T0293277

The Energy of Data and Distance Correlation

Energy distance is a statistical distance between the distributions of random vectors, which characterizes the equality of distributions. The name energy derives from Newton's gravitational potential energy, and there is an elegant relation to the notion of potential energy between statistical observations. Energy statistics are functions of distances between statistical observations in metric spaces. The authors hope this book will spark the interest of most statisticians who so far have not explored E-statistics and would like to apply these new methods using R. *The Energy of Data and Distance Correlation* is intended for teachers and students looking for dedicated material on energy statistics but can serve as a supplement to a wide range of courses and areas, such as Monte Carlo methods, U-statistics or V-statistics, measures of multivariate dependence, goodness-of-fit tests, nonparametric methods, and distance-based methods.

- E-statistics provides powerful methods to deal with problems in multivariate inference and analysis.
- Methods are implemented in R, and readers can immediately apply them using the freely available energy package for R.
- The proposed book will provide an overview of the existing state-of-the-art in development of energy statistics and an overview of applications.
- Background and literature review are valuable for anyone considering further research or application in energy statistics.

MONOGRAPHS ON STATISTICS AND APPLIED PROBABILITY

Editors: F. Bunea, R. Henderson, N. Keiding, L. Levina, N. Meinshausen, R. Smith

Recently Published Titles

Multistate Models for the Analysis of Life History Data
Richard J. Cook and Jerald F. Lawless 158

Nonparametric Models for Longitudinal Data
with Implementation in R
Colin O. Wu and Xin Tian 159

Multivariate Kernel Smoothing and Its Applications
José E. Chacón and Tarn Duong 160

Sufficient Dimension Reduction
Methods and Applications with R
Bing Li 161

Large Covariance and Autocovariance Matrices
Arup Bose and Monika Bhattacharjee 162

The Statistical Analysis of Multivariate Failure Time Data: A Marginal Modeling Approach
Ross L. Prentice and Shanshan Zhao 163

Dynamic Treatment Regimes
Statistical Methods for Precision Medicine
Anastasios A. Tsiatis, Marie Davidian, Shannon T. Holloway, and Eric B. Laber 164

Sequential Change Detection and Hypothesis Testing
General Non-i.i.d. Stochastic Models and Asymptotically Optimal Rules
Alexander Tartakovsky 165

Introduction to Time Series Modeling
Genshiro Kitigawa 166

Replication and Evidence Factors in Observational Studies
Paul R. Rosenbaum 167

Introduction to High-Dimensional Statistics, Second Edition
Christophe Giraud 168

Object Oriented Data Analysis
J.S. Marron and Ian L. Dryden 169

Martingale Methods in Statistics
Yoichi Nishiyama 170

The Energy of Data and Distance Correlation
Gabor J. Szekely and Maria L. Rizzo

For more information about this series please visit: https://www.crcpress.com/
Chapman--HallCRC-Monographs-on-Statistics--Applied-Probability/book-series/
CHMONSTAAPP

The Energy of Data and Distance Correlation

Gábor J. Székely
Maria L. Rizzo

CRC Press
Taylor & Francis Group
Boca Raton London New York

CRC Press is an imprint of the
Taylor & Francis Group, an **informa** business

A CHAPMAN & HALL BOOK

First edition published 2023
by CRC Press
6000 Broken Sound Parkway NW, Suite 300, Boca Raton, FL 33487-2742

and by CRC Press
4 Park Square, Milton Park, Abingdon, Oxon, OX14 4RN

CRC Press is an imprint of Taylor & Francis Group, LLC

ISBN: 978-1-482-24274-4 (hbk)
ISBN: 978-1-032-43379-0 (pbk)
ISBN: 978-0-429-15715-8 (ebk)

Typeset in Latin Modern
by KnowledgeWorks Global Ltd.

DOI: 10.1201/9780429157158

Publisher's note: This book has been prepared from camera-ready copy provided by the authors.

Access the companion website: http://cran.us.r-project.org/web/packages/energy/index.html
Access the Supplementary Material: github.com/mariarizzo/energy

Contents

Preface

This book summarizes over 30 years of research on what is now referred to as "energy statistics." While we designed it as a reference for the research, many applications, extensions, and related ideas are also discussed. The title of the book suggests that the fundamental ideas are related to the notion of energy in physics, and this is absolutely true. To avoid any confusion, perhaps we should first emphasize what this book is *not* about. Energy of data can easily be confused with data of energy, or data on energy, or simply energy statistics, which may refer to collecting and analyzing data on coal, crude oil, gas, electricity, nuclear energy, clean energy, etc. This book is not about this, not about the world energy resources; this book is about a new type of energy that was introduced by one of the co-authors of this book in several lectures in 1985. This new type of energy was inspired by physics and is called the energy of data, or simply data energy. In this context our use of the term "energy statistic" always refers to the sample coefficient or test statistic based on energy distance. For some insight into the name "energy" for this body of work, see Chapter 1, explaining the connection between our energy of data and physics.

This book is designed for researchers, graduate students, or advanced undergraduates with preparation in calculus, linear algebra, probability, and mathematical statistics. The text will be suitable for an introductory course in energy-based inference and data analysis, and may also be used for independent study. Although there is a strong connection between the energy approach and classical physics, knowledge of physics is not a prerequisite to read and apply the methods we discuss.

Most of the energy methods developed by the authors have been implemented for the statistical software R [R Core Team, 2022] in the *energy* package [Rizzo and Székely, 2022]. This package is open source with General Public License GPL $>=$ 2.0 and is available at `github.com/mariarizzo/energy` as well as `https://cran.r-project.org/package=energy`. For RStudio users it can be installed directly from the Packages tab. Refer to the relevant chapters and sections for examples or details about the functions available in the *energy* package, which is fully documented in the manual [Rizzo and Székely, 2022].

The book is organized as follows. Part I introduces energy distance, applications, and extensions. Topics such as one-sample goodness-of-fit tests, two-sample nonparametric tests for equal distributions, multi-sample tests and methods including a nonparametric extension of ANOVA, hierarchical clustering, and k-groups clustering. In this part, we also discuss in detail the energy test of multivariate normality for the simple or composite hypothesis and cover methods for computing and estimating the eigenvalues of the asymptotic distribution of the test statistics.

Part II is almost entirely devoted to distance covariance, distance correlation, and its extensions and applications. Here we also include a full chapter on energy U-statistics and the Hilbert space of U-centered distance matrices.

Selected data files and some supplementary materials are available at github.com/mariarizzo.

We hope that all readers can appreciate the energy perspective of statistics, through energy, through "physics of data."

Acknowledgments

The authors would like to recognize the contributions of our co-author on several research articles, Nail K. Bakirov (1952–2010) of the Russian Academy of Sciences.

We also thank all colleagues and co-authors who have contributed through their discussions and research to the growing body of research on energy statistics.

The authors thank their families for their constant support during the many years of research on energy and distance correlation.

Gábor J. Székely
National Science Foundation (ret.)
Email: gabor.j.szekely@gmail.com

Maria L. Rizzo
Department of Mathematics and Statistics
Bowling Green State University
Email: mrizzo@bgsu.edu

Authors

Gábor J. Székely graduated from Eötvös Loránd University, Budapest, Hungary (ELTE) with MS in 1970 and Ph. D. in 1971. He joined the Department of Probability Theory of ELTE in 1970. In 1989 he became the funding chair of the Department of Stochastics of the Budapest Institute of Technology (Technical University of Budapest). In 1995 Székely moved to the United States. Before that, in 1990–1991, he was the first distinguished Lukacs Professor at Bowling Green State University, Ohio. Székely had several visiting positions, e.g., at the University of Amsterdam in 1976 and at Yale University in 1989. Between 1985 and 1995, he was the first Hungarian director of Budapest Semesters in Mathematics. Between 2006 and 2022, until his retirement, he was program director of statistics of the National Science Foundation (USA). Székely has almost 250 publications, including six books in several languages. In 1988 he received the Rollo Davidson Prize from Cambridge University, jointly with Imre Z. Ruzsa for their work on algebraic probability theory. In 2010 Székely became an elected fellow of the Institute of Mathematical Statistics for his seminal work on physics concepts in statistics like energy statistics and distance correlation. Székely was invited speaker at several Joint Statistics Meetings and also organizer of invited sessions on energy statistics and distance correlation. Székely has two children, Szilvia and Tamás, and six grandchildren: Elisa, Anna, Michaël and Lea, Eszter, Avi who live in Brussels, Belgium and Basel, Switzerland. Székely and his wife, Judit, live in McLean, Virginia, and Budapest, Hungary.

Maria L. Rizzo is Professor in the Department of Mathematics and Statistics at Bowling Green State University in Bowling Green, Ohio, where she teaches statistics, actuarial science, computational statistics, statistical programming, and data science. Prior to joining the faculty at BGSU in 2006, she was a faculty member of the Department of Mathematics at Ohio University in Athens, Ohio. Her main research area is energy statistics and distance correlation. She is the software developer and maintainer of the energy package for R, and author of textbooks on statistical computing: "Statistical Computing with R" 1st and 2nd editions, "R by Example" (2nd edition in progress) with Jim Albert, and a forthcoming textbook on data science. Dr. Rizzo has eight PhD students and one current student, almost all with dissertations on energy statistics. Outside of work, she enjoys spending time with her family including her husband, daughters, grandchildren, and a large extended family.

Notation

Symbols

X'	An independent copy of X; that is, X and X' are iid.	$\mathcal{V}_n, \mathcal{R}_n$	Distance covariance, distance correlation statistics
X''	An independent copy of X'; that is, X' and X'' are iid.	$\mathcal{V}_n^*, \mathcal{R}_n^*$	Unbiased squared distance covariance, bias-corrected squared distance correlation statistics
$|x|_d$	Euclidean norm of $x \in \mathbb{R}^d$.		
$|x|$	Euclidean norm of x when dimension is clear in context; norm of a vector; complex norm if argument is complex.	$R_{x,y}^*$	Bias-corrected squared distance correlation statistic [Ch. 17]
$\|x\|$	Weighted L_2 norm of a complex function.	$R_{x,y;z}^*$	Partial distance correlation statistic
$A^\top$	Transpose of A.	$\mathcal{R}^*(X, Y; Z)$	Partial distance correlation
$< x, y >$	Dot product (scalar product)	$E[X]$	Expected value of the random variable X
(x, y)	Inner product (scalar product)	$I(A)$	Indicator function on the set A: $I(x) = 1$ if $x \in A$ and $I(x) = 0$ if $x \notin A$
$\hat{f}_X$	Characteristic function of X	I_d	The $d \times d$ identity matrix
$\hat{f}_X^n$	Empirical characteristic function of X	$\log x$	Natural logarithm of x
$\mathcal{E}, \mathcal{V}, \mathcal{R}$	Energy distance, distance covariance, distance correlation population coefficients	$\mathbb{R}$	The one dimensional field of real numbers
		$\mathbb{R}^d$	The d-dimensional real coordinate space
$\mathcal{W}$	Brownian covariance	$\Gamma(\cdot)$	Complete gamma function
$\mathcal{E}_n$	Energy statistic	$\overline{X}$	Sample mean or vector of sample means

Abbreviations

BCov	Brownian covariance
Cov, Cor	Pearson (linear) covariance, correlation
dCor	Distance correlation
dCov	Distance covariance
$\mathrm{dCov}_n, \mathrm{dCor}_n$	dCov, dCor statistics
disco	Distance Components
dVar	Distance variance
eCor	Earth mover's correlation
iid	independent and identically distributed
pdcor,pdCor	Partial distance correlation

Part I

The Energy of Data

1

Introduction

CONTENTS

> There are so many ideas in physics begging for mathematical inspira-
> tion . . . for new mathematical concepts [Ulam, 1976].

Data Science, and in a sense, the world as we know it started almost 6,000 years ago when, "in the beginning," Man created integers. The accountants of Uruk in Mesopotamia invented the first numerals: symbols encoding the concept of one-ness, two-ness, three-ness, etc. abstracted from any particular entity. Before that, for many thousand years jars of oil were counted with ovoid, measures of grain were counted with cones, tally marks, and pebbles (pebble is "calculus" in Latin) that helped people in counting [Rudman, 2007, Schmandt-Bessarat, 1992]. Abstract numerals, that are dissociated from the objects being counted, revolutionized our civilization: They expressed abstract thoughts, after all, "two" does not exist in nature, only two fingers, two people, two sheep, two apples, two oranges. One, two, three, . . . exist in our Mind only. After this abstraction, one could not tell from the numerals what the objects were; seeing the symbols of 1, 2, 3, . . . one could not see or smell oranges, apples, etc., but comparisons are possible. We could do accounting, "statistics," "statistical inference," "data science," and "exact science." The idea to divide objects into equal parts is responsible for important bases of number systems, the least common multiples of the first integers: 2, 6, 12, 60. The numbers that followed them, 3, 7, 13 became unlucky and then mystical [Székely, 1981]. On the interesting story of number names, see Menninger [2015]. Data gradually became more general than numbers; today data might mean vectors, functions, graphs, networks, and many other abstract notions. Data science is a continuation of some of the data analysis fields such as statistics, machine learning, data mining, and predictive analytics. It was John Tukey who started to transform academic statistics in the direction of data science in his classical work "The Future of Data Analysis" [Tukey, 1962]. See also "50 Years of Data Science" [Donoho, 2017]. The *energy of data* is a new concept that was introduced in the 1980's [Székely, 1989] (see also Székely [2002]) to help us to work with real numbers even if the data is complex objects like matrices, graphs, functions, etc. Instead of working with these objects themselves, we can work with their real-valued distances. For this, all we need is that distances between the data are defined. Mathematically this means that the data are in a metric space. In this way we can go back to the Paradise of real numbers.

1.1 Distances of Data

Data energy is a real number (typically a nonnegative number) that depends on the distances between data. This concept is based on the notion of Newton's gravitational potential energy, which is also a function of the distance between bodies. The idea of data energy or energy statistics is to consider statistical observations (data) as heavenly bodies governed by the potential energy of data, which is zero if and only if an underlying statistical hypothesis is true.

Following Székely [1989], we can represent the three pairs of dualities between energy, matter, and mind with the help of a triangle (Figure 1.1) whose vertices are energy, matter, and mind, and the connecting sides represent the equivalence/duality/dichotomy between these notions. Manifestations of matter are mass (m), disorder (measured by entropy, S), etc. Manifestations of the "immaterial" mind are memory, information (I), observation, data, etc., inputs passed on by the sensory organs.

The duality between E and m is Einstein's famous $E = mc^2$ [Einstein, 1905]. The duality between matter and mind is Szilárd's idea, which first appeared in his 1922 dissertation; namely, that in a closed material system, the decrease of uncertainty/entropy (S) equals the increase of information (I) in our mind about the system [Szilárd, 1929, Schrödinger, 1944, Brillouin, 2004]. To use Szilárd's words, it is possible to reduce the "entropy of a thermodynamic system by the intervention of intelligent beings," for example, a "Maxwell's demon." Thus, Szilárd eradicated the ancient dichotomy of matter and mind just as Einstein eradicated the dichotomy of energy and matter. *This book is about the third side of the triangle, the connection between energy and mind in terms of data distance D defined below.* Our mind regulates the flow of information (I) and data distance (D), the source of statistical energy to help achieve "mental harmony." It is interesting to note that the formula $S = k \log W$ where k is the Boltzmann constant and W denotes probability (the German word for Probability is Wahrscheinlichkeit) is considered by many the second most influential formula in physics, next to Einstein's $E = mc^2$. The formula $S = k \log W$ is carved on Boltzmann's gravestone. It is also interesting to mention that about a hundred years ago, another triangle mesmerized the scientific world in the book written by Hermann Weyl. It tracked the development of relativity physics. The book *Raum, Zeit, Materie (Space, Time, Matter)* had five editions between 1918 and 1922 [Weyl, 1922].

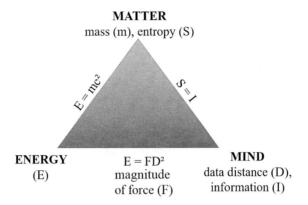

FIGURE 1.1
Connections between Energy, Matter, and Mind

In our book and in our Energy-Matter-Mind triangle, the single most frequently used *energy function of data* or *energy of data* or *data energy* is the square of the *energy distance* or *data distance* D defined below. First, we define D for data in Euclidean spaces.

For data $\mathbf{X} = \{X_1, \ldots, X_n\}$ and $\mathbf{Y} = \{Y_1, \ldots, Y_n\}$ in Euclidean space $\mathbb{R}^d$, $d \geq 1$, define the data distance of $\mathbf{X}$ and $\mathbf{Y}$ as the square root of

$$D^2 := \frac{1}{n^2} \left[2 \sum_{i=1}^{n} \sum_{j=1}^{n} |X_i - Y_j| - \sum_{i=1}^{n} \sum_{j=1}^{n} |X_i - X_j| - \sum_{i=1}^{n} \sum_{j=1}^{n} |Y_i - Y_j| \right],$$

where $|\cdot|$ denotes the Euclidean norm.

We will see that D^2 is always nonnegative. Moreover, D is always a metric in the space of data of length n. The source of data energy is D. This can also be viewed as energy in physics if we multiply D^2 by a constant force of magnitude F; that is, energy

$$E = F \cdot D^2.$$

This energy of data formula is the counterpart of Einstein's energy of matter formula, $E = m \cdot c^2$, when matter is replaced by mind. The signs in D^2 resemble the computation of electrostatic potential where the signs depend on the charges. For statistical inference, we do not need to know F. In statistics, the value of F is irrelevant, and we can simply take $F = 1$. But F can also be viewed from the physics perspective as a physical constant like the speed of light. This F would play an important role if we wanted to "free" data energy as a counterpart to nuclear energy. If we can decrease D in our brain, then this can free energy there. Thus, F can play an important role in brain research, especially in Alzheimer and attention deficit/hyperactivity disorder (ADHD) research. On the mathematics of freeing energy, see Section 3.8.

We have arrived at an interesting philosophical point. When we say "knowledge is power," we always think of this in an abstract sense. The first attempt to quantify this sentence was a lecture series in 1985 by Székely at MIT, Yale, and Columbia University. Power, say electric power, in physics is the energy per unit time, but how can we quantify knowledge? A huge part of knowledge can be translated to data distance (in face recognition, voice recognition, pattern recognition); we will see many further examples in this book. Of course, not all knowledge is data distance, as not all information can be expressed via bits. In this book, we work with knowledge that is related to distances between data points. Information theory showed us that we can measure abstract notions like information via an exact formula like Shannon's entropy, and John Tukey introduced a unit for that. The name "bit" (a portmanteau of "binary digit") was coined by Tukey while working with John von Neumann in the 1940s. We can do something similar with knowledge. In this book, we work with knowledge that is related to distances between data points. *The abstract/general notion of knowledge can be measured by data distance via our exact formula D.*

We can also introduce a unit for Knowledge and call it *Hokhma*, which is the Biblical Hebrew word for "Wisdom." Its physical dimension or unit is the square root of the dimension of distance (square root of the unit of distance). Quantifying knowledge, that is, mapping the mind (via computational neuroscience) is the road to a kind of immortality: Knowledge can be mapped; soul cannot. In our Energy-Matter-Mind triangle, we can replace "data distance" by "knowledge" that can be measured by D. Data energy in our mind can explain some of the "magic" of extrasensory perception (ESP), also called the "sixth sense" of our mind. Einstein's $E = mc^2$ can be used for freeing nuclear energy, while $E = FD^2$ can be used for freeing data energy by decreasing segregation, pooling of data; in effect increasing "colorblindness" in our mind.

If we choose $F = 1$ and the sample sizes n, m are not necessarily equal, then the *energy of data* formula becomes

$$\mathcal{E}_{n,m}(\mathbf{X}, \mathbf{Y}) = \frac{2}{nm} \sum_{i=1}^{n} \sum_{j=1}^{m} |X_i - Y_j|$$

$$- \frac{1}{n^2} \sum_{i=1}^{n} \sum_{j=1}^{n} |X_i - X_j| - \frac{1}{m^2} \sum_{i=1}^{m} \sum_{j=1}^{m} |Y_i - Y_j|. \qquad (1.1)$$

If $n = m$, and $Y_1 = Y_2 = \cdots = Y_n = y$, and we minimize the energy $\mathcal{E}_{n,m}(\mathbf{X}, \mathbf{Y})$ in y, then if the minimum is taken at $y = C$ we can call C the *energy center* of the data $\{X_1, X_2, \ldots, X_n\}$. This center is not necessarily unique. For example, let $X_1, X_2, \ldots, X_n$ be real numbers and order them increasingly: $X_{(1)}, X_{(2)}, \ldots, X_{(n)}$; if $n = 2k + 1$ is an odd number, then

$C = X_{(k+1)}$ because if $y - C = d$ then the energy increases by $|d|$. In case $n = 2k$, an even number, C can be any number between $X_{(k)}$ and $X_{(k+1)}$. This is the same as the *median* that separates the upper half and the lower half of real-valued data. The energy center, however, is defined not only for real numbers but for any data in higher dimensional spaces or in any space where a distance is defined.

An interesting example is due to P. Fermat, who asked E. Torricelli about the following problem in a 1629 letter. Given a triangle with vertices X_1, X_2, X_3, find a point C such that the sum of its distances from the vertices is minimum. Torricelli's answer is that if each angle of the triangle is less than $2\pi/3$ then the three segments CX_1, CX_2, CX_3 should form three angles equal to $2\pi/3$; otherwise, C is the vertex whose angle is not less than $2\pi/3$. The energy center C in geometry is called the median, or the Fermat point, or the Torricelli point of the triangle.

It was the French mathematician, Maurice Fréchet, who defined distances and thus *metric spaces* endowed with distances, in complete generality in his 1906 dissertation [Fréchet, 1906] and he is also responsible for the definition of the *Fréchet mean* [Fréchet, 1948], which is equivalent to our definition of energy center. These general centers of data are important in many areas of applications when the data are complex objects or shapes. The value of the energy minimum is clearly a measure of dispersion.

In this book, statistics are always based on distances between observations. The observations can be numbers, vectors, functions, graphs, etc., as long as the distance between them is defined; in other words, as long as the possible observations come from a metric space.

A formal definition of Fréchet mean is the following. If (M, d) is a metric space, $x_1, x_2, \ldots, x_n$ and p are points in M, and d denotes the distance, then the *Fréchet mean* is the point m in M that minimizes the sum of the squared distances

$$V(p) = \sum_{i=1}^{n} d^2(p, x_i)$$

if such a global minimizer $p = m$ exists. The corresponding $V(m)$ is the *Fréchet variance*. If we minimize the sum of the distances (we do not square them), then the location of the minimum is the *Fréchet median*.

Energy inference based on distances between data in metric spaces can be considered a type of method in topological data analysis (TDA) that helps to find structure in data, typically geometric structure, using methods such as clustering, dimension reduction, etc.

For statistical purposes, we can also work with powers of distances like $|X_i - Y_j|^\alpha$ where $0 < \alpha < 2$, because in this range of exponents

$$D_\alpha^2 := 2 \sum_{i=1}^{n} \sum_{j=1}^{m} |X_i - Y_j|^\alpha - \sum_{i=1}^{n} \sum_{j=1}^{n} |X_i - X_j|^\alpha - \sum_{i=1}^{m} \sum_{j=1}^{m} |Y_i - Y_j|^\alpha$$

remains the square of a metric (see Theorem 3.3). The square root of D_α^2 is a metric while $(D_\alpha)^{1/\alpha}$ is measured in the same units as $|X_i - Y_j|$ and thus $E_\alpha := F(D_\alpha)^{2/\alpha}$ can be considered a generalized statistical energy. Newton's gravitational potential is proportional to the reciprocal of the distance, while in elasticity the energy is proportional to the second power of the extension or displacement (Hooke's law), hence we have that the exponents relevant for data science are between the exponents relevant in gravitation and in elasticity. The difference between the gravitational energy of Matter and the data energy of Mind is the exponent of the data distance. The source of data energy is the dissimilarity between data, which is independent of the nature of the data (color, music, temperature, blood pressure, etc.).

The term

$$\frac{1}{n^2} \sum_{i=1}^{n} \sum_{j=1}^{n} |X_i - X_j|$$

in $\mathcal{E}_{n,m}$ is also known as *Gini's mean difference*. This was introduced in 1912 by Corrado Gini as an alternative to variance to measure the dispersion of data [Gini, 1912]. For more details, see Yitzhaki [2003] where "Gini is out of the bottle."

The general exponent version of Gini's mean difference,

$$\frac{1}{n^2} \sum_{i=1}^{n} \sum_{j=1}^{n} |X_i - X_j|^\alpha,$$

was applied in Riesz [1938, 1949] for $\alpha < 0$ to define α-energy for fractional calculus: differentiating or integrating fractional times.

The two-sequence version of energy, that is, the relative energy of one sequence to another one, $\mathcal{E}_{n,m}$, was initiated in 1985 (see History, Appendix A and Székely [1989], Székely [2002]). This energy is a double-centered version of the bivariate Riesz energy.

TABLE 1.1
Many methods of classical statistics have their counterparts in the energy world.

	CLASSICAL APPROACH	ENERGY APPROACH
Dependence	Pearson's correlation (can be 0 for dependent variables)	Distance correlation (equals 0 iff the variables are independent)
Goodness-of-fit & homogeneity	EDF tests Likelihood ratio tests	Energy goodness-of-fit & homogeneity characterizes equality of distributions Nonparametric energy tests
Tests for normality Multivariate normality	Shapiro-Wilk, Anderson-Darling (lack natural multivariate extensions) Multivariate skewness & kurtosis (test is not consistent)	Energy test for normality (arbitrary dimension) Multivariate energy test of normality (test is consistent)
Skewness	Third central moment (can be 0 for nonsymmetric distributions)	Distance skewness (equals 0 iff the probability dist. is centrally symmetric)
Multi-sample problems	ANOVA (tests if normally distributed variables have equal means)	DISCO (tests if random vectors have the same distribution)
Cluster analysis	Ward's minimum variance method k-means clustering (sensitive to cluster centers)	Hierarchical energy clustering k-groups (energy) clustering (sensitive to cluster distributions)

1.2 Energy of Data: Distance Science of Data

By now, energy inference is a huge field of statistics that is applicable to general metric space valued data. This is the topic of our book. Measures of dependence *Distance Covariance and Distance Correlation* and the Distance Variance are discussed extensively in this book, corresponding to the covariance, correlation, and variance of double centered distances. For details, see the relevant chapters on distance correlation in Part II.

In this book, *data energy* in the narrow sense means a real-valued (typically nonnegative valued) function of the energy distance D in Euclidean spaces. A simple generalization is when we replace D by D_α, $0 < \alpha < 2$. Then we can replace Euclidean spaces with Hilbert spaces or with general metric spaces where the metric is negative definite. In most cases we use the energy of data in this sense, but if our data are in metric spaces where the metric is not negative definite, then we need more general real-valued functions of distances of data. An example of this general notion of energy is the Fréchet variance or Gini's mean. The arithmetic mean of real numbers does not look like data energy, but it is because the arithmetic mean of real numbers is their Fréchet mean, which is a real number defined via distances between data. On the other hand, the arithmetic mean or centroid of vectors (with a dimension bigger than 1) is not an energy because it is not real valued. The same refers to statistical functions whose values are functions, matrices, networks, or geometrical objects. In these cases, energy inference means that we rely on real functions of their distances. Since energy inference is based on distances between data, energy inference is always invariant with respect to distance preserving transformations (also called isometries or congruences), which is not true for many commonly applied statistical methods such as those based on Cramér-distances (L_2-distances of multivariate distribution functions) that are typically not rotation invariant and thus not isometry-invariant.

Any data has energy depending on the distances or dissimilarities between data points. In this respect, the nature of the data is irrelevant; it can be blood pressure, temperature, stock price, petroleum price, brain image, or a genetic marker. The source of statistical energy or data energy is the distance or dissimilarity between observations. This is similar to Newton's potential energy, which is also a function of distances between objects. Data energy is simply a function of distances between observations, typically Euclidean distances.

Objective distances or dissimilarities can have reflections in our minds. These reflections can be called perceived distances. For example, red and green lights are different and their distance can be measured by difference between their wave lengths. This is objective. The reflection in our minds, i.e., the way we sense them, can be very different. For example, if one is color blind, the perceived distance can be zero. Distances or metrics in our mind are typically not

Euclidean. Psycho-physics studies the relation between stimuli and sensation or perception. Typically, the perception or sensation is a monotone function of the stimulus. According to the classical Weber-Fechner law, this function in many cases is logarithmic; that is, if a stimulus varies as a geometric progression, the corresponding perception is altered in an arithmetic progression. The energy of perception data in this case is the energy of the logarithm of the stimuli. This includes stimuli to all senses: vision, hearing, taste, touch, and smell. *Manifestations of data energy in our mind include happiness, increased willpower, focus, productivity, and confidence.* Measures of intelligence, like IQ, are also related to mental energy. We will provide a formula to compute the energy of data, first, when the distances are Euclidean and then when the "metrics" are general. A given function of statistical observations (data) is usually called a statistic, therefore we call functions of distances between data or between statistical observations "energy statistics." In this book, however, this notion should not be confused with statistics on energy because we simply do not even touch on this topic here. Instead, we focus on the quantitative formula for energy of data. This type of energy becomes more important by the day. Those who own and understand data own the future. Not so long ago, the most valuable commodities were gold, land, stocks, etc. Today, increasingly, the most valuable commodity is to own and understand data: Its energy.

Thus far, the most frequent application of this new type of energy is the distance correlation. Instead of working with the observations themselves, like in Pearson's correlation, we work only with the distances between observations. However, it is important to understand that our "distance correlation" does not mean "correlation of distances." That is, distance correlation does not apply to Pearson correlation of distances, which is a method sometimes applied but it is not an energy coefficient and does not have the same properties. Our distance correlation approach has two advantages: (i) We can easily measure the dependence of objects from completely different spaces. For example, we can measure the degree of dependence between functions and three-dimensional objects, say, because all that matters is their distances within their spaces. Another advantage is more shocking. (ii) While Pearson's correlation can easily be zero for dependent variables, our distance correlation is zero if and only if they are independent. This answers one of the most important questions in science: how to tell if two "things," that is, two random variables are dependent or independent. In other words, we can tell if there is a relationship between two quantities or they are independent. This is the topic of Part II "Distance Correlation and Dependence," where we define a new way of quantifying dependence.

Other applications of the data energy concept include testing for homogeneity, goodness-of-fit, hierarchical clustering, distance component analysis (DISCO), etc. We discuss these topics in separate chapters.

The theme of this book is "distance science" where we study data based only on the distances between them. Distance geometry is a special case; the two thousand year old *Heron's formula* to compute the area A of a triangle from the distances between the three vertices (side lengths) a, b, c is

$$A = \sqrt{s(s-a)(s-b)(s-c)}$$

where $s := (a+b+c)/2$.

The d dimensional generalization of Heron's formula is a formula for the square root of the d-dimensional volume of a simplex spanned by $d+1$ points; the formula is the so-called Cayley-Menger determinant of all pairwise distances between the vertices of the simplex. Tao [2019] posted an interesting note on the spherical Cayley-Menger determinant and the radius of the Earth. Another important area of distance science is the 3-dimensional "reconstruction" of protein molecules based on the pairwise distances of its atoms.

In this book, however, we focus on the statistical aspects of distance science. This is the first book on the application of energy distance, what we can call *distance statistics*, that analyzes the distances between observations rather than the observations themselves. We will narrow down our attention to (non-negative) real valued functions of distances between data. These functions will be called *energy of data* especially when the "energy equals zero" has a special statistical meaning that a certain "null hypothesis" is true. An example for energy functions is the distance correlation. This is zero if and only if the variables are independent. Our general notion of energy, of course, is not always Newtonian in the sense that it does not always correspond to the potential energy of existing physical objects.

Our book offers a new way of teaching statistics based on the notion of data energy. Students and researchers can easily use the *energy* package in R [Rizzo and Székely, 2022, R Core Team, 2022] for their data to experiment, teach, learn, and do research. Although there are mathematically challenging proofs in this book, most readers can skip them and focus on the main trend that needs limited mathematical background and can focus on interesting applications in classical statistics like testing independence, testing normality, or on machine learning, a tool for data mining and artificial intelligence.

2

Preliminaries

CONTENTS

Although this chapter is titled "Preliminaries," it is not necessarily a prerequisite to understanding the first few chapters.

In this chapter we give an introduction to V-statistics and U-statistics. Most of the statistics introduced in this book are V-statistics, and for some applications, U-statistics are needed. See Chapter 16 for much more detail and further examples.

We also introduce our notation and a fundamental Lemma that is essential for deriving many of our results.

2.1   Notation

The following notation is used consistently throughout.

- The Euclidean norm for random vectors $X = (X_1, \ldots, X_d)^\top \in \mathbb{R}^d$, $d \geq 1$, is

$$|X|_d = \left(\sum_{j=1}^{d} X_j^2 \right)^{1/2}.$$

  Whenever the dimension d is clear in context, we omit the subscript. (If the argument of $|\cdot|$ is complex, this denotes the complex norm.) We reserve $\|\cdot\|$ for the weighted L_2 norm to be defined later.

DOI: 10.1201/9780429157158-2

- The symbol X' indicates that X' is an "iid copy" of the random variable X. That is, X and X' are iid, Y, Y' are iid, etc. We use $X^\top$ to indicate the transpose of X.

- The dot product or scalar product of two vectors $a \in \mathbb{R}^d$, $b \in \mathbb{R}^d$ is

$$< a, b >= \sum_k a_k b_k = a \cdot b^\top,$$

  also denoted by (a, b). For complex vectors a, b the dot product is

$$< a, b >= \sum_k a_k \overline{b_k},$$

  where $\overline{b_k}$ is the complex conjugate of b_k. If the argument of $|\cdot|$ is complex, $|a| = \sqrt{< a, a >}$ is the complex norm.

- The Fourier-transform (characteristic function) of the probability density functions f is denoted by $\hat{f}$. If X is a random variable with density $f(x) = f_X(x)$, its characteristic function is $\hat{f}_X$. If X and Y are jointly distributed with density $f_{X,Y}$ then the joint characteristic function of (X, Y) is denoted $\hat{f}_{X,Y}$.

- If $X \in \mathbb{R}^d$ is supported on A, $Y \in \mathbb{R}^d$ is supported on B, $\xi \in \mathbb{R}^d$ is a constant, and X, Y are independent, the expected distances $E|X - Y|$, $E|X - X'|$, $E|\xi - X|$ are

$$E|X - Y| = \int_A \int_B |x - y| dF_X(x)\, dF_Y(y),$$

$$E|X - X'| = \int_A \int_A |s - t| dF_X(s)\, dF_X(t),$$

$$E|\xi - X| = \int_A |\xi - x| dF_X(x),$$

  provided these expectations exist. For discrete data, the definite integrals are replaced by sums and $dF_X(x) := P(X = x)$. Then if X is supported on A, Y is supported on B, ξ is constant, and X, Y are independent,

$$E|X - Y| = \sum_{x \in A} \sum_{y \in B} |x - y| P(X = x) P(Y = y),$$

$$E|X - X'| = \sum_{a \in A} \sum_{b \in A} |a - b| P(X = a) P(X = b),$$

$$E|\xi - X| = \sum_{x \in A} |\xi - x| P(X = x).$$

2.2 V-statistics and U-statistics

Energy statistics defined as in Chapter 3, Definition 3.1 are *V*-statistics. We can alternately work with *U*-statistics. Here we present an introduction to *V*-statistics and *U*-statistics. See Chapter 16 for more results, applications, and examples of *U*-statistics and *V*-statistics.

In a fundamental paper, von Mises [1947] developed asymptotic distribution theory of a class of statistics called *V*-statistics ("V" for von Mises). Readers may be more familiar with the closely related *U*-statistics ("U" for unbiased) introduced by Hoeffding [1948]. See Lee [2019], Koroljuk and Borovskich [1994], Kowalski and Tu [2008]. The asymptotic theory of *U*-statistics and *V*-statistics is essentially parallel, although derived by different approaches.

Statistics are functions of one or more samples of data. Let $F_n(x)$ denote the empirical cumulative distribution function (ecdf) of a sample from distribution F_X. Statistics that can be represented as *functionals* of the empirical distribution function F_n are called "statistical functionals." Differentiability of the functional plays a key role in von Mises' approach, thus the title of his 1947 paper "On the Asymptotic Distributions of Differentiable Statistical Functionals."

2.2.1 Examples

As an example of a statistical function, consider the sample variance. If $E(X) = \mu$, then $T(F) = \sigma_X^2 = \int (x - \mu)^2 dF_X(x)$ is the statistical functional. The corresponding statistical function substitutes $dF_n(x_i) = \frac{1}{n}$ for dF, and we get the sample variance $\hat{\sigma}^2 = \frac{1}{n} \sum_{i=1}^{n} (x_i - \overline{x})^2$.

Consider the Cramér–von-Mises and Anderson–Darling goodness-of-fit statistics for testing $H_0 : F = F_0$ vs $H_1 : F \neq F_0$ for some specified null distribution F_0. Both statistics are based on a functional of this form:

$$T(F) = \int (F(x) - F_0(x))^2 \, w(x; F_0) \, dF_0(x),$$

where $w(x; F_0)$ is a given weight function depending on F_0. When $w(x, F_0) = 1$, the statistical function is the well-known Cramér-von-Mises test statistic, substituting F_n for F. For the Anderson-Darling test, the weight function is $[F_0(x)(1 - F_0(x))]^{-1}$.

2.2.2 Representation as a V-statistic

Suppose that $x_1, \ldots, x_n$ is a sample. In typical applications, the statistical function can be represented as a *V*-statistic. If $h(x, y)$ is a symmetric kernel function, the corresponding V-statistic is

$$V_{n,2} = \frac{1}{n^2} \sum_{i=1}^{n} \sum_{j=1}^{n} h(x_i, x_j).$$

We can also work with symmetric kernels of higher degree r, $h(x_{i1}, x_{i2}, \dots, x_{ir})$. Then

$$V_{n,r} = \frac{1}{n^r} \sum_{i_1,\dots,i_r=1}^{n} h(x_{i1}, x_{i2}, \dots, x_{ir}).$$

See, e.g., [Serfling, 1980, Section 6.5] about how to find the kernel in practice. Kernels of U-statistics and V-statistics are discussed in Chapter 16.

An example of a degree-2 V-statistic is the maximum-likelihood estimator of the variance with kernel $h(x, y) = \frac{1}{2}(x-y)^2$:

$$V_{n,2} = \frac{1}{n^2} \sum_{i=1}^{n} \sum_{j=1}^{n} \frac{1}{2}(x_i - x_j)^2 = \frac{1}{n} \sum_{i=1}^{n} (x_i - \bar{x})^2.$$

Recall that $\hat{\sigma}^2 = V_{n,2}$ is a biased estimator of the variance. The corresponding U-statistic with the same kernel $h(x, y)$ is

$$U_{2,n} = \frac{1}{\binom{n}{2}} \sum_{1 \leq i < j \leq n} h(x_i, x_j),$$

which gives the unbiased estimator of variance $s^2 = \frac{1}{n-1} \sum_{i=1}^{n} (x_i - \bar{x})^2$.

For simplicity, below we drop the degree r from the subscript of V and U.

Example 2.1 (Energy goodness-of-fit statistics). Suppose that we have observed a sample $x_1, \dots, x_n$ from distribution F_X and we want to test $H : F_X = F_0$, where F_0 is a fully specified null distribution. Define

$$h(x_i, x_j) = E|x_i - Y| + E|x_j - Y| - E|Y - Y'| - |x_i - x_j|, \qquad (2.1)$$

where Y, Y' are iid with distribution F_0, $E|x_i - Y| = \int |x_i - y| dF_0(y)$, and $E|Y - Y'| = \int \int |y - t| dF_0(y) dF_0(t)$.

The V-statistic with kernel (2.1) is $\frac{1}{n^2} \sum_{i,j=1}^{n} h(x_i, x_j)$, which gives

$$\mathcal{E}_n = \frac{1}{n^2} \sum_{i,j=1}^{n} \left(E|x_i - Y| + E|x_j - Y| - E|Y - Y'| - |x_i - x_j| \right)$$

$$= \frac{1}{n^2} \left(2n \sum_{i=1}^{n} E|x_i - Y| - n^2 E|Y - Y'| - \sum_{i,j=1}^{n} |x_i - x_j| \right)$$

$$\frac{2}{n} \sum_{i=1}^{n} E|x_i - Y| - E|Y - Y'| - \frac{1}{n^2} \sum_{i,j=1}^{n} |x_i - x_j|. \qquad (2.2)$$

The U-statistic with kernel (2.1) is defined by the average of (2.1) over all $\binom{n}{2}$ pairs $x_i \neq x_j$. This gives the corresponding energy U-statistic

$$\mathcal{E}_n^* = \frac{2}{n}\sum_{i=1}^{n} E|x_i - Y| - E|Y - Y'| - \frac{1}{n(n-1)}\sum_{i,j=1}^{n}|x_i - x_j|. \qquad (2.3)$$

The test statistic is $nV_n = n\mathcal{E}_n$ (or $nU_n = n\mathcal{E}_n^*$).

Example 2.2 (A two-sample energy statistic). A two-sample energy statistic for testing if two independent random samples, $x_1, \ldots, x_n$ and $y_1, \ldots, y_m$ are drawn from equal distributions can be defined with the following two-sample kernel. Let

$$h(x_i, x_j; y_k, y_\ell) = \tfrac{1}{2}(|x_i - y_k| + |x_j - y_\ell| + |x_i - y_\ell| + |x_j - y_k|$$
$$- |x_i - x_j| - |x_j - x_i| - |y_k - y_\ell| - |y_\ell - y_j|). \qquad (2.4)$$

Then h is symmetric in its arguments (x_i, x_j) and (y_k, y_ℓ). The corresponding energy V-statistic for testing $H : F_X = F_Y$ is $\frac{1}{n^2 m^2}\sum_{i,j,k,\ell} h(x_i, x_j; y_k, y_\ell)$.

To simplify the equations, introduce distance matrices $D = (d_{ik}) = (|x_i - y_k|)$, $A = (a_{ij}) = (|x_i - x_j|)$, and $B = (b_{k\ell}) = (|y_k - y_\ell|)$. Let $d_{..}$ denote the sum of matrix D, and similarly define $a_{..}$ and $b_{..}$. Then the V-statistic can be written

$$\mathcal{E}_{n,m} = \frac{1}{n^2 m^2} h(x_i, x_j; y_k, y_\ell)$$
$$= \frac{1}{2n^2 m^2}(4\, nm\, d_{..} - 2m^2 a_{..} - 2n^2 b_{..})$$
$$= \frac{2}{nm}\sum_{i=1}^{n}\sum_{k=1}^{m}|x_i - y_k| - \frac{1}{n^2}\sum_{i=1}^{n}\sum_{j=1}^{n}|x_i - x_j| - \frac{1}{m^2}\sum_{k=1}^{m}\sum_{\ell=1}^{m}|y_k - y_\ell|.$$
$$\qquad (2.5)$$

The test statistic is $\frac{nm}{n+m}V_{n,m} = \frac{nm}{n+m}\mathcal{E}_{n,m}$. (The factor $\frac{nm}{n+m}$ is half the harmonic mean of n and m.)

Alternately, one could apply the corresponding U-statistic with the same kernel (2.4).

2.2.3 Asymptotic Distribution

Informally, the type of asymptotic distribution of a V-statistic (or a U-statistic) depends on the order of degeneracy. This order is determined by what is the first non-vanishing term of the Taylor expansion of the statistical functional. If it is the linear term, the limit distribution is normal. Otherwise, under suitable conditions, there is a hierarchy of cases. For the variance example, the asymptotic distribution is related to a chi-squared distribution. Limit theory for U-statistics and V-statistics is parallel in this hierarchy; it may be helpful to refer to any of the references on U-statistics, such as Serfling [1980], Chapters 5–6 or Lee [2019], Chapter 3.

Energy statistics have a degenerate kernel of order 2. We have a degenerate kernel of order 2 if $E(h^2(X,Y)) < \infty$, $E(h(X,Y)) < \infty$, and $E(h(x,Y)) \equiv 0$. In this case, nV_n converges in distribution to a weighted sum of independent chi-squared variables, ˜

$$\sum_{i=1}^{\infty} \lambda_i Z_i^2,$$

where Z_i are iid standard normal, and λ_i are constants that depend on the underlying distributions F and the functional. The Cramér-von-Mises (CvM) example has a degenerate kernel with this type of asymptotic distribution, a quadratic form of centered Gaussian random variables. (See [Lee, 2019, Chapter 3] for the CvM kernel function.)

When an energy statistic is formulated as a U-statistic the asymptotic distribution has the form

$$\sum_{i=1}^{\infty} \lambda_i (Z_i^2 - 1),$$

where Z_i are iid standard normal, and λ_i are constants.

See Examples 2.1 and 2.2 for the kernels of the energy goodness-of-fit statistics and the two-sample energy statistic.

2.2.4   E-statistics as V-statistics vs U-statistics

Energy statistics have typically been defined as V-statistics. Why not work with U-statistics, which are unbiased?

The energy V-statistics have the advantage of a nice interpretation as a statistical distance, because energy distance is always nonnegative. Another advantage is that because energy distance $\mathcal{E}(X,Y) \geq 0$, we can take its square root, and the square root is a metric.

However, all theory and applications could also be defined in terms of U-statistics, and many authors have taken this approach to extend or apply energy methods. The asymptotic theory of U-statistics and V-statistics are parallel. An advantage of U-statistics is that a U-statistic is unbiased.

More examples of energy V-statistics are introduced starting in Chapter 3 and some energy U-statistics appear in later chapters.

In general *Energy statistics* or *E-statistics* are arbitrary (typically non-negative) real-valued functions of distances between data. They are not necessarily U-statistics nor V-statistics.

Energy statistics generally compare two or more distributions for equality; in a one-sample problem, the second distribution is a hypothesized null distribution. Each distribution is characterized by its characteristic function. Population coefficients can be formulated by evaluating a corresponding integral involving a weighted norm of the difference in characteristic functions. See e.g. Chapter 12 and 15 definitions of distance covariance.

2.3   A Key Lemma

To simplify the integral expression of an energy coefficient we need to work with complex functions and the complex norm. Considering the operations on characteristic functions involved in evaluating the integrand in these expressions, we need the following important lemma. It is applied for results in several of our chapters.

Lemma 2.1. *If $0 < \alpha < 2$, then for all x in $\mathbb{R}^d$*

$$\int_{\mathbb{R}^d} \frac{1 - \cos\langle t, x \rangle}{|t|_d^{d+\alpha}}\, dt = C(d, \alpha)|x|_d^\alpha,$$

where

$$C(d, \alpha) = \frac{2\pi^{\frac{d}{2}}\, \Gamma(1 - \frac{\alpha}{2})}{\alpha 2^\alpha \Gamma(\frac{d+\alpha}{2})},$$

and $\Gamma(\cdot)$ is the complete gamma function. The integrals at 0 and ∞ are meant in the principal value sense: $\lim_{\varepsilon \to 0} \int_{\mathbb{R}^d \setminus \{\varepsilon B + \varepsilon^{-1} B^c\}}$, *where B is the unit ball (centered at 0) in $\mathbb{R}^d$ and B^c is the complement of B.)*

The following proof of Lemma 2.1 is given in Székely and Rizzo [2005a]. See Appendix A for some notes on the history of this and related results.

Proof. First let us see the proof for $\alpha = 1$. Apply an orthogonal transformation $t \mapsto z = (z_1, z_2, \ldots, z_d)$ with $z_1 = (t, x)/|x|$ followed by a change of variables: $s = |x| \cdot z$ to get

$$\int_{\mathbb{R}^d} \frac{1 - \cos(z_1 |x|)}{|z|^{d+1}}\, dz = |x| \int_{\mathbb{R}^d} \frac{1 - \cos s_1}{|s|^{d+1}}\, ds,$$

where $s = (s_1, s_2, \ldots, s_d)$. Then for $s = (s_1, s_2, \ldots, s_d)$,

$$c_d := C(d, 1) = \int_{\mathbb{R}^d} \frac{1 - \cos s_1}{|s|^{d+1}}\, ds = \frac{\pi^{(d+1)/2}}{\Gamma\left(\frac{d+1}{2}\right)}.$$

For $d = 1$ see Prudnikov et al. [1986] p. 442. Note that $2c_d = \omega_{d+1}$ is the area of the unit sphere in $\mathbb{R}^{d+1}$.

In the general case, when both d and α can differ from 1, more technical steps are needed. Applying formulas 3.3.2.1, p. 585, 2.2.4.24 p. 298 and 2.5.3.10 or 2.5.3.13 p. 387 of Prudnikov et al. [1986], we obtain

$$A := \int_{\mathbb{R}^{d-1}} \frac{dz_2\, dz_3 \ldots dz_d}{(1 + z_2^2 + z_3^2 + \cdots + z_d^2)^{\frac{d+\alpha}{2}}}$$

$$= \frac{2\pi^{\frac{d-1}{2}}}{\Gamma(\frac{d-1}{2})} \int_0^\infty \frac{x^{d-2} dx}{(1 + x^2)^{\frac{d+\alpha}{2}}} = \frac{\pi^{\frac{d-1}{2}} \Gamma(\frac{\alpha+1}{2})}{\Gamma(\frac{d+\alpha}{2})};$$

$$\frac{d}{da}\left(\int_0^\infty \frac{1-\cos au}{u^{1+\alpha}}\,du\right) = a^{\alpha-1}\int_0^\infty \frac{\sin v}{v^\alpha}\,dv = a^{\alpha-1}\frac{\sqrt{\pi}\Gamma(1-\frac{\alpha}{2})}{2^\alpha\Gamma(\frac{\alpha+1}{2})}.$$

Introduce new variables $s_1 := z_1$, and $s_k := s_1 z_k$ for $k = 2\dots,d$. Then

$$C(d,\alpha) = A \times \int_{-\infty}^\infty \frac{1-\cos z_1}{|z_1|^{1+\alpha}}\,dz_1$$

$$= \frac{\pi^{\frac{d-1}{2}}\Gamma(\frac{\alpha+1}{2})}{\Gamma(\frac{d+\alpha}{2})} \times \frac{2\sqrt{\pi}\Gamma(1-\frac{\alpha}{2})}{\alpha 2^\alpha\Gamma(\frac{\alpha+1}{2})} = \frac{2\pi^{\frac{d}{2}}\Gamma(1-\frac{\alpha}{2})}{\alpha 2^\alpha\Gamma(\frac{d+\alpha}{2})},$$

and this was to be proved. $\qquad\qquad\qquad\qquad\qquad\qquad\qquad\qquad\square$

2.4 Invariance

Classical statistics operate with real-valued data like height, weight, blood pressure, etc., and it is supposed that typical data are (approximately) normally distributed so that one can apply the theory of normal (Gaussian) distributions for inference. Even if the observations were not Gaussian, in case of "big data" when the number of observations, n, was large, one could often refer to central limit theorems to claim that for large n the normal approximation is valid and classical methods are applicable.

What happens if the data are not real numbers but vectors, functions, graphs, etc.? In this case, even addition or multiplication of data might be a problem if for example, the observed vectors have different dimensions. We can overcome this difficulty if the observations are elements of a metric space. In this case, instead of working with the observations themselves, we can work with their (nonnegative) real valued distances. This brings us back to the "real world" where we can work with real numbers. We will call this type of inference *energy inference*.

The energy approach to testing a hypothesis is built on comparing distributions for equality, in contrast to normal theory methods like a two-sample t-test for comparing means. Energy tests can detect any difference between distributions. An attractive property of working with distances is that energy statistics have invariance with respect to any distance preserving transformation of the full data set to be analyzed (rigid motion invariance), which includes translation, reflection, and angle-preserving rotation of coordinate axes. We will have an energy counterpart of the variance, covariance and correlation, called respectively, distance variance, distance covariance, and distance correlation. An important advantage of applying them is that the distance correlation coefficient (Chapter 12) equals zero if *and only if* the variables are independent. This characterization of independence typically does not hold for Pearson's correlation; for example, for nonnormal data.

Energy coefficients have scale equivariance; distance correlation has scale invariance.

In Part I of this book, we focus on energy V-statistics. In Part II, we also work with energy U-statistics. An important invariance property in Part II is the jackknife invariance, which characterizes whether a statistic is a U-statistic (Section 16.4).

2.5 Exercises

Exercise 2.1. In Example 2.1, if the null hypothesis is true, $F_X = F_0$. Thus, $\mathcal{E}(X,Y) = 2E|X - Y| - E|X - X'| - E|Y - Y'| = 0$. Show that for the U-statistic, $E(\mathcal{E}_n^*) = 0$. That is, show that the U-statistic given by (2.3) is unbiased for the energy distance $\mathcal{E}(X,Y)$.

Exercise 2.2. In Example 2.2, if X, Y are independent and $F_X = F_Y$ then $\mathcal{E}(X,Y) = 2E|X - Y| - E|X - X'| - E|Y - Y'| = 0$. However, the test statistic $\frac{nm}{n+m}V_{n,m} = \frac{nm}{n+m}\mathcal{E}_{n,m}$ is not unbiased for $\mathcal{E}(X,Y)$. Show that under the null hypothesis, for independent random samples, the bias in the test statistic $\frac{nm}{n+m}\mathcal{E}_{n,m}$ is

$$E\left(\frac{nm}{n+m}\mathcal{E}_{n,m}\right) = E|X - Y|.$$

Hint: First consider the case when $n = m$.

3

Energy Distance

CONTENTS

3.1 Introduction: The Energy of Data

Let $X \in \mathbb{R}^d$, $Y \in \mathbb{R}^d$, and $E|X| + E|Y| < \infty$, where $|\cdot|$ denotes the Euclidean norm. We denote a sample of size n from distribution X using $\mathbf{X}$, an $n \times d$ data matrix with sample observations in rows. Data matrix $\mathbf{Y}$ denotes a sample from distribution Y.

Definition 3.1 (Two-sample energy statistic)**.** For data $\mathbf{X} = (X_1, X_2, \ldots, X_n)$ and $\mathbf{Y} = (Y_1, Y_2, \ldots, Y_m)$ the (potential) data energy between $\mathbf{X}$ and $\mathbf{Y}$ is defined as follows:

$$\mathcal{E}_{n,m}(\mathbf{X}, \mathbf{Y}) = \frac{2}{nm} \sum_{i=1}^{n} \sum_{j=1}^{m} |X_i - Y_j|$$

$$- \frac{1}{n^2} \sum_{i=1}^{n} \sum_{j=1}^{n} |X_i - X_j| - \frac{1}{m^2} \sum_{i=1}^{m} \sum_{i=1}^{m} |Y_i - Y_j|. \qquad (3.1)$$

We will show that the statistic (3.1) is always nonnegative and equals 0 if and only if (iff) the sets $\{X_1, X_2, \ldots, X_n\}$ and $\{Y_1, Y_2, \ldots, Y_m\}$ are equal (this of course implies that $n = m$ and the samples are identical up to a permutation of the sample indices).

DOI: 10.1201/9780429157158-3

The two-sample energy statistic (3.1) is an empirical measure of the distance between X and Y. A common one-sample problem is to measure the distance between X and a hypothesized distribution Y.

Notation: A primed random variable is an iid copy of the random variable; that is, X' is independent and identically distributed as X, and Y, Y' are iid.

Definition 3.2 (Energy statistic for goodness-of-fit). For data $\mathbf{X} = (X_1, X_2, \ldots, X_n)$ and random variable Y with finite expectation, the data energy between $\mathbf{X}$ and Y is

$$\mathcal{E}_n(\mathbf{X}, Y) = \frac{2}{n} \sum_{i=1}^n E|X_i - Y| - E|Y - Y'| - \frac{1}{n^2} \sum_{i=1}^n \sum_{j=1}^n |X_i - X_j|. \quad (3.2)$$

We will see that (3.2) is always nonnegative and equals 0 if and only if Y takes the values $X_1, X_2, \ldots, X_n$ with uniform probability $1/n$.

(Note that in $E|X_i - Y|$, X_i is a constant, an observed sample value. Refer to Section 2.1 for explicit definitions of the expected differences above.)

Example 3.1 (Two-parameter Exponential Distribution). Suppose that one wishes to test whether a variable T has a two-parameter exponential distribution, with density

$$f_T(t) = \lambda e^{-\lambda(t-\mu)}, \qquad t \geq \mu.$$

It is easy to verify that

$$E|t - T| = t - \mu + \frac{1}{\lambda}[1 - 2(1 - e^{-\lambda(t-\mu)})], \qquad t \geq \mu;$$

$$E|t - T| = \frac{1}{\lambda} + \mu - t, \qquad t < \mu;$$

$$E|T - T'| = \frac{1}{\lambda}.$$

Then, if $t_1, \ldots, t_n$ is a random sample, the energy distance between the sampled distribution and the hypothesized Exponential(μ, λ) distribution follows Definition 3.2, which is

$$\mathcal{E}_n(\mathbf{X}; T) = \frac{2}{n} \sum_{i=1}^n \left[t_i - \mu + \frac{1 - 2(1 - e^{-\lambda(t_i-\mu)})}{\lambda} \right] - \frac{1}{\lambda} - \frac{1}{n^2} \sum_{i,j=1}^n |t_i - t_j|.$$

See Section 8.2 an application of the two-parameter exponential test.

Definition 3.3 (Population energy). If $X \in \mathbb{R}^d, Y \in \mathbb{R}^d$ are independent random variables with finite expectation, then the energy between X and Y is defined as follows:

$$\mathcal{E}(X, Y) = 2E|X - Y| - E|X - X'| - E|Y - Y'|. \quad (3.3)$$

We will see that (3.3) is always nonnegative and equals 0 iff X and Y are identically distributed.

What can we do if in the above definitions the random variable X or Y does not have finite expected value? In this, case we can work with the following extension using a fractional exponent $0 < \alpha < 2$ on distance that we call α-energy.

Choose $\alpha \in (0, 2)$ such that α-moments exist: $E|X|^\alpha < \infty$ and $E|Y|^\alpha < \infty$. Then Definitions 3.1, 3.2, and 3.3 are modified replacing $|\cdot|$ with $|\cdot|^\alpha$.

Definition 3.4 (Population α-energy). If $X \in \mathbb{R}^d, Y \in \mathbb{R}^d$ are independent random variables such that $E|X|^\alpha < \infty$, $E|Y|^\alpha < \infty$, then the α-energy between X and Y is defined as follows:

$$\mathcal{E}^{(\alpha)}(X, Y) = 2E|X - Y|^\alpha - E|X - X'|^\alpha - E|Y - Y'|^\alpha. \tag{3.4}$$

The potential α-energy between two samples is defined as follows for $0 < \alpha < 2$.

Definition 3.5 (Two-sample potential α-energy).

$$\mathcal{E}_{n,m}^{(\alpha)}(\mathbf{X}, \mathbf{Y}) = \frac{2}{nm} \sum_{i=1}^{n} \sum_{j=1}^{m} |X_i - Y_j|^\alpha$$

$$- \frac{1}{n^2} \sum_{i=1}^{n} \sum_{j=1}^{n} |X_i - X_j|^\alpha - \frac{1}{m^2} \sum_{i=1}^{m} \sum_{j=1}^{m} |Y_i - Y_j|^\alpha. \tag{3.5}$$

The one-sample statistic is defined for general $0 < \alpha < 2$ as follows.

Definition 3.6 (One-sample potential α-energy).

$$\mathcal{E}_n^{(\alpha)}(\mathbf{X}, Y) = \frac{2}{n} \sum_{i=1}^{n} E|X_i - Y|^\alpha - E|Y - Y'|^\alpha - \frac{1}{n^2} \sum_{i=1}^{n} \sum_{j=1}^{n} |X_i - X_j|^\alpha. \tag{3.6}$$

Then for $0 < \alpha < 2$ if $E|X|^\alpha < \infty$ and $E|Y|^\alpha < \infty$ the α-energy distances (3.4), (3.5), and (3.6) are always nonnegative, with equality to zero under the same conditions for $\alpha - 1$.

At the endpoint, $\alpha = 2$, we lose the important property that $\mathcal{E}^{(\alpha)}(X, Y) = 0$ if and only if X and Y are identically distributed. With $\alpha = 2$ we would have

$$\mathcal{E}^{(2)}(X, Y) = 2E|X - Y|^2 - E|X - X'|^2 - E|Y - Y'|^2,$$

which is zero under the weaker condition that $E(X) = E(Y)$.

Therefore, α-energy distance is only defined for $0 < \alpha < 2$, with α strictly less than 2. However, in Chapter 9 we will see that "energy" distances with $\alpha = 2$ lead us to methods based on analysis of variance, such as ANOVA or Ward's minimum variance clustering, which are methods for finding differences in means.

3.2   The Population Value of Statistical Energy

The population value of data energy is the energy between random variables.

The (potential) energy of X with respect to Y, or vice versa, or simply the energy of (X, Y), is defined as follows.

Definition 3.7 (Energy $\mathcal{E}(X, Y)$). The (potential) energy of the independent random variables X, Y that take their values in a metric space with distance function δ is defined as

$$\mathcal{E}(X, Y) = 2E[\delta(X, Y)] - E[\delta(X, X')] - E[\delta(Y, Y')],$$

provided that these expectations are finite. This can also be viewed as the (potential) energy between the distributions of X and Y.

In Chapter 10 are theorems stating that many important metric spaces share the following properties:

(1) $\mathcal{E}(X, Y) \geq 0$ and

(2) $\mathcal{E}(X, Y) = 0$ if and only if X and Y are identically distributed.

A necessary and sufficient condition for (1) is the conditional negative definiteness of δ. (See Chapter 10 for the definition.) See also Schoenberg [1938a,b] and Horn [1972], Steutel and van Harn [2004], Berg [2008]. Metric spaces with properties (1) and (2) include all Euclidean spaces, all separable Hilbert spaces, all hyperbolic spaces, and many graphs with geodesic distances. For more details, see Theorems 3.2, 3.3, 10.1, and Chapter 10. These theorems make it possible to develop clustering methods that depend not only on cluster centers but on cluster distributions (Chapter 9) and to develop a simple new dependence measure, the distance correlation, that is zero if and only if the random variables are independent (see Chapter 12). A simple example where (1) holds but (2) fails is the taxicab metric or Manhattan distance. By the way, $E[\delta(X, X')]$ can be viewed as the δ-energy of X or of the distribution of X. For a special case, see Riesz [1938, 1949]. The potential energy of conservative vector fields in physics is always a harmonic function. In the world of data energy or statistical energy "harmonic" is replaced by "conditionally negative definite."

Before continuing with the consequences of the energy notion in metric spaces, let us see a simple example where at least one of the properties {(1), (2)} above do not hold.

Take a circle with the shortest arc length as a metric. This is also called the geodesic distance. If X is uniformly distributed on two points, the two endpoints of a diameter, and Y is uniformly distributed on 4 points, the four endpoints of two orthogonal diameters, then it is easy to compute that $\mathcal{E}(X, Y) = 0$ thus property (2) does not hold. In situations like this we can try to find a function $D = D(\delta)$ such that the corresponding energy satisfies (1) and (2). In every finite metric space, we can always find such a D. One can show that if D is zero whenever $\delta = 0$ and $D := 1 + \delta/K$ where K is a sufficiently large number, then D always satisfies (1) and (2). (The proof follows the additive constant argument in Chapter 17 on partial distance correlation.) This new metric space is a small perturbation of a simplex. If the metric space is not finite, then we can try to find a positive bounded function b and a large enough constant K such that the energy defined by $D = 0$ whenever $\delta = 0$ and $D := 1 + b(\delta)/K$ otherwise satisfies (1) and (2). In the case of the unit circle example above, one can show that the energy defined via $D := \delta(2\pi - \delta)$ satisfies (1) and (2). The same holds if D is the length of the chord that connects the points on the circle.

3.3 A Simple Proof of the Inequality

The energy inequality in Definition 3.1 when $n = m$ claims that

$$2 \sum_{i=1}^{n} \sum_{j=1}^{n} |X_i - Y_j| - \sum_{i=1}^{n} \sum_{j=1}^{n} |X_i - X_j| - \sum_{i=1}^{n} \sum_{j=1}^{n} |Y_i - Y_j| \geq 0.$$

The heart of the energy inequality is that "twice the number of *mixed* minus the number of *pure* is always nonnegative." We can prove such an inequality very easily if we replace the distances by counting.

Suppose there are n red cities x_i, $i = 1, 2, \ldots, n$, located on two sides of a line L called a river, k of them on the left side, and $n - k$ on the right side. Similarly, m green cities y_i are on the left side of the same line, and $n - m$ are on the right side. We connect two cities if they are on different sides of the river. Red cities are connected with red, greens with green, and mixed with blue. Now count the number of red, green, and blue lines.

Claim: $2 \times blue - red - green \geq 0$ and equality to 0 holds iff $k = m$.
Proof:

$$k(n - m) + m(n - k) - k(n - k) - m(n - m) = (k - m)^2 \geq 0.$$

From this very elementary proof, we can get a proof for the energy inequality if the line (river) is considered a random line and, thus, the chance that the river crosses a connecting line segment between cities is proportional to the length of the line segment.

See Morgenstern [2001] on the combinatorics idea and also Crofton [1868] on the integral geometric formula on random lines to get the energy distance inequality. Other proofs of the energy inequality will be discussed later.

Energy statistics in general are statistical functions of distances of the data. In a narrower sense, energy statistics are U-statistics or V-statistics with a symmetric kernel function of *distances* between data, or between data and a distribution. In an even narrower sense, the kernel of these U-statistics or V-statistics are energy kernels that generate data energy as we defined them before. For a brief introduction to V-statistics and U-statistics, see Chapter 2. Energy U-statistics are applied in Chapters 16 and 17. For more details, see e.g., Lee [2019].

Energy statistics so far have been defined as V-statistics. Why not work with U-statistics, which are unbiased? See the discussion in Section 2.2.4 and also in Section 14.2.4.

3.4   Energy Distance and Cramér's Distance

Many types of distances can be defined between statistical objects. The L_2 distance is one well-known and widely applied distance. Let F be the cumulative distribution function (cdf) of a random variable, and let F_n denote the empirical cdf or ecdf of a sample of size n from distribution F. The squared L_2 distance between F and F_n

$$\int_{-\infty}^{\infty} (F_n(x) - F(x))^2 \, dx \qquad (3.7)$$

was introduced in Cramér [1928]. However, Cramér's distance is not distribution-free; thus, to apply this distance for testing goodness-of-fit, the critical values must depend on F. This disadvantage was easily rectified by replacing dx in Cramér's distance by $dF(x)$, to obtain the Cramér-von Mises-Smirnov distance defined

$$\int_{-\infty}^{\infty} (F_n(x) - F(x))^2 \, dF(x). \qquad (3.8)$$

There remains, however, another important disadvantage of both the Cramér L_2 distance [Cramér, 1928] and Cramér-von-Mises-Smirnov distance. For samples from a d-dimensional space where $d > 1$, neither distance is rotation invariant. For important problems, such as testing for multivariate normality, we require rotation invariance. Below we will show how to overcome this difficulty.

Theorem 3.1 (Cramér–Energy equality on the real line). *Let X and Y be independent real-valued random variables with finite expectations and with cdf's F and G, respectively. Let X' denote an independent and identically distributed (iid) copy of X, and let Y' denote an iid copy of Y. Then (see e.g., Székely [1989, 2002])*

$$2 \int_{-\infty}^{\infty} (F(x) - G(x))^2 \, dx = 2E|X - Y| - E|X - X'| - E|Y - Y'|.$$

This observation was the starting point of energy inference.

Proof. The geometric idea of the proof is the following. If we have 4 points on the real line (X, X', Y, Y'), we connect any two of them and after that we connect the remaining two points. Then we get two intervals such that the sum of their lengths is equal to either the sum or the difference of the two farthest points and the distance of the two closest points. Now, two points of the realizations of X, X', Y, Y' are neighbors if they both are smaller than the other two or bigger than the other two. If two points are neighbors, then the remaining two are also neighbors. If X, X' and thus Y, Y' are neighbors then let Z be twice the distance of the middle two points. If X, Y and thus X', Y' are neighbors, then Z is twice the negative distance of the middle two points. In all other cases, $Z = 0$.

Thus, if $I(\cdot)$ is the indicator function, $\vee$ and $\wedge$ denote the maximum and the minimum, resp., then

$$
\begin{aligned}
Z = {} & 2I(X \vee X' < Y \wedge Y')(Y \wedge Y' - X \vee X') \\
& + 2I(Y \vee Y' < X \wedge X')(X \wedge X' - Y \vee Y') \\
& - 2I(X \vee Y < X' \wedge Y')(X' \wedge Y' - X \vee Y) \\
& - 2I(X' \vee Y' < X \wedge Y)(X \wedge Y - X' \vee Y') \\
= {} & 2[Z(1) + Z(2) - Z(3) - Z(4)],
\end{aligned}
$$

where

$$
Z(1) = \int_{-\infty}^{\infty} I(X \vee X' \le t < Y \wedge Y')dt = \int_{-\infty}^{\infty} I(X \le t < Y)I(X' \le t < Y')dt.
$$

Thus by the independence of the random variables

$$
E(Z(1)) = \int_{-\infty}^{\infty} P(X \le t < Y)P(X' \le t < Y')dt = \int_{-\infty}^{\infty} P(X \le t < Y)^2 dt.
$$

We get similar formulae for the other three Z's, thus

$$
\begin{aligned}
E(Z) = {} & 2 \int_{-\infty}^{\infty} [P(X \le t < Y)^2 + P(Y \le t < X)^2 \\
& - 2P(X \le t < Y)P(Y \le t < X)]dt \\
= {} & 2 \int_{-\infty}^{\infty} [P(X \le t < Y) - P(Y \le t < X)]^2 \ge 0.
\end{aligned}
$$

Finally,

$$
P(X \le t < Y) - P(Y \le t < X)] = P(X \le t - P(Y \le t) = F(t) - G(t),
$$

and this proves our equation. $\qquad\square$

Remark 3.1. Energy distance in one dimension can easily be rewritten in terms of order statistics. Let us see the details.

$$\sum_{i,j=1}^{n} |X_i - X_j| = 2\sum_{k=1}^{n} ((2k-1) - n)X_{(k)} \tag{3.9}$$

where $X_{(1)}, \dots, X_{(n)}$ denotes the ordered sample (see Proposition 20.1). Now denote by $U_{(k)}$ the ordered sample of the union of the X and Y samples (pooled samples). Then

$$\sum_{i=1}^{n}\sum_{j=1}^{m} |X_i - Y_j| = \sum_{k=1}^{n+m} ((2k-1) - (n+m))U_{(k)} -$$

$$- \sum_{k=1}^{n} ((2k-1) - n)X_{(k)} - \sum_{k=1}^{m} ((2k-1) - m)Y_{(k)}.$$

Since the ordering of n elements needs $O(n \log n)$ steps, we can say that the computation of energy distance in one dimension for $n \geq m$ requires $O(n \log n)$ operations.

Example 3.2 (Energy distance for exponential distributions). Suppose that X and Y are independent exponential distributions, with density functions $f_X(x) = \lambda e^{-\lambda x}$ and $f_Y(y) = \beta e^{-\beta y}$ for nonnegative x, y. The parameters λ, β are positive. Then for any constant $c \geq 0$,

$$E|c - X| = c + \frac{1}{\lambda}(1 - 2F_X(c)) = c + \frac{1}{\lambda} - \frac{2(1 - e^{-\lambda c})}{\lambda},$$

hence

$$E|X - Y| = \int_0^\infty \int_0^\infty |x - y| f_X(x) f_Y(y) \, dx \, dy$$

$$= \int_0^\infty \left(y + \frac{1}{\lambda} - \frac{2(1 - e^{-\lambda y})}{\lambda} \right) \beta e^{-\beta y} \, dy$$

$$= \frac{\lambda^2 + \beta^2}{\lambda \beta (\lambda + \beta)}.$$

Then $E|X - X'| = \frac{1}{\lambda}$, $E|Y - Y'| = \frac{1}{\beta}$ and the energy distance between random variables X and Y is

$$\mathcal{E}(X, Y) = 2E|X - Y| - E|X - X'| - E|Y - Y'|$$

$$= \frac{2(\lambda^2 + \beta^2)}{\lambda \beta (\lambda + \beta)} - \frac{1}{\lambda} - \frac{1}{\beta} = \frac{(\lambda - \beta)^2}{\lambda \beta (\lambda + \beta)} \geq 0, \tag{3.10}$$

and $\mathcal{E}(X,Y) = 0$ if and only if $\lambda = \beta$. Alternately, if we express the energy distance in terms of the means $\mu_X = E(X) = 1/\lambda$ and $\mu_Y = E(Y) = 1/\beta$, we have

$$\mathcal{E}(X,Y) = \frac{|\mu_X - \mu_Y|^2}{\mu_X + \mu_Y}.$$

It is easy to see from (3.10) that $\mathcal{E}(X,Y)$ is increasing as the difference between parameters λ and β increases. Figures 3.1(a) and 3.1(b) show the energy distance $\mathcal{E}(X,Y)$ as a function of the rate parameter λ and as a function of the mean μ_X for the case when Y is standard exponential ($\beta = \mu_Y = 1$). In both plots we see that the energy distance is zero when the parameters are equal (at $\lambda = 1$ or $\mu_x = 1$), and increasingly positive as we move away from equality.

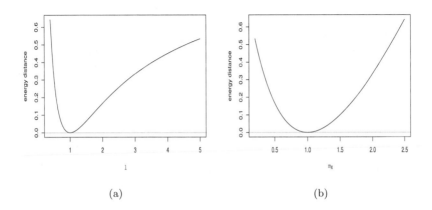

(a)   (b)

FIGURE 3.1
Energy distance $\mathcal{E}(X,Y)$ between two exponential distributions: X is exponential with rate λ (mean μ_X) and Y is standard exponential.

It is interesting to compare energy distance with Cramér's distance whose square is defined as

$$\int_{-\infty}^{\infty} [F(x) - G(x)]^2 dx,$$

where F, G are the cumulative distribution functions of X, Y, resp.

Working with the exponential distributions above we have that this integral is

$$\int_0^\infty [e^{-\lambda x} - e^{-\beta x}]^2 dx = \frac{1}{2\lambda} + \frac{1}{2\beta} - \frac{2}{\lambda + \beta} = \frac{(\lambda - \beta)^2}{2\lambda\beta(\lambda + \beta)}.$$

This is exactly half of the energy distance. Having seen Theorem 3.1 this coincidence is of course no surprise.

3.5   Multivariate Case

A rotation invariant natural extension of $2E|X - Y| - E|X - X'| - E|Y - Y'|$ for higher dimensions $d \geq 1$ is

$$2E|X - Y|_d - E|X - X'|_d - E|Y - Y'|_d, \tag{3.11}$$

where $X, Y \in \mathbb{R}^d$ are independent. Proof of the rotational invariance of this expression is straightforward, but it is *not trivial at all* that expression (3.11) is nonnegative and equals zero if and only if X and Y are identically distributed. See Theorem 3.2 below.

Definition 3.8. The energy distance between the d-dimensional independent random variables X and Y is defined as

$$\mathcal{E}(X, Y) = 2E|X - Y|_d - E|X - X'|_d - E|Y - Y'|_d, \tag{3.12}$$

where $E|X|_d < \infty$, $E|Y|_d < \infty$, X' is an iid copy of X, and Y' is an iid copy of Y. We omit the subscript d whenever it is clear in context.

Denote the Fourier-transform (characteristic function) of the probability density functions f and g by $\hat{f}$ and $\hat{g}$, respectively. Then, according to the Parseval-Plancherel formula,

$$2\pi \int_{-\infty}^{\infty} (f(x) - g(x))^2 \, dx = \int_{-\infty}^{\infty} |\hat{f}(t) - \hat{g}(t)|^2 \, dt.$$

The Fourier transform of the cdf $F(x) = \int_{-\infty}^{x} f(u) \, du$ is $\hat{f}(t)/(\imath t)$, where $\imath = \sqrt{-1}$. Thus we have

$$2\pi \int_{-\infty}^{\infty} (F(x) - G(x))^2 \, dx = \int_{-\infty}^{\infty} \frac{|\hat{f}(t) - \hat{g}(t)|^2}{t^2} \, dt. \tag{3.13}$$

The pleasant surprise is that the natural multivariate generalization of the right-hand side of (3.13) is rotation invariant and it is exactly a constant multiple of (3.12).

Theorem 3.2. *If the d-dimensional random variables X and Y are independent with $E|X|_d + E|Y|_d < \infty$, and $\hat{f}, \hat{g}$ denote their respective characteristic functions, then their energy distance is*

$$2E|X - Y|_d - E|X - X'|_d - E|Y - Y'|_d = \frac{1}{c_d} \int_{R^d} \frac{|\hat{f}(t) - \hat{g}(t)|^2}{|t|_d^{d+1}} \, dt, \tag{3.14}$$

where

$$c_d = \frac{\pi^{(d+1)/2}}{\Gamma\left(\frac{d+1}{2}\right)}, \tag{3.15}$$

and $\Gamma(\cdot)$ is the complete gamma function. Thus $\mathcal{E}(X, Y) \geq 0$ with equality to zero if and only if X and Y are identically distributed.

In view of (3.14), the square root of energy distance $\mathcal{E}(X,Y)^{1/2}$ is a metric on the set of d-variate distribution functions.

A proof of this theorem can be found in Székely and Rizzo [2005a,b]. Rather than include the proof here, we refer to the proof of a more general result, Theorem 3.3, which contains Theorem 3.2 as a special case. On the history of this Theorem, see Appendix A.

Theorem 3.3. *Let X and Y be independent d-dimensional random variables with characteristic functions $\hat{f}, \hat{g}$. If $E|X|^\alpha < \infty$ and $E|Y|^\alpha < \infty$ for some $0 < \alpha \leq 2$, then*

(i) For $0 < \alpha < 2$,

$$\mathcal{E}^{(\alpha)}(X,Y) = 2E|X - Y|^\alpha - E|X - X'|^\alpha - E|Y - Y'|^\alpha$$

$$= \frac{1}{C(d,\alpha)} \int_{R^d} \frac{|\hat{f}(t) - \hat{g}(t)|^2}{|t|^{d+\alpha}} \, dt, \tag{3.16}$$

where

$$C(d,\alpha) = 2\pi^{d/2} \frac{\Gamma(1 - \alpha/2)}{\alpha 2^\alpha \Gamma\left(\frac{d+\alpha}{2}\right)}. \tag{3.17}$$

(ii) $\mathcal{E}^{(2)}(X,Y) = 2|E(X) - E(Y)|^2$.

Statements (i) and (ii) show that for all $0 < \alpha < 2$, we have $\mathcal{E}^{(\alpha)}(X,Y) \geq 0$ with equality to zero if and only if X and Y are identically distributed; but this characterization does not hold for $\alpha = 2$ since we have equality to zero in (ii) whenever $E(X) = E(Y)$. For a generalization to metric spaces, see Theorem 10.1.

The following proof is from Székely and Rizzo [2005a].

Proof of Theorem 3.3. Statement (ii) is obvious. For (i), let $\overline{f(\cdot)}$ denote the complex conjugate of $f(\cdot)$. Notice that

$$|\hat{f}(t) - \hat{g}(t)|^2 = [\hat{f}(t) - \hat{g}(t)]\overline{[\hat{f}(t) - \hat{g}(t)]}$$

$$= [1 - \hat{f}(t)\overline{\hat{g}(t)}] + [1 - \overline{\hat{f}(t)}\hat{g}(t)] - [1 - \hat{f}(t)\overline{\hat{f}(t)}] - [1 - \hat{g}(t)\overline{\hat{g}(t)}]$$

$$= E\{[2 - \exp\{i(t, X - Y)\} - \exp\{i(t, Y - X)\}]$$

$$[1 - \exp\{i(t, X - X')\}] - [1 - \exp\{i(t, Y - Y')\}]\}$$

$$= E\left\{2[1 - \cos(t, X - Y)] - [1 - \cos(t, X - X')] - [1 - \cos(t, Y - Y')]\right\},$$

thus

$$\int_{\mathbb{R}^d} \frac{|\hat{f}(t) - \hat{g}(t)|^2}{|t|^{d+\alpha}} \, dt$$

$$= E\left[\int_{\mathbb{R}^d} \frac{2[1 - \cos(t, X - Y)] - [1 - \cos(t, X - X')] - [1 - \cos(t, Y - Y')]}{|t|^{d+\alpha}} \, dt\right].$$

Therefore, for (i), all we need is the following lemma. $\qquad\square$

Lemma 3.1. *For all $x \in \mathbb{R}^d$, if $0 < \alpha < 2$, then*

$$\int_{\mathbb{R}^d} \frac{1 - \cos(t, x)}{|t|_d^{d+\alpha}} \, dt = C(d, \alpha)|x|_d^\alpha,$$

where (t, x) represents inner product, $C(d, \alpha)$ is a constant defined in (3.17), $t \in \mathbb{R}^d$. (The integrals at $t = 0$ and $t = \infty$ are meant in the principal value sense: $\lim_{\varepsilon \to 0} \int_{\mathbb{R}^d \setminus \{\varepsilon B + \varepsilon^{-1}\overline{B}\}}$, where B is the unit ball (centered at 0) in $\mathbb{R}^d$ and $\overline{B}$ is the complement of B.)

See Section 2.3 for the proof of Lemma 3.1.

For the goodness-of-fit problem we want to compare a distribution F with a hypothesized null distribution F_0. The corresponding energy distance is a special case of Theorem 3.3.

Corollary 3.1. *Let $X \in \mathbb{R}^d$ with cdf F_X and characteristic function $\hat{f}_X$. Let $Y \in \mathbb{R}^d$ have a specified null distribution with cdf F_0 and characteristic function $\hat{f}_0$. If $E|Y|^\alpha < \infty$ for some exponent $0 < \alpha < 2$, then the α-energy distance between the distribution F_X and F_0 is*

$$\mathcal{E}^{(\alpha)}(F_X, F_0) = \frac{1}{C(d, \alpha)} \int_{\mathbb{R}^d} \frac{|\hat{f}_X(t) - \hat{f}_0(t)|^2}{|t|^{d+\alpha}} \, dt, \qquad (3.18)$$

where $C(d, \alpha)$ is the constant (3.17).

The potential α-energy between two multivariate samples is defined for $0 < \alpha < 2$ when the α moments of the variables are finite. For the sample coefficients, we replace the characteristic functions with their respective empirical characteristic functions, and apply Lemma 3.1.

Definition 3.9 (Multivariate two-sample potential α-energy). If $\mathbf{X} = \{X_1, \dots, X_n\}$ and $\mathbf{Y} = \{Y_1, \dots, Y_m\}$ are independent $\mathbb{R}^d$-valued random samples, $E|X|_d^\alpha + E|Y|_d^\alpha < \infty$, $0 < \alpha < 2$, then

$$\begin{aligned}
\mathcal{E}_{n,m}^{(\alpha)}(\mathbf{X}, \mathbf{Y}) &= \frac{1}{C(d, \alpha)} \int_{\mathbb{R}^d} \frac{|\hat{f}_n(t) - \hat{g}_n(t)|^2}{|t|^{d+\alpha}} \, dt \\
&= \frac{2}{nm} \sum_{i=1}^n \sum_{j=1}^m |X_i - Y_j|_d^\alpha - \frac{1}{n^2} \sum_{i=1}^n \sum_{j=1}^n |X_i - X_j|_d^\alpha \\
&\quad - \frac{1}{m^2} \sum_{i=1}^m \sum_{j=1}^m |Y_i - Y_j|_d^\alpha, \qquad (3.19)
\end{aligned}$$

where $C(d, \alpha)$ is the constant (3.17).

Then for $0 < \alpha < 2$, if $E|X|_d^\alpha < \infty$ and $E|Y|_d^\alpha < \infty$, $\mathcal{E}_{n,m}^{(\alpha)}$ is always non-negative, with equality to zero if and only if $\mathbf{X} = \mathbf{Y}$; that is, $n = m$ and there is a permutation of indices $\{1, \ldots, n\}$ of $\mathbf{X}$ such that the samples are identical.

Similarly, the one-sample statistic is defined for $0 < \alpha < 2$ by replacing $\hat{f}_X$ with the empirical characteristic function $\hat{f}_n$ of the sample in (3.18).

Definition 3.10 (Multivariate one-sample potential α-energy). Let F_0 be the cdf of a specified $\mathbb{R}^d$-valued random variable, with c.f. $\hat{f}_0$. If $\mathbf{X} = \{X_1, \ldots, X_n\}$ is a d-variate random sample, the empirical c.f. of the sample is $\hat{f}_n$, Y has the null distribution F_0 and $E|Y|^\alpha < \infty$, then

$$\mathcal{E}_n^{(\alpha)}(\mathbf{X}, F_0) = \frac{1}{C(d, \alpha)} \int_{\mathbb{R}^d} \frac{|\hat{f}_n(t) - \hat{f}_0(t)|^2}{|t|^{d+\alpha}} \, dt$$

$$= \frac{2}{n} \sum_{i=1}^n E|X_i - Y|_d^\alpha - E|Y - Y'|_d^\alpha - \frac{1}{n^2} \sum_{i=1}^n \sum_{j=1}^n |X_i - X_j|_d^\alpha, \quad (3.20)$$

where $C(d, \alpha)$ is the constant (3.17).

3.6 Why is Energy Distance Special?

Energy distance (3.14) is a weighted L_2 distance between characteristic functions, with weight function $w(t) = \text{const}/|t|^{d+1}$. Suppose that the following three technical conditions on the weight function hold: $w(t) > 0$, $w(t)$ is continuous, and

$$\int |\hat{f}(t) - \hat{g}(t)|^2 w(t) \, dt < \infty. \quad (3.21)$$

We claim that under these conditions, if the weighted L_2 distance between $\hat{f}$ and $\hat{g}$ is rotation invariant and scale equivariant, then $w(t) = \text{const}/|t|^{d+1}$. In other words, rotation invariance and scale equivariance (under some technical conditions) imply that the weighted L_2 distance between characteristic functions is the energy distance.

Why do we have this characterization? One can show that if two weighted L_2 distances of the type (3.21) are equal for all characteristic functions $\hat{f}$ and $\hat{g}$, then the (continuous) weight functions are also equal (for proof of a similar claim see Székely and Rizzo [2012]).

Scale equivariance and rotation invariance imply that for all real numbers a,

$$\int |\hat{f}(at) - \hat{g}(at)|^2 w(t) \, dt = |a| \times \int |\hat{f}(t) - \hat{g}(t)|^2 w(t) \, dt.$$

Introduce $s = at$. We can see that if $a \neq 0$ then

$$\int |\hat{f}(s) - \hat{g}(s)|^2 \frac{w(s/a)}{|a|} \, ds = |a| \times \int |\hat{f}(t) - \hat{g}(t)|^2 w(t) \, dt.$$

Thus $w(s/a)/|a| = |a|w(s)$. That is, if $c := w(1)$ then $w(1/a) = ca^2$, implying that $w(t) = \text{const}/|t|^{d+1}$.

Interestingly, this weight function appears in Feuerverger [1993], where it is applied for testing bivariate dependence. Although this singular weight function is "special" from the equivariance point of view, other weight functions are also applied in tests based on characteristic functions. See e.g., Gürtler and Henze [2000], Henze and Zirkler [1990], or Matsui and Takemura [2005].

If in the definition of equivariance above, we replace $|a|$ by $|a|^\alpha$ where $0 < \alpha < 2$ then we get a more general weight function which is $w(t) = \text{const}/|t|^{d+\alpha}$. Ratios of the corresponding statistics remain scale invariant.

3.7   Infinite Divisibility and Energy Distance

The fact that $w_\alpha(t) = c/|t|^{d+\alpha}$, $0 < \alpha < 2$ is a good choice for a weight function is not completely out of the blue if we know some basic facts from the theory of infinitely divisible distributions. They were introduced by the Italian statistician, Bruno de Finetti in 1929 to give a flexible model for probability and statistics. The definition is as follows: A probability distribution F is *infinitely divisible* if, for every positive n, there exist n iid random variables whose sum has the same distribution as F. Important examples are the normal, Poisson, and exponential distributions and the symmetric stable distributions with characteristic function $\exp\{-|t|^\alpha\}$, $0 < \alpha < 2$. (Note: In this context, stability is not related to any kind of physical stability. It is about the stability of iid sums.) A fundamental result is the complete description of infinitely divisible characteristic functions, i.e., characteristic functions of infinitely divisible distributions. This theorem is due to the French Paul Lévy and the Russian Alexandr Khinchin. In their theorem, called the Lévy–Khinchin representation of infinitely divisible distributions, they describe the logarithm of infinitely divisible characteristic function with the help of a so-called spectral measure. Now, our $w_\alpha(dt)$ is the spectral measure of symmetric stable distributions because the negative logarithm of the symmetric stable characteristic function, $\exp\{-|t|^\alpha\}$, can be written in the following form:

$$|t|^\alpha = \int_{\mathbb{R}^d} [1 - cos(t, x)]w_\alpha(dx).$$

Thus our weight function $w_\alpha(dx)$ plays the role of Lévy spectral measure in the Lévy–Khinchin representation of symmetric stable distributions. The representation formula above can be proved directly; see, e.g., Székely and Rizzo [2013b]. Symmetric stable characteristic functions are "the next in line" in terms of simplicity after the Gaussian. This is why our choice of w_α is not completely out of the blue. By the way, the Gaussian characteristic function would correspond to stability index $\alpha = 2$; in this case, however, the integral representation above does not hold (for more explanation, see below).

Negative logarithms of infinitely divisible characteristic functions have the following characterization due to Schoenberg [1938b].

A continuous complex function $\psi(t)$ such that $\psi(0) = 0$ is the negative logarithm of an infinitely divisible characteristic function if and only if

(i) $\psi(-t) = -\bar{\psi}(t)$ for all $t \in \mathbb{R}^d$ where the bar denotes the complex conjugate,

(ii) $\psi(t)$ is conditionally negative definite: for every finite subset A of the $\mathbb{R}^d$ such that $\sum_{t \in A} h(t) = 0$ we have

$$\sum_{s,t \in A} \psi(s - t)h(s)\bar{h}(t) \leq 0.$$

Now, there is an intimate relationship between negative definite functions and energy distance. Let P be a probability measure on the Borel sigma-algebra of $\mathbb{R}^d$ and let X, Y be two $\mathbb{R}^d$ valued (Borel measurable) random variables. If $\psi : \mathbb{R}^d \to \mathbb{R}$ is symmetric ($\psi(-t) = \psi(t)$) then

$$\mathcal{E}(X, Y) = 2E\psi(X - Y) - E\psi(X - X') - E\psi(Y - Y') \geq 0 \qquad (3.22)$$

holds whenever the expectations are finite if and only if ψ is negative definite. A proof is the following.

The definition of negative definiteness shows that

$$\int_{\mathbb{R}^d} \int_{\mathbb{R}^d} \psi(s - t)h(s)h(t)dQ(s)dQ(t) \leq 0 \qquad (3.23)$$

for any probability measure Q on the Borel sets of $\mathbb{R}^d$ where h is an integrable function such that $\int_{\mathbb{R}^d} h(t)dQ(t) = 0$. Now define $\mu(B) := P(X \in B)$ and $\nu(B) = P(Y \in B)$ for Borel sets B. Then the energy inequality (3.22) can be rewritten as

$$2 \int_{\mathbb{R}^d} \int_{\mathbb{R}^d} \psi(s - t)\, d\mu(s)d\nu(t) - \int_{\mathbb{R}^d} \int_{\mathbb{R}^d} \psi(s - t)\, d\mu(s)d\mu(t)$$

$$- \int_{\mathbb{R}^d} \int_{\mathbb{R}^d} \psi\, d\nu(s)d\nu(t) \geq 0.$$

The notation $h_1 = d\mu/dQ^*$, $h_2 = d\nu/dQ^*$ where Q^* dominates μ and ν otherwise arbitrary, and $h := h_1 - h_2$ gives

$$\int_{\mathbb{R}^d} \int_{\mathbb{R}^d} \psi(s - t)h(s)h(t)dQ^*(s)dQ^*(t) \leq 0$$

whenever $\int_{\mathbb{R}^d} h(s)dQ^*(s) = 0$. The arbitrariness of X, Y imply the arbitrariness of Q^* and h.

This also shows that there is an intimate relationship between infinitely divisible real-valued characteristic functions $\exp\{-\psi(t)\}$ and energy distance. Instead of $\psi(t) = |t|^\alpha$ we can work with other negative logarithms $\psi(t)$ of real valued infinitely divisible characteristic functions. There is a caveat, however. For the energy inequality above, if we want to be sure that equality holds if *and only if* X and Y are identically distributed (in other words, if we want the energy to be a distance between probability distributions), then we need to suppose that the negative definiteness is strict; that is $\sum_{s,t \in A} \psi(s - t)h(s)\bar{h}(t) \leq 0$ implies $\psi = 0$.

Strict negative definiteness does not hold for $\psi(t) = |t|^2$. Take for example, $d = 1, n = 3, A = \{-1, 0, 1\}$ and $h(-1) = 1, h(0) = -2, h(1) = 1$. That is why in the Gaussian case we will choose a strictly negative definite kernel instead of $|t|^2$, namely $\psi(t) = 1 - \exp\{-|t|^2\} = |t|^2 - |t|^4/2 + |t|^6/3! - \ldots$. Because $|t|^2$ is NOT *strictly* negative definite it is not surprising that in (3.22) with $\psi(t) = |t|^2$ we can have equality even if the distributions of X and Y are not equal, namely iff $E(X) = E(Y)$. If we want to guarantee that equality to 0 in the inequality (3.22) implies that X and Y are identically distributed, then we need to suppose that if in (3.23) equality holds then h is identically 0 Q-almost surely. This is called *strong negative definiteness* and, in this case, ψ is called strongly negative definite.

Remark 3.2. One might suspect (incorrectly) that strict negative definiteness is the same as strong negative definiteness. For an example of a metric space of strict negative type but not of strong negative type, see Lyons [2013] Remark 3.3. In this example, the probability measures have countable support.

See also Chapter 10 and Theorem 10.1 on energy distance

$$2E\delta(X, Y) - E\delta(X, X') - E\delta(Y, Y') \geq 0$$

for random variables X, Y taking values in a metric space $(\mathcal{X}, \delta)$.

Non-Stable kernels

A simple infinitely divisible characteristic function that is not stable is the characteristic function of the Laplace distribution, which is $1/(1 + t^2)$. According to Schoenberg's theorem above, the negative logarithm of $1/(1 + t^2)$, which is $\psi(t) = \log(1 + |t|^2)$, is conditionally negative definite and thus can be used in our energy inference. We just need to check the *strong* negative definite property. This follows from the fact that the spectral density is $dw(x) = \exp\{-|x|\}dx > 0$ everywhere. If this spectral density is $dw(x) = |x| \exp\{-|x|\}dx$ then $\psi(t) = |t|^2/(1+|t|^2)$ and this can also be applied in energy distance and energy inference. We can go even further: Bochner's classical theorem on the positive definiteness of all characteristic functions [Bochner, 1933] shows that if a random variable X is symmetric ($-X$ and X have the same distribution), then its real valued characteristic function subtracted from 1 is always conditionally negative definite, takes the value 0 at 0 and is symmetric, thus can be used as a kernel $\psi(t)$ for energy inference if the negative definiteness is strong. For further extensions via Bernstein functions see Guella [2022].

Finally, a few words about the machine learning language where they use reproducing kernel Hilbert spaces with positive definite kernels. This positive definiteness corresponds to the negative definiteness of our ψ. Interested readers can refer to Section 10.8 for more on machine learning.

3.8   Freeing Energy via Uniting Sets in Partitions

In a probability space $(\Omega, \mathcal{A}, P)$ take two independent partitions $\mathcal{P} = \{A_1, A_2, \ldots, A_n\}$ and $\mathcal{Q} = \{B_1, B_2, \ldots, B_n\}$ of Ω where $A_i \in \mathcal{A}, B_i \in \mathcal{A}, i = 1, 2, \ldots, n$. The independence of partitions means that A_i, B_j are independent for all $i, j = 1, 2, \ldots, n$; that is, $P(A_i \cap B_j) = P(A_i)P(B_j)$ for all $i, j = 1, , 2, \ldots, n$. Denote the indicator function of an event $A \in \mathcal{A}$ by $I(A)$ and introduce $X := [I(A_1), I(A_2), \ldots, I(A_n)]$ and $Y := I(B_1), B_2), \ldots, I(B_n)]$, two n-dimensional independent vectors. Define the energy distance of two partitions as follows:

$$\mathcal{E}(\mathcal{P}, \mathcal{Q}) := \mathcal{E}(X, Y) = 2E(|X - Y| - E|X - X'| - E|Y - Y'|$$

where (X', Y') and (X, Y) are iid.

Proposition 3.1. *$\mathcal{E}(\mathcal{P}, \mathcal{Q})$ is minimal if the probabilities of $A's$ and $B's$ are ordered the same way; that is, both $P(A_i)$ and $P(B_i)$ $i = 1, 2, \ldots, n$, are increasing (to be more precise: non-decreasing).*

Proof. Let us start with computing $\mathcal{E}(\mathcal{P}, \mathcal{Q})$ in terms of $P(A_i), P(B_i)$. Denote by I the indicator function. Then $[I(A_i) - I(B_i)]^2 = I(A_i) + I(B_i) - 2I(A_i \cap B_i)$, thus

$$E|X - Y| = E(\sum_{i=1}^{n}[I(A_i) + I(B_i) - 2I(A_i \cap B_i)])^{1/2}$$

$$= E(2 - 2\sum_{i=1}^{n} I(A_i \cup B_i))^{1/2}$$

$$= \sqrt{2}E(1 - I((A_i \cap B_i)))^{1/2}.$$

But in the big bracket there is an indicator function whose square root can be omitted, thus

$$E|X - Y| = \sqrt{2}[1 - P(\cup_{i=1}^{n}(A_i \cap B_i))] = \sqrt{2}(1 - \sum_{i=1}^{n} P(A_i)P(B_i)).$$

Hence, we get a very nice formula for the energy:

$$\mathcal{E}(\mathcal{P},\mathcal{Q}) = \sqrt{2}\sum_{i=1}^{n}[P(A_i)^2+P(B_i)^2-2P(A_i)P(B_i)] = \sqrt{2}\sum_{i=1}^{n}[P(A_i)-P(B_i)]^2.$$

Here $\sum_{i=1}^{n} P(A_i)^2$ and $\sum_{i=1}^{n} P(B_i)^2$ do not change if we permute the $A's$ and $B's$, thus the minimum of the energy $\mathcal{E}(\mathcal{P},\mathcal{Q})$ with respect to permutations of $A's$ and $B's$ is equivalent to the well-known claim that $\sum_{i=1}^{n} P(A_i)P(B_i)$ is maximal if $P(A_i)$ and $P(B_i)$, $i=1,2,\ldots,n$, are ordered the same way; that is, increasingly (non-decreasingly). This is true because if we fix the order of the $A's$ and permute the $B's$ and somewhere there is a pair of $B's$ where the $P(A)'s$ and $P(B)'s$ are ordered in the opposite direction then we can increase the energy via changing the order of these two $B's$. □

For the next Proposition, denote by $\mathcal{P}'$ and $\mathcal{Q}'$ the two partitions where A_1, A_2 are replaced by their union $A_1 \cup A_2$ and B_1, B_2 are also replaced by their union $B_1 \cup B_2$:

$$\mathcal{P}' = \{A_1 \cup A_2, \ldots,, A_n\}$$

and

$$\mathcal{Q}' = \{B_1 \cup B_2, \ldots,, B_n\}.$$

Proposition 3.2. *By suitable permutations of $A's$ and $B's$, we can always get A_1, A_2, B_1, B_2 such that*

$$\mathcal{E}(\mathcal{P},\mathcal{Q}) \geq \mathcal{E}(\mathcal{P}',\mathcal{Q}').$$

This proposition means that we can always free energy via uniting two pairs of suitable sets. Thus in n steps of forming suitable unions, we can always decrease the energy to 0.

Proof. Using the formula for $\mathcal{E}(\mathcal{P},\mathcal{Q})$ above we get

$$\mathcal{E}(\mathcal{P}',\mathcal{Q}') - \mathcal{E}(\mathcal{P},\mathcal{Q}) = \sqrt{2}([P(A_1) - P(B_1)] + [P(A_2) - P(B_2)])^2$$
$$- \sqrt{2}[P(A_1) - P(B_1)]^2 - \sqrt{2}[P(A_2) - P(B_2)]^2$$
$$= 2\sqrt{2}[P(A_1) - P(B_1)][P(A_2) - P(B_2)].$$

This is always at most 0 if we permute the $A's$ and $B's$ such that in the resulting permutation $[P(A_1) - P(B_1)][P(A_2) - P(B_2)] \leq 0$. Such a permutation always exists because one can always find A_i, B_i, A_j, B_j such that $[P(A_i)-P(B_i)][P(A_j)-P(B_j)] \leq 0$ otherwise $\sum_{i=1}^{n} P(A_i) = \sum_{i=1}^{n} P(B_i) = 1$ cannot hold. □

Remark 3.3. The information theoretic Kullback-Leibler divergence or relative entropy for two discrete distributions: $\{p_1, p_2, \ldots, p_n\}$ and $\{q_1, q_2, \ldots, q_n\}$ is defined as follows

$$D(p_1, p_2, \ldots, p_n \| q_1, q_2, \ldots, q_n) := \sum_{i=1}^{n} p_i \log(p_i/q_i) \geq 0.$$

(Without loss of generality we can assume all $q_i > 0$ and we can interpret $0 \log 0 = 0$. For more information on Kullbach-Leibler divergence, see Section 10.10.)

One can easily show that

$$D(p_1, p_2, \ldots, p_n || q_1, q_2, \ldots, q_n) \geq D(p_1 + p_2, p_3, \ldots, p_n || q_1 + q_2, q_3, \ldots, q_n).$$

This inequality is the information theoretic counterpart of our energy inequality above for partitions. Thus we can "free" both energy and information via uniting disjoint sets (or adding their probabilities). Our "Mind" obeys similar principles.

3.9 Applications of Energy Statistics

In this book, we will see how powerful energy distance is for statistical applications. These applications include, for example,

1. Consistent one-sample goodness-of-fit tests [Székely and Rizzo, 2005b, Rizzo, 2009, Yang, 2012, Rizzo and Haman, 2016, Haman, 2018, Móri et al., 2021].

2. Consistent multi-sample tests of equality of distributions [Székely and Rizzo, 2004a, Rizzo, 2002, 2003, Baringhaus and Franz, 2004, Rizzo and Székely, 2010].

3. Hierarchical clustering algorithms [Székely and Rizzo, 2005a] that extend and generalize the Ward's minimum variance algorithm.

4. K-groups clustering [Li, 2015a, Li and Rizzo, 2017, França et al., 2020], a generalization of k-means clustering.

5. Distance components (*DISCO*) [Rizzo and Székely, 2010], a nonparametric extension of analysis of variance for structured data.

6. Characterization and test for multivariate independence [Feuerverger, 1993, Székely et al., 2007, Székely and Rizzo, 2009a].

7. Change point analysis based on Székely and Rizzo [2004a] [Kim et al., 2009, Matteson and James, 2014].

8. Feature selection based on distance correlation [Li et al., 2012, Chen et al., 2017].

9. Uplift modeling [Jaroszewitz and Lukasz, 2015].

Several of these applications, and many more, are discussed in the following chapters. Software for energy statistics applications is available under General Public License in the *energy* package for R [Rizzo and Székely, 2022, R Core Team, 2022].

3.10 Exercises

Exercise 3.1. For the two-sample energy statistic, prove that under the null hypothesis $H_0 : X \overset{D}{=} Y$, the expected value of the test statistic $\frac{nm}{n+m}\mathcal{E}_{n,m}(\mathbf{X}, \mathbf{Y})$ is $E|X - Y|$.

Exercise 3.2. Prove that the two-sample energy test for equal distributions is consistent. Hint: Prove that under an alternative hypothesis,

$$E\left[\frac{nm}{n+m}\mathcal{E}_{n,m}(\mathbf{X}, \mathbf{Y})\right] > \frac{nm}{n+m}\mathcal{E}(X, Y).$$

Exercise 3.3. Refer to the energy goodness-of-fit test of the two-parameter exponential family $T \sim Exp(\mu, \lambda)$ described in Section 3.1. Prove that under the null hypothesis, the expected distances are

$$E|t - T| = t - \mu + \frac{1}{\lambda}(1 - 2F_T(t)), \qquad t \geq \mu;$$

$$E|T - T'| = \frac{1}{\lambda}.$$

Exercise 3.4. Using Theorem 3.2 and equation (3.13) give another proof for the Cramér-Energy equality on the real line (Theorem 3.1).

Hint: If $d = 1$ then $c_d = c_1 = \pi$ and thus if we compare the left-hand sides of (3.13) and Theorem 3.1 we get

$$2\int_{-\infty}^{\infty} (F(x) - G(x))^2 dx = \mathcal{E}(X, Y)$$

which is exactly Theorem 3.1.

Exercise 3.5. Using software, check that formula (3.9) holds for a numerical example, say `aircondit` data in the *boot* package or other small data set. If using R, see R functions `sort`, `seq`, and `dist`.

Exercise 3.6. Extend Lemma 3.1 (Lemma 2.1) for $\alpha \notin (0, 2)$. In particular, is it true that for $-d < \alpha < 0$ we have

$$\int_{\mathbb{R}^d} \frac{e^{i(t,x)}}{|t|_d^{d+\alpha}} = (d + \alpha)C(d, -\alpha)|x|_d^{\alpha} \quad ?$$

For reference, Lemma 2.1 is proved in Section 2.3.

(Notice that if we want to apply a result or lemma of this kind for energy inference then we need to assume $E|X - X'|^\alpha < \infty$ and this cannot be true if $\alpha < 0$ and X is discrete or has at least one atom (a value taken with positive probability).)

4

Introduction to Energy Inference

CONTENTS

4.1 Introduction

This chapter provides a basic introduction to hypothesis testing based on energy statistics. Later chapters develop theory, applications, and extensions of these methods. We begin with the two-sample problem of testing for equal distributions. Then we discuss the one-sample problem of goodness-of-fit, where we test whether an observed sample was drawn from a specified distribution or family of distributions.

We have introduced two types of energy distances in Chapter 3.

(i) The energy $\mathcal{E}_{n,m}(\mathbf{X}, \mathbf{Y})$ between two samples measures the distance between two sampled populations. The statistic $\frac{nm}{n+m}\mathcal{E}_{n,m}(\mathbf{X}, \mathbf{Y})$ can be applied to test $H_0 : F = G$ vs $H_1 : F \neq G$, where F and G are the cdf's of the two sampled distributions; equality means equal distributions. Usually F and G are unknown, so we implement a nonparametric test.

(ii) A goodness-of-fit energy statistic $n\mathcal{E}_n(\mathbf{X}; F_0)$ measures the distance between the sampled distribution F and a hypothesized distribution F_0. In Example 3.1, we may want to test the simple hypothesis $H_0 : f_T(t) = \lambda e^{-\lambda(t-\mu)}$, $t \geq \mu$ with parameters $\{\lambda, \mu\}$ specified. More typically, we want to test a composite hypothesis; for example, $H_0 : X$ has an exponential distribution vs $H_1 : X$ does not have an exponential distribution.

For a test criterion, we need to refer to the sampling distribution of the test statistic under the condition that H_0 is true. We know that the limit of our null sampling distribution is a quadratic form of centered Gaussian

DOI: 10.1201/9780429157158-4

variables, $Q \overset{D}{=} \sum_{i=1}^{\infty} \lambda_i Z_i^2$, where $\{Z_i\}$ are iid standard normal, and $\{\lambda_i\}$ are positive constants that depend on the distributions of $X \sim F_0$ (one-sample) or X and Y (two-sample problems). If the eigenvalues λ_i can be derived, we can obtain a numerical estimate of the cdf (applying Imhof [1961] as in Chapter 7). However, solving the integral equations for the eigenvalues is difficult and not a feasible approach except in special cases. See Chapter 7 for a discussion of computing eigenvalues for testing normality, and other numerical approaches to estimating eigenvalues.

A more practical and general approach to both problems (i) and (ii) above is to apply Monte Carlo methods to obtain a reference distribution for the test decision. In the *energy* package for R [Rizzo and Székely, 2022], we apply a nonparametric *permutation* resampling method for testing equal distributions, and a parametric Monte Carlo approach for goodness-of-fit, sometimes called a *parametric bootstrap*. Note that a parametric bootstrap does not apply resampling as in ordinary bootstrap, but a parametric simulation from a specified distribution.

4.2   Testing for Equal Distributions

Consider the problem of testing whether two random variables, X and Y, have the same cumulative distribution function. Suppose that the cdf's of X and Y are F and G, respectively. We consider the nonparametric problem, where F and G are unknown.

Recall that if $E|X| < \infty$, $E|Y| < \infty$, and X, Y are independent, the energy distance $\mathcal{E}(X, Y) = D^2(F, G)$ is

$$\mathcal{E}(X, Y) = 2E|X - Y| - E|X - X'| - E|Y - Y'|, \qquad (4.1)$$

where X, X' are iid with cdf F and Y, Y' are iid with cdf G. We have seen that $\mathcal{E}(X, Y)$ characterizes equality of distributions because $\mathcal{E}(X, Y) \geq 0$ with equality to 0 if and only if X and Y are identically distributed. The sample energy distance corresponding to (4.1) is as follows.

For independent random samples $x_1, \ldots, x_n$ and $y_1, \ldots, y_m$ from X and Y, respectively, the sample energy distance for testing $H_0 : F = G$ is

$$\mathcal{E}_{n,m}(X, Y) := 2A - B - C,$$

where A, B, and C are averages of pairwise distances between and within samples:

$$A = \frac{1}{nm} \sum_{i=1}^{n} \sum_{j=1}^{m} |x_i - y_j|, \ B = \frac{1}{n^2} \sum_{i=1}^{n} \sum_{k=1}^{n} |x_i - x_k|, \ C = \frac{1}{m^2} \sum_{i=1}^{m} \sum_{\ell=1}^{m} |y_i - y_\ell|.$$

The statistic $\mathcal{E}_{n,m}$ is always non-negative, with equality to 0 only if $n = m$ and the two samples are identical ($x_i = y_{\pi(i)}$, $i = 1, \ldots, n$, for some permutation π of the indices.) Therefore, $\mathcal{E}_{n,m}$ is a statistical distance.

When the null hypothesis of equal distributions is true, the test statistic

$$T = \frac{nm}{n+m} \mathcal{E}_{n,m}(\mathbf{X}, \mathbf{Y}).$$

converges in distribution to a quadratic form of independent standard normal random variables. Under an alternative hypothesis, the statistic T tends to infinity stochastically as sample sizes tend to infinity, so the energy test for equal distributions that rejects the null for large values of T is consistent [Székely and Rizzo, 2004a].

Example 4.1 (Birth weights). This example uses the `birthwt` data "Risk Factors Associated with Low Infant Birth Weight" from the R package *MASS* [Venables and Ripley, 2002]. We can test whether there is a significant difference in birth weight (`bwt`) for the risk factor "smoking status during pregnancy" (`smoke`). Histograms of the data are shown in Figure 4.1.

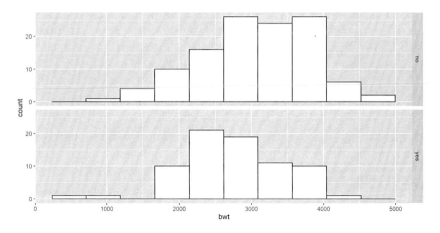

FIGURE 4.1
Birth weights (grams) by risk factor smoking status for the birthwt data.

In this example, the energy distance is $\mathcal{E}_{n,m}(\mathbf{X}, \mathbf{Y}) = 83.25031$, and the weight $\frac{nm}{n+m} = \frac{115 \cdot 74}{115+74} = 45.02646$, so the test statistic is 3748.466.

We can compute the energy two-sample test statistic using the `eqdist.e` function in the *energy* package. The R code to prepare the data and compute the energy distance is:

```
library(MASS)
library(energy)
```

```
o <- order(birthwt$smoke)   #for sorting by smoke variable
#next reorder the rows of the data frame by smoke
birthwt1 <- birthwt[o,]
sizes <- as.vector(table(birthwt$smoke))
sizes
## [1] 115   74
```

```
eqdist.e(birthwt1$bwt, sizes)
## E-statistic
##     3748.466
```

The inputs are the pooled data sample in order $X_1, \ldots, X_n, Y_1, \ldots, Y_m$ and the vector `sizes` of sample sizes, which are 115 and 74. We compute the energy statistic $\frac{nm}{n+m}\mathcal{E}_{n,m} = 3748.466$. However, the value 3748.466 of the statistic does not lead to a test decision without a frame of reference or critical value to understand whether it is statistically significant. Let's have a look at the permutation test `eqdist.etest` before continuing. The argument R is the number of permutation replicates. (The permutation test is explained in detail in the following section.)

```
eqdist.etest(birthwt1$bwt, sizes, R=999)
##
##  Univariate 2-sample E-test of equal distributions
##
## data:   sample sizes 115 74, replicates 999
## E-statistic = 3748.5, p-value = 0.004
```

At 5% significance, the test decision is to reject the null hypothesis (p-value $< .05$) and conclude that the distributions are different.

4.3   Permutation Distribution and Test

For simplicity, the following details are given specifically for a two-sample energy test of equal distributions. Permutation tests apply for many other problems, such as e.g., multi-sample tests (Chapter 9), and testing independence (Chapter 13).

Let Z be the ordered set $\{X_1, \ldots, X_n, Y_1, \ldots, Y_m\}$, indexed by

$$\nu = \{1, \ldots, n, n+1, \ldots, n+m\} = \{1, \ldots, N\}.$$

Now if we re-index Z as Z^* in a random order and partition it so that the first n elements are X^* and the last m elements are Y^*, under the null hypothesis $H_0 :$ $F = G$, the sampling distribution of $\mathcal{E}_{n,m}(\mathbf{X}, \mathbf{Y})$ is identical to the sampling distribution of $\mathcal{E}_{n,m}(\mathbf{X}^*, \mathbf{Y}^*)$. We use this property repeatedly, reordering Z by random permutations of $1, \ldots, N$ and computing the corresponding statistics $\mathcal{E}_{n,m}(\mathbf{X}^*, \mathbf{Y}^*)$ to obtain a reference distribution for our test decision.

Let $Z^* = (X^*, Y^*)$ represent a partition of the pooled sample $Z = X \cup Y$, where X^* has n elements and Y^* has $N - n = m$ elements. Then Z^* corresponds to a permutation π of the integers ν, where $Z_i^* = Z_{\pi(i)}$. There are $\binom{N}{n}$ ways to partition the pooled sample Z into two subsets of size n and m.

The Permutation Lemma [Efron and Tibshirani, 1993, p. 207] states that under $H_0 : F_X = F_Y$, a randomly selected Z^* has probability

$$\frac{1}{\binom{N}{n}} = \frac{n! \, m!}{N!}$$

of equaling any of its possible values; i.e., if $F_X = F_Y$ then all permutations are equally likely.

If $M = \binom{N}{n}$, the *permutation distribution* of $\frac{nm}{n+m} \mathcal{E}_{n,m}(\mathbf{X}^*, \mathbf{Y}^*)$ is the distribution of $T^* = T^{(1)}, \ldots, T^{(M)}$, where $T^{(j)} = \frac{nm}{n+m} \mathcal{E}_{n,m}(\mathbf{X}^{(j)}, \mathbf{Y}^{(j)})$, $\pi_j(\nu)$ is a permutation of ν and $(\mathbf{X}^{(j)}, \mathbf{Y}^{(j)})$ is the corresponding partition of $Z^{(j)} = Z_{\pi_j(1)}, \ldots, Z_{\pi_j(N)}$, $j = 1, \ldots, N$. If large values of T are significant, then the achieved significance level (ASL) of the test is

$$P(T^* \geq T_0) = \binom{N}{n}^{-1} \sum_{j=1}^{N} I(T^{(j)} \geq T_0).$$

Above we considered *all* possible permutations. In this case, the permutation test is exact. However, unless the sample size is tiny, evaluating the test statistic for all of the $\binom{N}{n}$ permutations requires an excessive and impractical amount of computing time. In practice, one applies an approximate permutation test by randomly drawing a large number R of samples without replacement. The test is significant at level α if

$$\hat{p} = \frac{1 + \sum_{r=1}^{R} I(T^{(r)} > T_0)}{1 + R} \leq \alpha.$$

The formula for $\hat{p}$ is given by Davison and Hinkley [Davison and Hinkley, 1997, p. 159], who state that "at least 99 and at most 999 random permutations should suffice." We usually choose R such that $(R + 1)\alpha$ is an integer.

To visualize the permutation distribution for the birth weight data of Example 4.1 a probability histogram of the 999 replicates generated for our test is shown in Figure 4.2. On the histogram, with the observed statistic marked by the vertical line, it is easy to see that the test is significant.

The shape of the histogram in Figure 4.2 is typical of the null sampling distribution of any energy V-statistic, although the exact shape and the cdf depend on the type of test and the underlying random variables.

Permutation tests are implemented in the *boot* package provided with R [Davison and Hinkley, 1997], by the function `boot` with `sim="permutation"`.

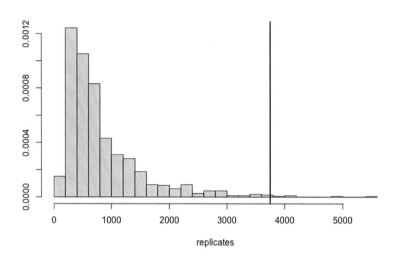

FIGURE 4.2
Permutation distribution generated for the energy test of equal distribution in Example 4.1. The observed test statistic is marked by the vertical line.

In the *energy* package, our `eqdist.etest` is coded internally in compiled C. (For the birth weight data in Example 4.1, *microbenchmark* [Mersmann, 2021] times the energy implementation about ten times faster than the *boot* version.)

Remark 4.1. Both *energy* and *boot* packages generate a random set of permutations, so in general, different sets of random permutations are generated each time the function is called. This means that the set of replicates will vary, hence the estimated p-value of the test can vary (although $\hat{p}$ should not vary greatly). In a simulation to estimate Type 1 error rates or power, $\hat{p}$ is random but the expected value of $\hat{p}$ is constant.

Remark 4.2 $(\mathcal{E}(F_n, G_m))$. The empirical cdf (ecdf) of a univariate sample $x_1, \ldots, x_n$ is

$$F_n(x) = \frac{1}{n} \sum_{i=1}^{n} I(x \leq x_i),$$

where $I(\cdot)$ is the indicator function taking values 1, 0 when its argument is true, false, respectively. Let F_n be the ecdf of the X sample, and G_m be the ecdf of the Y sample. Both F_n and G_m are cdf functions on finite sets of sizes n and m, respectively. What is the energy distance between the two distributions F_n and G_m?

The joint cdf of (X, Y) is $F_{X,Y}(x, y) = F_X(x)F_Y(y)$ by independence. Let us denote the joint ecdf of the samples by $F_{n,m}(x, y)$. The samples are independent, so we have $F_{n,m}(x, y) = F_n(x)G_m(y)$. For simplicity suppose that the sampled distributions are continuous, so that $F_n(x_i) = 1/n$, $G_m(y_j) = 1/m$, and $F_{n,m}(x_i, y_j) = 1/(nm)$.

Now define X^* as the random variable uniformly distributed on the n points of the first sample, and define Y^* as the random variable uniformly distributed on the m points of the second sample. Then

$$
\begin{aligned}
\mathcal{E}(F_n, G_m) &= 2E|X^* - Y^*| - E|X^* - X^{*\prime}| - E|Y^* - Y^{*\prime}| \\
&= 2\sum_{i,j}|X_i^* - Y_j^*|\frac{1}{nm} - \sum_{i,j}|X_i^* - X_j^*|\frac{1}{n^2} - \sum_{i,j}|Y_i^* - Y_j^*|\frac{1}{m^2} \\
&= \frac{2}{nm}\sum_{i,j}|x_i - y_j| - \frac{1}{n^2}\sum_{i,j}|x_i - x_j| - \frac{1}{m^2}\sum_{i,j}|y_i - y_j| \\
&= \mathcal{E}_{n,m}(\mathbf{X}, \mathbf{Y}).
\end{aligned}
$$

4.4 Goodness-of-Fit

Goodness-of-fit is a one-sample problem, but there are two distributions to consider: one is the hypothesized distribution and the other is the underlying distribution from which the observed sample has been drawn. Energy distance applies to compare these two distributions with a variation of the two sample energy distance.

The energy distance for this problem must be the same as $\mathcal{E}(X, Y)$ where one of the variables now represents the unknown sampled distribution. Suppose that a random sample $x_1, \ldots, x_n$ is observed and the problem is to test whether the sampled distribution F is equal to the hypothesized distribution F_0. The energy goodness-of-fit statistic is

$$
n\mathcal{E}_n = n\left(\frac{2}{n}\sum_{i=1}^{n}E|x_i - X| - E|X - X'| - \frac{1}{n^2}\sum_{i=1}^{n}\sum_{j=1}^{n}|x_i - x_j|\right), \quad (4.2)
$$

where X and X' are iid with distribution F_X, or more generally

$$
n\mathcal{E}_n = n\left(\frac{2}{n}\sum_{i=1}^{n}E|x_i - X|^\alpha - E|X - X'|^\alpha - \frac{1}{n^2}\sum_{i=1}^{n}\sum_{j=1}^{n}|x_i - x_j|^\alpha\right), \quad (4.3)
$$

where $0 < \alpha < 2$. Here the expectations are with respect to the hypothesized distribution F_0 as defined in Section 2.1.

The statistic is defined in arbitrary dimension and is not restricted by sample size. The only required condition is that X has finite α moment under the null hypothesis. Under the null hypothesis $E[n\mathcal{E}_n] = E|X - X'|^\alpha$, and the asymptotic distribution of $n\mathcal{E}_n$ is a quadratic form of centered Gaussian random variables. The rejection region is in the upper tail. Under an alternative hypothesis, $n\mathcal{E}_n$ tends to infinity stochastically as $n \to \infty$, and therefore, $n\mathcal{E}_n$ determines a consistent goodness-of-fit test.

For most applications the exponent $\alpha = 1$ (Euclidean distance) can be applied, but smaller exponents have been applied for testing distributions with heavy tails including Pareto, Cauchy, and stable distributions [Rizzo, 2009, Yang, 2012]. In addition to the examples of this chapter, see Chapter 5 for several goodness-of-fit tests that apply (4.2) and Chapter 8 for examples where (4.3) is applied. Chapter 6 covers the important special case of testing multivariate normality [Rizzo, 2002, Székely and Rizzo, 2005b], which is fully implemented in the *energy* package for R.

Example 4.2 (Exponential Distribution). In Example 3.1 we considered a goodness-of-fit test for a two-parameter exponential distribution with density

$$f(x) = \lambda e^{-\lambda(x-\mu)}, \qquad x \geq \mu, \lambda > 0.$$

The expected distances for that example were

$$E|x - X| = x - \mu + \frac{1}{\lambda}(1 - 2F_0(x)), \qquad x \geq \mu;$$

$$E|X - X'| = \frac{1}{\lambda},$$

where $F_0(x)$ is the cumulative distribution function (cdf) of X under H_0.

Suppose that we have observed a random sample and want to test the simple hypothesis that the sampled distribution is Exponential with rate λ and $\mu = 0$. Then

$$F_0(x) = 1 - e^{-\lambda x}, \qquad x \geq 0.$$

If $x_1, \ldots, x_n$ is a random sample, the energy distance between the sampled distribution and the hypothesized Exponential(λ) distribution F_0 is

$$\mathcal{E}_n(\mathbf{X}; \lambda) = \frac{2}{n} \sum_{i=1}^{n} \left[x_i + \frac{1}{\lambda}(1 - 2F_0(x_i)) \right] - \frac{1}{\lambda} - \frac{1}{n^2} \sum_{i,j=1}^{n} |x_i - x_j|.$$

The test statistic is $n\mathcal{E}_n(\mathbf{X}; \lambda)$. Here we have a univariate sample so by Proposition 20.1, $\sum_{i,j=1}^{n} |x_i - x_j| = 2 \sum_{k=1}^{n} (2k - 1 - n)x_{(k)}$, where $x_{(1)} \leq \cdots \leq x_{(n)}$ is the ordered sample.

To implement an energy goodness-of-fit test, we use a parametric simulation from the hypothesized distribution to obtain a reference distribution. The test procedure is much simpler by comparison than the permutation test for equal distributions in Section 4.3.

Algorithm 4.1 Simulation Method for an Energy Goodness-of-Fit Test, Simple Hypothesis

---

To test the simple hypothesis $H_0 : F_X = F_0$ vs $H_1 : F_X \neq F_0$ for an observed sample $\mathbf{X} = (X_1, \ldots, X_n)$ at significance level α with R replicates:
1. Compute the test statistic $T_0 = n\mathcal{E}_n(\mathbf{X}; F_0)$.
2. For $j = 1$ to R:
   (a) Generate a random sample $\mathbf{X}^* = (X_1^*, \ldots, X_n^*)$ from F_0.
   (b) Compute $T^{(j)} = n\mathcal{E}_n(\mathbf{X}^*; F_0)$.
3. Find the $100(1 - \alpha)$ percentile c_α (or $1 - \alpha$ quantile) of $T^{(1)}, \ldots, T^{(R)}$ and reject H_0 if $T_0 \geq c_\alpha$.

Algorithm 4.1 is for a test of the simple hypothesis, with a fully specified null distribution F_0, all parameters known. The algorithm must be modified to test a composite hypothesis. Any Monte Carlo approach to implementing a goodness-of-fit test (for any test statistic) depends in part on the family of distributions of interest. A modified algorithm for the energy test of normality with estimated parameters (composite hypothesis) is outlined in Section 4.5.

In Section 4.5, we focus on the energy goodness-of-fit test of normality and discuss two methods of implementing the test.

4.5 Energy Test of Univariate Normality

For a test of univariate normality, we apply the statistic (4.2). Suppose that $x_1, \ldots, x_n$ is a random sample and the null hypothesis is that the sampled distribution is normal with mean μ and variance σ^2. Then it can be derived that for any fixed x, under H_0,

$$E|x - X| = 2(x - \mu)F(x) + 2\sigma^2 f(x) - (x - \mu), \qquad (4.4)$$

where F and f denote the cdf and density of the hypothesized $N(\mu, \sigma^2)$ distribution, and

$$E|X - X'| = \frac{2\sigma}{\sqrt{\pi}}.$$

The energy goodness-of-fit statistic for testing the simple hypothesis $H : X \sim N(\mu, \sigma)$ is therefore

$$n\mathcal{E}_n(\mathbf{X}; \mu, \sigma) = n\left[\frac{2}{n}\sum_{i=1}^{n}\left\{2(x_i - \mu)F(x_i) + 2\sigma^2 f(x_i) - (x_i - \mu)\right\}\right.$$

$$\left. - \frac{2\sigma}{\sqrt{\pi}} - \frac{1}{n^2}\sum_{i,j=1}^{n}|x_i - x_j|\right]. \qquad (4.5)$$

The univariate normal distribution has the property that if X is normally distributed, then $Y = (X - c)/s$ is normally distributed, for any constants c and $s > 0$. For the simple hypothesis, we can test if $Y = (X - \mu)/\sigma$ has a standard normal distribution.

The last sum in the statistic $\mathcal{E}_n$ can be linearized in terms of the ordered sample, which allows computation in $O(n \log n)$ time:

$$\sum_{i,j=1}^{n} |X_i - X_j| = 2 \sum_{k=1}^{n} (2k - 1 - n) X_{(k)},$$

where $X_{(1)}, \dots, X_{(n)}$ denotes the ordered sample (see Proposition 20.1). Then if $y_{(1)}, \dots, y_{(n)}$ is the ordered standardized sample, the test statistic is

$$n\mathcal{E}_n(\mathbf{Y}) = n \left[\frac{2}{n} \sum_{i=1}^{n} \{ 2y_i \Phi(y_i) + 2\phi(y_i) - y_i \} \right.$$
$$\left. - \frac{2\sigma}{\sqrt{\pi}} - \frac{2}{n^2} \sum_{k=1}^{n} (2k - 1 - n) y_{(k)} \right], \qquad (4.6)$$

where $\Phi(\cdot)$ and $\phi(\cdot)$ are the standard normal cdf and density, respectively.

In practice, the parameters μ and σ are unknown, and we need to test a composite hypothesis. If both parameters are unknown, we test whether the distribution belongs to the family of normal distributions. In this case we condition on the sample. Center and scale the sample using the sample mean and the sample standard deviation, so that $Y = (X - \overline{X})/S$, where $\overline{X}$ is the sample mean, and S^2 is the unbiased estimator of σ^2. Although Y is not normal, the distribution of $Y | (\mu = \overline{X}, \sigma = S)$ is normal. The test statistic is formally the same as (4.6), but with $y_i = (x_i - \overline{x})/s$.

The estimated parameters change the critical values of the distribution but not the general shape, and the rejection region is in the upper tail. See Rizzo [2002] for a detailed proof that the test with estimated parameters is statistically consistent (also for the multivariate case).

For testing normality one can consider four types of hypotheses.

1. Simple Hypothesis: μ, σ are known and specified.
2. Composite Hypothesis: μ is unknown, σ is known.
3. Composite Hypothesis: μ is known, σ is unknown.
4. Composite Hypothesis: Both μ and σ are unknown.

As we will see in Chapter 7, the eigenvalues λ_i of the asymptotic distribution of the test statistics under H_0 are quite different for each of the four cases above. Therefore, we know that the probabilities and percentage points or critical values for test criteria will be different in each case. This shows that we cannot use the same Monte Carlo simulation for testing a simple hypothesis when we want to test a composite hypothesis.

Eigenvalues for these four cases are derived in Móri et al. [2021]. See Chapter 7 for details. A list of the largest 10 eigenvalues is given in Table 7.1. A more complete table with the first 125 eigenvalues is found as data in the *energy* package. The largest few eigenvalues are:

```
> head(EVnormal)
      i      Case1         Case2         Case3         Case4
[1,]  1 0.59454782  0.18450123  0.59454782  0.11310754
[2,]  2 0.18450123  0.11310754  0.08967162  0.08356687
[3,]  3 0.08967162  0.05300453  0.08356687  0.03911317
[4,]  4 0.05300453  0.03911317  0.03501530  0.03182242
[5,]  5 0.03501530  0.02485880  0.03182242  0.01990113
[6,]  6 0.02485880  0.01990113  0.01856336  0.01697827
```

Remark 4.3. As we should expect seeing the very different eigenvalues, these four cases have very different critical values or percentage points. For example, in Case 1 the 90th percentile $Q = \sum_i \lambda_i Z_i^2$ is at $q = 2.2$, with upper tail probability 0.09964207. If we are in Case 4, the upper tail probability at the same point 2.2 is less than .0001. In fact, for Case 4, the 90th percentile of Q is closer to 0.7. For the same reason one cannot use results of a simulation from Case 1 for testing the composite hypothesis in Case 4. The test statistics have very different null sampling distributions for each case, whether we apply the asymptotic distribution or the finite sample simulated distribution. The simulation algorithm for Case 4 is outlined below.

Remark 4.4. The eigenvalues were obtained by numerical solution to a system of integral equations [Móri et al., 2021]. In general, we cannot easily derive or compute eigenvalues; in that case we have a few other options. *Our method of simulation applies in general to testing the composite hypothesis of normality in **any dimension** $d \geq 1$.* Details for $d > 1$ are in Chapter 6. For $d = 1$ we can optionally apply the limit distribution using the pre-computed eigenvalues discussed above. Alternately for $d \geq 2$ (simple hypothesis) one can estimate eigenvalues from the sample data. See Chapter 7 for details of this method.

For the simulation method, composite hypothesis, we need to modify the simulation algorithm given in Section 4.4. To test the *composite* hypothesis of normality $H_0 : X \in \mathcal{N}$ vs $H_1 : X \notin \mathcal{N}$, where $\mathcal{N}$ is the family of univariate normal distributions, the modified simulation is outlined in Algorithm 4.2.

Example 4.3 (Testing Univariate Normality). Suppose that we want to test the iris data for normality. We are in Case 4 because the mean and variance are unknown. Case 4 is implemented two ways in the *energy* package. The `mvnorm.test` function uses parametric simulation, and it handles testing normality in any dimension. The *energy* function `normal.test` is provided for testing univariate normality by either simulation (default) or the limit distribution. Below we test the univariate iris setosa sepal length data for normality.

The first two tests are equivalent. Both apply the Monte Carlo approach, and if we use the same random number seed, the result is identical for both functions.

Algorithm 4.2 Simulation for Testing Normality Case 4: Estimated Parameters

---

1. Compute $\bar{x} = \frac{1}{n}\sum_{i=1}^{n} x_i$ and $s^2 = \frac{1}{n-1}\sum_{i=1}^{n}(x_i - \bar{x})^2$.
2. Center and scale the data: $y_i = \frac{x_i - \bar{x}}{s}$, $i = 1, \ldots, n$.
3. Compute the observed test statistic $T_0 = n\mathcal{E}_n(\mathbf{Y})$ from equation (4.6).
4. For $j = 1$ to R:

   (a) Generate $\mathbf{X}^* = (x_1^*, \ldots, x_n^*)$ from $N(\mu = 0, \sigma = 1)$.
   (b) Compute $\bar{x}^* = \frac{1}{n}\sum_{i=1}^{n} x_i^*$ and $s^{2*} = \frac{1}{n-1}\sum_{i=1}^{n}(x_i^* - \bar{x}^*)^2$.
   (c) Center and scale the data: $y_i^* = \frac{x_i^* - \bar{x}^*}{s^*}$, $i = 1, \ldots, n$.
   (d) Compute $T^{(j)} = n\mathcal{E}_n(\mathbf{Y}^*)$ from equation (4.6).

5. Find the $100(1-\alpha)$ percentile c_α (or $1-\alpha$ quantile) of $T^{(1)}, \ldots, T^{(R)}$ and reject H_0 if $T_0 \geq c_\alpha$.

```
library(energy)
x <- iris[1:50, 1]
set.seed(1)
normal.test(x, R=200)

##  Energy test of normality: estimated parameters
##
## data:  x, sample size 50, dimension 1, replicates 200
## E-statistic = 0.46503, p-value = 0.33
## sample estimates:
##       mean          sd
## 5.0060000 0.3524897

set.seed(1)
mvnorm.etest(x, R=200)

##  Energy test of multivariate normality: estimated parameters
##
## data:  x, sample size 50, dimension 1, replicates 200
## E-statistic = 0.46503, p-value = 0.33
```

Finally, our third test applies the limit distribution based on the pre-computed eigenvalues for Case 4.

```
normal.test(x, method="limit")
##
##  Energy test of normality: limit distribution
##
## data:  Case 4: composite hypothesis, estimated parameters
## statistic = 0.46503, p-value = 0.2869
## sample estimates:
##       mean          sd
## 5.0060000 0.3524897
```

We do not reject normality for this sample.

Note: One reason that we see only approximately equal p-values rather than exactly equal p-values in the two methods `method="mc"` (default) and `method="limit"` of `normal.test` is that the simulation involves random number generation. Another reason is that the eigenvalues determine the asymptotic distribution ($n = \infty$) and here our sample size is only 50, so the finite sample distribution may not have precise agreement with the asymptotic distribution, especially if n is small. The `"mc"` method, however, generates a sampling distribution that depends on n, so it does reflect the finite sample size n.

Remark 4.5. Recall that in Chapter 3, we proved that in one dimension

$$2E|X - Y| - E|X - X'| - E|Y - Y'| = 2 \int_{-\infty}^{\infty} (G(t) - F(t))^2 dt. \quad (4.7)$$

(The equality (4.7) cannot hold in higher dimensions because the right-hand side is not rotation invariant.) In the multivariate case, the energy goodness-of-fit test is a rotation invariant multivariate version of a Cramér-von Mises type test, and the power of our univariate test is similar to the power of a suitable Cramér-von Mises type test. From the right-hand side of (4.7), we get the classical "distribution free" Cramér-von Mises goodness-of-fit formula if dt is replaced by $dF(t)$ (and G is replaced by the empirical distribution F_n). When the weight function is $[F(t)(1 - F(t))]^{-1}$, we have the Anderson-Darling distance [Anderson and Darling, 1954]. In case of standard normal null F, the shape of the curve $[\psi(t)]^{-1} = F(t)(1 - F(t))$ is similar to the shape of the density $F'(t)$: their ratio is close to a constant c (empirically $c \approx 0.67$.) That is, if we replace $dF(t)/[F(t)(1 - F(t))]$ by $c^{-1} dt$, we can see that our univariate test hardly differs from the powerful Anderson-Darling test of univariate normality. In fact our simulations show that in one dimension the power of our test, even for small samples sizes, is almost the same as that of Anderson-Darling, which has extremely good 0.96 Bahadur local index for Gaussian null and location alternatives (Nikitin 1995, p. 80).

4.6   Multivariate Normality and other Energy Tests

The above energy test of univariate normality is a special case of the energy test of multivariate normality. To apply the test of multivariate normality, the function `mvnorm.etest` is applied in the same way, with the $n \times d$ data matrix (observations in rows) as the first argument. Although our next chapter is on other energy goodness-of-fit tests, interested readers could alternately jump to Chapter 6 "Testing Multivariate Normality" next.

Multisample tests for equal distributions are covered in Chapter 9. There are several applications of multisample statistics also covered in Chapter 9 including nonparametric extensions of ANOVA and two types of energy based cluster analysis.

4.7   Exercises

Exercise 4.1. Implement a goodness-of-fit test for the Exponential distribution (see Example 4.2) in R. Hint: See the help topics for `rexp` (random exponential generator), `dist` (Euclidean distances) and `replicate` (utility to construct the loop for the replicates).

Exercise 4.2. In the goodness-of-fit test procedure of Section 4.5, instead of the estimated critical value c_α, give a formula for a an estimated p-value. Hint: Source code of the *energy* goodness-of-fit test functions.

Exercise 4.3. Derive formula (4.4) for the expected distance $E|x - X|$ if X is Normal(μ, σ^2). Hint: Write the integrals in the expression as $F(x)$ or $1 - F(x)$.

5

Goodness-of-Fit

CONTENTS

5.1 Energy Goodness-of-Fit Tests

A one-sample goodness-of-fit statistic is designed to measure the distance between a hypothesized distribution F_0 and the distribution F from which an observed d-dimensional random sample $x_1, \ldots, x_n$ is drawn, $d \geq 1$. Suppose that the characteristic function of F_0 is $\hat{f}_0$, the empirical characteristic function of the sample is $\hat{f}_n$, $X, X' \in \mathbb{R}^d$ are iid with cdf F_0 and $E|X|_d < \infty$. The energy test statistic for the goodness-of-fit test $H_0 : F = F_0$ vs $H_1 : F \neq F_0$ is

$$n \, \mathcal{E}_{n,d} := \frac{n}{c_d} \int_{\mathbb{R}^d} \frac{|\hat{f}_n(t) - \hat{f}_0(t)|^2}{|t|^{d+1}} \, dt =$$

$$= n \left(\frac{2}{n} \sum_{j=1}^{n} E|x_j - X|_d - E|X - X'|_d - \frac{1}{n^2} \sum_{j,k=1}^{n} |x_j - x_k|_d \right), \quad (5.1)$$

where

$$c_d = \frac{\pi^{\frac{d+1}{2}}}{\Gamma\left(\frac{d+1}{2}\right)}. \quad (5.2)$$

If the first moment of F_0 does not exist but an $\alpha > 0$ moment is finite then we can apply the generalized α-energy distance and test statistic:

$$n \, \mathcal{E}_{n,d}^{(\alpha)} := \frac{n}{C(d,\alpha)} \int_{\mathbb{R}^d} \frac{|\hat{f}_n(t) - \hat{f}_0(t)|^2}{|t|^{d+\alpha}} \, dt =$$

$$= n \left(\frac{2}{n} \sum_{j=1}^{n} E|x_j - X|_d^{\alpha} - E|X - X'|_d^{\alpha} - \frac{1}{n^2} \sum_{j,k=1}^{n} |x_j - x_k|_d^{\alpha} \right), \quad (5.3)$$

where $0 < \alpha < 2$ and

$$C(d,\alpha) = 2\pi^{d/2} \frac{\Gamma(1 - \alpha/2)}{\alpha 2^{\alpha} \Gamma\left(\frac{d+\alpha}{2}\right)}. \quad (5.4)$$

The expectations above are taken with respect to the null distribution F_0. The derivation of formulas (5.1) and (5.3) apply Lemma 2.1.

$\mathcal{E}_n$ is a V-statistic, and the corresponding unbiased statistics are U-statistics. The kernel function for the energy goodness-of-fit V, or U-statistic is (7.3), discussed in Section 7.2. See Example 16.1 for a comparison of the energy goodness-of-fit V-statistic and U-statistic.

Under the null hypothesis, $H_0 : F = F_0$, the test statistic $n\mathcal{E}_n$ has a nondegenerate asymptotic distribution as $n \to \infty$ (see Section 7.2). Under an alternative hypothesis $H_1 : F \neq F_0$, the statistic $n\mathcal{E}_n$ tends to infinity stochastically. Hence, the energy goodness-of-fit test that rejects the null hypothesis for large values of $n\mathcal{E}_n$ is consistent against general alternatives.

Energy goodness-of-fit tests based on (5.1) and (5.3) have been implemented for testing the composite hypothesis of multivariate normality [Székely and Rizzo, 2005b], Pareto family [Rizzo, 2009], stable [Yang, 2012], and other distributions.

Let us begin by illustrating the application of energy goodness-of-fit tests with a few univariate examples.

5.2 Continuous Uniform Distribution

If a random variable has the continuous uniform distribution over an interval (a, b), then for any fixed $x \in [a, b]$, we have

$$E|x - X| = \frac{(x - a)^2}{b - a} - x + \frac{b - a}{2}$$

and $E|X - X'| = \frac{b-a}{3}$. The energy test statistic for a goodness-of-fit test of standard uniform distribution, that is, $H_0 : X \sim \text{Uniform}(0,1)$, is therefore given by

$$n\mathcal{E}_n = n \left(\frac{2}{n} \sum_{i=1}^{n} \left(X_i^2 - X_i + \frac{1}{2} \right) - \frac{1}{3} - \frac{2}{n^2} \sum_{k=1}^{n} (2k - 1 - n) X_{(k)} \right),$$

where $X_{(k)}$ denotes the k-th order statistic of the sample. Note that the linearization in the last sum, $\frac{1}{n^2} \sum_{i,j=1}^{n} |X_i - X_j| = \frac{2}{n^2} \sum_{k=1}^{n} (2k - 1 - n) X_{(k)}$, simplifies the energy statistic for any univariate sample, reducing the computational complexity to $O(n \log n)$ in the one-dimensional case.

5.3 Exponential and Two-Parameter Exponential

The energy goodness-of-fit test statistic is given in Examples 3.1 and 4.2. The energy distance between two exponential variables is discussed in Example 3.2. See also Section 8.2 for an application of the test.

5.4 Energy Test of Normality

The energy test for normality and multivariate normality is a rotation invariant, consistent goodness-of-fit test that has excellent power relative to commonly applied tests for normality. Theory is developed in Rizzo [2002] and Székely and Rizzo [2005b], and further results are in Móri and Székely [2020]. Testing normality is an important application; it is covered in detail in other chapters. The energy test of univariate normality is covered in Section 4.5. The energy test of multivariate normality is covered separately in Chapter 6, with further results in Chapter 7.

5.5 Bernoulli Distribution

The Bernoulli random variable takes two values: 1 with probability p and 0 with probability $1-p$. This can be identified with a simple biased coin flipping experiment where the Head is 1 and the Tail is 0. A sample of n elements is a sequence from $\{0,1\}$ of length n. Here, $E|k - X| = 1 - p$ if $k = 1$ and $E|k - X| = p$ if $k = 0$.

Thus if we have a random sample $Y_i \in \{0,1\}$, $i = 1, \ldots, n$ with h 1's and $n - h$ 0's, then

$$\frac{2}{n} \sum_{i=1}^{n} E|Y_i - X| = \frac{2}{n}[h(1 - p) + (n - h)p],$$

and

$$E|X - X'| = 2p(1 - p).$$

Finally

$$\frac{1}{n^2} \sum_{i=1}^{n} \sum_{j=1}^{n} |Y_i - Y_j| = \frac{2h(n - h)}{n^2}.$$

Thus the energy is

$$\mathcal{E}_n(X, \mathbf{Y}) = \frac{2}{n}[h(1 - p) + (n - h)p] - 2p(1 - p) - \frac{2h(n - h)}{n^2}.$$

This energy takes its minimum at $p = h/n$ which is of course the usual estimator of p. We can also apply the energy formula to test if the unknown $p = p_0$. We just need to generate many iid Bernoulli samples with the given p_0 and compare the simulation results with our original sample of h 1's and $n - h$ 0's.

5.6 Geometric Distribution

If X has a Geometric(p) distribution, with probability function

$$p(x) = p(1 - p)^x, \quad x = 0, 1, 2, \ldots,$$

and cdf $F(x) = 1 - q^{x+1}$ then $E(X) = q/p$ where $q = 1 - p$. In this parameterization, X is the number of failures before the first success. For any $k = 0, 1, 2, \ldots$,

$$E|k - X| = k - \frac{1 - p}{p} + \frac{2(1 - p)^{k+1}}{p} = k + 1 + \frac{1}{p}(1 - 2F(k)).$$

Alternately,

$$E|k - X| = k + \theta - 2\theta F(k - 1)$$

where $\theta = E(X) = (1 - p)/p$. We also need

$$E|X - X'| = \frac{2(1 - p)}{1 - (1 - p)^2} = \frac{2q}{1 - q^2}.$$

Geometric distributions can be applied to test if a given $0 - 1$ sequence is a realization of an iid sequence. The explanation is very simple. If the sequence is iid and the probability of 1 is $0 < p < 1$, say, then the pure blocks ("runs") have lengths that are geometrically distributed on 1, 2, ... (if the previous geometric variable is X then this variable is $Y = X + 1$). In case of odd blocks the parameter of this geometric distribution is $q = 1 - p$; in case of even blocks the parameter is p. Thus if we want to test if the $0 - 1$ sequence is iid with parameter p then we just need to check if the pure blocks have geometric distributions. There is a similar approach if we want to test if the $0 - 1$ sequence is independent and the probability that the k-th digit is 1 is p_k, $k = 1, 2, \ldots, n$.

The simplest case is when $p = 1/2$ and we want to check if a $0 - 1$ sequence is the result of fair coin flipping. The length of the longest run can easily help us to identify the sequences that are not random, but they just try to imitate randomness. According to the Erdős-Rényi law of large numbers (Erdős and Rényi [1970]) in a (random) coin flipping sequence of length n there must be a run of length $\log_2 n$ with probability tending to 1. Thus, e.g., if $n = 100$ and we do not see a run of length 6 (only 5 or fewer), then it is likely that the $0 - 1$ sequence is not the result of a fair coin flipping. For an example in R, see Example 1.6 (Run length encoding) in Rizzo [2019].

5.7 Beta Distribution

The Beta distribution with shape parameters $\alpha > 0, \beta > 0$ has pdf

$$f(x) = \frac{x^{\alpha-1}(1 - x)^\beta}{B(\alpha, \beta)} \frac{1}{}, \qquad 0 < x < 1,$$

where $B(\alpha, \beta)$ is the complete beta function. It is easy to show that the expected distance of a sample point x_i to X is

$$E|x_i - X| = 2x_i F(x_i) - x_i + \frac{\alpha}{\alpha + \beta} - 2\frac{B(\alpha + 1, \beta)}{B(\alpha, \beta)}G(x_i),$$

where F is the CDF of Beta(α, β) and G is the CDF of the Beta$(\alpha + 1, \beta)$ distribution. This expression is easy to evaluate using vectorized operations in R, and the CDF function `pbeta`.

Let's see how to evaluate this first mean for the energy statistic (5.1).

```
EyX <- function(y, a, b) {
  t1 <- 2*y*pbeta(y, a, b) - y + a/(a+b) -
          2*pbeta(y, a+1, b)*beta(a+1,b)/beta(a,b)
}
```

We can use `sapply` to compute the vector $E|x_i - X|$ for the whole sample. Here is a toy example with sample size 5.

```
set.seed(1)
n <- 5
a <- .5; b <- 1.5
x <- rbeta(5, a, b)
sapply(x, EyX, a=a, b=b)
## [1] 0.4836536 0.1969334 0.2023728 0.5861122 0.2130562
```

An expression for $E|X - X'|$ is a complicated formula involving Gauss hypergeometric functions. Instead, let us use numerical integration to compute it from our formula for $E|y - X|$.

```
integrate(EyX, 0, 1, a=a, b=b)$value
## [1] 0.375
```

The test statistic is $n\mathcal{E}_n$ where $\mathcal{E}_n$ is (5.1). The third mean can be linearized as

$$\frac{1}{n^2} \sum_{i,j=1}^{n} |X_i - X_j| = \frac{2}{n^2} \sum_{k=1}^{n} (2k - 1 - n) X_{(k)}.$$

For a test of the composite hypothesis, one can estimate parameters by method of moments or maximum likelihood. (For the MLEs see the *Rfast* package function `beta.mle` [Papadakis et al., 2022].) For estimated parameters, we apply a Monte Carlo algorithm similar to Algorithm 4.2 for the composite hypothesis in testing univariate normality.

The Dirichlet distribution is a multivariate beta distribution. There is an easy to apply nonparametric distance correlation goodness-of-fit test for Dirichlet based on an independence characterization; see Section 14.5.6 for details.

5.8 Poisson Distribution

In this section we discuss two energy-type tests for the Poisson distribution. The 'E-test' is the usual energy goodness-of-fit test which applies the statistic (5.1). The 'M-test' is a test based on deriving Poisson probabilities in terms of mean distances.

5.8.1 The Poisson E-test

For an energy test statistic (5.1) we need the expected distances for $X \sim$ Poisson(λ). If X has a Poisson distribution with mean λ, then for integers $k = 0, 1, 2, \ldots,$

$$E|k - X| = 2nP(X \leq k) - 2\lambda P(X \leq k - 1) + \lambda - k. \qquad (5.5)$$

We also need the expected distance $E|X - X'|$ under the null hypothesis. If X is a Poisson random variable with expected value λ then

$$E|X - X'| = \sum_{i=0}^{\infty} \sum_{j=0}^{\infty} |i - j| \frac{\lambda^{i+j}}{i! j!} e^{-2\lambda}$$

$$= e^{-2\lambda} \sum_{m=0}^{\infty} \lambda^m \sum_{j=0}^{m} \frac{|m - 2j|}{j!(m - j)!}.$$

Here the inner sum is

$$\sum_{j=0}^{m} \frac{|m - 2j|}{j!(m - j)!} = 2 \sum_{j=0}^{\lfloor m/2 \rfloor} \frac{m - 2j}{j!(m - j)!} = 2 \sum_{j=0}^{\lfloor m/2 \rfloor} \frac{(m - j) - j}{j!(m - j)!}$$

$$= 2 \sum_{j=0}^{\lfloor m/2 \rfloor} \frac{1}{j!(m - j - 1)!} - 2 \sum_{j=0}^{\lfloor m/2 \rfloor} \frac{1}{(j - 1)!(m - j)!}$$

$$= 2 \sum_{j=0}^{\lfloor m/2 \rfloor} \frac{1}{j!(m - j - 1)!} - 2 \sum_{j=0}^{\lfloor m/2 \rfloor - 1} \frac{1}{j!(m - j - 1)!}$$

$$= \frac{2}{\lfloor m/2 \rfloor!(m - 1\lfloor m/2 \rfloor)!}$$

$$= \frac{2}{\lfloor m/2 \rfloor!(m - 1 - \lfloor m/2 \rfloor)!} = \frac{2}{\lfloor m/2 \rfloor!\lfloor (m - 1)/2 \rfloor!}.$$

Thus

$$E|X - X'| = e^{-\lambda} \sum_{m=0}^{\infty} \frac{2}{\lfloor m/2 \rfloor!\lfloor (m - 1)/2 \rfloor!}$$

$$= 2e^{-2\lambda} \sum_{k=0}^{\infty} \left[\frac{\lambda^{2k+1}}{k!k!} + \frac{\lambda^{2k+2}}{k!(k + 1)!} \right]$$

$$= 2\lambda \sum_{k=0}^{\infty} p_k (p_k + p_{k+1}),$$

where $p_k = P(X = k)$. We have

$$E|X - X'| = 2\lambda \sum_{k=0}^{\infty} p_k (p_k + p_{k+1}).$$

Note that this immediately implies $E|X - X'| \le 4\lambda = 4Var(X)$, although the bound is not sharp.

Using the Bessel functions defined as

$$I_\alpha(x) := \frac{(x/2)^\alpha}{\Gamma(\alpha + 1)} {}_0F_1(; \alpha + 1; (1/4)x^2), \qquad (5.6)$$

we have the following computing formula:

$$E|X - X'| = 2\lambda e^{-2\lambda}[I_0(2\lambda) + I_1(2\lambda)]. \qquad (5.7)$$

The Bessel I_α function is implemented in the GNU Scientific Library [Galassi et al., 2003] and other software such as the R package *gsl* [Hankin, 2006], and R function `besselI`.

Then with equations (5.5) and (5.7) we have a computing formula for an energy goodness-of-fit test for the Poisson distribution. The energy statistic for testing whether a sample $x_1, \ldots, x_n$ is drawn from $X \sim$ Poisson(λ) with CDF $F(x)$ is

$$
\begin{aligned}
n\mathcal{E}_n = n &\left\{ \frac{2}{n} \sum_{i=1}^n |x_i - X| - E|X - X'| - \frac{1}{n^2}|x_i - x_j| \right\} \\
= n &\left\{ \frac{2}{n} \sum_{i=1}^n [2x_i F(x_i) - 2\lambda F(x_i - 1) + \lambda - x_i] \right. \\
&\left. - 2\lambda e^{-2\lambda}[I_0(2\lambda) + I_1(2\lambda)] - \frac{2}{n^2} \sum_{k=1}^n (2k - 1 - n)x_{(k)} \right\},
\end{aligned}
$$

where $x_{(1)}, \ldots, x_{(n)}$ denotes the ordered sample. Note: The Poisson cdf is easily computed in R as `ppois(x, lambda)`.

5.8.2   Probabilities in Terms of Mean Distances

Consider the mean distance $E_X|k - X|$, the weighted average distance of a fixed real number k to the possible values $0, 1, 2, \ldots$ of a discrete random variable X. That is, for each fixed k, $E_X|k - X| = \sum_{j=0}^\infty |k - j|f_X(j)$, where $f_X(\cdot)$ is the probability mass function (pmf) of X.

Theorem 5.1 (Székely and Rizzo [2004b]). *Suppose X and Y are discrete random variables taking values on the nonnegative integers, $E[X] < \infty$, and $E[Y] < \infty$. Then X and Y are identically distributed if and only if*

$$E_X|k - X| = E_Y|k - Y| \qquad (5.8)$$

holds for every nonnegative integer k.

Proof. Assume that X and Y are nonnegative discrete random variables with finite expectations. Clearly (5.8) holds for all real k whenever X and Y are identically distributed. To prove the converse, it is sufficient to show that the set of distances $\{E_X|k - X| : k = 0, 1, \ldots\}$ uniquely determines the pmf of X. Let $F_X(k) = \sum_{j=0}^{k} f_X(j)$ be the cumulative distribution function (cdf) of X. Define $m_k = E_X|k - X|$ and $\mu = E[X]$. Then $m_0 = \mu$, and for positive integers k,

$$m_k = \sum_{j=0}^{\infty} |k - j| f_X(j) = 2 \sum_{j=0}^{k-1} (k - j) f_X(j) + \mu - k$$

$$= 2kF_X(k - 1) - 2 \sum_{j=0}^{k-1} j f_X(j) - (k - \mu).$$

Let $G_X(k) = \sum_{j=0}^{k} j f_X(j)$. Then

$$m_k = 2kF_X(k - 1) - 2G_X(k - 1) - (k - \mu)$$
$$= 2(kF_X(k - 2) - G_X(k - 2)) + 2f_X(k - 1) - (k - \mu).$$

Therefore, $f_X(0) = (m_1 + 1 - m_0)/2$, and for $k = 1, 2, \ldots$,

$$f_X(k - 1) = (m_k - 2(kF_X(k - 2) - G_X(k - 2)) + k - m_0)/2.$$

Simplifying, we obtain the recursive formula

$$f_X(k - 1) = (m_k - 2 \sum_{j=0}^{k-2} (k - j) f_X(j) + k - m_0)/2, \qquad (5.9)$$

which uniquely determines the distribution of X. Therefore, by induction, X and Y are identically distributed if $E_X|k - X| = E_Y|k - Y|$ for all integers $k \geq 0$. $\qquad\square$

Remark 5.1. Theorem 5.1 holds for arbitrary random variables with finite first moment if k is replaced by an arbitrary real number x. Since the proof of this general result is more technical, we proved only the special case above.

5.8.3   The Poisson M-test

If X has a Poisson distribution with mean λ, then $m_0 = \lambda$ and

$$m_k = E|k - X| = 2nP(X \leq k) - 2\lambda P(X \leq k - 1) + \lambda - k, \qquad (5.10)$$

$k = 1, 2, \ldots$.

In the proof of Theorem 5.1, an equivalent expression for $m_k = E|k - X|$ is

$$E|k - X| = 2kP(X \leq k) + \lambda - k - 2 \sum_{j=1}^{k} j f(j). \qquad (5.11)$$

For the Poisson(λ) distribution, $k = 1, 2, \ldots$, we have

$$\sum_{j=1}^{k} j f(j) = \sum_{j=0}^{n} \frac{j e^{-\lambda} \lambda^j}{j!} = \sum_{j=1}^{n} \frac{e^{-\lambda} \lambda^x}{(j-1)!}$$

$$= \lambda e^{-\lambda} \sum_{j=0}^{k-1} \frac{\lambda^j}{j!} = \lambda \sum_{j=1}^{k-1} f(j) = \lambda F(k-1).$$

Substituting into (5.11) we obtain (5.10). An equivalent formula is

$$m_k = E|k - X| = 2(k - \lambda)F(k-1) + \lambda - k + 2kf(k). \tag{5.12}$$

Székely and Rizzo [2004b] applied Theorem 5.1 for a Poisson goodness-of-fit test. Under Poisson(λ) distribution, $m_0 = E|X| = \lambda$ and $m_1 = 2f_X(0) - (1 - \lambda)$. Hence, $F_X(0) = f_X(0) = (m_1 + 1 - \lambda)/2$. Solving for $f(k)$ in equation (5.12) yields the recursive formula

$$f_X(k) = \frac{m_{k+1} - (k + 1 - \lambda)(2F_X(k-1) - 1)}{2(k+1)}. \tag{5.13}$$

This test uses sample estimates $\hat{m}_k$ to estimate the CDF of the Poisson distribution. It is not the *energy* goodness-of-fit test although it applied expected distances.

5.8.4   Implementation of Poisson Tests

The above Poisson M-test is implemented as a Monte Carlo test (parametric bootstrap) in the *energy* package for R as `poisson.mtest`. With the estimated CDF (from the $\{\hat{m}_k\}$) one can apply e.g. any of the quadratic EDF statistics, such as Cramér-von Mises (CvM) or Anderson-Darling (AD). The *energy* function `poisson.etest` implements the E-test. To get all tests use `poisson.tests`, which is convenient for a power comparison, because the parametric bootstrap computes all test statistics on each replicated sample. It returns the test results in a data.frame. See the *energy* manual for details [Rizzo and Székely, 2022]. The following shows the output of a test on sample size 30. The sampled data is Poisson with mean 2.

	estimate	statistic	p.value	method
M-CvM	1.966667	0.002107634	0.95	M-CvM test
M-AD	1.966667	0.011806386	0.97	M-AD test
E	1.966667	0.305291965	0.78	Energy test

To compute just the test statistics see the `poisson.m` and `poisson.e` functions.

Our benchmarks find that the M-test is considerably faster than the E-test. Power studies reported in Székely and Rizzo [2004a] suggest that the Poisson M-test is usually superior to EDF tests.

Note that there is also an R function `poisson.test`, which is an exact binomial test not related to the tests described in this chapter. See the documentation [R Core Team, 2022] for details.

5.9   Energy Test for Location-Scale Families

A location-scale family $\mathcal{F}$ of distributions has the property that for every cdf $F \in \mathcal{F}$, the distribution $G = a + bF$ is a member of $\mathcal{F}$ for all real numbers a and $b > 0$. If $X \in \mathcal{F}$, $E(X) = 0$ and $Var(X) = 1$, then $Y = a + bX$ has mean a and standard deviation b, with cdf $F_Y(y) = F_X((y-a)/b)$, so the distribution of Y is parameterized by location parameter a and scale parameter b. In the multivariate case, X is a d-dimensional random vector with zero mean vector and covariance matrix equal to the d-dimensional identity matrix. If Σ is a $d \times d$ symmetric positive-semidefinite matrix (covariance matrix) and $a \in \mathbb{R}^d$, then the cdf of $Y = \Sigma^{-1/2}(X - a)$ belongs to $\mathcal{F}$.

Examples of location-scale families include normal, Cauchy, Laplace, Student t, normal, and other univariate and multivariate distributions.

Above we have discussed energy goodness-of-fit tests including the important special case of the normal distribution. An energy goodness-of-fit test can be developed for any location-scale family along similar lines.

Suppose that one wants to test if the random sample $x_1, x_2, \ldots, x_n$ comes from a given location-scale family of distribution. That is, test whether the sampled distribution is of the form $F_0(A + Bx)$ where F_0 is a cdf that is fully specified except for the unknown constants A, B. Let $y_1, y_2, \ldots, y_n$ be the transformed sample $y_i = (x_i - \bar{x})/s$ where $\bar{x}$ is the sample mean and s is the sample standard deviation of $x_1, \ldots, x_n$.

Suppose that the characteristic function of F_0 is $\hat{f}_0$, the empirical characteristic function of y_n is $\hat{f}_n$, X, X' are iid with cdf F_0. For estimation of scale we assume that $E|X|_d^2 < \infty$, the matrix $Cov(X, X)$ is non-degenerate, and for $d > 1$, $E|X - X'|_d^{-1} < \infty$ (for $d = 1$ $|X - X'|$ has continuous pdf f such that $f(0) \neq 0$).

Then a test of the hypothesis that $F_X \in \mathcal{F}$ applies the energy statistic (5.1) to the sample $y_1, \ldots, y_n$.

If the first moment of F_0 does not exist but an α moment is finite for some $0 < \alpha < 1$, then we can apply the generalized energy distance (5.3).

A parametric simulation similar to Algorithm 4.2 for testing the composite hypothesis of normality can be applied for a test decision.

5.10 Asymmetric Laplace Distribution

The Laplace distribution and its generalizations in some cases can be a useful alternative to the normal law. It is often applied to model data from heavy-tailed distributions; for example, stock market returns, currency exchange rates, or non-Gaussian noise in signal analysis. Energy tests are attractive for this type of goodness-of-fit problem because energy tests typically perform quite well when the null or alternative distribution is heavy tailed. An energy test for the Laplace and asymmetric Laplace family of distributions is developed in Rizzo and Haman [2016] and Haman [2018]. Here we present selected results originally published in Rizzo and Haman [2016].

The Laplace distribution with location parameter $\theta \in \mathbb{R}$ and scale parameter $s > 0$ is the distribution with probability density function

$$f(x; \theta, s) = \frac{1}{2s} e^{-|x-\theta|/s}, \qquad -\infty < x < \infty. \tag{5.14}$$

The standard Laplace distribution is the special case $\theta = 0$ and $s = 1$, which has variance 2.

Asymmetric Laplace is a skewed family of distributions that contains the Laplace distribution with density (5.14) as a special case. Below, our parameterization follows Definition 3.1.1 of Kotz et al. [2001].

A random variable Y has an asymmetric Laplace distribution $\mathcal{AL}(\theta, \mu, \sigma)$ if there exist parameters $\theta \in \mathbb{R}$, $\mu \in \mathbb{R}$, and $\sigma \geq 0$ such that the characteristic function (c.f.) of Y has the form

$$\psi(t) = \frac{e^{i\theta t}}{1 + \frac{1}{2}\sigma^2 t^2 - i\mu t}, \qquad -\infty < t < \infty. \tag{5.15}$$

When $\mu = 0$ (5.15) is the c.f. of a symmetric Laplace distribution.

5.10.1 Expected Distances

For deriving our results, we work with the following alternate parameterization. We write $Y \sim \mathcal{AL}^*(\theta, \kappa, \sigma)$ (see Kotz et al. [2001, Sec. 3.1]) if Y has the density function

$$f_{\theta, \kappa, \sigma}(y) = \frac{\sqrt{2}}{\sigma} \frac{\kappa}{1+\kappa^2} \begin{cases} \exp\left(-\frac{\sqrt{2}\kappa}{\sigma}|y-\theta|\right), & y \geq \theta; \\ \exp\left(-\frac{\sqrt{2}}{\sigma\kappa}|y-\theta|\right), & y < \theta. \end{cases} \tag{5.16}$$

The skewness parameters μ and κ are related by

$$\kappa = \frac{\sqrt{2}\,\sigma}{\mu + \sqrt{2\sigma^2 + \mu^2}} = \frac{\sqrt{2\sigma^2 + \mu^2} - \mu}{\sqrt{2}\,\sigma}, \qquad \mu = \frac{\sigma}{\sqrt{2}}\left(\frac{1}{\kappa} - \kappa\right). \tag{5.17}$$

If $\mu = 0$ ($\kappa = 1$) the density (5.16) is a symmetric Laplace density. The standard Laplace case is $\mathcal{AL}^*(\theta = 0, \kappa = 1, \sigma = \sqrt{2})$.

The cumulative distribution function (c.d.f.) of $Y \sim \mathcal{AL}^*(\theta, \kappa, \sigma)$ is

$$F_{\theta,\kappa,\sigma}(y) = \begin{cases} 1 - \frac{1}{1+\kappa^2} \exp\left(-\frac{\sqrt{2}\kappa}{\sigma}|y - \theta|\right), & y \geq 0; \\ \frac{\kappa^2}{1+\kappa^2} \exp\left(-\frac{\sqrt{2}}{\sigma\kappa}|y - \theta|\right), & y < 0. \end{cases} \tag{5.18}$$

Introduce

$$p_\kappa := Pr(Y > \theta) = \frac{1}{1+\kappa^2}, \qquad q_\kappa := Pr(Y \leq \theta) = \frac{\kappa^2}{1+\kappa^2},$$

and

$$\lambda := \frac{\sqrt{2}\kappa}{\sigma}, \qquad \beta := \frac{\sqrt{2}}{\kappa\sigma}.$$

In this notation, the density is

$$f_{\theta,\kappa,\sigma}(y) = \begin{cases} \lambda p_\kappa \exp\left(-\lambda|y - \theta|\right), & y \geq 0; \\ \lambda p_\kappa \exp\left(-\beta|y - \theta|\right), & y < 0, \end{cases} \tag{5.19}$$

and the cdf is

$$F_{\theta,\kappa,\sigma}(y) = \begin{cases} 1 - p_\kappa \exp\left(-\lambda|y - \theta|\right), & y \geq 0; \\ q_\kappa \exp\left(-\beta|y - \theta|\right), & y < 0. \end{cases} \tag{5.20}$$

The mean and variance of $Y \sim \mathcal{AL}^*(\theta, \kappa, \sigma)$ are

$$E[Y] = \theta + \frac{\sigma}{\sqrt{2}}\left(\frac{1}{\kappa} - \kappa\right) = \theta + \mu, \qquad Var[Y] = \frac{\sigma^2}{2}\left(\frac{1}{\kappa^2} + \kappa^2\right) = \mu^2 + \sigma^2.$$

We need to derive $E|y - Y|$, the expected distance of Y to an arbitrary constant real number y.

The expected distance of $Y \sim \mathcal{AL}^*(\theta, \kappa, \sigma)$ to the parameter θ is

$$E|Y - \theta| = \frac{\sigma}{\sqrt{2}\kappa} \frac{1 + \kappa^4}{1 + \kappa^2}.$$

[Kotz et al., 2001, (3.1.26)]

Proposition 5.1. *If* $Y \sim \mathcal{AL}^*(\theta, \kappa, \sigma)$ *then for any constant real number* y,

$$E|y - Y| = \begin{cases} y - \theta - \mu + \frac{2p_\kappa}{\lambda} \exp\left(-\lambda|y - \theta|\right), & y \geq 0; \\ -y + \theta + \mu + \frac{2q_\kappa}{\beta} \exp\left(-\beta|y - \theta|\right), & y < 0, \end{cases} \tag{5.21}$$

where $\lambda = \frac{\sqrt{2}\kappa}{\sigma}$, $\beta = \frac{\sqrt{2}}{\kappa\sigma}$, $\mu = \frac{\sigma}{\sqrt{2}}(\frac{1}{\kappa} - \kappa)$, $p_\kappa = \frac{1}{1+\kappa^2}$, *and* $q_\kappa = 1 - p_\kappa$.

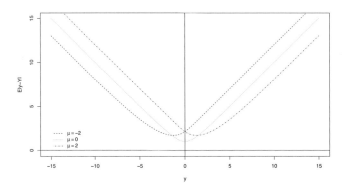

FIGURE 5.1
The expected distances $E|y - Y|$ of a point y to the distribution $Y \sim \mathcal{AL}(\theta = 0, \mu, \sigma = \sqrt{2})$ as a function of y for $\mu = -2, 0, 2$ ($\kappa \doteq 2.41, 1, 0.41$).

A proof of Proposition 5.1 is given in Section 5.14. Figure 5.1 illustrates the expected distance $E|y - Y|$ from a point y to the distribution $Y \sim \mathcal{AL}(\theta, \mu, \sigma)$ as a function of y for $\theta = 0$ and $\sigma = \sqrt{2}$.

In (5.21), μ could be re-expressed as κ using (5.17). However, we expressed (5.21) using μ for better interpretation because $E[Y] = \mu + \theta$. In the symmetric case,

$$E|y - Y| = |y - \theta| + \frac{\sigma}{\sqrt{2}} \exp\left(-\frac{\sqrt{2}}{\sigma}|y - \theta|\right).$$

As a corollary to Proposition 5.1, we obtain the mean absolute deviation of Y around $E[Y]$ and of Y around median(Y).

Corollary 5.1. *If* $Y \sim \mathcal{AL}^*(\theta, \kappa, \sigma)$ *then*

$$(i) \quad E|Y - E[Y]| = \begin{cases} \frac{\sqrt{2}\sigma\kappa^3}{(1+\kappa^2)} \exp\{(1-\kappa^2)/\kappa^2\}, & \kappa \le 1; \\ \frac{\sqrt{2}\sigma}{\kappa(1+\kappa^2)} \exp\{\kappa^2 - 1\}, & \kappa > 1. \end{cases} \quad (5.22)$$

$$(ii) \quad E|Y - \text{median}(Y)| = \frac{\sigma}{\sqrt{2}} \begin{cases} \kappa + \frac{1}{\kappa}\left[1 + \log\left(\frac{2}{1+\kappa^2}\right)\right], & \kappa \le 1; \\ \frac{1}{\kappa} + \kappa\left[1 + \log\left(\frac{2\kappa^2}{1+\kappa^2}\right)\right], & \kappa > 1. \end{cases} \quad (5.23)$$

Remark 5.2. Corollary 5.1(i) provides a correction to an error in Kotz et al. [2001, Equation (3.1.27)].

Corollary 5.1(i) is obtained by evaluating $E|y - Y|$ with $y = E[Y] = \mu + \theta$.

Corollary 5.1(ii) follows by evaluating $E|y - Y|$ at the median:

$$\text{median}(Y) = \begin{cases} \theta + \frac{\sigma}{\sqrt{2}\kappa} \log\left(\frac{2}{1+\kappa^2}\right), & \kappa \le 1; \\ \theta - \frac{\kappa\sigma}{\sqrt{2}} \log\left(\frac{2\kappa^2}{1+\kappa^2}\right), & \kappa > 1. \end{cases}$$

We derive three equivalent expressions for the expected distance $E|Y - Y'|$.

Proposition 5.2. *If $Y \sim \mathcal{AL}^*(\theta, \kappa, \sigma)$ or $Y \sim \mathcal{AL}(\theta, \mu, \sigma)$ where $\mu = \frac{\sigma}{\sqrt{2}}(\frac{1}{\kappa} - \kappa)$, then*

$$E|Y - Y'| = \frac{p_\kappa}{\beta} + \frac{q_\kappa}{\lambda} + \frac{p_\kappa^2}{\lambda} + \frac{q_\kappa^2}{\beta}. \tag{5.24}$$

$$= \frac{\sigma}{\sqrt{2}} \left(k + \frac{1}{k} - \frac{1}{k + \frac{1}{k}} \right) \tag{5.25}$$

$$= \frac{\sigma}{\sqrt{2}} \left(\sqrt{4 + \frac{2\mu^2}{\sigma^2}} - \frac{1}{\sqrt{4 + \frac{2\mu^2}{\sigma^2}}} \right), \tag{5.26}$$

where $\lambda = \frac{\sqrt{2}\kappa}{\sigma}$, $\beta = \frac{\sqrt{2}}{\kappa\sigma}$, $\mu = \frac{\sigma}{\sqrt{2}}(\frac{1}{\kappa} - \kappa)$, $p_\kappa = \frac{1}{1+\kappa^2}$, and $q_\kappa = 1 - p_\kappa$.

Proposition 5.2 is proved in Section 5.14.

5.10.2 Test Statistic and Empirical Results

Finally, we can state the computing formula for the energy goodness-of-fit statistic for testing the $\mathcal{AL}(\theta, \kappa, \sigma)$ hypothesis. If $y_1, \ldots, y_n$ is a random sample,

$$n\mathcal{E}_n = n \left(\frac{2}{n} \sum_{i=1}^{n} E|y_i - Y| - E|Y - Y'| - \sum_{i,j=1}^{n} |y_i - y_j| \right)$$

$$= n \left(\frac{2}{n} \sum_{i=1}^{n} E|y_i - Y| - E|Y - Y'| - \frac{2}{n^2} \sum_{k=1}^{n} (2k - 1 - n)y_{(k)} \right),$$

where $E|y_i - Y|$ is given by (5.21), $E|Y - Y'|$ is given by (5.24), and $y_{(1)} \le y_{(2)} \le \cdots \le y_{(n)}$ is the ordered sample.

Example 5.1. In this example, we test $H_0 : Y \sim \mathcal{AL}^*(\theta = 0, \kappa = 1, \sigma = \sqrt{2})$ (symmetric Laplace) against alternatives with varying skewness parameter κ. The power of the energy test (E) is compared with three tests: Cramér-von Mises (CVM), Anderson-Darling (AD), and Kolmogorov-Smirnov (KS). The KS test decisions were based on the p-values returned by the `ks.test` in R, and the others were implemented as Monte Carlo tests.

Power performance for sample size $n = 30$ is summarized in Figure 5.2. The proportion of significant tests at $\alpha = 0.05$ significance level is plotted for several values of κ. Power was estimated from 10,000 tests for each alternative value of κ.

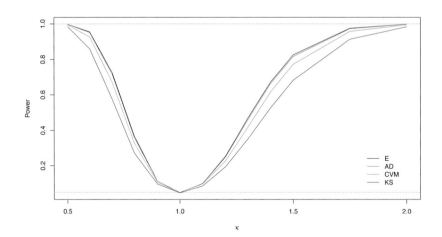

FIGURE 5.2
Power comparison of the energy test (E) vs three EDF tests. The null distribution is $\mathcal{AL}(0, \kappa = 1, \mu = 0)$ (symmetric) and the alternatives have varying skewness κ. Sample size is $n = 30$, and the significance level is 0.05.

At $\kappa = 1$ (the null hypothesis is true) observe that the Type 1 error rate is controlled at 5%. As κ decreases or increases away from 1, the power of each test is increasing. The energy test and the AD test are similar, with better power in this study than the CVM and KS tests. One advantage that the energy test has over AD is that the energy test can be extended to test bivariate or multivariate Laplace. See Haman [2018] for more details.

5.11 The Standard Half-Normal Distribution

The standard half-normal (SHN) distribution has density $f(x) = \frac{2}{\sqrt{2\pi}}e^{-x^2/2}$, $x \in \mathbb{R}$. If X has the SHN distribution, then the expected distance between a sample observation $x_i > 0$ and X is

$$E|x_i - X| = 4x_i\Phi(x_i) + 4\phi(x_i) - 3x_i + \sqrt{\frac{2}{\pi}}. \tag{5.27}$$

where $\Phi(\cdot)$ and $\phi(\cdot)$ are the standard normal cdf and density, respectively. If X and X' are iid with SHN distribution, then

$$E|X - X'| = \frac{2(2 - \sqrt{2})}{\sqrt{\pi}}. \tag{5.28}$$

An energy goodness-of-fit test for the SHN distribution has test statistic (5.1) with the mean distances given by (5.27) and (5.28). We apply this test in Section 5.12.

5.12 The Inverse Gaussian Distribution

One of the available distribution families applied to model size, lifetime, loss, etc. in finance, science, and other fields is the Inverse Gaussian or Wald Distribution.

Schrödinger [1915] derived the distribution of the first passage time for a Wiener process with positive drift. Tweedie discovered that the cumulant generating function of the first passage distribution is the inverse of the cumulant generating function of the Gaussian distribution and adopted the name 'inverse Gaussian' for the first passage time distribution [Tweedie, 1957a,b]. Wald [1947] derived the same class of distributions, and the name 'Wald distribution' is also used for the inverse Gaussian (IG) family.

The IG distribution with location parameter $\mu > 0$ and scale parameter $\lambda > 0$ has density function

$$f(x; \mu, \lambda) = \sqrt{\frac{\lambda}{2\pi x^3}} \exp\left\{\frac{-\lambda(x - \mu)^2}{2\mu^2 x}\right\}, \qquad x > 0,$$

and cdf

$$F(x; \mu, \lambda) = \Phi\left\{\sqrt{\frac{\lambda}{x}}\left(\frac{x}{\mu} - 1\right)\right\} + \exp\left\{\frac{2\lambda}{\mu}\right\}\Phi\left\{-\sqrt{\frac{\lambda}{x}}\left(\frac{x}{\mu} - 1\right)\right\},$$

where $\Phi(\cdot)$ is the standard normal cdf. The inverse Gaussian distribution has several interesting properties analogous to the properties of normal distributions: the sample mean of an IG sample is IG; $\frac{\lambda}{\mu^2 x}(x - \mu)^2$ has the $\chi^2(1)$ distribution; the sample mean and the MLE for λ are independent; the family is closed under a change of scale and the sum of iid IG variables is IG.

Ofosuhene [2020] developed an energy goodness-of-fit test for the Inverse Gaussian (IG) distribution.

If $Y \sim IG(\mu, \lambda)$, the expected distance of a fixed sample point x to Y is

$$E|x - Y| = 2xF_Y(x) + (\mu - x)$$

$$- 2\left(\mu\Phi\left[\sqrt{\frac{\lambda}{x}}\left(\frac{x}{\mu} - 1\right)\right] - \mu e^{2\lambda/\mu}\Phi\left[-\sqrt{\frac{\lambda}{x}}\left(\frac{x}{\mu} + 1\right)\right]\right). \quad (5.29)$$

See Ofosuhene [2020], Proof of Proposition 3.4 for the proof.

If X and Y are iid $IG(\mu, \lambda)$, then

$$E|X - Y| = \mu \int_0^\infty \frac{8e^{-y^2}\operatorname{erf}(y)}{\sqrt{\pi}\sqrt{y^2 + 2\phi^2}}\,dy, \quad (5.30)$$

where $\phi = \mu/\lambda$ [Girone and D'Uggento, 2016]. The closed form expression of $E|X - Y|$ is not known but its values can be numerically computed.

Then the energy test of $H_0 : X \sim IG(\mu, \lambda)$ is given by (5.1), with the mean distance formulas (5.29) and (5.30).

It is known that if $X \sim IG(\mu, \lambda)$ then

$$R = \left|\sqrt{\frac{\lambda}{x}}(x - \mu)/\mu\right|$$

has the standard half-normal (SHN) distribution with density $f(r) = \frac{2}{\sqrt{2\pi}}e^{-r^2/2}$, $r > 0$. Thus, an alternate test can be developed using the energy test for SHN distribution (see Section 5.11).

Suppose that a random sample $X_1, \ldots, X_n$ has been observed and we wish to test the composite hypothesis that $X \sim IG$. If $X \sim IG(\mu, \lambda)$ then $Y = |\sqrt{\lambda/X}(X - \mu)/\mu| \sim SHN$. To implement the test of the composite hypothesis, we estimate μ, λ with the maximum likelihood estimators $\hat{\mu} = \overline{X}$ and $\hat{\lambda} = n/\sum(X_i^{-1} - \overline{X}^{-1})$. Our test statistic is

$$n\mathcal{E}_n = n\left(\frac{2}{n}\sum_{i=1}^n\left[4y_i\Phi(y_i) - 3y_i - \sqrt{\frac{2}{\pi}} + 4\phi(y_i)\right]\right.$$

$$\left. - \frac{2(2 - \sqrt{2})}{\sqrt{\pi}} - \frac{2}{n^2}\sum_{k=1}^n(2k - 1 - n)y_{(k)})\right), \quad (5.31)$$

where $y_1, \ldots, y_n$ is the transformed sample and $y_{(1)}, \ldots, y_{(n)}$ is the ordered Y sample.

To implement the test, one can apply a parametric simulation with estimated parameters, similar to the algorithm implemented for testing univariate normality in Section 4.5; Algorithm 4.2. The estimation and transformation steps must be repeated for each random sample generated from IG, $j = 1, \ldots, R$. See Ofosuhene [2020] for the methodology and empirical results including power comparisons with competing tests.

5.13   Testing Spherical Symmetry; Stolarsky Invariance

Spherical symmetry around the origin (around 0) holds if the projection of the variable X onto the surface of the unit sphere is uniform. That is, for $X \neq 0$, $X/|X|$ is uniformly distributed on the surface of the unit sphere. With the following result, we can test for uniform distribution on the sphere.

Suppose that we want to test if a random sample $\{x_1, \ldots, x_n\}$ of points on the surface of a d-dimensional unit sphere (whose radius is 1) is from a uniform distribution on the d-dimensional unit sphere. Under the null hypothesis, the rotational invariance implies that $E|X - x| = E|X - X'|$ for every x on the unit sphere. If $d = 2$ (unit circle) then $E|X - X'| = 4/\pi$, thus the energy goodness-of-fit test statistic (5.1) is

$$n\mathcal{E}_n = \frac{4n}{\pi} - \frac{1}{n} \sum_{\ell=1}^{n} \sum_{m=1}^{n} |x_\ell - x_m|.$$

For $d = 3$ we have $E|X - X'| = 4/3$. For arbitrary dimension d, the energy test statistic for uniform distribution of points on the d-dimensional unit sphere is

$$n\mathcal{E}_n = nE|X - X'| - \frac{1}{n} \sum_{\ell=1}^{n} \sum_{m=1}^{n} |x_\ell - x_m|,$$

where

$$E|X - X'| = \frac{\left(\Gamma\left(\frac{d}{2}\right)\right)^2 2^{d-1}}{\Gamma\left(d - \frac{1}{2}\right)\Gamma\left(\frac{1}{2}\right)}. \tag{5.32}$$

For a proof see Panagiotis [2014]. The sequence (5.32) is monotone, increasing in d with limit $\sqrt{2}$. Since energy is always non-negative, we have that $E|X-X'|$ is maximum if (and in fact, only if) X and X' have uniform distributions on the sphere. (Notice that if we want to maximize the squared distances then because of $E|X - X'|^2 = 2(1 - E^2(X))$ we have that the maximum is attained if and only if $E(X) = 0$.)

If the center of symmetry is unknown, then apply the same test for the centered sample when the sample mean is subtracted from all sample elements and then projected onto the surface of the unit sphere.

The difference

$$\mathcal{E}_n = E|X - X'| - \frac{1}{n^2} \sum_{\ell=1}^{n} \sum_{m=1}^{n} |x_\ell - x_m|,$$

can be viewed as the deviation between the theoretical energy of the uniform distribution on the unit sphere and the empirical energy of the points $x_1, x_2, \ldots, x_n$ on the unit sphere. It was proved in Stolarsky [1973] that this deviation is the same as the "spherical cap discrepancy" of the points (for the exact mathematical definition see the paper). Thus if we move $x_1, x_2, \ldots, x_n$ on the unit sphere then their (empirical) energy decreases/increases if and only if their discrepancy increases/decreases and by the exact same value. This is a beautiful invariance. Because discrepancy of sequences is a well investigated area of mathematics (see, e.g., Schmidt [1969], Beck [1984]), we can translate discrepancy theorems into energy theorems. Extrema of the Riesz α-energy

$$\frac{1}{n^2} \sum_{\ell=1}^{n} \sum_{m=1}^{n} |x_\ell - x_m|^\alpha$$

has lots of interesting applications. For example,

$$\frac{1}{n^2} \sum_{\ell=1}^{n} \sum_{m=1}^{n} |x_\ell - x_m|^{-1}$$

represents the energy of n charged particles that repel each other according to Coulomb's law. Thus energy minimum can explain the location of electrons on the surface of a sphere. Similar problems on the plane are much easier to solve. In the plane, the centers and vertices of regular hexagonal tiling solve many interesting energy optimum problems. For more details, see Saff and Kuijlaars [1997].

A related problem is the so called Fekete problem: given a natural number n and a real $\alpha > 0$ find the points $x_1, \ldots, , x_n$ on the 2-sphere (in dimension 3) for which the energy

$$\sum_{1 \leq i < j \leq n} |x_i - x_j|^{-\alpha}$$

for $\alpha > 0$ is minimal. The optimal points $x_1, \ldots, x_n$ are called Fekete points. One can consider the same problem on higher dimensional spheres or on Riemannian manifolds in which case the Euclidean norm $|x_i - x_j|$ is replaced with the Riemannian distance between x_i and x_j. The first problem of this type was solved by Michael Fekete answering a question of Issai Schur. For more details, see Fekete [1923]. See also Saff and Kuijlaars [1997] and Smale [1998]. Two related problems are described in Fejes-Tóth and Fejes-Tóth [1980].

Problem 1: On a sphere distribute n points so as to maximize the least distance between pairs of them.

Problem 2: On a sphere mark n points so as to minimize the greatest distance between a point of the sphere and the mark nearest to it.

There is a jocular interpretation of these problems. On a planet n dictators face the problem of assigning the sites of their residences. The first problem is the problem of inimical dictators who want to get as far from one another as possible. The second problem is the problem of the allied dictators who want to control the planet as firmly as possible.

Typically the solution of the two problems are different: *minmax* > *maxmin* (the only known cases when they are equal is $n = 2, 3, 4, 5, 6$ and

12) but von Neumann's minimax theorem suggest that if we take the points $X_1, X_2, \ldots, X_n$ randomly according to the Haar (uniform) probability measure of the sphere then there is a "saddle point":

$$E[\min_i \max_j |X_i - X_j|] = E[\max_j \min_i |X_i - X_j|].$$

In other words, the Haar measure is an optimal (mixed/random) strategy for dictators on the planet.

Another type of interesting application of minimum energy points is the discretization of manifolds [Hardin and Saff, 2004].

5.14 Proofs

The following proofs relate to goodness-of-fit for the asymmetric Laplace family of distributions in Section 5.10.

Proof of Proposition 5.1. Suppose that $Y \sim \mathcal{AL}^*(\theta, \kappa, \sigma)$ and $x \in R$ is constant. Then

$$
\begin{aligned}
E|x - Y| &= \int_{x \leq y} (y - x) f_Y(y) dy + \int_{x > y} (x - y) f_Y(y) dy \\
&= x(2F_Y(x) - 1) - E(Y) + 2 \int_x^\infty y f_Y(y) dy \qquad (5.33) \\
&= x(2F_Y(x) - 1) + E(Y) - 2 \int_{-\infty}^x y f_Y(y) dy. \qquad (5.34)
\end{aligned}
$$

Case 1: $x \geq \theta$. In this case $y \geq x \geq \theta$ in the integrand in (5.33), so that

$$\int_x^\infty y f_Y(y) dy = \int_x^\infty y p_\kappa \lambda e^{-\lambda|y - \theta|} dy = \frac{p_\kappa}{\lambda} e^{-\lambda|x - \theta|}(\lambda x + 1).$$

Thus (5.33) can be simplified to

$$
\begin{aligned}
E|x - Y| &= x(2(1 - p_\kappa e^{-\lambda|x - \theta|}) - 1) - E(Y) - \frac{p_\kappa}{\lambda} e^{-\lambda|x - \theta|}(\lambda x + 1) \\
&= x - \theta - \mu + \frac{2p_\kappa}{\lambda} e^{-\lambda|x - \theta|}, \qquad x \geq \theta.
\end{aligned}
$$

Case 2: $x < \theta$. In this case $y \leq x < \theta$ in the integrand in (5.34), so that

$$\int_{-\infty}^x y f_Y(y) dy = \int_{-\infty}^x y p_\kappa \lambda e^{-\beta|y - \theta|} dy = \frac{p_\kappa \lambda}{\beta^2} e^{-\beta|x - \theta|}(x\beta - 1).$$

Hence using (5.34) we have

$$E|x - Y| = x(2F_Y(x) - 1) + E(Y) - 2\int_{-\infty}^{x} y f_Y(y) dy$$

$$= x\left(2q_\kappa e^{-\beta|x-\theta|} - 1\right) + \theta + \mu - \frac{2p_\kappa \lambda}{\beta^2} e^{-\beta|x-\theta|}(x\beta - 1)$$

$$= -x + \theta + \mu - 2xe^{-\beta|x-\theta|}\left(q_\kappa - \frac{p_\kappa \lambda}{\beta}\right) + \frac{2\lambda p_\kappa}{\beta^2} e^{-\beta|x-\theta|}$$

$$= -x + \theta + \mu + \frac{2q_\kappa}{\beta} e^{-\beta|x-\theta|}, \qquad x < \theta.$$

In the last step we used the identities $q_\kappa \beta = p_\kappa \lambda$ and $\frac{\lambda p_\kappa}{\beta^2} = \frac{q_\kappa}{\beta}$.

$\square$

Proof of Proposition 5.2. Since $E|(Y - \theta) - (Y' - \theta)| = E|Y - Y'|$, without loss of generality we can suppose that $\theta = 0$. Then if $Y \sim \mathcal{AL}^*(0, \kappa, \sigma)$ we have $E|Y - Y'| = L + U$, where, by Proposition 5.1,

$$L = \int_{-\infty}^{0}\left(-y + \mu + \frac{2q_\kappa}{\beta} e^{-\beta|y|}\right) \cdot p_\kappa \lambda e^{-\beta|y|} dy$$

$$= \frac{q_\kappa}{\beta} + \mu q_\kappa + \frac{q_\kappa \lambda}{\beta^2} = \frac{q_\kappa}{\beta} + \mu q_\kappa + \frac{q_\kappa^2}{\beta},$$

and

$$U = \int_{0}^{\infty}\left(y - \mu + \frac{2p_\kappa}{\lambda} e^{-\lambda|y|}\right) \cdot p_\kappa \lambda e^{-\lambda|y|} dy = \frac{p_\kappa}{\lambda} - \mu p_\kappa + \frac{p_\kappa^2}{\lambda}.$$

Observe that $\mu = \frac{\sqrt{2}}{\sigma}(k - \frac{1}{k})$ implies $\mu = \frac{1}{\lambda} - \frac{1}{\beta}$. Substituting, we obtain

$$E|Y - Y'| = \frac{q_\kappa}{\beta} + \frac{p_\kappa}{\lambda} + \frac{q_\kappa^2}{\beta} + \frac{p_\kappa^2}{\lambda} + \mu(q_\kappa - p_\kappa)$$

$$= \frac{q_\kappa}{\beta} + \frac{p_\kappa}{\lambda} + \frac{q_\kappa^2}{\beta} + \frac{p_\kappa^2}{\lambda} + \frac{q_\kappa}{\lambda} - \frac{q_\kappa}{\beta} - \frac{p_\kappa}{\lambda} + \frac{p_\kappa}{\beta}$$

$$= \frac{p_\kappa}{\beta} + \frac{q_\kappa}{\lambda} + \frac{p_\kappa^2}{\lambda} + \frac{q_\kappa^2}{\beta}.$$

To prove (5.26) we substitute $p_\kappa = \frac{\beta}{\lambda + \beta}$, $q_\kappa = \frac{\lambda}{\lambda + \beta}$, and apply several identities of the type $\frac{\lambda}{\beta} = \kappa^2$, $\lambda + \beta = \frac{\sqrt{2}}{\sigma}(k + \frac{1}{k})$, etc. After lengthy algebraic manipulation we obtain (5.25):

$$\frac{p_\kappa}{\beta} + \frac{q_\kappa}{\lambda} + \frac{p_\kappa^2}{\lambda} + \frac{q_\kappa^2}{\beta} = \frac{\sigma}{\sqrt{2}}\left(k + \frac{1}{k} - \frac{1}{k + \frac{1}{k}}\right)$$

and the right hand side equals (5.26) using another identity

$$\sqrt{4 + \frac{2\mu^2}{\sigma^2}} = k + \frac{1}{k}.$$

$\square$

5.15 Exercises

Exercise 5.1. Derive equation (5.7). That is, prove that if $0 < y < 1$ and $X \sim Beta(\alpha, \beta)$ then

$$E|y - X| = 2yF(y) - y + \frac{\alpha}{\alpha + \beta} - 2\frac{B(\alpha + 1, \beta)}{B(\alpha, \beta)}G(y),$$

where G is the CDF of Beta($\alpha + 1, \beta$).

Exercise 5.2. Do a small power study to investigate whether the Poisson M-test and the energy test (E-test) have different power performance. Some alternatives to consider could be Poisson mixtures (different λ), negative binomial, zero-modified or zero-inflated Poisson, or other counting distributions.

Exercise 5.3. Run a benchmark comparison to compare the running times of the Poisson M-test and E-test. Is there a noticeable difference? The *microbenchmark* package for R is convenient for this [Mersmann, 2021].

Exercise 5.4. Derive formulas (5.27) and (5.28).

Exercise 5.5. Write an R function to compute the energy test statistic for a goodness-of-fit test of the simple hypothesis $H_0 : X \sim Beta(\alpha, \beta)$, where α, β are specified. How would you modify the function for the composite hypothesis?

6

Testing Multivariate Normality

CONTENTS

6.1   Energy Test of Multivariate Normality

The energy goodness-of-fit test for multivariate normality is developed in Rizzo [2002] and Székely and Rizzo [2005b]. As we have seen in Chapter 3, an energy goodness-of-fit test is a type of L_2 test. See Ebner and Henze [2020] for a review of L_2 tests of multivariate normality, including the energy test, and the comments in Székely and Rizzo [2020]. The asymptotic distribution of the test statistic is the topic of Móri, Székely, and Rizzo [2021].

The energy test of multivariate normality is a consistent, rotation-invariant test that applies in arbitrary dimension. Tests for the simple hypothesis with all parameters specified, and the composite hypothesis with estimated parameters, are each discussed below. Results of a comparative power study vs seven other tests of multivariate normality are summarized, as well as a projection pursuit type of test. See Chapter 5 for a detailed discussion of the energy test of univariate normality. Chapter 7 covers the asymptotic distribution and solution to the integral equations that determine the eigenvalues of the distribution.

DOI: 10.1201/9780429157158-6

Considering the close relation (see Chapters 2 and 4) between $\widehat{\mathcal{E}}$ and Anderson-Darling (AD), multivariate $\widehat{\mathcal{E}}_{n,d}$ can be viewed as a natural way to lift AD, a powerful EDF test, to arbitrarily high dimension.

For testing goodness-of-fit, we apply the kernel function

$$h(x, y; \alpha) = E|x - X|_d^\alpha + E|y - X|_d^\alpha - E|X - X'|_d^\alpha - |x - y|_d^\alpha, \qquad (6.1)$$

where X has the hypothesized null distribution, for arbitrary exponents $0 < \alpha < 2$ and arbitrary dimension $d \geq 1$. Then the α-energy goodness-of-fit statistic, $\mathcal{E}_{n,d,\alpha}$, is the V-statistic with kernel $h(x, y; \alpha)$, which is

$$\mathcal{E}_{n,d}^{(\alpha)} := \frac{2}{n} \sum_{j=1}^{n} E|x_j - X|_d^\alpha - E|X - X'|_d^\alpha - \frac{1}{n^2} \sum_{j,k=1}^{n} |x_j - x_k|_d^\alpha,$$

where $0 < \alpha < 2$. For a test of normality or multivariate normality, X has the hypothesized normal distribution, and $E|X| < \infty$ so exponent $\alpha = 1$ is the simplest choice. The test statistic is $n\mathcal{E}_{n,d}$.

Remark 6.1. The choice $\alpha = 1$ (Euclidean distance) is the simplest case and it is applicable for testing normality in arbitrary dimension. We have found no advantage to choose $\alpha \neq 1$ for this problem, but the formulas below are given for general $0 < \alpha < 2$ because they may be applicable in other problems. Intermediate results in the derivation of $E|y - Z|_d^\alpha$ include fractional moments of non-central Chi, for example.

We derive the expected distances in the following sections, and break the testing procedure into four cases, starting with the simple hypothesis (all parameters are known).

6.1.1   Simple Hypothesis: Known Parameters

The energy test of multivariate normality is rigid motion invariant and consistent. When we apply the test to standardized samples, it is affine invariant.

First we develop the test for the case when the parameters are specified. The null hypothesis is $H_0 : F_X = F_0$ where F_0 is the cdf of a $N_d(\mu, \Sigma)$ distribution with mean vector μ and covariance matrix Σ completely specified. In this case, the data can be transformed to standard multivariate normal $N_d(0, I_d)$, where I_d is the $d \times d$ identity matrix. For an observed sample $\mathbf{X} = (X_1, \ldots, X_n)$, let $Y_i = \Sigma^{-1/2}(X_i - \mu)$, $i = 1, \ldots, n$. Then the problem is to test $H_0 : Y \sim N_d(0, I_d)$.

For standard multivariate normal $Z \in \mathbb{R}^d$ with mean vector 0 and identity covariance matrix,

$$E|Z - Z'|_d = \sqrt{2}E|Z|_d = 2 \frac{\Gamma\left(\frac{d+1}{2}\right)}{\Gamma\left(\frac{d}{2}\right)}.$$

Then, if $\{y_1, \ldots, y_n\}$ denotes the standardized sample elements, the energy test statistic for standard d-variate normal distribution is

$$n\,\mathcal{E}_{n,d} = n\left(\frac{2}{n}\sum_{j=1}^{n} E|y_j - Z|_d - 2\frac{\Gamma\left(\frac{d+1}{2}\right)}{\Gamma\left(\frac{d}{2}\right)} - \frac{1}{n^2}\sum_{j,k=1}^{n}|y_j - y_k|_d\right)$$

where

$$E|a - Z|_d = \frac{\sqrt{2}\,\Gamma\left(\frac{d+1}{2}\right)}{\Gamma\left(\frac{d}{2}\right)} + \sqrt{\frac{2}{\pi}}\sum_{k=0}^{\infty}\frac{(-1)^k}{k!\,2^k}\frac{|a|_d^{2k+2}}{(2k+1)(2k+2)}\frac{\Gamma\left(\frac{d+1}{2}\right)\Gamma\left(k+\frac{3}{2}\right)}{\Gamma\left(k+\frac{d}{2}+1\right)}.$$
$$(6.2)$$

This is the original series formula in Székely and Rizzo [2005b]. See Section 6.3 for details. We can also write the series using a hypergeometric function.

An equivalent expression for $E|a - Z|_d$ is

$$E|a - Z|_d = \sqrt{2}\,\frac{\Gamma\left(\frac{d+1}{2}\right)}{\Gamma\left(\frac{d}{2}\right)}\,{}_1F_1(-1/2; d/2; -|a|_d^2/2),\qquad(6.3)$$

where ${}_1F_1$ is Kummer's confluent hypergeometric function

$$M(a; b; z) = {}_1F_1(a; b; z) = \sum_{n=0}^{\infty}\frac{a^{\overline{n}}}{b^{\overline{n}}}\frac{z^n}{n!},\qquad(6.4)$$

and $a^{\overline{n}} = \Gamma(n+a)/\Gamma(a)$ denotes the ascending factorial.

Formula (6.3) is a special case $\alpha = 1$ of the general result stated in Theorem 6.1 below. Formula (6.3) or (6.5) is easily evaluated using the GNU Scientific Library. See, for example, the function `hyperg_1F1` in the *gsl* package for R [Hankin, 2006] for an implementation of the hypergeometric function ${}_1F_1$. See Section 6.3 for the derivation of (6.2) and (6.3). See also Székely and Rizzo [2020].

In the univariate case, if $X \sim N(\mu, \sigma^2)$, then for real exponents $\alpha > -1$ the raw absolute moments are

$$E|X|^{\alpha} = \sigma^{\alpha}2^{\alpha/2}\frac{\Gamma\left(\frac{\alpha+1}{2}\right)}{\sqrt{\pi}}\cdot M\left(-\frac{\alpha}{2}; \frac{1}{2}; \frac{\mu^2}{2\sigma^2}\right).\qquad(6.5)$$

For arbitrary dimension d, one way to obtain the expression (6.2) for $E|a - Z|_d$ follows from a result in [Zacks, 1981, p. 55], which states that if $Z \in \mathbb{R}^d$ is standard normal, then the variable $|a - Z|_d^2$ has a noncentral chisquare distribution $\chi^2[\nu; \lambda]$ with degrees of freedom $\nu = d + 2\psi$, and noncentrality parameter $\lambda = |a|_d^2/2$, where ψ is a Poisson random variable with mean λ. See Section 6.3 for details.

The following result from Székely and Rizzo [2020] holds for arbitrary dimension $d \geq 1$.

Theorem 6.1. *If Z is a standard d-dimensional multivariate normal random vector, and $y \in \mathbb{R}^d$ is fixed, then for all $\alpha > 0$*

$$E|y - Z|_d^\alpha = \frac{2^{\alpha/2}\Gamma\left(\frac{d+\alpha}{2}\right)}{\Gamma\left(\frac{d}{2}\right)} \, {}_1F_1(-\alpha/2; d/2; -|y|_d^2/2), \qquad (6.6)$$

where $_1F_1(a; b; z) = M(a; b; z)$ is Kummer's confluent hypergeometric function (6.4).

The proof of Theorem 6.1 is in Section 6.3.

When $y = 0$ we have $_1F_1(-\alpha/2, d/2; -|y|^2/2) = {}_1F_1(-\alpha/2, d/2; 0) = 1$, so

$$E|Z|_d^\alpha = \frac{2^{\alpha/2}\Gamma\left(\frac{d+\alpha}{2}\right)}{\Gamma\left(\frac{d}{2}\right)},$$

and

$$E|Z - Z'|_d^\alpha = \frac{2^\alpha\Gamma\left(\frac{d+\alpha}{2}\right)}{\Gamma\left(\frac{d}{2}\right)}$$

follows by applying the formula to $\frac{1}{2}(Z - Z')$.

Corollary 6.1. *A computing formula for the energy statistic for a generalized energy goodness-of-fit test of univariate or multivariate normality with kernel $h(x, y; \alpha)$ is*

$$n\mathcal{E}_{n,d}^{(\alpha)} = \qquad\qquad\qquad\qquad\qquad\qquad\qquad\qquad\qquad\qquad (6.7)$$

$$n\left[2^{\alpha/2}\kappa_{d,\alpha} \cdot \frac{2}{n}\sum_{i=1}^n {}_1F_1\left(\frac{-\alpha}{2}, \frac{d}{2}; \frac{-|y_i|_d^2}{2}\right) - 2^\alpha\kappa_{d,\alpha} - \frac{1}{n^2}\sum_{i,j=1}^n |y_i - y_j|_d^\alpha\right],$$

where $y_1, \ldots, y_n$ is the standardized sample, $\kappa_{d,\alpha} = \Gamma((d+\alpha)/2)/\Gamma(d/2)$, and $_1F_1(a, b; z)$ is given by (6.4).

If $\alpha = 1$, $\kappa_{d,1} = \kappa_d = \Gamma(\frac{d+1}{2})/\Gamma(\frac{d}{2})$ and a computing formula is

$$n\mathcal{E}_{n,d} = \qquad\qquad\qquad\qquad\qquad\qquad\qquad\qquad\qquad\qquad (6.8)$$

$$n\left[\sqrt{2}\kappa_d \cdot \frac{2}{n}\sum_{i=1}^n {}_1F_1\left(-\frac{1}{2}, \frac{d}{2}; \frac{-|y_i|_d^2}{2}\right) - 2\kappa_d - \frac{1}{n^2}\sum_{i,j=1}^n |y_i - y_j|_d\right],$$

A U-statistic with kernel $h(x, y; \alpha)$ has the same form, simply replacing the divisor n^2 of the third mean with $n(n-1)$ in (6.7) and (6.8).

Remark 6.2 (Computation). For $E|a - Z|_d$ where $d > 1$, we apply formulas (6.6) and (6.8) with $\alpha = 1$ in the current version of the *energy* package [Rizzo and Székely, 2022]. One reason to prefer formula (6.6) is that the function `hyperg_1F1` in the *gsl* package computes the series with better precision than a partial sum in (6.2).

6.1.2 Composite Hypothesis: Estimated Parameters

In practice, the parameters of the hypothesized normal distribution are usually unknown. If $\mathcal{N}_d(\mu, \Sigma)$ denotes the family of d-variate normal distributions with mean vector μ and covariance matrix $\Sigma > 0$, and F_X is the distribution of a d-dimensional random vector X, the problem is to test if $F_X \in \mathcal{N}_d(\mu, \Sigma)$. In this case, parameters μ and Σ are estimated from the sample mean vector $\overline{X}$ and sample covariance matrix

$$S = (n-1)^{-1} \sum_{j=1}^{n} (X_j - \overline{X})(X_j - \overline{X})^\top,$$

and the transformed sample vectors are

$$Y_j = S^{-1/2}(X_j - \overline{X}), \qquad j = 1, \ldots, n.$$

The joint distribution of $Y_1, \ldots, Y_n$ does not depend on unknown parameters, but $Y_1, \ldots, Y_n$ are dependent. As a first step, we will ignore the dependence of the standardized sample, and use the computing formula given for testing the simple hypothesis.

Denote by $n\widehat{\mathcal{E}}_{n,d}$ the version of the statistic (6.8) obtained by standardizing the sample with estimated parameters. The test rejects the hypothesis of multivariate normality for large values of $n\widehat{\mathcal{E}}_{n,d}$. The modified test statistic $n\widehat{\mathcal{E}}_{n,d}$ has the same type of limit distribution as $n\mathcal{E}_{n,d}$, with rejection region in the upper tail, but with different critical values. Percentiles of the finite sample null distribution of $n\widehat{\mathcal{E}}_{n,d}$ can be estimated by parametric simulation. (See, e.g., Table 6.1 in Section 6.1.4.) The test based on $n\widehat{\mathcal{E}}_{n,d}$ is clearly affine invariant, but since the standardized sample is dependent, an alternate proof of consistency is needed. The proof of consistency for a test based on $n\widehat{\mathcal{E}}_{n,d}$ is given in Rizzo [2002] and Székely and Rizzo [2005b].

In testing the composite hypothesis of normality, a Monte Carlo test implementation can be applied. The statistic $n\widehat{\mathcal{E}}_{n,d}$ is compared to replicates of the energy statistic for standardized d-dimensional normal samples of the same size and dimension, computing

$$Y_j^* = S^{*-1/2}(X_j^* - \overline{X}^*), \qquad j = 1, \ldots, n.$$

for each of the generated samples $\mathbf{X}^*$. Algorithm 6.1 below outlines the Monte Carlo test procedure. This is the method of implementation in function mvnorm.test in the *energy* package for R if $d > 1$. Theory for the case of estimated parameters is derived in Rizzo [2002] and Székely and Rizzo [2005b].

The energy test of multivariate normality is applicable for arbitrary dimension (not constrained by sample size) and it practical to apply. Monte Carlo power comparisons in the references suggest that energy is a powerful competitor to other tests of multivariate normality. Several examples appear in Rizzo [2002], Székely and Rizzo [2005b] to illustrate the power of the test of

---

Algorithm 6.1 Simulation for Testing Multivariate Normality Case 4: Estimated Parameters

---

1. Compute the sample mean vector $\overline{X}$ and the sample covariance matrix S.
2. Center and scale the data: $Y_j = S^{-1/2}(X_j - \overline{X})$, $j = 1, \ldots, n$.
3. Compute the observed test statistic $T_0 = n\mathcal{E}_n(\mathbf{Y})$ from equation (6.8).
4. For $j = 1$ to R:
    (a) Generate $\mathbf{X}^* = (X_1^*, \ldots, X_n^*)$ from $N_d(\mu = 0, \Sigma = I_d)$.
    (b) Compute $\overline{X}^*$, $S^* = (n-1)^{-1} \sum_{j=1}^{n} (X_j^* - \overline{X}^*)(X_j^* - \overline{X}^*)^\top$, and the transformed vectors

$$Y_j^* = S^{*-1/2}(X_j^* - \overline{X}^*), \qquad j = 1, \ldots, n.$$

    (c) Compute $T^{(j)} = n\mathcal{E}_n(\mathbf{Y}^*)$ from equation (6.8).
5. Find the $100(1-\alpha)$ percentile c_α (or $1-\alpha$ quantile) of $T^{(1)}, \ldots, T^{(R)}$ and reject H_0 if $T_0 \geq c_\alpha$ (where α is the significance level).

---

multivariate normality against various alternatives, compared with competing tests like Mardia's test and the Henze-Zirkler test [Mardia, 1970, Henze and Zirkler, 1990]. Overall, the energy test is a powerful test of multivariate normality, consistent against all alternatives with relatively good power compared with other commonly applied tests.

6.1.3   On the Asymptotic Behavior of the Test

If $X \in \mathbb{R}^d$ and Z, Z' are iid standard d-variate normal random vectors, the kernel for testing if X is sampled from multivariate normal distribution is

$$h(x, y) = E|x - Z| + E|y - Z| - E|Z - Z'| - |x - y|. \qquad (6.9)$$

The V-statistics $\mathcal{E}_{n,d}$ based on the kernel (6.9) are degenerate kernel V-statistics and it is known in this case, nV_n converges in distribution to a Gaussian quadratic form

$$\sum_{k=0}^{\infty} \lambda_k Z_k^2$$

where $\{Z_k\}$ is a sequence of independent standard normal random variables, and $\{\lambda_k\}$ are the eigenvalues determined by the integral equation of the form

$$\int \psi(x) h(x, y) dF_X(x) = \lambda \psi(y). \qquad (6.10)$$

For testing normality, the density $dF_X(x)$ in (6.10) is the standard normal (or multivariate normal) density of the null hypothesis.

See Chapter 2 for a brief introduction; Serfling [1980] or Van der Vaart [2000] for asymptotic distribution theory of degenerate kernel V-statistics, and Chapter 7 for the fully specified integral equations corresponding to the V-statistics for testing normality.

Integral equations of the form (6.10) typically do not have nice analytical solutions. A remarkable exception is the Cramér-von Mises case when F is linear in $[0,1]$ and $\psi(t)$ is identically 1. For other exceptions see Thas and Ottoy [2003].

Móri et al. [2021] solved the integral equations for testing normality, obtaining numerical solutions to the eigenvalues λ_i in dimension $d = 1$ and empirical estimates for $d > 1$. The test based on these numerical estimates is implemented in the *energy* package for R in function `normal.test`. See Chapter 7 for details.

Remark 6.3. In the *energy* package, we have implemented a Monte Carlo test in function `mvnorm.test`. For testing univariate normality, the function `normal.test` includes an option to use pre-computed eigenvalues (see Chapter 7). In the following simulations, we used the parametric simulation approach because (i) our Monte Carlo test implementation does not require the limit distribution, only its existence, and (ii) according to Götze [1979] and Bickel et al. [1986], the rate of convergence is $o(n^{-1})$ for degenerate kernel U-statistics (and V-statistics), compared with $O(n^{-1/2})$ rate of convergence to the normal limit in the non-degenerate case. For this reason, as Henze and Wagner [1997] observed for their degenerate kernel BHEP V-statistics, "the test is practically sample size independent." We observed the same behavior for our degenerate kernel; see Table 6.1. This fact decreases the practical importance of large sample analysis.

6.1.4   Simulations

The results of a power comparison of several tests of multivariate normality are summarized below. These results first appeared in Székely and Rizzo [2005b].

TABLE 6.1
Empirical percentiles of $\widehat{\mathcal{E}}_{n,d}$.

	$n = 25$		$n = 50$		$n = 100$	
d	.90	.95	.90	.95	.90	.95
1	0.686	0.819	0.695	0.832	0.701	0.840
2	0.856	0.944	0.872	0.960	0.879	0.969
3	0.983	1.047	1.002	1.066	1.011	1.077
4	1.098	1.147	1.119	1.167	1.124	1.174
5	1.204	1.241	1.223	1.263	1.234	1.277
10	1.644	1.659	1.671	1.690	1.685	1.705

- Sample sizes were small to moderate, so we used empirical critical values for all test statistics in our power study. These critical values were computed from results on 20,000 samples generated under the null hypothesis (40,000 samples for $n = 25$). See Table 6.1 for critical values of $\widehat{\mathcal{E}}_{n,d}$ for selected values of d and n.

- $\widehat{\mathcal{E}}_{n,d}$ is compared with seven multivariate tests for $d = 2, 3, 5, 10$, $n = 25, 50, 100$, at significance level $\alpha = 0.05$.

- Details on the normal mixture alternatives: $pN_d(\mu_1, \Sigma_1) + (1 - p)N_d(\mu_2, \Sigma_2)$ denotes a normal mixture, where the sampled populations is $N_d(\mu_1, \Sigma_1)$ with probability p, and $N_d(\mu_2, \Sigma_2)$ with probability $1 - p$. The multivariate normal mixtures have a wide variety of types of departures from normality depending on the parameters and mixing probabilities. For example, a 50% normal location mixture is symmetric with light tails, and a 90% normal location mixture is skewed with heavy tails. A normal location mixture with $p = 1 - \frac{1}{2}(1 - \frac{\sqrt{3}}{3}) = .7887$ is skewed with normal kurtosis [Henze, 1994]. The scale mixtures are symmetric with heavier tails than normal.

Results for tests of bivariate normality ($n = 25, 50, 100$) are in Table 2 of Székely and Rizzo [2005b]. These results suggest that $\widehat{\mathcal{E}}_{n,d}$ test is more powerful than Mardia's skewness or kurtosis tests against symmetric mixtures with light or normal tails, and more powerful than the kurtosis test against skewed heavy tailed mixtures. In this comparison, the Henze–Zirkler test is very sensitive against the alternatives with light or normal tails, but less powerful against the heavy tail alternatives.

Results in dimensions $d = 3, 5$, and 10 for $n = 50$ in Table 3 of Székely and Rizzo [2005b] are similar to the bivariate case, and all eight tests are practical and effective in higher dimensions. See the figures below for a visual comparison of power for $d = 5; n = 25, 50, 100$.

Figure 6.1 shows the empirical power comparison across sample sizes in $d = 5$ for a 90% normal location mixture. Here, the $\widehat{\mathcal{E}}_{n,d}$ test was superior to the multivariate skewness, kurtosis, and HZ tests. In Figure 6.2 power is compared in $d = 5$ for a 79% normal location mixture, where $\widehat{\mathcal{E}}_{n,d}$ and HZ have similar performance, both tests superior to the skewness and kurtosis tests. In Figure 6.3 are results for the 50% normal location–scale mixture. In this comparison $\widehat{\mathcal{E}}_{n,d}$ is at least as powerful as the HZ and skewness tests. Performance of fixed β BHEP tests against different classes of alternatives depends on the parameter β. Henze and Wagner [1997] give a graphical illustration and discussion of this dependence. Our study agrees with conclusions of Henze and Zirkler [1990] and Henze and Wagner [1997] that small β is a good choice against symmetric heavy tailed alternatives, while large β is better against the symmetric light tailed alternatives. The parameter β in the Henze–Zirkler test is between 1 and 2, and although not the optimal choice of β for most alternatives, provides a test that is effective against a wide class of alternatives.

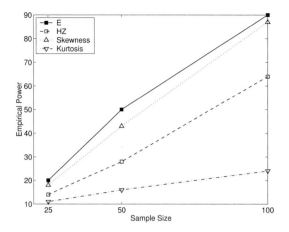

FIGURE 6.1
Empirical power of tests of multivariate normality (d = 5) against normal location mixture $.9N_5(0, I) + .1N_5(2, I)$: percent of significant tests of 2000 Monte Carlo samples at $\alpha = .05$. E denotes $\widehat{\mathcal{E}}_{n,5}$ and HZ denotes the Henze–Zirkler test.

Our empirical results summarized in Tables 2 and 3 of Székely and Rizzo [2005b] illustrate that while none of the tests compared are universally superior, some general aspects of relative power performance are evident. Overall, the relative performance of $\widehat{\mathcal{E}}_{n,d}$ was good against heavy tailed alternatives, and better than a skewness test against symmetric light tailed alternatives. Among the four tests compared, $\widehat{\mathcal{E}}_{n,d}$ was the only test that never ranked below all other tests in power.

6.2   Energy Projection-Pursuit Test of Fit

In this section, we summarize methods and results first published in Székely and Rizzo [2005b] on the energy projection pursuit approach to testing multivariate normality. A projection pursuit method is more computationally expensive, but may have better power than the energy test against certain alternatives.

6.2.1   Methodology

The projection pursuit (PP) approach to testing multivariate normality is based on the characterization property that a d-dimensional random variable

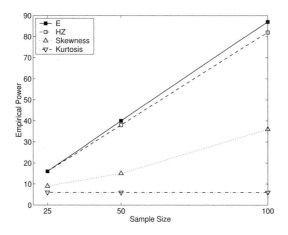

FIGURE 6.2
Empirical power of tests of multivariate normality (d = 5) against normal location mixture $.7887N_5(0, I) + .2113N_5(2, I)$: percent of significant tests of 2000 Monte Carlo samples at $\alpha = .05$. E denotes $\widehat{\mathcal{E}}_{n,5}$ and HZ denotes the Henze–Zirkler test.

X with mean vector μ and covariance matrix Σ has a multivariate normal $N_d(\mu, \Sigma)$ distribution if and only if the distribution of $a^\top X$ is univariate normal with mean vector $a^\top \mu$ and covariance matrix $a^\top \Sigma a$ for all vectors $a \in \mathbb{R}^d$. The PP method tests the "worst" one-dimensional projection of the multivariate data according to a univariate goodness-of-fit index. This index could be an EDF test statistic, such as Anderson-Darling, or any other univariate statistic for testing normality, in particular, the energy statistic.

Suppose C_n is a statistic for testing univariate normality that rejects normality for large C_n. For a d-dimensional random sample $X_1, \ldots, X_n$, define

$$C_n{}^*(X_1, \ldots, X_n) = \sup_{a \in \mathbb{R}^d, |a|=1} \{C_n(a^\top X_1, \ldots, a^\top X_n)\}.$$

The PP test based on the index C_n rejects multivariate normality for large values of $C_n{}^*$. To implement a PP test based on C_n, one can approximate $C_n{}^*$ by finding the maximum value of C_n over a finite set of projections. The set of projections is determined by a suitable "uniformly scattered" net of vectors in $\mathbb{R}^d$. For example, in $d = 2$ a suitable net of K points in $\mathbb{R}^2$ is given by $\{a_k = (cos(\theta_k), sin(\theta_k)), \ k = 1, \ldots, K\}$, where $\theta_k = 2\pi(2k - 1)/(2K)$. See Fang and Wang [1994] (Chapters 1 and 4) for details of implementation in $d > 2$.

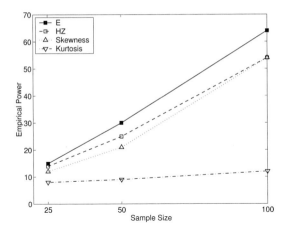

FIGURE 6.3
Empirical power of tests of multivariate normality (d = 5) against normal mixture $.5N_5(0, C) + .5N_5(2, I)$: percent of significant tests of 2000 Monte Carlo samples at $\alpha = .05$. Covariance matrix C has 1 on diagonal and .5 off diagonal. E denotes $\widehat{\mathcal{E}}_{n,5}$ and HZ denotes the Henze–Zirkler test.

6.2.2 Projection Pursuit Results

How does the power performance of various PP tests compare with each other? One can also ask whether the PP version of a multivariate test like energy is better or worse than the test based on the multivariate statistic. To evaluate the power performance, we applied PP tests based on five univariate tests of normality: univariate statistic $n\mathcal{E}_n$, modified Cramér-von Mises W^2, Kolmogorov-Smirnov D, univariate skewness $\sqrt{b_1}$, and univariate kurtosis b_2 (kurtosis is a two-tailed test). If $\{x_1, \ldots, x_n\}$ is the observed d-dimensional random sample, and $a \in \mathbb{R}^d$, let $y_{aj} = a^\top S^{-1/2}(x_j - \bar{x})$, $j = 1, \ldots, n$ be the univariate projection corresponding to a. Let $y_{a(1)}, \ldots, y_{a(n)}$ denote the ordered standardized univariate sample corresponding to a. The univariate $\mathcal{E}$ statistic has a very simple form, implemented as

$$\mathcal{E}^a = n\left(\frac{2}{n}\sum_{j=1}^{n}[2y_{aj}\Phi(y_{aj}) + 2\phi(y_{aj})] - \frac{2}{\sqrt{\pi}} - \frac{2}{n^2}\sum_{j=1}^{n}(2j - 1 - n)y_{a(j)}\right),$$

where $\Phi(\cdot)$ and $\phi(\cdot)$ are the standard normal cdf and pdf, respectively.

Let $F_n(j) = n^{-1}\sum_{k=1}^{n} I\{x_k \leq x_j\}$, where $I(\cdot)$ is the indicator function, denote the empirical distribution function (edf) of the sample, and let $F_n^a(\cdot)$ denote the edf of the projected sample $y_{a1}, \ldots, y_{an}$. The statistics $W^2, D, \sqrt{b_1}$, and b_2 corresponding to a are $W^{2^a} = n\sum_{j=1}^{n}(F_n^a(y_{aj}) - \Phi(y_{aj}))^2$, $D^a = \sqrt{n}\max_{1\leq j\leq n}|F_n^a(y_{aj}) - \Phi(y_{aj})|$, $\sqrt{b_1}^a = n^{-1}\sum_{j=1}^{n}(y_{aj})^3$, and $b_2{}^a = n^{-1}\sum_{j=1}^{n}(y_{aj})^4$.

We used empirical critical values for all test statistics in the PP power comparisons. That is, we estimated critical values of tests by simulation of 20,000 samples under the null hypothesis (40,000 for $n = 25$).

The PP tests were implemented by projecting each standardized sample in 15,000 directions: $\{a_k = (cos(\theta_k), sin(\theta_k)), \; k = 1, \ldots, 15,000\}$, where $\theta_k = 2\pi(2k-1)/(30,000)$.

The five PP tests were compared for $d = 2$, and sample sizes $n = 25, 50, 100$, at significance level $\alpha = 0.05$. Power of each test was estimated from a simulation of 2,000 random samples from the alternative distribution. The alternatives considered were normal mixtures, as described above.

Empirical results for the PP tests are summarized in Table 6.2. In these simulations, the PP-$\mathcal{E}$ test was comparable to or better than the multivariate $\mathcal{E}$, depending on the alternative. For sample sizes $n \geq 50$, there was little or no difference in power between PP-$\mathcal{E}$ and PP-W^2, but overall, the PP-$\mathcal{E}$ test was more powerful than the PP EDF tests W^2 and D against the normal mixture alternatives. Although the multivariate skewness test was more powerful than the PP-skewness test against the 90% scale mixture alternative, in the other scenarios, both versions of skewness tests were comparable in power. For the symmetric location mixture, the PP-kurtosis test was considerably more sensitive than the multivariate version, but against other alternatives such as the 90% scale mixture, the PP-kurtosis was considerably weaker than the multivariate test.

Projection pursuit tests are more computationally expensive. In case of the PP-$\mathcal{E}$ statistic, the PP-$\mathcal{E}$ test had better power against some alternatives without sacrificing power against others. Neither PP-skewness nor PP-kurtosis appear to offer a clear advantage over Mardia's multivariate tests, which are computationally less intensive.

6.3 Proofs

6.3.1 Hypergeometric Series Formula

The following is a proof of (6.3):

$$E|y - Z|_d^r = \frac{2^r \, \Gamma\left(\frac{d+1}{2}\right)}{\Gamma\left(\frac{d}{2}\right)} {}_1F_1(-r; d/2; -|y|_d^2/2),$$

Proof. If Z is standard d-dimensional normal, then $|y - Z|_d^2$ has a non-central chi-squared distribution with d degrees of freedom and non-centrality parameter $|y|_d^2$. If $X \sim \chi^2(d, \lambda)$, the density of X can be written as a Poisson weighted mixture of central chi-squared distributions:

$$f(x; d, \lambda) = \sum_{n=0}^{\infty} \frac{e^{-\lambda/2}(\lambda/2)^n}{n!} g(x; d + 2n), \qquad (6.11)$$

TABLE 6.2
Percentage of significant tests of bivariate normality, by projection pursuit method, of 2000 Monte Carlo samples at $\alpha = 0.05$.

Alternative	n	$\mathcal{E}$	$\sqrt{b_1}$	b_2	W^2	D
$.5N_d(0, I) + .5N_d(3, I)$	25	36	3	48	28	22
$.79N_d(0, I) + .21N_d(3, I)$	25	60	24	10	60	48
$.9N_d(0, I) + .1N_d(3, I)$	25	47	49	21	46	36
$.5N_d(0, B) + .5N_d(0, I)$	25	20	17	15	19	15
$.9N_d(0, B) + .1N_d(0, I)$	25	20	18	14	20	14
$.5N_d(0, I) + .5N_d(3, I)$	50	88	2	88	84	61
$.79N_d(0, I) + .21N_d(3, I)$	50	96	55	8	96	87
$.9N_d(0, I) + .1N_d(3, I)$	50	82	87	35	80	67
$.5N_d(0, B) + .5N_d(0, I)$	50	40	22	26	36	26
$.9N_d(0, B) + .1N_d(0, I)$	50	40	20	25	37	25
$.5N_d(0, I) + .5N_d(3, I)$	100	100	2	100	100	97
$.79N_d(0, I) + .21N_d(3, I)$	100	100	94	7	100	100
$.9N_d(0, I) + .1N_d(3, I)$	100	99	100	62	98	94
$.5N_d(0, B) + .5N_d(0, I)$	100	75	24	53	73	54
$.9N_d(0, B) + .1N_d(0, I)$	100	77	24	52	73	53

where $g(x; d+2n)$ is the density of a central chi-squared random variable with $d + 2n$ degrees of freedom (see [Johnson et al., 1995, Chapter 29] or [Zacks, 1981, p. 55]). Thus, if $V \sim \chi^2(d, \lambda)$, and $r > 0$,

$$E[V^r] = \int_0^\infty v^r \sum_{n=0}^\infty \left[\frac{e^{-\lambda/2}(\lambda/2)^n}{n!} \times \frac{v^{(d+2n)/2-1}e^{-v/2}}{2^{(d+2n)/2}\Gamma(\frac{d+2n}{2})} \right] dv$$

$$= \sum_{n=0}^\infty \left[\frac{e^{-\lambda/2}(\lambda/2)^n}{n!} \int_0^\infty v^r \frac{v^{d/2+n-1}e^{-v/2}}{2^{d/2+n}\,\Gamma(d/2+n)} dv \right].$$

If X has a gamma distribution with shape parameter β and scale parameter θ, then $E[X^r] = \theta^r \Gamma(\beta + r)/\Gamma(\beta)$. Therefore, the integrals in the sum are

$$\frac{2^r \, \Gamma\left(\frac{d}{2} + r + n\right)}{\Gamma\left(\frac{d}{2} + n\right)}, \qquad n = 0, 1, \ldots.$$

We have

$$E[V^r] = 2^r \, e^{-\lambda/2} \sum_{n=0}^\infty \frac{(\lambda/2)^n}{n!} \cdot \frac{\Gamma\left(\frac{d}{2} + n\right)}{\Gamma\left(\frac{d}{2} + r + n\right)}. \tag{6.12}$$

The ascending factorial is $a^{\overline{n}} := \Gamma(a+n)/\Gamma(a)$. If $a = d/2+r$ and $b = d/2$, then

$$E[V^r] = \frac{2^r \,\Gamma\left(\frac{d}{2}+r\right)}{\Gamma\left(\frac{d}{2}\right)} e^{-\lambda/2} \sum_{n=0}^{\infty} \frac{a^{\overline{n}}}{b^{\overline{n}}} \frac{(\lambda/2)^n}{n!}$$

$$= \frac{2^r \,\Gamma\left(\frac{d}{2}+r\right)}{\Gamma\left(\frac{d}{2}\right)} e^{-\lambda/2}\,_1F_1(d/2+r;d/2;\lambda/2),$$

where

$$_1F_1(a;b;z) = M(a;b;z) = \sum_{n=0}^{\infty} \frac{a^{\overline{n}}}{b^{\overline{n}}} \frac{z^n}{n!}$$

is Kummer's confluent hypergeometric series. Apply Kummer's transformation: $e^z M(b-a,b;-z) = M(a,b;z)$ to obtain

$$E[V^r] = \frac{2^r \,\Gamma\left(\frac{d}{2}+r\right)}{\Gamma\left(\frac{d}{2}\right)}\,_1F_1(-r;d/2;-\lambda/2). \tag{6.13}$$

Let $V = |y-Z|_d^2$, so that $\lambda = |y|_d^2$. Then with $r=1/2$ in (6.13) we have

$$E|y-Z|_d = \frac{\sqrt{2}\,\Gamma\left(\frac{d+1}{2}\right)}{\Gamma\left(\frac{d}{2}\right)}\,_1F_1(-1/2;d/2;-|y|_d^2/2), \tag{6.14}$$

and for $r>0$

$$E|y-Z|_d^r = \frac{2^r \,\Gamma\left(\frac{d+1}{2}\right)}{\Gamma\left(\frac{d}{2}\right)}\,_1F_1(-r;d/2;-|y|_d^2/2), \tag{6.15}$$

$\square$

6.3.2 Original Formula

The original formula (6.2) published in Székely and Rizzo [2005b] is derived directly from (6.11). The following is an alternate way to see that the two formulas are equivalent.

Proof. To obtain (6.2), let us start with (6.14).
We have

$$E|y-Z|_d = \sqrt{2}\,\frac{\Gamma\left(\frac{d+1}{2}\right)}{\Gamma\left(\frac{d}{2}\right)}\,_1F_1(-1/2;d/2;-|y|_d^2/2). \tag{6.16}$$

The ascending factorial is $a^{\overline{n}} = \Gamma(n+a)/\Gamma(a)$, so

$$
{}_1F_1\left(-\frac{1}{2};\frac{d}{2};-\frac{|y|^2}{2}\right) = \sum_{n=0}^{\infty} \frac{\Gamma(n-\frac{1}{2})}{\Gamma(-\frac{1}{2})}\left(\frac{\Gamma(n+\frac{d}{2})}{\Gamma(\frac{d}{2})}\right)^{-1}\frac{(-|y|^2/2)^n}{n!}
$$

$$
= \sum_{n=0}^{\infty} \frac{(-1)^n|y|^{2n}}{2^n\,n!}\frac{\Gamma(n-\frac{1}{2})}{\Gamma(-\frac{1}{2})}\frac{\Gamma(\frac{d}{2})}{\Gamma(n+\frac{d}{2})}
$$

$$
= 1 + \sum_{n=1}^{\infty} \frac{(-1)^n|y|^{2n}}{2^n\,n!}\cdot\frac{\Gamma(n-\frac{1}{2})}{\Gamma(-\frac{1}{2})}\frac{\Gamma(\frac{d}{2})}{\Gamma(n+\frac{d}{2})}.
$$

Now we re-index the sum, substitute $\Gamma(-1/2) = -2\sqrt{\pi}$, and apply the identity $\Gamma(k+3/2) = (k+1/2)\Gamma(k+1/2)$ to get

$$
{}_1F_1\left(-\frac{1}{2};\frac{d}{2};-\frac{|y|^2}{2}\right) = 1 + \sum_{k=0}^{\infty} \frac{(-1)^{k+1}|y|^{2(k+1)}}{2^{k+1}\,(k+1)!}\cdot\frac{\Gamma(k+\frac{1}{2})}{-2\sqrt{\pi}}\frac{\Gamma(\frac{d}{2})}{\Gamma(k+1+\frac{d}{2})}
$$

$$
= 1 + \frac{1}{\sqrt{\pi}}\sum_{k=0}^{\infty} \frac{(-1)^k||y|^{2(k+1)}}{2^k\,k!\,(2k+1)(2k+2)}\cdot\frac{\Gamma(k+\frac{3}{2})\Gamma(\frac{d}{2})}{\Gamma(k+1+\frac{d}{2})}.
$$

Hence

$$
E|y-Z|_d = \sqrt{2}\,\frac{\Gamma\left(\frac{d+1}{2}\right)}{\Gamma\left(\frac{d}{2}\right)}\,{}_1F_1\left(-\frac{1}{2};\frac{d}{2};-\frac{|y|^2}{2}\right)
$$

$$
= \sqrt{2}\,\frac{\Gamma\left(\frac{d+1}{2}\right)}{\Gamma\left(\frac{d}{2}\right)}\left(1+\frac{1}{\sqrt{\pi}}\sum_{k=0}^{\infty}\frac{(-1)^k|y|^{2(k+1)}}{2^k\,k!\,(2k+1)(2k+2)}\cdot\frac{\Gamma(k+\frac{3}{2})\Gamma(\frac{d}{2})}{\Gamma(k+1+\frac{d}{2})}\right)
$$

$$
= \sqrt{2}\,\frac{\Gamma\left(\frac{d+1}{2}\right)}{\Gamma\left(\frac{d}{2}\right)} + \sqrt{\frac{2}{\pi}}\sum_{k=0}^{\infty}\frac{(-1)^k|y|^{2(k+1)}}{2^k\,k!\,(2k+1)(2k+2)}\cdot\frac{\Gamma(k+\frac{3}{2})\Gamma(\frac{d+1}{2})}{\Gamma(k+\frac{d}{2}+1)},
$$

which is formula (6.2). □

6.4 Exercises

Exercise 6.1. Prove that if $Z_1, Z_2, \ldots Z_n$ are d-dimensional iid standard normal random variables,

$$
\overline{Z} = \frac{1}{n}\sum_{i=1}^{n}Z_i, \quad S_n = \frac{1}{n-1}\sum_{i=1}^{n}(Z_i-\overline{Z})(Z_i-\overline{Z})^T,
$$

and $Y_i = S_n^{-1/2}(Z_i-\overline{Z})$, $i = 1,2,\ldots,n$ is the standardized sample, $n > d$, then

$$
E|Y_1-Y_2| > E|Z_1-Z_2|,
$$

and the ratio $\frac{E|Y_1-Y_2|}{E|Z_1-Z_2|}$ does not depend on d, only on n.

Hint: In the family of d-dimensional normal samples with positive definite covariance matrices, $(\overline{Z}, S_n)$ is a complete and sufficient statistic, $\xi := Y_1 - Y_2$ is ancillary and thus by Basu's theorem $(\overline{Z}, S_n)$ and ξ are independent. Introduce $f(w) := E(S_n^{1/2}w)$ where $w \in R^d$. Then $f(w) = C|w|$ where C is a constant because for arbitrary orthonormal $d \times d$ matrix U (whose transpose is its inverse), $U^T S_n^{1/2} U = (U^T S_n U)^{1/2}$ has the same distribution as $S^{1/2}$. Thus

$$f(Uw) = E|S^{1/2}Uw| = E|U^T S_n^{1/2}Uw| = E|S^{1/2}w| = f(w).$$

But then by the independence of ξ and S_n we have

$$E|Z_1 - Z_2| = E|S_n^{1/2}\xi = Ef(\xi) = CE|\xi|.$$

We just need to show that $C < 1$. Put $w := (1, 0, \dots, 0)^T$. Then $|S_n^{1/2}w|^2 = w^T S_n w$ is a univariate empirical variance whose distribution is $\chi_{n-1}^2/(n-1)$ and thus the expected value of its square root is $C = \Gamma(n/2)/\Gamma((n+1)/2) < 1$.

Exercise 6.2. Prove that under the hypothesis of normality for every fixed integer $d \geq 1$,

1. $E[\widehat{\mathcal{E}}_{n,d}]$ is bounded above by a finite constant that depends only on d.

2. For every fixed $\alpha \in (0, 1)$, the sequence of critical values $c_{\alpha,n,d}$ of $\widehat{\mathcal{E}}_{n,d}$ are bounded above by a finite constant $k_{\alpha,d}$ that depends only on α and d.

7

Eigenvalues for One-Sample E-Statistics

CONTENTS

In this chapter, several interesting connections can be seen between statistical potential energy of data and the corresponding energy of physics; the motivation for the name "energy statistics" should be clear by the end of this chapter.

7.1 Introduction

The form of the asymptotic null distribution of an energy statistic is known to be a quadratic form of centered Gaussian variables

$$\sum_{i=1}^{\infty} \lambda_i Z_i^2, \tag{7.1}$$

where Z_i are independent and identically distributed (iid) standard normal variables. The eigenvalues λ_i and associated eigenvectors/eigenfunctions are solutions to the corresponding integral equation.

In this chapter, we discuss how to compute the upper tail probabilities and critical values of the asymptotic distribution of an energy goodness-of-fit statistic. In order to compute probabilities or critical values for inference,

DOI: 10.1201/9780429157158-7

we need the values of the constants $\{\lambda_1, \lambda_2, \dots\}$. In general, these constants depend on the distributions of the random variables, and typically are difficult to derive analytically.

Methods of this chapter apply to energy goodness-of-fit test statistics. We will apply our results to find the asymptotic distributions of the statistics for the energy tests of univariate and multivariate normality introduced in Section 4.5 and Chapter 6. The main results were first published in Móri, Székely, and Rizzo [2021].

The one-sample V-statistic for testing $H_0 : F_X = F_0$ vs $H_1 : F_X \neq F_0$ has kernel

$$h(x, y) = E|x - X| + E|y - X| - E|X - X'| - |x - y|,$$

where the expected value is $E|y - X| = \int_{\mathbb{R}^d} |y - x| dF_0(x)$.

Suppose that $X_1, \dots, X_n$ are iid, d-dimensional random variables. Let $x_1, \dots, x_n$ denote the sample observations. Then the V-statistic for goodness-of-fit is

$$\mathcal{E}_n(X, F_0) = \frac{2}{n} \sum_{i=1}^{n} E|x_i - X| - E|X - X'| - \frac{1}{n^2} \sum_{\ell=1}^{n} \sum_{m=1}^{n} |x_\ell - x_m|, \quad (7.2)$$

where X and X' are iid with distribution F_0, and the expectations are taken with respect to the null distribution F_0. The energy goodness-of-fit test statistic is $nV_n = n\mathcal{E}_n(X, F_0)$.

The energy statistic (7.2) is a degenerate kernel V-statistic, so $n\mathcal{E}_n$ asymptotically has a sampling distribution that is a Gaussian quadratic form (7.1) under the null hypothesis. The parameters λ_i of the limit distribution are the eigenvalues of the integral operator determined by the energy distance. Although a Monte Carlo approach provides excellent approximations to the sampling distribution of the test statistic for finite samples, our goal here is to determine the asymptotic distribution.

For testing normality we derive the exact form of the integral equation for four cases, according to whether the population parameters are known or estimated. After solving the integral equations for the eigenvalues [Móri et al., 2021], one can obtain the probability distribution and compute the upper tail probabilities using a method for Gaussian quadratic forms [Imhof, 1961]. As the eigenvalues can be pre-computed, p-values for a large-sample test can be computed using any available software that implements Imhof's method or alternate methods.

The corresponding large-sample test implementation based on applying (7.1), as well as computational methods to evaluate or estimate the eigenvalues are discussed below in more detail. The asymptotic test for univariate normality is implemented in the *energy* package [Rizzo and Székely, 2022] in the function `normal.test` for the composite hypothesis.

In the following sections, we derive the explicit integral equations for the eigenvalue problem for the simple and composite hypotheses of normality. These equations can be solved by numerical methods. We will apply a variation of Nyström's method. For the simple hypothesis, with all parameters known, we also obtain the eigenvalues by an empirical approach.

7.2 Kinetic Energy: The Schrödinger Equation

The one sample energy statistic $\mathcal{E}_n$ in (7.2) is a V-statistic whose kernel is a double centered potential function defined by

$$h(x,y) = E|x - X| + E|y - X| - E|X - X'| - |x - y|, \qquad (7.3)$$

where $h : \mathbb{R}^d \times \mathbb{R}^d \to \mathbb{R}$. The statistic is $\mathcal{E}_n = \frac{1}{n^2} \sum_{i,j=1}^{n} h(x_i, y_j)$.

By the law of large numbers for V-statistics (see, e.g., Serfling [1980] or Koroljuk and Borovskich [1994]), we have

$$\lim_{n \to \infty} \mathcal{E}_n = E[h(X, X')]$$

with probability one. Applying Proposition 3.2, we see that $\mathcal{E}(F, F_0) > 0$ whenever $H_0 : F = F_0$ is false. Hence under an alternative hypothesis $n\mathcal{E}_n \to \infty$ with probability one, as $n \to \infty$.

On the other hand, if H_0 is true, then the kernel h is degenerate; that is, $E[h(x, X)] = 0$ for almost all $x \in \mathbb{R}^d$. Thus $n\mathcal{E}_n$ has a finite limit distribution under the extra condition $E[h^2(X, X')] < \infty$. This result combined with the property that $n\mathcal{E}_n \to \infty$ under the alternative, shows that tests can be constructed based on $\mathcal{E}_n$ that are consistent against general alternatives.

If $h(x, y)$ is (7.3) and $E[h^2(X, X')] < \infty$, under the null hypothesis the limit distribution of $n\mathcal{E}_n$ is a quadratic form

$$Q = \sum_{k=1}^{\infty} \lambda_k Z_k^2 \qquad (7.4)$$

of iid standard normal random variables Z_k, $k = 1, 2, \ldots$ [Koroljuk and Borovskich, 1994, Theorem 5.3.1]. The nonnegative coefficients $\{\lambda_k\}$ are eigenvalues of the integral operator with kernel $h(x, y)$, satisfying the Hilbert-Schmidt eigenvalue equation

$$\int_{\mathbb{R}^d} h(x, y)\psi(y) \, dF(y) = \lambda \psi(x). \qquad (7.5)$$

We will call the eigenvalues λ the *statistical potential energy levels*.

Our goal in this chapter is to find the eigenvalues λ_i for an energy goodness-of-fit test and thus, we can compute probabilities $P(Q > t)$ for any $t > 0$ to get a test decision.

The kernel h is symmetric ($h(x, y) = h(y, x)$), hence the eigenvalues are real. Since $|x - y|$ is conditionally negative definite, one can easily see that $h(x, y)$ is positive semidefinite, and thus all eigenvalues in (7.5) are nonnegative. It is also known that their sum is finite and equal to $E|X - X'|$ under H_0.

The kernel h is degenerate; that is,

$$\int h(x, y)\, dF(y) = 0.$$

Thus $\psi_0 = 1$ is an eigenfunction with eigenvalue 0. Since eigenfunctions with different eigenvalues are orthogonal we have for any ψ corresponding to a nonzero λ that

$$\int \psi(y)\, dF(y) = 0.$$

For such a ψ in (7.5) the y-independent terms in $h(x, y)$ integrate to 0 and thus (7.5) simplifies to

$$\int (E|y - X| - |x - y|)\psi(y)\, dF(y) = \lambda \psi(x).$$

In the one dimensional case, if we differentiate with respect to x we get

$$-\int_a^b sign(x - y)\psi(y)\, dF(y) = \lambda \psi'(x),$$

where (a, b) is the support of dF. (Note that a, b can be infinite). Now letting $x \to a$,

$$\lambda \psi'(a) = -\int_a^b sign(a - y)\psi(y)\, dF(y) = \int_a^b \psi(y)\, dF(y) = 0.$$

We get a similar equation for b. Therefore, the boundary conditions are

$$\psi'(a) = \psi'(b) = 0,$$

and this also holds for $\psi_0 = 1$. One more differentiation leads to

$$-2f\psi = \lambda \psi'',$$

where $f = F'$. This means that for $f \neq 0$ and for $\mu := 1/\lambda$, we have the simple eigenvalue equation

$$-\frac{1}{2f}\psi'' = \mu\psi. \tag{7.6}$$

Now let us recall the time independent (stationary) Schrödinger equation of quantum physics [Schrödinger, 1926]:

$$-\frac{\psi''(x)}{2m} + V(x)\psi(x) = \mathcal{E}\psi(x).$$

Here ψ is the standing wave function, m is the mass of a particle, $V(x)$ is the potential function, and $\mathcal{E}$ denotes the energy level. The left hand side of (7.6) corresponds to pure kinetic energy because the $V(x)\psi(x)$ term is missing in (7.6). We can thus call μ in (7.6) the *statistical kinetic energy level*.

We have just proved that in one dimension the statistical potential energy level λ is the exact reciprocal of the statistical kinetic energy level μ. This can be viewed as a counterpart of the *law of conservation of energy* in physics.

The derivation of this nice property relies on the fact that $(1/2)|x-y|$ is the fundamental solution of the one-dimensional Laplace equation

$$\frac{d^2}{dx^2}\frac{1}{2}|x-y| = -\delta(x-y),$$

where $\delta(\cdot)$ is the Dirac delta function, so in one dimension $|x-y|$ is a harmonic function. Thus in one dimension the inverse of the statistical potential energy operator

$$\mathbf{H}\psi(x) = \int_{\mathbb{R}^d} h(x,y)\psi(y)\,dF(y) \tag{7.7}$$

is the Schrödinger operator

$$\frac{d^2}{2f\,dx^2}.$$

It would be interesting to find the inverse of $\mathbf{H}$ in arbitrary dimension, or somewhat more generally, to find the inverse of the integral operator with kernel

$$h_\alpha(x,y) := E|x-X|^\alpha + E|y-X|^\alpha - E|X-X'|^\alpha - |x-y|^\alpha,$$

$0 < \alpha < 2$.

7.3 CF Version of the Hilbert-Schmidt Equation

We can rewrite the Hilbert-Schmidt eigenvalue equation (7.5) in terms of characteristic functions. If the characteristic function of X is $\widehat{f}$ and the empirical characteristic function of n sample observations of X is $\widehat{f}_n$, then provided that the variance of X exists, we have that under the null $\sqrt{n}(\widehat{f}_n(t) - \widehat{f}(t))$ tends to a (complex) Gaussian process with zero expected value and covariance function $\widehat{f}(s-t) - \widehat{f}(s)\overline{\widehat{f}(t)}$. Thus according to Székely and Rizzo [2013b], Section 2, under the null hypothesis the limit distribution of $n\mathcal{E}_n$ is a weighted integral of a Gaussian process. Therefore by the Karhunen-Loève theorem this limit distribution is the same as the distribution of a quadratic form of iid standard Gaussian random variables where the coefficients in this quadratic form are the eigenvalues of the characteristic function version of (7.5):

$$\frac{1}{c_d}\int_{\mathbb{R}^d}\frac{\widehat{f}(s-t)-\widehat{f}(s)\overline{\widehat{f}(t)}}{|s|^{(d+1)/2}|t|^{(d+1)/2}}\Psi(s)\,ds=\lambda\Psi(t),$$

where

$$c_d=\frac{\pi^{(d+1)/2}}{\Gamma\left(\frac{d+1}{2}\right)}. \qquad (7.8)$$

It is known that this eigenvalue equation has a countable spectrum; that is, we have a discrete set of solutions $\{\lambda_k : k = 1, 2, ...\}$.

For example, if $\widehat{f}$ is the standard normal characteristic function, then

$$\widehat{f}(s-t)-\widehat{f}(s))\overline{\widehat{f}(t)}=e^{-(|s|^2+|t|^2)/2}[e^{(s't)}-1],$$

where $(s't)$ is the inner product. Thus

$$\frac{1}{c_d}\int_{\mathbb{R}^d}\frac{e^{-(|s|^2+|t|^2)/2}[e^{(s't)}-1]}{|s|^{(d+1)/2}|t|^{(d+1)/2}}\Psi(s)\,ds=\lambda\Psi(t).$$

We consider four types of hypotheses for a test of normality (four cases):

1. Simple hypothesis; known parameters.
2. Estimated mean, known variance/covariance.
3. Known mean, estimated variance/covariance.
4. Estimated mean and estimated variance/covariance.

For a test of the simple hypothesis of normality, the parameters are specified, so it is equivalent to test whether the standardized sample was drawn from a standard normal or standard multivariate normal distribution. Hence we have Theorem 7.1 for Case 1, $d \geq 1$.

Theorem 7.1. *The eigenvalue equation for the energy goodness-of-fit test of standard d-dimensional multivariate normal distribution is*

$$\frac{1}{c_d}\int_{\mathbb{R}^d}\frac{e^{-(|s|^2+|t|^2)/2}[e^{(s't)}-1]}{|s|^{(d+1)/2}|t|^{(d+1)/2}}\Psi(s)\,ds=\lambda\Psi(t),$$

where c_d is given by (7.8). These equations are valid for statistics of the standardized sample; that is, for $y_i = (x_i - \mu)/\sigma$, $\mu = E(X), \sigma^2 = Var(X)$ if $d = 1$, or $y_i = \Sigma^{-1/2}(x_i - \mu)$ if $d > 1$, where $\mu = E(X)$ and Σ is population covariance matrix.

If the parameters are known (simple hypothesis) then the characteristic function version is equally good as the original Hilbert-Schmidt eigenvalue equation. But if the parameters are unknown then we simply do not know how to change the kernel in (7.5). On the other hand, we will see that the characteristic function version has a nice extension to the unknown parameter case. We just need to compute the new covariance function for a studentized sample.

Let us see the details. If the parameters of the normal distribution are unknown, one can center and scale the sample using the sample mean and sample standard deviation. If the empirical characteristic function of the transformed sample $y_1, y_2, \ldots, y_n$ is

$$\widehat{g}_n(t) = \frac{1}{n} \sum_{k=1}^{n} e^{ity_k}$$

and $\widehat{f}(t)$ denotes the standard normal characteristic function, then

$$T_n := \frac{n}{c_d} \int_{\mathbb{R}^d} \frac{|\widehat{g}_n(t) - \widehat{f}(t)|^2}{|t|^{d+1}} dt$$

$$= 2 \sum_{i=1}^{n} E|y_i - Z| - nE|Z - Z'| - \frac{1}{n} \sum_{i,j=1}^{n} |y_i - y_j|.$$

Here $\sqrt{n}(\widehat{g}_n(t) - \widehat{f}(t))$ tends to a (complex) Gaussian process with zero expected value, but the covariance function of this Gaussian process is different for each of the four types of hypotheses in Cases 1-4. As we shall prove in Section 7.7, if $d = 1$ the covariance function of this Gaussian process in Case 4 (μ, σ unknown) is

$$e^{-(s^2+t^2)/2}(e^{st} - 1 - st - (st)^2/2),$$

and the eigenvalue equation becomes

$$\frac{1}{c_1} \int_{\mathbb{R}} \frac{e^{-(s^2+t^2)/2}(e^{st} - 1 - st - (st)^2/2)}{|s||t|} \Psi(s) \, ds = \lambda\Psi(t), \qquad (7.9)$$

where $c_1 = \pi$ from equation (7.8).

Hence the limiting distribution of T_n is $Q = \sum_{k=1}^{\infty} \lambda_k Z_k^2$, where (for Case 4, $d = 1$), λ_k, $k = 1, 2, \ldots$ are the eigenvalues of the eigenvalue equation (7.9). The four types of hypotheses Cases 1–4 correspond to four eigenvalue equations.

Theorem 7.2. *The eigenvalue equations for energy tests of univariate normality are*

$$\frac{1}{\pi} \int_{\mathbb{R}} \frac{e^{-(s^2+t^2)/2}g(s,t)}{|s||t|} \Psi(s) \, ds = \lambda\Psi(t), \qquad (7.10)$$

where $g(s,t)$ is given by

1. $e^{st} - 1$ *(μ, σ known).*

2. $e^{st} - 1 - st$ *(μ unknown, σ known).*

3. $e^{st} - 1 - (st)^2/2$ *(μ known, σ unknown).*

4. $e^{st} - 1 - st - (st)^2/2$ *(μ, σ unknown)*

for Cases 1, 2, 3, 4, respectively.

These equations are valid for test statistics of the standardized sample $y_i = (x_i - \mu)/\sigma$ in Case 1, and the studentized samples $y_i = (x_i - \bar{x})/\sigma$, $y_i = (x_i - \mu)/s$, and $y_i = (x_i - \bar{x})/s$ in Cases 2–4, respectively.

See Section 7.7 (i) for the proof.

Csörgő [1986] proved a similar result on a similar limiting covariance function. However, the test suggested in Csörgő's paper is not affine invariant, because it is a function of the standardized sample only, but not of the Mahalanobis distances. The affine invariant energy test of normality is based on a limiting distribution of a similar form, namely a quadratic form of iid normal distributions $Q = \sum_{k=1}^{\infty} \lambda_k Z_k^2$, but the test function and the eigenvalue coefficients are different, because the eigenvalue equation is different.

The proof of Theorem 7.2 is valid for all four cases if $d = 1$, and Theorem 1 applies for the simple hypothesis if $d \geq 1$. The proof in Section 7.7 (i) does not apply for Case 4 if $d > 1$.

The covariance kernel for Case 4 (all parameters unknown, $d > 1$) is given by Henze and Wagner [1997], Theorem 2.1. The covariance kernels for d-dimensional normality in each of the four cases are

$$k(s,t) = \exp\{-(|s|_d^2 + |t|_d^2)/2\}\, g(s,t),   \tag{7.11}$$

where $g : \mathbb{R}^d \times \mathbb{R}^d \to \mathbb{R}$ is defined for each case by

$$g(s,t) = \begin{cases} (e^{s't} - 1), & \text{Case 1: } \mu, \Sigma \text{ known;} \\ (e^{s't} - 1 - s't), & \text{Case 2: } \mu \text{ unknown, } \Sigma \text{ known;} \\ (e^{s't} - 1 - (s't))^2/2, & \text{Case 3: } \mu \text{ known, } \Sigma \text{ unknown;} \\ (e^{s't} - 1 - s't - (s't))^2/2, & \text{Case 4: } \mu, \Sigma \text{ unknown,} \end{cases}$$

$s \in \mathbb{R}^d$, $t \in \mathbb{R}^d$.

The corresponding integral equations are

$$\frac{1}{c_d} \int_{\mathbb{R}^d} \frac{k(s,t)}{|s|_d^{(d+1)/2} |t|_d^{(d+1)/2}} \Psi(s)\, ds = \lambda \Psi(t),$$

where $s \in \mathbb{R}^d$, $t \in \mathbb{R}^d$, $k(s,t)$ is given by (7.11), and c_d is given by (7.8).

Henze and Wagner [1997] derived the general integral equation for the BHEP test in $d \geq 1$ with estimated parameters; however, no eigenvalues were derived or estimated. Instead, in Henze and Wagner [1997] the asymptotic distribution for the BHEP test statistic is estimated by fitting a lognormal distribution.

7.4 Implementation

To apply the energy test of normality for the simple or composite hypotheses, we have several approaches. Five methods are described below.

 (i) *Parametric simulation.*
 (ii) *Chi-square criterion* for the simple hypothesis, typically a conservative test criterion.
(iii) *Direct numerical solution for eigenvalues* to obtain the null limit distribution.
(iv) *Kernel matrix or Gram matrix eigenvalues* to estimate the eigenvalues of the null limit distribution.
 (v) *Schrödinger differential equation for eigenvalues* to obtain the null limit distribution.

The focus here is primarily on Methods (iii), (iv), and (v); a brief description of all five methods is outlined below.

Method (i): *Parametric simulation:* In this approach, we do not obtain the analytical form of the limit distribution, so the finite sampling distribution is obtained by parametric simulation and a Monte Carlo approach to testing is applied (see Chapter 6 and Algorithm 6.1). This method is applied in function `mvnorm.test` in the *energy* package for R (Case 4).

Method (ii): *Conservative chi-square criterion:*

The following theorem relates the distribution of the quadratic form Q to the chisquare distribution. Let $T_n = nV_n/E[nV_n]$ be the normalized statistic with expected value 1, $Q \overset{D}{=} \lim_{n\to\infty} nV_n/E[nV_n]$.

Theorem 7.3 (Székely and Bakirov [2003]). *Suppose $T_\alpha(T_n)$ is the test that rejects H_0 if*

$$T_n > \Phi^{-1}(1 - \alpha/2)^2,$$

where $\Phi(\cdot)$ denotes the standard normal cumulative distribution function. Let ASL_n denote the achieved significance level of the test $T_\alpha(T_n)$. Then for all $0 < \alpha \leq 0.215$

 1. $\lim_{n\to\infty} ASL_n \leq \alpha$.
 2. $\sup_{\Theta_0} \{\lim_{n\to\infty} ASL_n\} = \alpha$, where Θ_0 is the null parameter space.

Note that the chi-square test criterion applies to all statistics which have the same type of limiting distribution under the null, the quadratic form Q, if normalized to have mean 1.

The above theorem provides a simple test criterion for many of our energy tests. For example, in the two-sample energy test, if we divide the statistic $n\mathcal{E}_n$ by S_2, for large n we could apply this theorem. Unfortunately, using the percentiles of $\chi^2(1)$ leads to a very conservative test, which is typically quite low in power. Based on our empirical results, the chisquare criterion is not recommended except in certain special cases.

Method (iii): *Direct numerical solution for eigenvalues:* The following limiting result can be applied to the simple or composite hypothesis problems.

We have seen that under the null hypothesis, if $E[h^2(X, X')] < \infty$, the limit distribution of $n\mathcal{E}_n$ is is known to be a quadratic form (7.4), where the nonnegative coefficients λ_k are eigenvalues of the integral operator with kernel $h(x, y)$, satisfying the Hilbert–Schmidt eigenvalue equation

$$\int_{\mathbb{R}^d} h(x, y)\psi(y)\, dF_0(y) = \lambda\psi(x).$$

We call the eigenvalues λ of the above integral equation the *statistical potential energy levels*. The corresponding integral equation is a Fredholm equation of the second kind. For a proof, see, e.g., Koroljuk and Borovskich [1994].

Method (iv): *Kernel matrix method for eigenvalues:* For the simple hypothesis in dimensions $d \geq 1$, the eigenvalues of the kernel matrix of the sample (or Gram matrix) approximate the eigenvalues $\{\lambda_k : k = 1, 2, \dots\}$. See Section 7.5 for details.

Method (v): *Schrödinger equation for eigenvalues:* We derive a Schrödinger type differential equation on the eigenvalues, which leads to a set of simultaneous equations for the eigenvalues. The equations are derived separately for four cases; the simple hypothesis, and three types of composite hypotheses. After solving these equations the asymptotic distribution and test decision can then be obtained as in Method (iii). The details are in Section 7.2.

If the eigenvalues are known, then the asymptotic distribution of the test statistic can be computed using one of several methods, such as Imhof's method [Imhof, 1961] for evaluating the upper tail probabilities of a quadratic form of standard normals. Under the null hypothesis, the eigenvalues $\lambda_1 > \lambda_2 > \cdots > 0$ must sum to $E[n\mathcal{E}_n] = E|X - X'|$, so the series can be truncated when the sum has converged to within a specified tolerance of $\sum_{k=1}^{\infty} \lambda_k = E|X - X'|$. For univariate $N(\mu, \sigma^2)$, this sum is $E|X - X'| = 2\sigma/\sqrt{\pi}$ and for standard d-dimensional normal

$$\sum_{k=1}^{\infty} \lambda_k = E|Z - Z'| = \frac{2\Gamma\left(\frac{d+1}{2}\right)}{\Gamma\left(\frac{d}{2}\right)},$$

where $\Gamma(\cdot)$ is the complete gamma function.

In summary, Method (i) is a general finite sample approach, applying to all types of hypotheses and dimensions $d \geq 1$. The Monte Carlo test has correct size equal to the nominal significance level. Method (iii) is a general approach, but the numerical solution may be difficult for $d > 1$. Methods (ii) and (iv) apply to the simple hypothesis in all dimensions $d \geq 1$; however, the chi-square criterion is less powerful than the other four methods because it is a conservative test. Method (v) is based on the exact integral equations derived for testing normality for all four types of hypotheses. Thus, in case of the simple hypothesis there are several options to implement the energy goodness-of-fit test. The composite hypothesis leaves us with fewer options. In Section 7.6, the accuracy and computational efficiency of methods based on the asymptotic distribution are compared with the general Monte Carlo method.

7.5 Computation of Eigenvalues

Nyström's Method

We apply a version of Nyström's method (see, e.g., Sloan [1981]). Suppose that we have an integral operator T from $L^2(\mathbb{R}^d) \to L^2(\mathbb{R}^d)$:

$$T : \psi(y) \mapsto \int_{\mathbb{R}^d} K(x,y)p(x)\psi(x)dx.$$

Here, K is a symmetric kernel and $p(x)$ is a probability density function of a random variable on $\mathbb{R}^d$. In this approach, N points are sampled from the density $p(x)$ and the integral is estimated as

$$E[K(X,y)\psi_i(X)] \approx \frac{1}{N}\sum_{j=1}^{N} K(x_j,y)\psi_i(x_j), \qquad i = 1,2,\ldots$$

Let $\Psi_i = (\psi_i(x_1),\ldots,\psi_i(x_N))^\top$ and define the matrix $\widetilde{K} := K(x_i,x_k)$. Then we have the system of equations

$$\widetilde{K}\Psi_i = N\lambda_i\Psi_i, \quad i = 1,2,\ldots,$$

and the eigenvalues of $\widetilde{K}$ are $N\lambda_i$ with associated eigenvectors Ψ_i. As an approximation for $d = 1$, we can take $p(x)$ to be uniform on some interval $[-a,a]$ where a is chosen such that $|K(x,y)|$ is less than a specified tolerance outside of $[-a,a]$. Let $p(x) = \frac{1}{2a}I(-a < x < a)$. Then as an approximation we solve

$$2a\int_{-a}^{a} K(x,y)\frac{1}{2a}\psi_i(x)dx = \lambda_i\psi_i(y), \quad i = 1,2,\ldots$$

We can absorb the constant $2a$ into the kernel function, or divide the resulting eigenvalues by $2a$.

Kernel matrix method

Define the kernel matrix (or Gram matrix) H_n by expanding the kernel for each pair of sample observations. That is, $H_n(i,j) = \frac{1}{n}h(x_i,x_j), i,j = 1,\ldots,n$. The sum of the entries of H_n equals the energy test statistic $n\mathcal{E}_n$. The matrix H_n is positive definite and the eigenvalues of H_n are estimates of the n largest eigenvalues of the integral equation. Here $E[h^2(X, X')] < \infty$ so H_n is Hilbert-Schmidt, and the eigenvalues of H_n converge a.s. to $\{\lambda_k : k = 1, 2, ...\}$ as $n \to \infty$. See Koltchinskii and Giné [2000] for proof of convergence and other results on this approach. This method is discussed and implemented in Section 7.6.

Intuitively, it may seem that the limits of the eigenvalues of the corresponding $n \times n$ matrix $\widehat{H}_n$ may be approximately equal if one replaced the original sample values by the standardized sample centered and scaled by estimated parameters. That is, let $\widehat{H}_n(i, j) = \frac{1}{n}h(y_i, y_j)$, where $y_1, \ldots, y_n$ is the transformed sample, However, this method does not apply when the sample observations are dependent. Results in Section 7.6 show that in the location-scale composite hypothesis case the eigenvalues of H_n and $\widehat{H}_n$ are not even approximately equal.

Method (v) seems to be the only large sample, asymptotically exact result if we can compute the eigenvalues $\{\lambda_k\}$ of the limiting distribution. And we can!

For testing a simple hypothesis with specified parameters, one can apply the original Hilbert–Schmidt integral equation and also the characteristic function version for computing the eigenvalues. The kernel matrix method also applies in dimensions $d \geq 1$, but for simple hypotheses only. For example the kernel matrix approach to computing eigenvalues would not apply if the null hypothesis is that the distribution belongs to a given class such as a location-scale family of distributions, with estimated parameters; in particular, it does not apply for testing a composite hypothesis of normality or multivariate normality.

7.6 Computational and Empirical Results

We computed the eigenvalues for energy tests of univariate normality for each of the four types of null hypotheses, and for the simple hypothesis in $d > 1$.

7.6.1 Results for Univariate Normality

Using Nyström's method described in the previous section on the interval $[-a, a]$ with $a = 25$, and $N = 2000$ collocation points, we computed the eigenvalues for the energy test of univariate normality in the four cases:
1. Known parameters (simple hypothesis);
2. Known variance;
3. Known mean;
4. Unknown mean and variance (composite hypothesis),
Calculations were carried out in R [R Core Team, 2022]. In each case, the eigenvalues were essentially zero within the first 125. Integrating over $[-a, a]$

rather than $(-\infty, \infty)$ causes the estimates to slightly underestimate the eigenvalues. The ten largest eigenvalues are shown in Table 7.1. In Case 1, for example, we know that $\sum_{k=1}^{\infty} \lambda_k = 2/\sqrt{\pi} \doteq 1.128379$, while the computed eigenvalues sum to 1.102914, an error of about 0.025. When the corresponding upper tail probabilities are computed using Imhof's method, the fit to the sampling distribution is quite good. A complete table of 125 eigenvalues is provided in a supplementary file to Móri et al. [2021].

The 125 eigenvalues for each of the four cases are also available as a data set EVnormal in the *energy* package. From the last line below we can see that the smallest λ computed is nearly 0 in each case.

```
> head(energy::EVnormal)
        i      Case1       Case2       Case3       Case4
[1,] 1 0.59454782 0.18450123 0.59454782 0.11310754
[2,] 2 0.18450123 0.11310754 0.08967162 0.08356687
[3,] 3 0.08967162 0.05300453 0.08356687 0.03911317
[4,] 4 0.05300453 0.03911317 0.03501530 0.03182242
[5,] 5 0.03501530 0.02485880 0.03182242 0.01990113
[6,] 6 0.02485880 0.01990113 0.01856336 0.01697827
> tail(energy::EVnormal, 1)
          i        Case1        Case2        Case3        Case4
[125,] 125 6.07968e-15 3.962931e-15 3.916436e-15 3.662914e-15
```

TABLE 7.1

Largest eigenvalues for the energy test of normality $(d = 1)$, computed by Nyström's method

Eigenvalue	Case 1 μ, σ known	Case 2 μ unknown	Case 3 σ unknown	Case 4 μ, σ unknown
1	0.594548	0.184501	0.594548	0.113108
2	0.184501	0.113108	0.089672	0.083567
3	0.089672	0.053005	0.083567	0.039113
4	0.053005	0.039113	0.035015	0.031822
5	0.035015	0.024859	0.031822	0.019901
6	0.024859	0.019901	0.018563	0.016978
7	0.018563	0.014392	0.016978	0.012069
8	0.014392	0.012069	0.011485	0.010599
9	0.011485	0.009378	0.010599	0.008105
10	0.009378	0.008105	0.007803	0.007259

Although we do not need the estimated eigenfunctions for the test of normality, we obtain good estimates of the functions at each of the nodes and the functions can be evaluated by interpolation between the nodes. See Figure 7.1 and the supplementary file.

After computing the eigenvalues, we obtain the upper tail probabilities of the asymptotic distribution in each case by Imhof's method [Imhof, 1961].

Remark 7.1 (Imhof's Method). We apply the `imhof` function of the *Com-pQuadForm* package for R [Duchesne and de Micheaux, 2010]. This package offers several methods (Davies, Ruben/Farebrother, Imhof, and Liu's method) which could also be applied, but our results comparing very large scale, large sample parametric simulations found the best accuracy in the upper tail using Imhof's method.

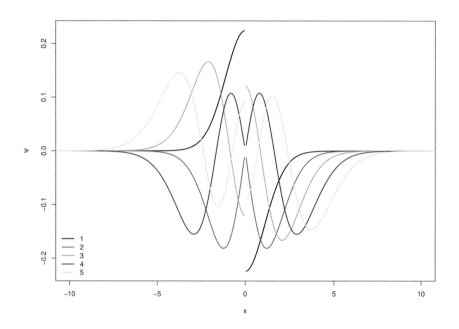

FIGURE 7.1
Eigenfunctions corresponding to the five largest eigenvalues of Case 1.

Remark 7.2. Closer inspection of Table 7.1 reveals an interesting structure connecting the four cases. Let us denote the eigenvalues for Case (i) as $\{\lambda_k^{(i)}\}$, $k = 1, 2, \ldots$. Then all of the eigenvalues of Case 1 appear as solutions of Case 2 or 3. Specifically, $\lambda_{2k-1}^{(2)} = \lambda_{2k}^{(1)}$, $k = 1, 2, \ldots$, and $\lambda_1^{(3)} = \lambda_1^{(1)}$, $\lambda_{2k}^{(3)} = \lambda_{2k+1}^{(1)}$ $k = 1, 2, \ldots$. Thus, Cases 2 and 3 each inherit half of their solutions from Case 1, corresponding to which parameter is known, with new eigenvalues corresponding to the estimated parameter. Moreover, Case 4 inherits the "new" eigenvalues of Cases 2 and 3. Given solutions to Cases 2 and 3, we have all of the solutions to Case 4. The structure observed in the extended table is that $\lambda_{2k-1}^{(4)} = \lambda_{2k}^{(2)}$ and $\lambda_{2k}^{(4)} = \lambda_{2k+1}^{(3)}$, $k = 1, 2, \ldots$. This structure in a sense reflects the structure of the integral equations among the four cases in Theorem 7.2. Although we do not have explicit formulae for the eigenvalues, these identities can be checked empirically with a few lines of code. These coincidences of eigenvalues are not completely mystical if we notice that the symmetry properties of the kernels imply that the eigenfunctions are either even or odd and they are alternating (see Figure 7.1). Whenever the eigenfunctions are even functions if we multiply them by an odd term in a kernel then the integral of this product term on the left hand side of the eigenvalue equation becomes zero, thus the corresponding eigenvalues coincide. This explains the coincidences of eigenvalues between Case 3 and Case 1 and also the coincidences of eigenvalues between Case 2 and Case 4.

In a simulation study, we compared the finite sample performance of the energy test for Case 4 (composite hypothesis, where μ and σ^2 are estimated by the sample mean and sample variance, respectively) by two methods:

(a) Monte Carlo method (MC) as implemented in the *energy* package [Rizzo and Székely, 2022].

(b) Application of the asymptotic distribution obtained by computing the Case 4 eigenvalues (EV).

Type 1 error rates for nominal significance levels 5% and 10% are reported in Table 7.2. The number of replications was 10000 for each sample size.

TABLE 7.2
Type 1 error rates for the energy test of normality by the Monte Carlo (MC) test method, and asymptotic distribution method (EV) at nominal significance levels 5% and 10% ($se < 0.005$).

n	MC 5%	EV 5%	MC 10%	EV 10%
10	0.051	0.042	0.102	0.092
25	0.051	0.053	0.101	0.107
50	0.049	0.054	0.102	0.113
100	0.053	0.058	0.103	0.114
200	0.047	0.052	0.099	0.114
500	0.048	0.054	0.095	0.109

We observe that while the Monte Carlo test controls the Type 1 error rate at the nominal level, the asymptotic test criterion also does so, or is perhaps only slightly more liberal than the Monte Carlo test at the 10% level (se < 0.005). Thus, we have a more computationally efficient method for obtaining the test decision as the eigenvalues are computed in advance and stored within our package. See Section 7.6.3 for more details.

Using the *energy* package, the following are the lines of code that compute the upper tail probability for the observed test statistic t0 from the precomputed eigenvalues:

```
## load pre-computed eigenvalues
ev <- energy::EVnormal[, "Case4"]
p <- CompQuadForm::imhof(t0, ev)$Qq
```

The *CompQuadForm* package [Duchesne and de Micheaux, 2010] can be installed from the RStudio Package menu or downloaded from https://cran.r-project.org/package=CompQuadForm.

The following example illustrates the *energy* package implementation of the asymptotic test criterion. Here it is applied to a simulated sample from a Student $t(5)$ distribution.

```
> x <- rt(100, df=5)
> energy::normal.test(x, method="limit")

        Energy test of normality: limit distribution

data:  Case 4: composite hypothesis, estimated parameters
statistic = 1.7989, p-value = 0.0004226
sample estimates:
      mean           sd
-0.1469036   1.1967225
```

7.6.2   Testing Multivariate Normality

To test (Case 1) the simple hypothesis $H : X \sim N_d(\mu, \Sigma)$ one first transforms the sample data vectors $x_1, \ldots, x_n$ to standard multivariate normal given the specified mean vector μ and specified covariance matrix Σ. In this case we can alternately apply the kernel matrix approach to computing eigenvalues by expanding the kernel function

$$h(x_j, x_k) = E|x_j - Z| + E|x_k - Z| - E|Z - Z'| - |x_j - x_k|,$$

$j, k = 1, \ldots, n$, to obtain the kernel matrix $H_n = \frac{1}{n}(h(x_j, x_k))$. Then the eigenvalues of the integral operator are approximately equal to the eigenvalues of the matrix H_n.

In Case 1, the sum of the eigenvalues equals the expected value of the test statistic:

$$\sum_{i=1}^{\infty} \lambda_i = E|Z - Z'| = \frac{2\Gamma\left(\frac{d+1}{2}\right)}{\Gamma\left(\frac{d}{2}\right)}. \tag{7.12}$$

The sum of estimated eigenvalues should converge to approximately $E|Z - Z'|$ as $n \to \infty$, but we necessarily truncate the sum when $\lambda \approx 0$ so that the matrix H is not too large to compute eigenvalues.

Estimated eigenvalues (Case 1) for $d = 1, 2, 3, 4, 5, 10$ are shown in Table 7.3. The length $n(\widehat{\lambda})$ of the sequence of eigenvalues is given, as well as the smallest eigenvalue computed. The sum of the full sequence of eigenvalues is also given with the exact value $\sum \lambda_i = E|Z - Z'|$ for comparison.

Asymptotic critical values can be derived using the estimated eigenvalues, and inverting the probability distribution function obtained with Imhof's method. These critical values for Case 1 (simple hypothesis) are summarized in Table 7.4.

TABLE 7.3
Estimated eigenvalues for Case 1 test of multivariate normality, simple
hypothesis.

d	1	2	3	4	5	10
$\widehat{\lambda}_1$	0.594550	0.461009	0.391909	0.346714	0.321561	0.238768
$\widehat{\lambda}_2$	0.184504	0.458829	0.385644	0.343556	0.313020	0.230383
$\widehat{\lambda}_3$	0.089675	0.134116	0.376981	0.333892	0.306473	0.226789
$\widehat{\lambda}_4$	0.053008	0.072919	0.109592	0.333120	0.303544	0.224884
$\widehat{\lambda}_5$	0.035019	0.071009	0.048642	0.090079	0.297535	0.222645
$\widehat{\lambda}_6$	0.024863	0.046139	0.047470	0.036053	0.084029	0.220443
$\widehat{\lambda}_7$	0.018567	0.045289	0.046580	0.034563	0.027431	0.218542
$\widehat{\lambda}_8$	0.014396	0.025966	0.044759	0.034093	0.026930	0.215544
$\widehat{\lambda}_9$	0.011489	0.024761	0.043898	0.033501	0.026837	0.208594
$\widehat{\lambda}_{10}$	0.009382	0.024675	0.030117	0.033302	0.026404	0.204424
$n(\widehat{\lambda})$	1000	2000	3000	4000	5000	5000
$\min(\widehat{\lambda})$	1.23e-06	1.04e-06	2.39e-06	1.37e-05	1.94e-05	9.43e-05
$\sum \widehat{\lambda}_i$	1.12803	1.77233	2.24710	2.64000	3.02414	4.35869
$\sum \lambda_i$	1.12838	1.77245	2.25676	2.65868	3.00901	4.36190

TABLE 7.4
Asymptotic critical values for testing multivariate normality (dimension d)
in Case 1, simple hypothesis, based on eigenvalues of the kernel matrix (see
Table 7.3).

d	10%	5%
1	2.22176	2.88781
2	3.01669	3.65446
3	3.53191	4.13539
4	3.94865	4.53036
5	4.35259	4.92071
10	5.69520	6.21063

7.6.3 Computational Efficiency

When the eigenvalues of the quadratic form are known or can be estimated
with sufficient accuracy, we have the option of applying a more computationally efficient procedure for testing normality. Our original Monte Carlo test
procedure in $d = 1$ is $M \cdot O(n \log n)$, where $M - 1$ is the number of replicates. Using the asymptotic distribution based on pre-computed eigenvalues,
$M = 1$ and we have only the additional computing time to apply Imhof's
method for the computed upper tail probability. This reduces the complexity
to $O(n \log n) + C$, where the constant C does not depend on n. In case $d > 1$

we have the original $M \cdot O(d \cdot n^2)$ complexity vs the faster $O(d \cdot n^2) + C$ of the eigenvalue method, where the constant C for computing the upper tail probabilities from pre-computed eigenvalues does not depend on n or d.

Table 7.5 reports a comparison of computing times of the two methods, Monte Carlo approach (MC) and asymptotic distribution (EV) for testing the composite hypothesis for selected sample sizes n in $d = 1$, where $M = 1000$ for the Monte Carlo method. The reported times are based on median timings for 100 independent samples. The table also reports the median computing times for the test statistic alone, and the Imhof computation alone for comparison. We applied the `imhof` function in the *CompQuadForm* package for R [Duchesne and de Micheaux, 2010] to compute the upper tail probabilities. Results in Table 7.5 were extracted from reports generated by the `microbenchmark` function in the *microbenchmark* package for R [Mersmann, 2021].

TABLE 7.5
Comparisons of computing times for the Monte Carlo test (MC) and the test based on the asymptotic distribution computed from the eigenvalues (EV), for the composite hypothesis (Case 4) in $d = 1$. The median times roughly in nanoseconds (10000s) for tests on 100 independent samples size n are reported.

n	MC	EV	Statistic	Imhof
10	23619	2106	26	1091
20	24380	1927	27	1133
50	26229	1656	29	1089
100	28705	2007	32	1118
200	33847	1734	34	1161

While the time for the step of computing the upper tail probability by Imhof's method is not negligible (see the last column of Table 7.5), the time does not scale with n. The asymptotic test procedure here is 10 to 20 times faster than the Monte Carlo test. The advantage will be more dramatic in $d > 1$ as the $O(n \log n)$ complexity increases to $O(dn^2)$ per statistic.

Remark 7.3 (Open Problems). Eigenvalues for testing multivariate normality $(d \geq 2)$ in Cases 1, 2, 3, 4 remain an open problem. For the simple hypothesis (Case 1) there is a method of estimating the eigenvalues in dimensions $d \geq 2$. In this book we discuss several energy goodness-of-fit tests, and to date there do not seem to be results other than Móri et al. [2021] for eigenvalues, leaving many interesting open problems.

7.7 Proofs

(i) Proof of Theorem 2, Case 4

Let $\mathbf{X} = (X_1, X_2, \ldots, X_n)^\top$ be an n-dimensional standard Gaussian vector, and, as usual, let

$$\overline{X} = \frac{1}{n} \sum_{k=1}^{n} X_k \quad \text{and} \quad S^2(\mathbf{X}) = \frac{1}{n} \sum_{k=1}^{n} (X_k - \overline{X})^2$$

denote the sample mean and sample variance, respectively. Consider the studentized sample $Y_k = \frac{X_k - \overline{X}}{S(\mathbf{X})}$, $k = 1, 2, \ldots, n$, and let

$$\widehat{g}_n(t) = \frac{1}{n} \sum_{k=1}^{n} e^{itY_k}$$

be the empirical characteristic function of the studentized sample, where t is real and i is the imaginary unit. We are to compute the limit

$$K(s, t) = \lim_{n \to \infty} n \operatorname{Cov} \left(\widehat{g}_n(s), \ \widehat{g}_n(t) \right).$$

(ii) Proof of Theorem 7.2

Theorem 7.2, Case 4 asserts that

$$K(s, t) = \exp \left(-\tfrac{1}{2} \left(s^2 + t^2 \right) \right) \left(e^{st} - 1 - st - \tfrac{1}{2} (st)^2 \right). \tag{7.13}$$

Proof. Since the random variables Y_k are exchangeable, we have

$$K(s, t) = \lim_{n \to \infty} \left(\operatorname{Cov} \left(e^{it_1 Y_1}, \ e^{it_2 Y_1} \right) + (n-1) \operatorname{Cov} \left(e^{it_1 Y_1}, \ e^{it_2 Y_2} \right) \right). \tag{7.14}$$

Let $\mathcal{P}$ denote the orthogonal projection onto the hyperplane $\{ \mathbf{x} : x_1 + \cdots + x_n = 0 \}$. That is,

$$\mathcal{P}\mathbf{x} = \mathbf{x} - \frac{\mathbf{x}^\top \mathbf{e}}{|\mathbf{e}|^2} \mathbf{e} = \mathbf{x} - \overline{x}\mathbf{e}, \quad |\mathcal{P}\mathbf{x}|^2 = nS^2(\mathbf{x}),$$

where $\mathbf{e} = (1, \ldots, 1)^\top$. It is well known that the random vector $\mathbf{Y}$ is uniformly distributed on the $n-1$ dimensional sphere

$$\left\{ \mathbf{y} \in \mathbb{R}^n : y_1 + \cdots + y_n = 0, \ y_1^2 + \cdots + y_n^2 = n \right\}.$$

For every $\mathbf{a} \in \mathbb{R}^n$, we have $\mathbf{a}^\top \mathbf{Y} = \mathbf{a}^\top \mathcal{P}\mathbf{Y} = (\mathcal{P}\mathbf{a})^\top \mathbf{Y}$. Regarding the distribution of $(\mathcal{P}\mathbf{a})^\top \mathbf{Y}$, the direction of $\mathcal{P}\mathbf{Y}$ in the hyperplane perpendicular to $\mathbf{e}$ is clearly irrelevant, hence without loss of generality it could be $\mathcal{P}\mathbf{f}$, where $\mathbf{f} = (1, 0, \ldots, 0)^\top$. Therefore,

$$E \exp \left(i \mathbf{a}^\top \mathbf{Y} \right) = E \exp \left(i \frac{|\mathcal{P}\mathbf{a}|}{|\mathcal{P}\mathbf{f}|} \mathbf{f}^\top \mathbf{Y} \right)$$

$$= E \exp \left(i \sqrt{\frac{n}{n-1}} |\mathcal{P}\mathbf{a}| Y_1 \right) = \widehat{\varphi} \left(\sqrt{\frac{n}{n-1}} |\mathcal{P}\mathbf{a}| \right),$$

where $\widehat{\varphi}$ is the characteristic function of Y_1, real and even. By this we have

$$\text{Cov}\left(e^{isY_1},\ e^{itY_1}\right) = \widehat{\varphi}(s-t) - \widehat{\varphi}(s)\widehat{\varphi}(t), \qquad (7.15)$$

$$\text{Cov}\left(e^{isY_1},\ e^{itY_2}\right) = E\exp\left(i(s\mathbf{f} - t\mathbf{g})^{\top}\mathbf{Y}\right) - \widehat{\varphi}(s)\widehat{\varphi}(t)$$

$$= \widehat{\varphi}\left(\sqrt{\frac{n}{n-1}}\,|s\mathcal{P}\mathbf{f} - t\mathcal{P}\mathbf{g}|\right) - \widehat{\varphi}(s)\widehat{\varphi}(t), \qquad (7.16)$$

where $\mathbf{g} = (0, 1, 0, \ldots, 0)^{\top}$. Simple calculation gives

$$|s\mathcal{P}\mathbf{f} - t\mathcal{P}\mathbf{g}|^2 = \left|s\mathbf{f} - t\mathbf{g} - \frac{1}{n}(s-t)\mathbf{e}\right|^2 = s^2 + t^2 - \frac{1}{n}(s-t)^2.$$

All we need to do is to find the first three terms in the expansion of $\widehat{\varphi}$ into a series of negative powers of n (that is, with a remainder of order $o(1/n)$). First, let us compute the density function φ of Y_1. It follows from [Lehmann, 1983, (2.14)] that

$$\varphi(y) = \kappa_n\left(1 - \frac{y^2}{n-1}\right)^{\frac{n-4}{2}}, \qquad |y| < \sqrt{n-1}, \qquad (7.17)$$

where

$$\kappa_n = \frac{1}{\sqrt{(n-1)\pi}}\frac{\Gamma\left(\frac{n-1}{2}\right)}{\Gamma\left(\frac{n-2}{2}\right)}.$$

As a first step, we evaluate the log of the density function (7.17). For the constant κ_n, by Stirling's formula, we have

$$\log\Gamma(n+\alpha) - \log\Gamma(n+\beta)$$

$$= (\alpha - \beta)\log n + \left(\binom{\alpha}{2} - \binom{\beta}{2}\right)\frac{1}{n} + O\left(n^{-2}\right),$$

thus

$$\kappa_n = \frac{1}{\sqrt{2\pi}}\left(1 - \frac{3}{4n} + O\left(n^{-2}\right)\right).$$

The log of the non-constant factor in the density is

$$\frac{n-4}{2}\log\left(1 - \frac{y^2}{n-1}\right) = -\frac{y^2}{2} + \frac{3y^2}{2n} - \frac{y^4}{4n} + O\left(n^{-2}\right).$$

Hence

$$\varphi(y) = f(y)\left(1 - \frac{3}{4n}\right)\left(1 + \frac{6y^2 - y^4}{4n}\right)(1 + O\left(n^{-2}\right))$$

$$= f(y)\left(1 - \frac{3 - 6y^2 + y^4}{4n} + O\left(n^{-2}\right)\right),$$

where $f(y)$ is the standard normal density function. The expressions above imply that

$$\widehat{\varphi}(t) = \int_{-\sqrt{n-1}}^{\sqrt{n-1}} \varphi(y) e^{ity} \, dy$$

$$= \int_{-\infty}^{\infty} f(y) \left(1 - \frac{3 - 6y^2 + y^4}{4n}\right) e^{ity} \, dy + O\left(n^{-2}\right)$$

$$= e^{-t^2/2} - \frac{1}{4n} \left(3e^{-t^2/2} + 6\frac{d^2}{dt^2} e^{-t^2/2} + \frac{d^4}{dt^4} e^{-t^2/2}\right) + O\left(n^{-2}\right)$$

$$= e^{-t^2/2} \left(1 - \frac{t^4}{4n} + O\left(n^{-2}\right)\right).$$

Consequently,

$$\widehat{\varphi}\left(\sqrt{\frac{n}{n-1}} \, |s\mathcal{P}\mathbf{f} - t\mathcal{P}\mathbf{g}|\right)$$

$$= \exp\left(-\frac{n}{2(n-1)} \left(s^2 + t^2 - \frac{1}{n}(s - t)^2\right)\right)$$

$$\times \left((1 - \frac{1}{4n} \left(\frac{n}{n-1}\right)^2 \left(s^2 + t^2 - \frac{1}{n}(s - t)^2\right)^2 + O\left(n^{-2}\right)\right)$$

$$= \exp\left(-\frac{1}{2}\left(s^2 + t^2\right)\right) \left(1 - \frac{1}{n}st - \frac{1}{4n}\left(s^2 + t^2\right)^2 + O\left(n^{-2}\right)\right).$$

Furthermore,

$$\widehat{\varphi}(s)\widehat{\varphi}(t) = \exp\left(-\frac{1}{2}\left(s^2 + t^2\right)\right) \left(1 - \frac{1}{4n}\left(s^4 + t^4\right) + O\left(n^{-2}\right)\right). \quad (7.18)$$

Then equation (7.13) immediately follows from (7.14), (7.15), (7.16), and (7.18). $\qquad \square$

7.8   Exercises

Exercise 7.1. Solve Schrödinger equation (7.6) for $f(x) = \frac{1}{2}$ on $[-1, 1]$, when $\psi'(-1) = \psi'(1) = 0$.

8

Generalized Goodness-of-Fit

CONTENTS

8.1 Introduction

Many important families of distributions have heavy tails, and possibly non-finite expectation. When $E|X| < \infty$ does not hold, the energy distance based on Euclidean distances (5.1) does not apply for a goodness-of-fit test, but a generalized energy distance can be applied with a suitable exponent on Euclidean distance such that $E|x|^\alpha < \infty$.

Recall that if $0 < \alpha < 2$, the generalized energy distance with index α is

$$\mathcal{E}_n^\alpha(\mathbf{X}, F_0) = \frac{2}{n}\sum_{i=1}^{n} E|x_i - X|^\alpha - E|X - X'|^\alpha - \frac{1}{n^2}\sum_{\ell=1}^{n}\sum_{m=1}^{n}|x_\ell - x_m|^\alpha, \quad (8.1)$$

where $0 < \alpha < 2$ is chosen such that $E|X|^\alpha < \infty$.

Under the null hypothesis, $H_0 : F = F_0$, the test statistic $n\mathcal{E}_n^\alpha$ has the same type of asymptotic distribution as $n\mathcal{E}_n$, which is a quadratic form of centered Gaussian variables. Under an alternative hypothesis $H_1 : F \neq F_0$, the statistic $n\mathcal{E}_n^\alpha$ tends to infinity stochastically. Hence, a generalized energy goodness-of-fit test that rejects the null hypothesis for large values of $n\mathcal{E}_n^\alpha$ is consistent.

DOI: 10.1201/9780429157158-8

Energy goodness-of-fit tests based on (8.1) have been implemented for testing the Pareto family [Rizzo, 2009], Cauchy, and symmetric stable distributions Yang [2012].

For Pareto distributions and stable distributions, in most literature the symbol α is a parameter, which denotes the tail index. Therefore, throughout this chapter, we will use the symbol s rather than α as the exponent on Euclidean distance in (8.1).

8.2   Pareto Distributions

Pareto distributions are named for Vilfredo Pareto (1848-1923), who developed models for income inequality. Pareto models are often used as statistical size distributions for modeling income, losses or other random variables, especially when there may typically be some large values observed. Arnold [2015] is a comprehensive reference on Pareto distributions.

A Pareto Type I distribution has the survival function

$$\overline{F}(x) = \left(\frac{x}{\sigma}\right)^{-\alpha}, \qquad x \geq \sigma > 0, \tag{8.2}$$

and density function $f(x) = \frac{\alpha\sigma^\alpha}{x^{\alpha+1}}$, $x \geq \sigma > 0$, where $\sigma > 0$ is a scale parameter, and $\alpha > 0$ is a shape parameter (Pareto's index of inequality), which measures the heaviness in the upper tail.

The Pareto Type II distribution has the survival distribution

$$\overline{F}(x) = \left[1 + \frac{x - \mu}{\sigma}\right]^{-\alpha}, \qquad x \geq \mu, \tag{8.3}$$

where $\mu \in \mathbb{R}$ is a location parameter, $\sigma > 0$ is a scale parameter and $\alpha > 0$ is a shape parameter. Pareto type I and type II models are related by a simple transformation. If $Y \sim P(II)(\mu, \sigma, \alpha)$ then $Y - (\mu - \sigma) \sim P(I)(\sigma, \alpha)$.

8.2.1   Energy Tests for Pareto Distribution

Pareto densities have a polynomial upper tail with index $-(\alpha + 1)$. Small values of α correspond to heavier tails, and the kth moments exist only if $\alpha > k$. The moments of $X \sim P(I)(\sigma, \alpha)$ are given by

$$E[X^k] = \frac{\alpha\sigma^k}{(\alpha - k)}, \quad \alpha > k; \tag{8.4}$$

in particular, $E[X] = \frac{\alpha\sigma}{\alpha-1}, \alpha > 1$ and $Var(X) = \frac{\alpha\sigma^2}{(\alpha-1)^2(\alpha-2)}, \alpha > 2$. For $P(II)(\mu, \sigma, \alpha)$ distributions, $E[X] = \frac{\alpha\sigma}{\alpha-\mu-\sigma}, \alpha > 1$.

Thus, theoretical results that depend on existence of moments do not necessarily extend to Pareto distributions with arbitrary shape parameter α.

Energy statistics overcome this limitation with the generalized energy distance. Replacing exponent α with s, we have

$$\mathcal{E}_n^s(\mathbf{X}, F_0) = \frac{2}{n} \sum_{i=1}^{n} E|x_i - X|^s - E|X - X'|^s - \frac{1}{n^2} \sum_{\ell=1}^{n} \sum_{m=1}^{n} |x_\ell - x_m|^s, \quad (8.5)$$

where $0 < s < \alpha$ is chosen such that $E|X|^s < \infty$.

The exponent s is a stability index in the sense that when $s < \alpha/2$ the distribution of X^s has finite variance. Then $E|x_j - X|^s$ and $E|X - X'|^s$ are computed under the hypothesized Pareto model. Expressions for $E|x_j - X|^s$ and $E|X - X'|^s$ are derived below. This statistic will be denoted $Q_{n,s}$.

8.2.2 Test of Transformed Pareto Sample

Alternately, for goodness-of-fit tests of $P(\sigma, \alpha)$ models, it is equivalent to test the hypothesis that $T = \log(X)$ has a two-parameter exponential distribution,

$$H_0 : T \sim \text{Exp}(\mu, \alpha),$$

where $\mu = \log(\sigma)$ is the location parameter and α is the rate parameter. Here $\log(X)$ always refers to the natural logarithm. The density of T is

$$f_T(t) = \alpha e^{-\alpha(t-\mu)}, \qquad t \geq \mu.$$

The first and second moments of T are finite. For all $\alpha > 0$ we have $E[T] = \frac{1}{\alpha} + \mu$, and $Var(T) = \alpha^{-2}$. Hence we can alternately apply the test statistic

$$V_s = n \left\{ \frac{2}{n} \sum_{j=1}^{n} E|T_j - T|^s - E|T - T'|^s - \frac{1}{n^2} \sum_{j,k=1}^{n} |T_j - T_k|^s \right\}, \quad (8.6)$$

where $T_j = \log(X_j)$, $j = 1, \ldots, n$, and $E|T_j - T|^s$ and $E|T - T'|^s$ are computed under the hypothesized exponential (log Pareto) model. The energy goodness-of-fit test for the two-parameter exponential is covered in Section 5.3. See Rizzo [2009] for more details on the energy goodness-of-fit test for Pareto distributions.

For Pareto type II samples, $X_j \sim P(II)(\mu, \sigma, \alpha)$, let $Y_j = X_j - (\mu - \sigma)$. Then $Y_j \sim P(I)(\sigma, \alpha)$. Moreover, Q and V are invariant to this transformation, as $|X_j - X_k| = |Y_j - Y_k|$, etc. Thus the statistics developed for Pareto type I distributions can be applied to the corresponding transformed Pareto type II distributions.

The expected values are $E[Q_s] = E|X - X'|^s$ and $E[V_1] = E|\log(X) - \log(X')|$. When the Pareto hypothesis is true and $Var(X)$ is finite, Q_s converges in distribution to a quadratic form

$$\sum_{j=1}^{\infty} \lambda_j Z_j^2, \quad (8.7)$$

as sample size n tends to infinity, where λ_j are non-negative constants, and Z_j are iid standard normal random variables. Asymptotic theory of V-statistics can be applied to prove that tests based on Q_s (or V_s) are statistically consistent goodness-of-fit tests.

8.2.3   Statistics for the Exponential Model

Assume that $X \sim P(\sigma, \alpha)$, and $T = \log(X)$. Then $T \sim Exp(\mu, \alpha)$, where $\mu = \log \sigma$, α is the rate parameter, and $F_T(t) = 1 - e^{-\alpha(t-\mu)}$, $t \geq \mu$. Then the integrals in V_1 are

$$E|t - T| = s - \mu + \frac{1}{\alpha}(1 - 2F_T(t)), \qquad t \geq \mu; \tag{8.8}$$

$$E|T - T'| = \frac{1}{\alpha}. \tag{8.9}$$

A computing formula for the corresponding test statistic is derived as follows. The first mean in the statistic $V = V_1$ is

$$\frac{1}{n} \sum_{j=1}^{n} \left(T_j - \mu + \frac{1}{\alpha}(1 - 2F_T(T_j)) \right) = \overline{T} - \left(\mu + \frac{1}{\alpha} \right) + \frac{2e^{\alpha\mu}}{\alpha n} \sum_{j=1}^{n} e^{-\alpha T_j},$$

where $\overline{T} = \frac{1}{n} \sum_{j=1}^{n} T_j$ is the sample mean.

Also, for $s = 1$, the third sum can be expressed as a linear function of the ordered sample. If $T_{(j)}$ denotes the jth largest sample element, then

$$\sum_{j,k=1}^{n} |T_j - T_k| = 2 \sum_{j=1}^{n} ((2j - 1) - n)T_{(j)}.$$

Hence, the computational complexity of the statistic Q_1 or V_1 is $O(n \log n)$. The statistic $V = V_1$ is given by

$$V = n \left\{ 2 \left[\overline{T} - \mu - \frac{1}{\alpha} + \frac{2e^{\alpha\mu}}{\alpha n} \sum_{j=1}^{n} e^{-\alpha T_j} \right] - \frac{1}{\alpha} - \frac{2}{n^2} \sum_{j=1}^{n} (2j - 1 - n)T_{(j)} \right\}. \tag{8.10}$$

If parameters are estimated, the corresponding estimates are substituted in (8.10). (Formula (8.10) can be simplified further for computation.)

8.2.4   Pareto Statistics

In this section we develop the computing formula for Q_s. First we present two special cases, $s = 1$ and $s = \alpha - 1$.

1. If $X \sim P(\sigma, \alpha)$, $\alpha > 1$ and $s = 1$, then

$$E|y - X| = y - E[X] + \frac{2\sigma^\alpha}{(\alpha - 1)\, y^{\alpha - 1}}$$

$$= y + \frac{2\sigma^\alpha y^{1-\alpha} - \alpha\sigma}{\alpha - 1}, \qquad y \geq \sigma; \qquad (8.11)$$

$$E|X - X'| = \frac{2\alpha\sigma}{(\alpha - 1)(2\alpha - 1)} = \frac{E[X]}{\alpha - 1/2}. \qquad (8.12)$$

2. If $X \sim P(\sigma, \alpha)$, $\alpha > 1$ and $s = \alpha - 1$, then

$$E|y - X|^{\alpha - 1} = \frac{(y - \sigma)^\alpha + \sigma^\alpha}{y}, \qquad y \geq \sigma; \qquad (8.13)$$

$$E|X - X'|^{\alpha - 1} = \frac{2\alpha\sigma^{\alpha - 1}}{\alpha + 1}. \qquad (8.14)$$

The statements of cases 1 and 2 can be obtained by directly evaluating the integrals.

Although the special cases above are easy to apply, in general it may be preferable to apply s that is proportional to α. For this we need case 1 below.

The Pareto type I family is closed under the power transformation. That is, if $X \sim P(\sigma, \alpha)$ and $Y = X^r$, then $Y \sim P(\sigma^r, \alpha/r)$. It is always possible to find an $r > 0$ such that the second moments of $Y = X^r$ exist, and Q_1 can be applied to measure the goodness-of-fit of Y to $P(\sigma^r, \alpha/r)$. This goodness-of-fit measure will be denoted $Q^{(r)}$.

Proofs of the following statements are given in Rizzo [2009].

1. If $X \sim P(\sigma, \alpha)$ and $0 < s < \alpha < 1$, then

$$E|y - X|^s = (y - \sigma)^s - \frac{\sigma^\alpha [s B_{y_0}(s, 1 - \alpha) - \alpha B(\alpha - s, s + 1)]}{y^{\alpha - s}}, \qquad y \geq \sigma; \qquad (8.15)$$

$$E|X - X'|^s = \frac{2\alpha^2 \sigma^s B(\alpha - s, s + 1)}{2\alpha - s}, \qquad (8.16)$$

where $y_0 = \frac{y - \sigma}{y}$ and $B(a, b)$ is the complete beta function.

2. If $X \sim P(\sigma, \alpha = 1)$, $0 < s < 1$, and $y_0 = (y - \sigma)/y$, then

$$E|y - X|^s = (y - \sigma)^s - \sigma s y^{s-1} \left\{ \frac{y_0^s}{s} + \frac{y_0^{s+1}}{s + 1}\, {}_2F_1(1, s + 1; s + 2; y_0) \right\}$$

$$+ \sigma y^{s-1} B(s + 1, 1 - s), \qquad y \geq \sigma; \qquad (8.17)$$

$$E|X - X'|^s = \frac{2\sigma^s}{2 - s}\, B(1 - s, s + 1), \qquad (8.18)$$

where $_2F_1(a, b; c; z)$ denotes the Gauss hypergeometric function,

$$_2F_1(a, b; c; z) = \sum_{k=0}^{\infty} \frac{(a)_k (b)_k}{(c)_k} \frac{z^k}{k!},$$

and $(r)_k = r(r+1)\cdots(r+k-1)$ denotes the ascending factorial.

For $\alpha > 1$ the expressions for $E|y - X|^s$ are complicated and involve the Gauss hypergeometric function. The above two cases and formulas for the remaining cases are derived in Rizzo [2009]; see equations (3.9), (3.10), (3.11), and (3.12) for $\alpha > 1$. It is simpler and more computationally efficient to apply the $Q^{(r)}$ statistics (or V) if $\alpha > 1$.

For a formal goodness-of-fit test based on V or Q, the unknown parameters of the hypothesized $P(\sigma, \alpha)$ distribution can be estimated by a number of methods. See Arnold [2015, Ch. 5].

The joint maximum likelihood estimator (MLE) of (α, σ) is $(\hat{\alpha}, \hat{\sigma})$, where

$$\hat{\alpha} = n \left[\sum \log \frac{X_j}{\hat{\sigma}} \right]^{-1}, \qquad \hat{\sigma} = X_{(1)},$$

and $X_{(1)}$ is the first order statistic. By the invariance property of the MLE, substituting $\hat{\alpha}$ for α and $\hat{\sigma}$ for σ, or $\hat{\mu} = \log \hat{\sigma}$, we obtain the corresponding MLEs of the mean distances in the test statistics V or Q.

Alternately, unbiased estimators of the parameters can be derived from the MLEs. If both parameters are unknown,

$$\sigma^* = X_{1:n} \left(1 - \frac{1}{(n-1)\hat{\alpha}} \right), \qquad \alpha^* = \left(1 - \frac{2}{n} \right) \hat{\alpha}$$

are unbiased estimators of the parameters. If one parameter is known, then

$$\alpha^* = \left(1 - \frac{1}{n} \right) \hat{\alpha} \quad \text{or} \quad \sigma^* = X_{1:n} \left(1 - \frac{1}{n\alpha} \right)$$

is unbiased for the unknown parameter.

Then for a test of the composite hypothesis, one can apply the same type of parametric simulation as for the energy test of univariate normality in Chapter 5. See Rizzo [2009] for empirical results and power comparisons.

8.2.5   Minimum Distance Estimation

The proposed Pareto goodness-of-fit statistics provide a new approach to estimation of the tail index α of a $P(\sigma, \alpha)$ distribution. A minimum distance approach can be applied, where the objective is to minimize the corresponding goodness-of-fit statistic under the assumed model. For this application, the statistic V_1 can be normalized to mean 1 by dividing by the mean $E|T - T'|$, and Q_s can be normalized by dividing by $E|X - X'|^s$, where $E|T - T'|$ or $E|X - X'|^s$ are computed under the hypothesized model.

For example, one could minimize

$$V = n\left\{2\left[\overline{T} - \mu - \frac{1}{\alpha} + \frac{2e^{\alpha\mu}}{\alpha n}\sum_{j=1}^{n}e^{-\alpha T_j}\right] - \frac{1}{\alpha} - \frac{2}{n^2}\sum_{j=1}^{n}(2j - 1 - n)T_{(j)}\right\}$$

or $V/E|T - T'| = \alpha V$ with respect to α to obtain a minimum energy estimate α^* of α. Note that the third term in V is constant given the sample.

8.3   Cauchy Distribution

Yang [2012] implements the energy goodness-of-fit test for Cauchy distributions (and more generally, symmetric stable distributions, discussed in Section 8.5). Gürtler and Henze [2000] introduced an L_2 type of test for Cauchy distribution similar to energy with a different weight function.

Cauchy is a symmetric location-scale family of distributions. The Cauchy distribution with location parameter δ and scale parameter γ has density function

$$\frac{1}{\pi\gamma\left[1 + (\frac{x-\delta}{\gamma})^2\right]}, \qquad x \in \mathbb{R}. \tag{8.19}$$

Standard Cauchy ($\delta = 0, \gamma = 1$) has the central t distribution with 1 degree of freedom.

Although there is no "standardized" Cauchy (because we do not have finite mean and variance) there is a *standard form*. If X has pdf (8.19) then the pdf of $Y = A + BX$ has the same form, with location $\delta^* = A + B\delta$ and scale $\gamma^* = |B|\gamma$. Thus $Y = (X - \delta)/\gamma$ has the standard Cauchy distribution. Then to test the simple hypothesis $H_0 : X \sim \text{Cauchy}(\delta, \gamma)$ it is equivalent to test if $Y = (X - \delta)/\gamma \sim \text{Cauchy}(0, 1)$.

For an energy test we need finite expected distance, so we will apply the generalized energy goodness-of-fit test based on the energy statistic (8.1). However, Cauchy is a member of the stable family, for which the symbol α denotes the stability index, so we have conflicting notation in this case. (Cauchy stability index is $\alpha = 1$.) Let us replace the exponent α in (8.1) with exponent s, and choose $s \in (0, 1)$ so that $E|X|^s < \infty$.

Then if X is standard Cauchy ($\delta = 0, \gamma = 1$) we have

$$E|X - x_i|^s = (1 + x_i^2)^{s/2}\frac{\cos(s\arctan x_i)}{\cos\frac{\pi s}{2}}. \tag{8.20}$$

We also have that

$$\lim_{x\to\infty}\frac{E|x - X|^s}{|x|^s} \to 1,$$

so if $|x|$ is large we can approximate $E|X - x_i|^s \approx |x_i|^s$.

If X, X' are iid standard Cauchy and $0 < s < 1$ then

$$E|X - X'|^s = \frac{2^s}{\cos \frac{\pi s}{2}}. \tag{8.21}$$

See Yang [2012], Chapter 3 for derivation of the expected distances above.

The energy statistic for standard Cauchy is a V-statistic with kernel

$$h(x_1, x_2) = E|x_1 - X|^s + E|x_2 - X|^s - E|X - X'|^s - |x_1 - x_2|^s$$

$$= (1 + x_1^2)^{s/2} \frac{\cos(s \arctan x_1)}{\cos \frac{\pi s}{2}} + (1 + x_2^2)^{s/2} \frac{\cos(s \arctan x_2)}{\cos \frac{\pi s}{2}}$$

$$- \frac{2^s}{\cos \frac{\pi s}{2}} - |x_1 - x_2|^s. \tag{8.22}$$

The test statistic is n times the V-statistic:

$$Q_{n,s} = 2 \sum_{j=1}^{n} (1 + x_j^2)^{s/2} \frac{\cos(s \arctan x_j)}{\cos \frac{\pi s}{2}} - \frac{n 2^s}{\cos \frac{\pi s}{2}} - \frac{1}{n} \sum_{j,k-1}^{n} |x_j - x_k|^s. \tag{8.23}$$

The V-statistic $Q_{n,s}/n$ is a degenerate kernel V-statistic, which under the null hypothesis of standard Cauchy converges in distribution to a quadratic form of centered Gaussian variables $Q \overset{D}{=} \sum_{i=1}^{\infty} \lambda_i Z_i^2$. As in Chapters 3 and 6, one can implement a Monte Carlo test by parametric simulation, or estimate eigenvalues by one of the methods in Chapter 7. See Yang [2012] for details of finding eigenvalues by Nyström's method, some asymptotic critical values and simulation results.

For testing the composite hypothesis that X has a Cauchy distribution one can estimate γ and δ by e.g. maximum likelihood. This approach is covered in detail in Chapter 5 of Yang [2012], including computing formulas and simulation results.

8.4   Stable Family of Distributions

Stable distributions have several parameterizations in the literature. We will use the unified notation $S(\alpha, \beta, \gamma, \delta)$ suggested by Nolan [2020]. Here

α is the tail index or characteristic exponent; the power rate at which the tail(s) of the density function decay, also called the stability index.

β is the skewness index.

γ is the scale parameter.

δ is the location parameter.

A random variable X is stable if and only if $X \overset{D}{=} aZ + b$, where Z is a random variable with characteristic function

$$\varphi_X(t) = E(e^{itZ}) = \begin{cases} \exp(-|t|^\alpha[1 - i\beta \tan \frac{\pi\alpha}{2} \operatorname{sgn}(t)]), & \alpha \neq 1; \\ \exp(-|t|[1 - i\beta \frac{2}{\pi} \operatorname{sgn}(t) \log |t|]), & \alpha = 1, \end{cases}$$

$0 < \alpha \leq 2$, $-1 \leq \beta \leq 1$, $a > 0$, $b \in \mathbb{R}$. In this notation, $S(1, 0, \gamma, \delta)$ denotes a Cauchy distribution, and $S(\frac{1}{2}, 1, \gamma, \delta)$ denotes a Lévy distribution. The normal distribution has stability index 2, and it is the only member of the stable family that has a finite first moment. Thus, for testing stability, one would apply the generalized energy distance (8.1) where the exponent of (8.1) is replaced by the symbol s, to avoid confusion with the tail index α in $S(\alpha, \beta, \gamma, \delta)$.

If $\alpha \neq 1$, and $E|X|^s < \infty$ under $H_0 : X \sim S(\alpha, 0, \gamma = 1, \delta = 0)$, the energy test statistic is

$$Q_{n,s} = \frac{4}{\pi} \Gamma(1 + s) \sin \frac{\pi s}{2} \sum_{j=1}^{n} \int_0^\infty \frac{1 - e^{-t^\alpha} \cos(\beta t^\alpha \tan \frac{\pi\alpha}{2} - x_i t)}{t^{s+1}} dt$$

$$- n \frac{2^{1+s/\alpha}}{\pi} \Gamma\left(1 - \frac{s}{\alpha}\right) \Gamma(s) \sin \frac{\pi s}{2} - \frac{1}{n} \sum_{j,k=1}^{n} |x_j - x_k|^s.$$

If $\alpha = 1$, $0 < s < \alpha$, $\gamma = 1$, $\delta = 0$ the energy goodness-of-fit statistic is

$$Q_{n,s} = \frac{4}{\pi} \Gamma(1 + s) \sin \frac{\pi s}{2} \sum_{j=1}^{n} \int_0^\infty \frac{1 - e^{-t} \cos(\beta \frac{2}{\pi} t \log t + x_i t)}{t^{s+1}} dt$$

$$- n \frac{2^{1+s}}{\pi} \Gamma(1 - s) \Gamma(s) \sin \frac{\pi s}{2} - \frac{1}{n} \sum_{j,k=1}^{n} |x_j - x_k|^s.$$

For standard Cauchy, the expected value is simpler; see formula (8.23) from Section 8.3.

Note: In practice, choosing $s < \alpha/2$ seems to be numerically stable. See Yang [2012] for details on the energy test of Cauchy and stable distributions including formulae for expected distances, computational formulas, and methodology for implementing the tests.

8.5 Symmetric Stable Family

If $X \sim S(\alpha, \beta = 0, \gamma, \delta)$ then $Y = (X - \delta)/\gamma$ is a symmetric stable distribution. Thus, without loss of generality, we can suppose $X \sim S(\alpha, 0, 1, 0)$. If $\alpha = 1$,

X is standard Cauchy, with kernel function and test statistic given in Section 8.3. If $\alpha \neq 1$, and $s < \alpha$, the test statistic is

$$Q_{n,s} = 2\sum_{j=1}^{n}\int_{0}^{\infty}\frac{1-e^{-t^{\alpha}}\cos(x_j,t)}{t^{1+s}}\,dt - \frac{n}{\pi}2^{(s+\alpha)/\alpha}\Gamma\left(1-\tfrac{s}{\alpha}\right)\Gamma(s)\sin\tfrac{\pi}{2}s$$

$$-\frac{1}{n}\sum_{j=1}^{n}\sum_{k=1}^{n}|x_j-x_k|^s. \tag{8.24}$$

For numerical stability, choosing $s < \alpha/2$ is recommended.

Yang [2012], Table 4.12, derived asymptotic critical values for the energy test of $S(1.5,0)$ at significance level 10%, using Nyström's method. As we should expect, the critical values depend on the choice of exponent s. With $N = 2000$ collocation points, the Type I error of the test is well controlled using the computed critical values. See Chapter 7 for a discussion of Nyström's method applied to compute the asymptotic distribution of $Q_{n,1}$ for the normal distribution, which is symmetric stable with $\alpha = 2$. See also Yang [2012] for more details on computing the test statistic and methods for estimating the asymptotic critical values, as well as empirical results.

8.6   Exercises

Exercise 8.1. Prove that if X is symmetric stable and $E|x - X|^s < \infty$ for all $x \in \mathbb{R}$, then as $x \to \infty$ (or $-x \to \infty$),

$$\frac{E|x-X|^s}{|x|^s} \to 1.$$

Exercise 8.2. Yang [2012] computes $E|x - X|^s$ as $|x|^s$ for $|x| \geq 2000$. Compute the relative error in this approximation. That is, compute (true value − approximate value)/(true value) for the Cauchy distribution when $x = 2000$.

Exercise 8.3. Prove that if X is d-dimensional standard normal, then

$$\lim_{|x|_d \to \infty}\frac{E|x-X|_d}{|x|_d} \to 1.$$

9

Multi-sample Energy Statistics

CONTENTS

9.1   Energy Distance of a Set of Random Variables

In this chapter we focus on multi-sample problems, so we begin by defining the energy distance of a set of random variables. For simplicity, let us start with three independent random variables, $X \in \mathbb{R}^d, Y \in \mathbb{R}^d, Z \in \mathbb{R}^d$, where $d \geq 1$. Define their mutual energy by

$$\mathcal{E}(X, Y, Z) := \mathcal{E}(X, Y) + \mathcal{E}(X, Z) + \mathcal{E}(Y, Z).$$

This mutual energy coefficient is always non-negative and equals 0 if and only if X, Y, Z are identically distributed. If any two terms on the right hand side are zero, then X, Y, Z are identically distributed, so the third term must be zero, too.

DOI: 10.1201/9780429157158-9

If we have k random variables, then their mutual energy is the sum of their pairwise energy distances. That is,

$$\mathcal{E}(X_1,\ldots,X_k) = \sum_{1\leq i<j\leq k} \mathcal{E}(X_i,X_j).$$

We have $\binom{k}{2}$ terms on the right hand side, and $\mathcal{E}(X_1,\ldots,X_k) = 0$ if and only if all k random variables are identically distributed. For this to hold, it is enough to see that a well-chosen set of $k-1$ terms are 0.

This definition of mutual energy among several random variables motivates the definition of multi-sample energy statistics in the following section.

9.2   Multi-sample Energy Statistics

Recall from Chapter 3 that for two samples, $\mathbf{X} = (X_1, X_2, \ldots, X_n)$ and $\mathbf{Y} = (Y_1, Y_2, \ldots, Y_m)$ their sample energy distance is

$$\mathcal{E}_{n,m}(\mathbf{X},\mathbf{Y}) = \frac{2}{nm}\sum_{i=1}^{n}\sum_{j=1}^{m}|X_i - Y_j|$$

$$- \frac{1}{n^2}\sum_{i=1}^{n}\sum_{j=1}^{n}|X_i - X_j| - \frac{1}{m^2}\sum_{i=1}^{m}\sum_{j=1}^{m}|Y_i - Y_j|. \qquad (9.1)$$

To test the hypothesis $H_0 : F_X = F_Y$, our test statistic is

$$T_{n,m} = \frac{nm}{n+m}\mathcal{E}_{n,m}(\mathbf{X},\mathbf{Y}).$$

For a multisample statistic, we can consider two cases: a balanced design, where all sample sizes are equal, and an unbalanced design. For a balanced design with sample size n for each sample, we have the same weight $\frac{n}{2}$ for each pairwise test statistic. We can simply add the pairwise test statistics.

For example, in a balanced design the three sample test statistic would be

$$\mathcal{E}_n(\mathbf{X},\mathbf{Y},\mathbf{Z}) = \frac{n}{2}\left[\mathcal{E}_{n,n}(\mathbf{X},\mathbf{Y}) + \mathcal{E}_{n,n}(\mathbf{X},\mathbf{Z}) + \mathcal{E}_{n,n}(\mathbf{Y},\mathbf{Z})\right].$$

In an unbalanced design, however, the weights $\frac{n_i n_j}{n_i+n_j}$ are not all equal. Suppose that $\mathbf{X}$, $\mathbf{Y}$, and $\mathbf{Z}$ are independent random samples from the distributions F_X, F_Y, and F_Z, respectively, where $X, Y, Z \in \mathbb{R}^d$, $d \geq 1$. If their sample sizes are n_1, n_2, n_3, respectively, and $N = n_1 + n_2 + n_3$, then a multi-sample energy statistic for testing $H_0 : F_X = F_Y = F_Z$ is

$$T_N(\mathbf{X},\mathbf{Y},\mathbf{Z}) = \frac{n_1 n_2}{n_1+n_2}\mathcal{E}_{n_1,n_2}(\mathbf{X},\mathbf{Y}) + \frac{n_1 n_3}{n_1+n_3}\mathcal{E}_{n_1,n_3}(\mathbf{X},\mathbf{Z})$$

$$+ \frac{n_2 n_3}{n_2+n_3}\mathcal{E}_{n_2,n_3}(\mathbf{Y},\mathbf{Z}). \qquad (9.2)$$

Then $T_N(\mathbf{X}, \mathbf{Y}, \mathbf{Z}) \geq 0$. Under the alternative hypothesis, at least one pair of distributions differ. If $N \to \infty$ such that none of n_i/N tend to zero, then the test is consistent.

Our original k-sample energy test for equal distributions based on the k-sample version of (9.2) is implemented in the energy package as `eqdist.e` and `eqdist.etest`. The default method is "original," corresponding to (9.2).

Example 9.1. The familiar iris data set is a good example with data in $\mathbb{R}^4$, and three samples of size 50 each.

```
> library(energy)
> eqdist.etest(iris[,1:4], c(50,50,50), R = 199)

        Multivariate 3-sample E-test of equal distributions

data:  sample sizes 50 50 50, replicates 199
E-statistic = 357.71, p-value = 0.005
```

The test is implemented as a permutation test, where argument R specifies the number of permutation replicates. The hypothesis of equal distributions is rejected at $\alpha = .05$.

In (9.2) when sample sizes are not approximately equal, the unequal weights mean that the large sample pairs are weighted more heavily in the sum. In clustering or change-point analysis, for example, we may have varying sample sizes throughout the analysis, including some very unbalanced sample sizes.

However, as shown in Section 9.3, an alternate statistic can be formulated using a weighted average of the two-sample statistics based on sample sizes: Distance Components (disco). For the iris example above, with equal sample sizes, the *disco* "between-components" statistic is exactly equal to the average of the three pairwise statistics in (9.2).

```
> 3 * disco.between(iris[,1:4], iris$Species, R=0)
[1] 357.7119
```

9.3 Distance Components: A Nonparametric Extension of ANOVA

Rizzo and Székely [2010] introduced a nonparametric test for the K-sample hypothesis $(K \geq 2)$ of equal distributions: $H_0 : F_1 = \cdots = F_K$, vs $H_1 : F_i \neq F_j$ for some $i \neq j$. This problem could be compared to the multi-sample test for equal means, but with energy statistics we can test the more general hypothesis of equal distributions.

Analogous to the ANOVA decomposition of variance, we partition the total dispersion of the pooled samples into between and within components, called *DIStance COmponents* (DISCO). For two samples, the between-sample component is the two-sample energy statistic above. For several samples, the between component is a weighted combination of pairwise two-sample energy statistics.

Then to test the K-sample hypothesis $H_0 : F_1 = \cdots = F_K$, $K \geq 2$, one can apply the between-sample test statistic, or alternately a ratio statistic similar to the familiar F statistic of ANOVA. Both options are available in the *energy* package [Rizzo and Székely, 2022], and both are implemented as a randomization test.

Recall that the characterization of equality of distributions by energy distance also holds if we replace Euclidean distance by $|X-Y|^\alpha$, where $0 < \alpha < 2$. The characterization does **not** hold if $\alpha = 2$ because $2E|X - Y|^2 - E|X - X'|^2 - E|Y - Y'|^2 = 0$ whenever $E(X) = E(Y)$. We denote the corresponding two-sample energy distance by $\mathcal{E}_{n,m}^{(\alpha)}$.

Let $A = \{a_1, \ldots, a_{n_1}\}$, $B = \{b_1, \ldots, b_{n_2}\}$ be two samples, and define

$$g_\alpha(A, B) := \frac{1}{n_1 n_2} \sum_{i=1}^{n_1} \sum_{j=1}^{n_2} |a_i - b_j|^\alpha, \qquad (9.3)$$

for $0 < \alpha \leq 2$.

9.3.1   The DISCO Decomposition

Suppose that we have K independent random samples $A_1, \ldots, A_K$ of sizes $n_1, n_2, \ldots, n_K$, respectively, and $N = \sum_{j=1}^{K} n_j$. Let $A = \cup_{j=1}^{K} A_j$ be the pooled sample.

The within-sample dispersion statistic is

$$W_\alpha = W_\alpha(A_1, \ldots, A_K) = \sum_{j=1}^{K} \frac{n_j}{2} g_\alpha(A_j, A_j). \qquad (9.4)$$

and the total dispersion is

$$T_\alpha = T_\alpha(A_1, \ldots, A_K) = \frac{N}{2} g_\alpha(A, A), \qquad (9.5)$$

where A is the pooled sample. The between-sample statistic is

$$S_{n,\alpha} = \sum_{1 \leq j < k \leq K} \left(\frac{n_j + n_k}{2N} \right) \left[\frac{n_j n_k}{n_j + n_k} \, \mathcal{E}_{n_j, n_k}^{(\alpha)}(A_j, A_k) \right] \qquad (9.6)$$

$$= \sum_{1 \leq j < k \leq K} \left\{ \frac{n_j n_k}{2N} \left(2g_\alpha(A_j, A_k) - g_\alpha(A_j, A_j) - g_\alpha(A_k, A_k) \right) \right\},$$

See Section 20.1.3 for an $O(N \log N)$ computing algorithm for univariate samples.

Note that $S_{n,\alpha}$ weights each pairwise $\mathcal{E}$-statistic based on the proportion of data in the two samples. Analogous to the decomposition of between-sample variance and within-sample variance of ANOVA, we have the decomposition of distances.

Theorem 9.1 (disco). *The total pairwise distances T_α (9.5) between all elements of K samples can be decomposed into between-sample S_α (9.6) and within sample W_α (9.4) components as*

$$T_\alpha = S_\alpha + W_\alpha,$$

where both S_α and W_α are nonnegative.

See Rizzo and Székely [2010] or Section 9.9 for a proof.

We refer to the between-sample statistic S_α (9.6) as the disco (or disco-between) statistic. For every $0 < \alpha < 2$, the statistic (9.6) determines a consistent test of the multi-sample hypothesis of equal distributions [Rizzo and Székely, 2010]. In case all sample sizes are equal, S_α is simply the arithmetic average of the pairwise energy statistics.

Above we saw that the energy characterization of equal distributions does not hold for $\alpha = 2$. What happens at the endpoint $\alpha = 2$? In the special case where all F_j are univariate distributions and $\alpha = 2$, $S_{n,2}$ is the ANOVA between sample sum of squared error, and the decomposition $T_2 = S_2 + W_2$ is the ANOVA decomposition. The ANOVA test statistic measures differences in means, not distributions.

However, if we apply $\alpha = 1$ (Euclidean distance) or any $0 < \alpha < 2$ as the exponent on Euclidean distance, the corresponding energy test is consistent against all alternatives with finite α moments. The simplest choice is $\alpha = 1$. If the underlying distributions may have non-finite first moment, a suitable choice of α extends the energy test to this situation.

A multi-sample permutation test (randomization test) can be applied similar to the two-sample energy test described in Section 4.3.

Example 9.2. The penguin data [Horst et al., 2020] records size and other variables for three species of penguins on three islands in the Palmer Archipelago near Palmer Station, Antarctica. Here we analyze bill length, bill depth, and flipper length for the Adelie species by island. The sample sizes for Adelie on the three islands are

```
##     Biscoe    Dream Torgersen
##         44       55        47
```

and we are testing for equal distribution of measurements by island.

Let us see how to prepare the data for input into either `eqdist.etest` or `disco`. After reviewing summaries of the data, we see that there are two missing values. We remove those observations and subset the data for species

Adelie. Then `disco` is easiest to apply because we already have the factor `island` in the data. Note that the data can be in any order using `disco`, unlike `eqdist.etest`.

```
library(palmerpenguins)
p <- penguins[complete.cases(penguins), ]
x <- p[p$species == "Adelie", ]
disco(x[, 3:5], factors=x$island, R=999)
```

```
Distance Components: index   1.00
Source              Df    Sum Dist   Mean Dist    F-ratio    p-value
factors              2    10.18606    5.09303      1.178      0.233
Within             143   618.41111    4.32455
Total              145   628.59716
```

This runs a permutation test based on the disco statistic, and summarizes the result by a DISCO table that looks similar to an ANOVA table. We do not reject the null hypothesis of equal distributions.

The "F"-ratio reported in the table

$$F_{n,\alpha} = \frac{S_{n,\alpha}/(K-1)}{W_{n,\alpha}/(N-K)}$$

is applied (for the permutation test) if `method="discoF"`. Degrees of freedom are determined by the combined constraints on sums of distances.

If we only want the between sample statistic it is $S_1 = 2(5.09303) = 10.18606$, in the "Sum Dist" column; or directly from

```
disco.between(x[, 3:5], factors=x$island, R=0)
```

To use `eqdist.etest` we need to stack the data so that island 1 is in the first 44 rows, island 2 is in the next rows, etc. After subsetting the data above, we need to order the data by island.

```
o <- order(x$island)
x <- x[o, ]
eqdist.etest(x[, 3:5], sizes=table(x$island), R=999)
```

```
##
##  Multivariate 3-sample E-test of equal distributions
##
## data:   sample sizes 44 55 47, replicates 999
## E-statistic = 31, p-value = 0.244
```

One can obtain a table of pairwise energy statistics for any number of samples in one step using the `edist` function in the energy package. It can be used, for example, to display $\mathcal{E}$-statistics for the result of a cluster analysis. By default it gives the pairwise energy distances of the disco method:

```
edist(x[ ,3:5], sizes=table(x$island), method="disco")
```

```
##          1        2
## 2 1.757060
## 3 4.877647 3.551350
```

The pairwise statistics sum to $S_1 = 10.18606$ shown in the DISCO table above.

Example 9.3. Returning to Example 9.2, let us consider a second variable, sex, for the Adelie species of penguins. The sample sizes for the interaction term are

```
##          Biscoe Dream Torgersen
##   female    22    27        24
##   male      22    28        23
```

A test for significant interaction in this context is a test for independence of class labels and distances. We can apply distance components to test this hypothesis.

```
## disco(x = x[, 3:5], factors = x$island:x$sex, R = 999)
##
## Distance Components: index  1.00
## Source            Df    Sum Dist  Mean Dist   F-ratio   p-value
## factors            5    79.63883   15.92777     4.062     0.001
## Within           140   548.95834    3.92113
## Total
```

The test for interaction is significant, so we should analyze the island effect separately for female and male penguins.

Remark 9.1. On implementation of other permutational tests for multi-factor and multi-level designs based on partitions of distances or dissimilarities, see the vignettes and manuals for R packages *vegan* and *permute* [Oksanen et al., 2022, Simpson, 2022]. The adonis2 function in package *vegan* can perform permutation tests based on sequential decomposition of distances. To use vegan::adonis2 with Euclidean distances, it is necessary to specify method="euclidean" and sqrt.dist=TRUE. (With these options, the one-way decomposition matches disco.) On nonparametric MANOVA see also Anderson [2001] and Anderson [2014] describing the "PERMANOVA" permutational methods.

For comparison, the corresponding decomposition obtained by adonis2 is

```
library(vegan)
y <- x[,3:5]
adonis2(y ~ island : sex, data=x,
        method="euclidean", sqrt.dist=TRUE)$SumOfSq
## [1]   79.63883 548.95834 628.59716
```

and the permutation test with default number 999 of replicates based on the pseudo-F-ratio is significant.

The interaction contains the between sample distances for both variables, so it can be further decomposed. A sequential decomposition and permutation tests can be obtained using `vegan::adonis2` as follows.

```
## Permutation test for adonis under reduced model
## Terms added sequentially (first to last)
## Permutation: free
## Number of permutations: 199
##
## adonis2(formula = y ~ island * sex, data = x, permutations = 199,
##           method = "euclidean", sqrt.dist = TRUE)
##              Df SumOfSqs      R2       F Pr(>F)
## island        2    10.19 0.01620  1.2989  0.170
## sex           1    61.62 0.09803 15.7147  0.005 **
## island:sex    2     7.83 0.01246  0.9989  0.360
## Residual    140   548.96 0.87331
## Total       145   628.60 1.00000
```

Note: With the sequential decomposition, the decomposition depends on the order of the terms from left to right. That is, the `adonis2` partition for model `y ~ island*sex` differs from the partition for `y ~ sex*island`.

9.3.2   Application: Decomposition of Residuals

The following application and example originally appeared in Rizzo and Székely [2010]. Consider the residuals from a fitted linear model on a univariate response with one factor. Denote the fitted model L. Regardless of whether the hypothesis of equal means is true or false, the residuals do not reflect differences in means. If treatments differ in some way other than the mean response, then the differences can be measured on the residuals by distance components, $0 < \alpha < 2$. If we consider models of the type proposed by Akritas and Arnold [1994], we could regard the linear portion L for treatment effect as an "intercept" term. That is,

$$F_j(x) = L(x) + R_j(x), \qquad \sum_{j=1}^{a} R_j(x) = 0,$$

where F_j is the distribution function of x_{ij}, $i = 1, \ldots, n_j$. If all $R_j(x) = 0$, then $F_j = L$ for every j. One can test the hypothesis $H_0 : all\ R_j(x) = 0$ by testing the sample of residuals of L for equal distributions.

Example 9.4. (Gravity data)

The gravity data consist of 81 measurements in a series of eight experiments conducted by the National Bureau of Standards in Washington D.C. between May, 1934 and July, 1935, to estimate the acceleration due to gravity at Washington. Each experiment consisted of replicated measurements with a reversible pendulum expressed as deviations from 980 cm/sec^2. The data set is provided in the *boot* package for R [Davison and Hinkley, 1997]. Boxplots of the data [Rizzo and Székely, 2010, Figure 9.1] reveal non-constant variance of the measurements over the series of experiments.

The decompositions of gravity measurements by Series for $\alpha = 1$ and $\alpha = 2$ are shown in Table 9.1. Note that when index $\alpha = 2$ is applied, the DISCO decomposition is exactly equal to the ANOVA decomposition, also shown in Table 9.1. In fact, with our implementation as random permutation test, the F_2 test is actually a *permutation test* based on the ANOVA F statistic. In this example 999 permutation replicates were used to estimate the p-values.

TABLE 9.1
Comparison of DISCO and ANOVA decompositions in Example 9.4.

```
DISCO
Distance Components: index  1.00
Source          Df   Sum Dist  Mean Dist    F-ratio    p-value
Between:
  Series         7   100.62287  14.37470      2.781      0.001
Within          73   377.27836   5.16820


Distance Components: index  2.00
Source          Df   Sum Dist  Mean Dist    F-ratio    p-value
Between:
  Series         7  2818.62413  402.66059      3.568      0.002
Within          73  8239.37587  112.86816


ANOVA
Analysis of Variance Table
Response: Gravity
          Df Sum Sq Mean Sq F value    Pr(>F)
Series     7 2818.6   402.7  3.5675 0.002357 [0.002 by perm. test]
Residuals 73 8239.4   112.9
```

When we decompose residuals by Series using DISCO ($\alpha = 1$) as shown in Table 9.2, the DISCO F_1 statistic is significant (p-value < 0.05). We can conclude that the residuals do not arise from a common error distribution. (The ANOVA F statistic is zero on residuals.)

Example 9.5. (Iris data) The iris data set records four measurements (sepal length and width, petal length and width) for 50 flowers from each of three species setosa, versicolor, and virginica. The model is $Y \sim$ Species, where Y is a four dimensional response corresponding to the four measurements of each iris. In this example we apply the disco decomposition to the residuals of the linear model.

TABLE 9.2
Distance Components of ANOVA residuals in Example 9.4.

```
Distance Components: index  1.00
Source              Df   Sum Dist  Mean Dist    F-ratio    p-value
Between:
  Series             7    56.66334   8.09476      1.566      0.046
Within              73   377.27836   5.16820
```

The DISCO F_1 and MANOVA Pillai-Bartlett F test, implemented as permutation tests, each have p-value 0.001 based on 999 permutation replicates. The residuals from the fitted linear model are a 150×4 data set.

Results of the multivariate analysis are shown in Table 9.3. Based on the DISCO decomposition of the residuals and test for equality of distributions of residuals (p-value < 0.04), there are differences due to Species that are not explained by the linear component of the model.

TABLE 9.3
Analysis of iris data and residuals in Example 9.5.

```
DISCO analysis of multivariate iris data:
   Distance Components: index  1.00
   Source              Df   Sum Dist  Mean Dist    F-ratio    p-value
   Between:
     Species            2   119.23731  59.61865    124.597      0.001
   Within             147    70.33848   0.47849

MANOVA analysis of multivariate iris data:
           Df Pillai approx F num Df den Df    Pr(>F)
   Species  2  1.192   53.466       8    290 < 2.2e-16 ***
   Residuals 147
   [permutation test p = 0.001]

DISCO analysis of residuals of linear model for iris data:
   Distance Components: index  1.00
   Source              Df   Sum Dist  Mean Dist    F-ratio    p-value
   Between:
     Species            2    1.69845   0.84923      1.775      0.039
   Within             147   70.33848   0.47849
```

9.4 Hierarchical Clustering

Energy distance has been applied in hierarchical cluster analysis [Székely and Rizzo, 2005a]. It generalizes the well-known Ward's minimum variance method in a similar way that DISCO generalizes ANOVA. The DISCO decomposition has also been applied to generalize k-means clustering (see Section 9.6).

A clustering is defined to be a partition $P = \{C_1, \ldots, C_g\}$ of the set of n objects or observations into g disjoint non-empty classes or clusters. The number of classes g is not specified in advance. A dendrogram is a binary tree with $n - 1$ internal nodes. An agglomerative hierarchical clustering procedure determines a dendrogram, where at each step in the hierarchical clustering procedure, the distance between merging clusters is the height of the corresponding node in the tree connecting the clusters. A hierarchical clustering solution is the nested class structure represented by the dendrogram, which may reveal structure or groups in the data. The dendrogram can be cut at any level to obtain a clustering, with the level determining the number of classes.

In an *agglomerative* hierarchical clustering algorithm, starting with single observations or *singletons*, at each step we merge the pair of clusters that have minimum cluster distance. In the energy distance algorithm, the cluster distance is the two sample energy statistic.

There is a general class of hierarchical clustering algorithms uniquely determined by their respective Lance-Williams recursive formula for updating all cluster distances following each merge of two clusters. One can show [Székely and Rizzo, 2005a] that the energy clustering algorithm is also a member of this class and its recursive formula shows that it it is formally similar to Ward's method.

Proposition 9.1 (Recursive Formula for $\mathcal{E}$-distance). *Suppose $A, B,$ and C are disjoint nonempty finite subsets of $\mathbb{R}^d$ such that*

$$\mathcal{E}(A, B) \leq \min(\mathcal{E}(A, C), \mathcal{E}(B, C)).$$

Then a recursive formula for $\mathcal{E}(A \cup B, C)$ is given by

$$\mathcal{E}(A \cup B, C) = \frac{n_1 + n_3}{n_1 + n_2 + n_3}\, \mathcal{E}(A, C)$$
$$+ \frac{n_2 + n_3}{n_1 + n_2 + n_3}\, \mathcal{E}(B, C) - \frac{n_3}{n_1 + n_2 + n_3}\, \mathcal{E}(A, B). \qquad (9.7)$$

See Section 9.9 or Székely and Rizzo [2005a] for the proof.

Energy Lance-Williams Formula

Suppose that the disjoint clusters C_i, C_j are to be merged at the current step. If C_k is a disjoint cluster, then by Proposition 9.1 the new cluster distances can be computed by the following recursive formula:

$$d(C_i \cup C_j, C_k) = \frac{n_i + n_k}{n_i + n_j + n_k}\, d(C_i, C_k)$$
$$+ \frac{n_j + n_k}{n_i + n_j + n_k}\, d(C_j, C_k) - \frac{n_k}{n_i + n_j + n_k}\, d(C_i, C_j), \quad (9.8)$$

where $d(C_i, C_j) = \mathcal{E}_{n_i, n_j}(C_i, C_j)$, and n_i, n_j, n_k are the sizes of clusters C_i, C_j, C_k, respectively.

A Lance-Williams recursive formula has the form

$$d_{(ij)k} = \alpha_i d_{ik} + \alpha_j d_{jk} + \beta d_{ij} + \gamma |d_{ik} - d_{jk}|,$$

where $d_{ij} := d(C_i, C_j)$ and $d_{(ij)k} := d(C_i \cup C_j, C_k)$.

Thus the hierarchical $\mathcal{E}$-clustering algorithm has a Lance-Williams form

$$\begin{aligned} d_{(ij)k} &:= d(C_i \cup C_j, C_k) \\ &= \alpha_i\, d(C_i, C_k) + \alpha_j\, d(C_j, C_k) + \beta\, d(C_i, C_j) \\ &= \alpha_i d_{ik} + \alpha_j d_{jk} + \beta d_{ij} + \gamma |d_{ik} - d_{jk}|, \end{aligned}$$

with

$$\alpha_i = \frac{n_i + n_k}{n_i + n_j + n_k}; \qquad \beta = \frac{-n_k}{n_i + n_j + n_k}; \qquad \gamma = 0.$$

If we substitute squared Euclidean distances for Euclidean distances in this recursive formula, keeping the same parameters $(\alpha_i, \alpha_j, \beta, \gamma)$, then we obtain the Lance-Williams updating formula for Ward's minimum variance method. However, we know that Ward's method (with exponent $\alpha = 2$ on distances) is a geometrical method that separates clusters by their centers, not by their distributions. $\mathcal{E}$-clustering generalizes Ward because for every $0 < \alpha < 2$, the energy clustering algorithm separates clusters based on differences in distribution.

The clustering algorithm based on $\mathcal{E}$-distance has several desirable properties:

- Statistical consistency,
- Ultrametricity,
- Lance–Williams form,
- Space-dilating,
- Computational tractability.

See Székely and Rizzo [2005a] for details.

Overall in simulations and real data examples [Székely and Rizzo, 2005a] the characterization property of $\mathcal{E}$ is a clear advantage for certain clustering problems, without sacrificing the good properties of Ward's minimum variance method for separating spherical clusters.

Remark 9.2 (Implementation). The `hclust` hierarchical clustering function provided in R [R Core Team, 2022] implements its available clustering methods by Lance-Williams recursive formulas. In R (versions > 3.03) with `method="ward.D"` the hierarchical $\mathcal{E}$-clustering algorithm is applied, while `method="ward.D2"` applies Ward's minimum variance method.

The cluster heights determined by energy hierarchical clustering with $\alpha = 1$ are two times the heights from `hclust`; otherwise the implementations are identical. In the current releases of *energy* and R, the function `energy.hclust` optionally applies a user-specified exponent to distance, then passes the call to `hclust` with `method="ward.D"`. The return value from `energy.hclust` is class `hclust`, so all the `hclust` methods for analysis, such as `plot`, `print`, `cutree`, `dendrogram` are applicable.

Remark 9.3. Some software implementations (Matlab, for one) of "Ward's method" do not apply the Lance-Williams recursive formula (9.8) for updating cluster distances; we cannot use these algorithms to implement energy clustering. The reason that `hclust` provides (for exponent $\alpha = 1$) an algorithm for energy clustering is that method `ward.D` **does** apply the Lance-Williams recursion.

9.5 Case Study: Hierarchical Clustering

Classification of Human Tumors Based on Gene Expression Data

Comparison of gene expression levels of normal and diseased tissue can be used to help identify tumors and appropriate treatment. In Székely and Rizzo [2005a] we applied hierarchical $\mathcal{E}$-clustering to the NCI60 microarray data discussed in Chapter 14 of Hastie et al. [2009]. The raw data are expression levels from cDNA microarrays, in 60 cell cancer cell lines used in the screen for anti-cancer drugs by the National Cancer Institute. The data is an array of 6830 gene expression measurements for 64 human cancer samples. The gene expression levels in the NCI60 data are relative to a fixed common reference sample. In the raw data, x_{ij} is $\log_2(Cy5/Cy3)$ (the fluorescence ratio) for gene j in sample i. The data has been centered to row median zero and column median zero, and missing values were assigned the value of zero. To classify tumors, the genes are regarded as variables.

The samples include nine types of cancers: breast (7), central nervous system (CNS) (5), colon (7), leukemia (6), melanoma (8), non-small-cell-lung-carcinoma (NSCLC) (9), ovarian (6), prostate (2), renal (9), and unknown cancer (1). In our analysis, we have omitted 1884 variables with more than two missing values. The unknown cancer sample and the two prostate cancer samples are also omitted from this analysis, because of the small class size. The resulting data is a 61×4946 array of real numbers representing gene expression levels for 4946 genes in 61 cancer samples. Two of the cell lines are replicated in the data. The samples labeled "K562B-repro" are replicated leukemia samples, and the samples labeled "MCF7A-repro" are replicated breast cancer samples. In this analysis, replicated samples are treated as distinct samples, so the clustering solutions should cluster each pair of replicates together.

Hierarchical clustering solutions at the eight group level for three methods, e-clustering, Ward's method, and group average method are shown in Tables 5–7 of Székely and Rizzo [2005a]. Both $\mathcal{E}$-clustering and Ward's method appear to be more successful at recovering the correct classes than the group average method.

We computed an adjusted (corrected for chance) Rand statistic for each solution to measure agreement with the given classes. The adjusted Rand statistics for $\mathcal{E}$-clustering, Ward's method, and group average method are 0.45, 0.33, and 0.08, respectively (higher is better), which is a way to rank the methods in this example. The dendrogram for energy clustering is given in Figure 9.1. Here we can also observe that the replicated breast cancer and leukemia samples are each correctly grouped into a common cluster.

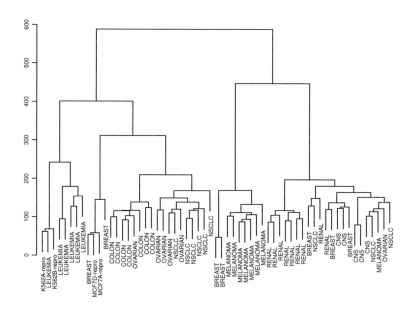

FIGURE 9.1
Cancer Classification by Hierarchical $\mathcal{E}$ Method

See Székely and Rizzo [2005a] for more examples and further analysis. See also Section 9.7 for a case study with comparison to energy k-groups clustering.

9.6   K-groups Clustering

Li and Rizzo [2017] proposed new clustering methods *k-groups* which generalize and extend the well known and widely applied *k*-means cluster analysis
method. See also Li [2015a], França et al. [2020]. Our cluster distance is based
on the energy characterization of equality between distributions, and it applies in arbitrary dimension. It generalizes *k*-means, which separates clusters
by differences in means. The *k*-groups cluster distance, which is based on
energy distance separates clusters by differences in distributions.

First we describe *k*-means clustering. It is a prototype-based algorithm
which uses the cluster mean as the centroid, and assigns observations to the
cluster with the nearest centroid. Let $D = \{x_1, \ldots, x_n\} \subset \mathbb{R}^m$ be the data set
to be clustered, and $P = \{\pi_1, \ldots, \pi_k\}$ be a k set partition of D, where k is
the number of clusters set by the user. Then $\cup_i \pi_i = D$, and $\pi_i \cap \pi_j = \emptyset$ if
$i \neq j$. Let n_i be the number of data objects assigned to cluster π_i, c_i be the
centroid of cluster π_i, $1 \leq i \leq k$ and $d(x, y)$ be a symmetric, zero-diagonal
dissimilarity function that measures the distance between data objects x and
y.

The distance or dissimilarity function $d(x, y)$ is an important factor that
influences the performance of *k*-means. The most commonly applied distance
functions are Euclidean quadratic distance, spherical distance, and Kullback-
Leibler Divergence. Each choice of distance function thus determines a cluster
distance function. In the energy approach, cluster distance is energy distance.
If $d(x, y) = (x - y)^2$, then the *k*-means clustering objective is

$$\min_{c_i, 1 \leq i \leq k} \sum_{i=1}^{k} \sum_{x_j \in \pi_i} (x_j - c_i)^2.$$

Implementing a *k*-means algorithm is equivalent to a global minimization
problem, which is computationally difficult (NP-hard). A computationally efficient iterative algorithm is usually applied. An early algorithm was proposed
by Lloyd [1982], and a more efficient version was proposed and published in
Fortran by Hartigan and Wong [1979], which is the default algorithm of the
R function `kmeans`.

In most available software and in applications "*k*-means" applies quadratic
distance, and the algorithm minimizes the variance within the clusters. For
k-groups we apply a weighted two-sample energy statistic to measure the dissimilarity between clusters, and modify the *k*-means algorithm given by Hartigan and Wong [1979] to implement *k*-groups clustering. The *energy* package
[Rizzo and Székely, 2022] provides the algorithm described below in the function `kgroups`. It returns an object of the same type as `kmeans`.

9.6.1   K-groups Objective Function

We define dispersion between two sets $A = \{a_1, \ldots, a_{n_1}\}$, $B = \{b_1, \ldots, b_{n_2}\}$ by

$$g_\alpha(A, B) = \frac{1}{n_1 n_2} \sum_{i=1}^{n_1} \sum_{m=1}^{n_2} |a_i - b_m|^\alpha,$$

where $0 < \alpha \leq 2$. K-groups is based on a distance components decomposition (see Section 9.3). At each step in the clustering solution, the K clusters can be regarded as K samples when we apply the disco decomposition.

Let $D = \{x_1, \ldots, x_N\} \subset \mathbb{R}^m$ be the data set to be clustered. At each step in the clustering algorithm, there is a partition $P = \{\pi_1, \ldots, \pi_k\}$ of the data of D into k sets, where k is the (pre-specified) number of clusters. Let $n_1, \ldots, n_k$ be the current cluster sizes. We define the total dispersion of the data as

$$T_\alpha(\pi_1, \ldots \pi_k) = \frac{N}{2} g_\alpha(\cup_{i=1}^k \pi_i, \cup_{i=1}^k \pi_i) = \frac{N}{2} g_\alpha(D, D).$$

The within-groups dispersion is defined by

$$W_\alpha(\pi_1, \ldots \pi_k) = \sum_{j=1}^k \frac{n_j}{2} g_\alpha(\pi_j, \pi_j).$$

The between-sample component is

$$B_\alpha(\pi_1, \ldots \pi_k) = \sum_{1 \leq i < j \leq k} \left\{ \frac{n_i n_j}{2N} \left(2g_\alpha(\pi_i, \pi_j) - g_\alpha(\pi_i, \pi_i) - g_\alpha(\pi_j, \pi_j) \right) \right\}.$$

When $0 < \alpha \leq 2$ we have the decomposition

$$T_\alpha(\pi_1, \ldots \pi_k) = W_\alpha(\pi_1, \ldots \pi_k) + B_\alpha(\pi_1, \ldots \pi_k),$$

where both $W_\alpha(\pi_1, \ldots \pi_k)$ and $B_\alpha(\pi_1, \ldots \pi_k)$ are nonnegative.

A good clustering solution should have large between-sample dispersion $B_\alpha(\pi_1, \ldots \pi_k)$. Since $T_\alpha(\pi_1, \ldots \pi_k)$ is constant, the problem is equivalent to minimizing $W_\alpha(\pi_1, \ldots \pi_k)$. Hence the objective is to find a partition which minimizes the within-cluster dispersion W_α. Therefore, the objective function for k-groups is

$$\min_{\pi_1, \ldots, \pi_k} \sum_{j=1}^k \frac{n_j}{2} g_\alpha(\pi_j, \pi_j) = \min_{\pi_1, \ldots, \pi_k} W_\alpha(\pi_1, \ldots \pi_k). \qquad (9.9)$$

Following Hartigan and Wong's algorithm, we implement an iterative approach, where we search for a locally optimal k-partition by moving points from one cluster to another.

A *variation* of a partition P is a partition P' obtained from P by removing a single point a from a cluster $\pi_i \in P$ and assigning this point to a different cluster π_j of P. For efficiency we do not want to recompute all components to find the updated within-cluster W component for any potential move of a point. It is enough to know the change in W.

The α-energy distance between a point a and a cluster π_c is

$$\mathcal{E}_\alpha(a, \pi_c) = \frac{2}{n_c} \sum_{x \in \pi_c} |x - a|^\alpha - \frac{1}{n_c^2} \sum_{x \in \pi_c} \sum_{y \in \pi_c} |x - y|^\alpha. \tag{9.10}$$

Our usual choice of exponent is $\alpha = 1$.

Theorem 9.2 (Li [2015a]). *Suppose that $P = \{\pi_1, \pi_2, ... \pi_k\}$ is a partition of the data, and $P^a = \{\pi_1^-, \pi_2^+, ..., \pi_k\}$ is the partition obtained by moving point a from π_1 to π_2. Then*

$$W_\alpha(P) - W_\alpha(P^a) = \frac{n_1}{2(n_1 - 1)} \mathcal{E}_\alpha(a, \pi_1) - \frac{n_2}{2(n_2 + 1)} \mathcal{E}_\alpha(a, \pi_2), \tag{9.11}$$

where $\mathcal{E}_\alpha(a, \pi_i)$ denotes the two-sample α-energy distance between point a and cluster π_i.

9.6.2 K-groups Clustering Algorithm

Similar to the Hartigan and Wong k-means algorithm, the k-groups algorithm moves point a from cluster π_1 to π_2 if

$$\frac{n_1}{2(n_1 - 1)} \mathcal{E}_\alpha(a, \pi_1) - \frac{n_2}{2(n_2 + 1)} \mathcal{E}_\alpha(a, \pi_2)$$

is positive. Otherwise point a remains in cluster π_1. For a new clustering problem, we initialize the partitions at random. Usually several random starts will be tried and the best solution will be kept. The k-groups clustering algorithm is as follows.

1. For each point I, $I = 1, ..., N$, randomly assign I to cluster $\pi_i, i = 1, ..., k$. Let $\pi(I)$ represent the cluster where I belongs, and $n(\pi(I))$ represent the size of cluster $\pi(I)$.

2. For each point I, $I = 1, ..., N$, compute

$$E_1 - \frac{n(\pi(I))}{2(n(\pi(I)) - 1)} \mathcal{L}_\alpha(I, \pi(I))$$

   and

$$E_2 = \left[\frac{n(\pi_i)}{2(n(\pi_i) + 1)} \mathcal{E}_\alpha(I, \pi_i) \right]$$

   for all clusters π_i, $\pi \neq \pi(I)$. If $E_1 \leq E_2$, observation I remains in cluster $\pi(I)$. Otherwise, move the point I to cluster π, and update clusters $\pi(I)$ and π.

3. Stop if there is no relocation in the last N steps.

Li [2015a] also provides algorithms with updating formulas for moving m points at each step, where $m \geq 1$.

9.6.3   K-means as a Special Case of K-groups

Theorem 9.3 shows that k-groups contains the k-means algorithm as a special case when $\alpha = 2$, by showing that Hartigan and Wong's k-means algorithm has the same objective function as k-groups when $\alpha = 2$.

We know that when $0 < \alpha < 2$, the α-energy distance is 0 if and only if the random variables are identically distributed, but for $\alpha = 2$, we have equality to 0 whenever $E(X) = E(Y)$. Thus k-groups is a distribution-based algorithm, while k-means is a prototype algorithm.

Theorem 9.3. *When $\alpha = 2$, the k-groups algorithm and the Hartigan and Wong k-means algorithm have the same objective function.*

Proof. For the k-groups algorithm with exponent $\alpha = 2$

$$\frac{n_i}{2}g_2(\pi_i, \pi_i) = \sum_{\ell=1}^{n_i} x_\ell^2 - n_i c_i^2,$$

where $c_i = \frac{1}{n_i}\sum_{j=1}^{n_i} x_j$, and $x_j \in \pi_i$, $j = 1,...n_i$. The objective function for k-means is

$$\min_{c_i, 1\leq i \leq k} \sum_{i=1}^{k} \sum_{x_j \in \pi_i} (x_j - c_i)^2,$$

and

$$\sum_{x_j \in \pi_i} (x_j - c_i)^2 = \frac{n_i}{2}g_2(\pi_i, \pi_i),$$

for all $i = 1, ..., k$. Hence, when $\alpha = 2$, k-groups and k-means have the same objective function.   □

9.7   Case Study: Hierarchical and K-groups Cluster Analysis

Diagnosis of Erythemato-Squamous Diseases in Dermatology

The dermatology data is available from the UCI Machine Learning Repository [Blake and Merz, 1998] at `https://archive.ics.uci.edu/ml/datasets/dermatology/`. For description of the data refer to the repository or Li and Rizzo [2017]. The clustering task is to cluster six types of erythemato-squamous diseases: psoriasis, seboreic dermatitis, lichen planus, pityriasis rosea, choronic dermatitis and pityriasis rubra pilaris. According to Güvenir et al. [1998], diagnosis is difficult since all these diseases share the similar clinical features of erythema and scaling.

The data has 366 observations with 34 attributes each. We standardize all the attributes to zero mean and unit standard deviation and delete any observations which have missing values. The effective data size is 358 in the cluster analysis.

We will compare the clustering results with the diagnosis provided in the data file for agreement. Four measures of cluster validation are computed: diagonal agreement, Kappa, Rand, and corrected Rand (cRand). In cluster validation, we are most interested in the corrected Rand statistic, because it corrects for agreement by chance. (See e.g. R package *e1071* [Meyer et al., 2022]; functions `classAgreement` and `matchClasses`.)

Table 9.4 shows the clustering result of k-means, k-groups, and Hierarchical $\mathcal{E}$. Hierarchical $\mathcal{E}$ is agglomerative hierarchical clustering by energy distance (see Section 9.4). The maximum Rand and cRand index values 0.9740 and 0.9188 are obtained by k-groups. Hierarchical $\mathcal{E}$ obtains the second largest Rand and cRand index values 0.9730 and 0.9159, and k-means obtained the smallest Rand and cRand index values among these algorithms: 0.9441 and 0.8390, respectively.

TABLE 9.4

Dermatology data results

Indices	k-means	k-groups	Hierarchical $\mathcal{E}$
Diagonal	0.8324	0.9553	0.9497
Kappa	0.7882	0.9440	0.9370
Rand	0.9441	0.9740	0.9730
cRand	0.8390	0.9188	0.9159

9.8 Further Reading

Diverse applications in science can be found in the literature. For example, the energy test was applied in chemometrics [Vaiciukynas et al., 2015], and Varina et al. [2009] applied energy hierarchical for clustering of chemical structures.

Change-point analysis: Kim et al. [2009] applied energy distance to identify change points in a multivariate streaming environment; Matteson and James [2014] proposed an energy approach to change point analysis. Energy change point analysis is closely related to DISCO analysis, a nonparametric version of MANOVA.

Alekseyenko [2016] proposed a distance-based version of a Welch t-test for two sample potentially unbalanced and heteroscedastic data, and applied it in the analysis of microbiome datasets.

9.8.1   Bayesian Applications

Approximate Bayesian computation (ABC) via energy distance (see e.g. [Nguyen, 2019, Chapter 7]) defines a psuedo-posterior by a comparison of the data with simulated data using a suitable discrepancy measure. An importance sampling algorithm is designed based on the two-sample energy statistic, such that ABC can be shown to have desirable asymptotic properties. Nguyen et al. [2020] found that the energy importance sampling IS-ABC method compares well with alternative discrepancy measures.

Energy distance has been applied for cross-validation of posterior distribution estimation with a new heteroscedastic version of Bayesian Additive Regression Trees (BART) by Pratola et al. [2017].

Markov Chain Monte Carlo (MCMC) is a frequently applied method to approximate multidimensional integrals, for example in Bayesian analysis. According to Mak and Joseph [2018a] the support point approximation enjoys an improved convergence rate. They propose a new method to compact a continuous probability distribution F into a set of representative points called support points. These points are obtained by minimizing the energy distance. They show that support points converge in distribution to F. The definition of support points P_i is the following:

$$\{P_i\}_{i=1}^n := \operatorname*{arg\,min}_{x_1, x_2, \ldots, x_n} \left\{ \frac{2}{n} \sum_{i=1}^n E|x_i - Y| - \frac{1}{n^2} \sum_{i=1}^n \sum_{j=1}^n |x_i - x_j| \right\}.$$

The minimization problem is handled by two algorithms to efficiently generate representative point sets. In simulation studies, Mak and Joseph [2018b] found that support points improved integration performance compared with Monte Carlo and a specific quasi-Monte Carlo method. Two important applications demonstrated are (1) To quantify the propagation of uncertainty in expensive simulations and (2) To optimally compact (MCMC) samples in Bayesian computation.

For details see Mak and Joseph [2018b], Mak and Joseph [2018a] and Huang et al. [2022]. This idea can also be used for optimal data splitting; see Joseph and Vakayil [2022].

9.9   Proofs

9.9.1   Proof of Theorem 9.1

One can obtain the DISCO decomposition by directly computing the difference between the total and within-sample dispersion. Given p-dimensional samples $A_1, \ldots, A_K$ with respective sample sizes $n_1, \ldots, n_K$ and $N = \sum_j n_j$, let $g_{jk} = g_\alpha(A_j, A_k)$ given by (9.3) and $G_{jk} = n_j n_k g_{jk}$, for $j, k = 1, \ldots, K$. Then for all $0 < \alpha \leq 2$ and $p \geq 1$,

$$
\begin{aligned}
T_\alpha - W_\alpha &= \frac{N}{2} g(A, A) - \frac{1}{2} \sum_j n_j g_{jj} \\
&= \frac{N}{2} \left(\sum_{j<k} \frac{2}{N^2} G_{jk} + \sum_j \frac{1}{N^2} G_{jj} \right) - \frac{1}{2} \sum_j \frac{1}{n_j} G_{jj} \\
&= \frac{1}{2N} \left(\sum_{j<k} 2G_{jk} + \sum_j G_{jj} \right) - \frac{1}{2} \sum_j \frac{1}{n_j} G_{jj} \\
&= \frac{1}{2N} \left(\sum_{j<k} n_j n_k (2g_{jk} - g_{jj} - g_{kk}) + \sum_{j<k} n_j n_k (g_{jj} + g_{kk}) \right) \\
&\quad + \frac{1}{2N} \sum_j n_j^2 g_{jj} - \frac{1}{2} \sum_j n_j g_{jj} \\
&= \sum_{j<k} \frac{n_j + n_k}{2N} \left(\frac{n_j n_k}{n_j + n_k} \right) (2g_{jk} - g_{jj} - g_{kk}) + \frac{1}{2N} \sum_{j<k} n_k(n_j g_{jj}) \\
&\quad + \frac{1}{2N} \sum_{j<k} n_j(n_k g_{kk}) + \frac{1}{2N} \sum_j n_j^2 g_{jj} - \frac{1}{2} \sum_j n_j g_{jj}.
\end{aligned}
$$

After simplification we have

$$
\begin{aligned}
T_\alpha - W_\alpha &= \sum_{j<k} \frac{n_j + n_k}{2N} \left(\frac{n_j n_k}{n_j + n_k} \right) (2g_{jk} - g_{jj} - g_{kk}) \\
&\quad + \frac{1}{2N} \sum_k \sum_j n_k(n_j g_{jj}) - \frac{1}{2} \sum_j n_j g_{jj} \\
&= \sum_{j<k} \frac{n_j + n_k}{2N} \left(\frac{n_j n_k}{n_j + n_k} \right) (2g_{jk} - g_{jj} - g_{kk}) \\
&\quad + \frac{N}{2N} \sum_j n_i g_{ii} - \frac{1}{2} \sum_j n_j g_{jj} \\
&= \sum_{j<k} \frac{n_j + n_k}{2N} \left(\frac{n_j n_k}{n_j + n_k} \right) (2g_{jk} - g_{jj} - g_{kk}) = S_\alpha,
\end{aligned}
$$

the between sample component. It is clear that $S_\alpha \geq 0$ because for each j, k, the term $2g_{jk} - g_{jj} - g_{kk} \geq 0$ because it is the two sample energy distance (9.1) between samples A_j and A_k.

9.9.2   Proof of Proposition 9.1

Proof. We need to prove that

$$
\mathcal{E}(A \cup B, C) = \frac{n_1 + n_3}{n_1 + n_2 + n_3} \, \mathcal{E}(A, C)
$$
$$
+ \frac{n_2 + n_3}{n_1 + n_2 + n_3} \, \mathcal{E}(B, C) - \frac{n_3}{n_1 + n_2 + n_3} \, \mathcal{E}(A, B). \qquad (9.12)
$$

Suppose that $A = \{a_1, \ldots, a_{n_1}\}$, $B = \{b_1, \ldots, b_{n_2}\}$ and $C = \{c_1, \ldots, c_{n_3}\}$ are disjoint, non-empty subsets of $\mathbb{R}^d$ (distinct clusters). Define the constants δ_{11}, δ_{22}, and δ_{12} by

$$
\delta_{11} = \frac{1}{n_1^2} \sum_{i=1}^{n_1} \sum_{j=1}^{n_1} |a_i - a_j|, \qquad \delta_{22} = \frac{1}{n_2^2} \sum_{i=1}^{n_2} \sum_{j=1}^{n_2} |b_i - b_j|,
$$

$$
\delta_{12} = \frac{1}{n_1 n_2} \sum_{i=1}^{n_1} \sum_{j=1}^{n_2} |a_i - b_j|.
$$

By definition,

$$
\mathcal{E}(A, B) = \frac{n_1 n_2}{n_1 + n_2} (2\delta_{12} - \delta_{11} - \delta_{22}).
$$

Similarly, if $\delta_{33} = \frac{1}{n_3^2} \sum_{i=1}^{n_3} \sum_{j=1}^{n_3} |c_i - c_j|$, $\delta_{13} = \frac{1}{n_1 n_3} \sum_{i=1}^{n_1} \sum_{j=1}^{n_3} |a_i - c_j|$, and $\delta_{23} = \frac{1}{n_2 n_3} \sum_{i=1}^{n_2} \sum_{j=1}^{n_3} |b_i - c_j|$, we have

$$
\mathcal{E}(A, C) = \frac{n_1 n_3}{n_1 + n_3} (2\delta_{13} - \delta_{11} - \delta_{33})
$$

and

$$
\mathcal{E}(B, C) = \frac{n_2 n_3}{n_2 + n_3} (2\delta_{23} - \delta_{22} - \delta_{33}).
$$

Consider the cluster $A \cup B$ formed by merging clusters A and B. Denote $A \cup B$ by subscript k, and define the corresponding constants

$$
\delta_{k3} = \frac{1}{(n_1 + n_2)n_3} \sum_{j=1}^{n_3} \left(\sum_{i=1}^{n_1} |a_i - c_j| + \sum_{i=1}^{n_2} |b_i - c_j| \right),
$$

$$
\delta_{kk} = \frac{1}{(n_1 + n_2)^2} \left(\sum_{i=1}^{n_1} \sum_{j=1}^{n_1} |a_i - a_j| + 2 \sum_{i=1}^{n_1} \sum_{j=1}^{n_2} |a_i - b_j| + \sum_{i=1}^{n_2} \sum_{j=1}^{n_2} |b_i - b_j| \right),
$$

so that in terms of the original constants we have

$$
\delta_{k3} = \frac{n_1 n_3 \delta_{13} + n_2 n_3 \delta_{23}}{(n_1 + n_2)n_3},
$$

$$
\delta_{kk} = \frac{n_1^2 \delta_{11} + n_2^2 \delta_{22} + 2 n_1 n_2 \delta_{12}}{(n_1 + n_2)^2}.
$$

Therefore, the $\mathcal{E}$-distance between the new cluster $A \cup B$ and disjoint cluster C is given by

$$\mathcal{E}(A \cup B, C) = \frac{(n_1 + n_2)n_3}{n_1 + n_2 + n_3}[2\delta_{k3} - \delta_{kk} - \delta_{33}]$$
$$= \frac{(n_1 + n_2)n_3}{n_1 + n_2 + n_3}\left[\frac{2n_1n_3\delta_{13} + 2n_2n_3\delta_{23}}{(n_1 + n_2)n_3}\right.$$
$$\left. - \frac{n_1^2\delta_{11} + n_2^2\delta_{22} + 2n_1n_2\delta_{12}}{(n_1 + n_2)^2} - \delta_{33}\right].$$

Simplify

$$-\frac{(n_1 + n_2)n_3}{n_1 + n_2 + n_3}\left[\frac{n_1^2\delta_{11} + n_2^2\delta_{22} + 2n_1n_2\delta_{12}}{(n_1 + n_2)^2}\right]$$
$$= \tfrac{1}{n_1+n_2+n_3}\left[-n_3\mathcal{E}(A, B) - n_1n_3\delta_{11} - n_2n_3\delta_{22}\right]$$

so that

$$(n_1 + n_2 + n_3)\,\mathcal{E}(A \cup B, C) = 2n_1n_3\delta_{13} + 2n_2n_3\delta_{23}$$
$$- n_3\mathcal{E}(A, B) - n_1n_3\delta_{11} - n_2n_3\delta_{22} - n_1n_3\delta_{33} - n_2n_3\delta_{33}$$
$$= n_1n_3[2\delta_{13} - \delta_{11} - \delta_{33}] + n_2n_3[2\delta_{23} - \delta_{22} - \delta_{33}] - n_3\mathcal{E}(A, B)$$
$$= (n_1 + n_3)\mathcal{E}(A, C) + (n_2 + n_3)\mathcal{E}(B, C) - n_3\mathcal{E}(A, B).$$

Therefore, the distance between new cluster $A \cup B$ and C is given in terms of $\mathcal{E}(A, C)$, $\mathcal{E}(B, C)$ and $\mathcal{E}(A, B)$ by recursive formula (9.12). □

9.10 Exercises

Exercise 9.1. Prove that in the univariate case, with $\alpha = 2$, the disco decomposition is the ANOVA decomposition of variance. Hint; Start by finding an identity between SST and S_2.

Exercise 9.2. Prove Theorem 9.2.

Exercise 9.3. Show that the update formula of k-groups and Hartigan and Wong's k-means algorithm are the same when $\alpha = 2$.

10

Energy in Metric Spaces and Other Distances

CONTENTS

10.1 Metric Spaces

10.1.1 Review of Metric Spaces

Data energy is a function of distances. If we want to see the boundaries of energy inference and make it as general as possible, then it is natural to work in metric spaces where all we need to assume is that a distance (metric) is defined. These spaces were introduced by Fréchet [1906].

Definition 10.1 (Metric space). A metric space is an ordered pair $(\mathcal{X}, \delta)$ where $\mathcal{X}$ is a set and δ is a metric on $\mathcal{X}$. That is, δ is a function, $\delta : \mathcal{X} \times \mathcal{X} \to R$, such that if x_1, x_2, x_3 are elements of $\mathcal{X}$, then

(i) $\delta(x_1, x_2) \geq 0$,

(ii) $\delta(x_1, x_1) = 0$

(iii) $\delta(x_1, x_2) = 0$ implies $x_1 = x_2$,

DOI: 10.1201/9780429157158-10

(iv) $\delta(x_1, x_2) = \delta(x_2, x_1)$ (symmetry),

(v) $\delta(x_1, x_3) \leq \delta(x_1, x_2) + \delta(x_2, x_3)$ (triangle inequality).

If only (i)–(iii) hold, then δ is called a *divergence*. A divergence can easily be made symmetric: $\delta^*(x_1, x_2) := \delta(x_1, x_2) + \delta(x_2, x_1)$ is a symmetric function such that properties (i)–(iv) hold for δ^*. If only (i)–(iv) hold, then δ is called a *semimetric*. If (i), (ii) and (iv) hold, then δ is a *dissimilarity*. See Cailliez [1983] on how to transform a dissimilarity to a metric. This idea will be applied in our discussion of partial distance correlation. If (iii) is not required, then δ is a *pseudometric*. A pseudometric can always be considered a metric on the set of clusters (equivalence classes) of elements in $\mathcal{X}$ whose distances are zero.

A *Hilbert space H* is a real or complex inner product space such that it is a complete metric space with respect to the distance function induced by the inner product. A Hilbert space is separable if it contains a dense countable subset.

10.1.2   Examples of Metrics

Suppose $x = (x_1, \ldots, x_d)^\top \in \mathbb{R}^d$ and $y = (y_1, \ldots, y_d)^\top \in \mathbb{R}^d$.

- The most frequently applied distance is the Euclidean distance in $\mathbb{R}^d$:

$$l_2(x, y) := \left(\sum_{i=1}^d |x_i - y_i|^2 \right)^{1/2}.$$

  This metric describes the shortest distance for birds if they want to fly from x to y.

- The Manhattan distance:

$$l_1(x, y) := \sum_{i=1}^d |x_i - y_i|$$

  is also a metric that is the shortest path for pedestrians who want to walk from x to y along a rectangular grid of streets. The Manhattan distance is also called the taxicab metric (although one-way streets might change the shortest taxicab path to a non-symmetric function).

- A common generalization of these distances is the Minkowski distance:

$$l_p(x, y) := \left(\sum_{i=1}^d |x_i - y_i|^p \right)^{1/p}.$$

  It is known that l_p is a metric if $p \geq 1$. At infinity we take the limit in the definition, which is $max_i |x_i - y_i|$. This is called the Chebyshev distance.

For $0 < p < 1$, l_p is not a metric because the triangle inequality fails to hold. Consider, for example, the points $x = (0,0), y = (1,1), z = (0,1)$. If $0 < p < 1$, then $l_p(x,y) = 2^{1/p} > 2$, with $l_p(x,y) + l_p(y,z) = 1 + 1 = 2$. Since this violates the triangle inequality, for $0 < p < 1$, l_p is not a metric, but one can show that l_p^p is a metric. As $p \to \infty$, l_p tends to $min_i|x_i - y_i|$.

- The l_0 norm is defined as $\lim_{p \to 0}(\sum_{i=1}^{d} |x_i - y_i|^p)^{1/p}$, which is the number of non-zero terms in the sum.

In high dimensional spaces Euclidean and other Minkowski distances with $p > 2$ can be misleading. From Agarwal et al. [2001]:

The Manhattan distance (l_1 distance) is consistently more preferable than the Euclidean distance metric (l_2) for high dimensional data mining applications. Using the intuition derived from our analysis, we introduce and examine a natural extension of l_p fractional distance metrics. We show that the fractional distance metric ($0 < p < 1$) provides more meaningful results both from the theoretical and empirical perspective. The results show that fractional distance metrics can significantly improve the effectiveness of standard clustering algorithms such as the k-means algorithm.

One can show that in high dimension the difference between the maximum and the minimum Euclidean distances to a given point does not increase as fast as the nearest distance to any point as the dimension tends to infinity. This is one reason to prefer the Manhattan distance in high dimension. On this topic see Yao et al. [2018].

In the following we will show that energy inference can be extended to metric spaces of strong negative type (defined below). The Manhattan distance itself is not of strongly negative type; it is only of negative type. However, its square root is strongly negative definite. This (or any other power $0 < r < 1$ of the Manhattan distance) is of strong negative type and thus it can be applied it for energy inference.

Other important metrics on $\mathbb{R}^d$ include the British Rail metric, which is $|x| + |y|$ for $x \neq y$ and is equal to 0 for $x = y$. The name alludes to the tendency of railway journeys to proceed via London irrespective of their final destination.

The radar screen metric is $min(1, |x - y|)$.

A systematic study of random variables taking values in metric spaces is Parthasarathy [1967].

10.2   Energy Distance in a Metric Space

Define energy $\mathcal{E}$ for all pairs of random variables X, Y that take their values in a metric space $(\mathcal{X}, \delta)$ with distance function δ:

$$\mathcal{E}(X, Y) = 2E[\delta(X, Y)] - E[\delta(X, X')] - E[\delta(Y, Y')],$$

provided that these expectations exist. However, if we replace Euclidean distance with a metric δ in an arbitrary metric space, then the claim that "$\mathcal{E}(X, Y) \geq 0$ *with equality to zero if and only if X and Y are identically distributed*" does not necessarily hold. It does hold for a large class of metric spaces, namely those of strong negative type. A necessary and sufficient condition that the above characterization of equal distributions holds is established in Theorem 10.1 below.

Definition 10.2 (negative type). A metric space $(\mathcal{X}, \delta)$ has *negative type* if for all $n \geq 1$ and all sets of n red points x_i and n green points x_i' in $\mathcal{X}$, we have

$$2 \sum_{i,j} \delta(x_i, x_j') - \sum_{i,j} \delta(x_i, x_j) - \sum_{i,j} \delta(x_i', x_j') \geq 0.$$

As a simple example consider the red points and the green points as red cities and green cities situated on both sides of a blue river (blue line). We connect two cities with a road if and only if they are on opposite sides of the river. The road is red if it connects two red cities, the road is green if it connects two green cities, otherwise the road is blue. Show that the number of blue roads is always bigger or equal to the sum of the number of red roads and green roads. Reformulate this inequality in terms of a δ function.

If we take repetitions of x_i and take limits, then we get a seemingly more general property called *conditional negative definiteness* of δ:

Definition 10.3 (conditional negative definiteness). A metric space $(\mathcal{X}, \delta)$ is conditionally negative definite if for all $n \geq 1$, $x_1, x_2, \ldots, x_n \in \mathcal{X}$, and $a_1, a_2, \ldots, a_n$ real numbers with $\sum_{i=1}^{n} a_i = 0$, we have

$$\sum_{i,j} a_i a_j \delta(x_i, x_j) \leq 0.$$

The metric space $(\mathcal{X}, \delta)$ has *strict negative type* if for every $n \geq 1$ and for all distinct points $x_1, x_2, \ldots, x_n$ equality holds in the inequality above only if $a_i = 0$ for all i.

For an interesting example, see the Variogram Example in the Exercises Section 10.12.

Some classical results on conditional negative definiteness are given in Schoenberg [1938a] and Schoenberg [1938b]. Now suppose that the Borel probability measures μ_i, $i = 1, 2$ on $\mathcal{X}$ have finite first moments; that is, $\int \delta(o, x) d\mu_i(x) < \infty$ for some (and thus for all) $o \in \mathcal{X}$. If we approximate μ_i by probability measures of finite support we arrive at an even more general version of the energy inequality:

Definition 10.4 (strong negative type). The metric space $(\mathcal{X}, \delta)$ has *strong negative type* if it has negative type and in the inequality

$$\int \delta(x_1, x_2) d(\mu_1 - \mu_2)^2(x_1, x_2) \leq 0.$$

equality holds if and only if $\mu_1 = \mu_2$.

Lyons [2013] in Remark 3.3 gives an example for a countably infinite metric space that is strictly negative definite but not strongly. On finite metric spaces of strict negative type see Hjorth et al. [1998]. On the application of the concept of negative definiteness to metric spaces of connected graphs where the distance between vertices is the length of the shortest path, see Deza and Laurent [1997] and Meckes [2013]. *So far we do not have a characterization of graphs whose metric space has strict negative type* but for example, weighted trees have negative type [Meckes, 2013, Theorem 3.6]. Another open problem is the characterization of Riemannian spaces that have strong negative type. On this topic, see e.g. Feragen et al. [2015]. It is easy to see that spheres do not have strict negative type. For a counterexample it is enough to check four points of the sphere: the north pole, the south pole and the middle points of the semicircles that connect them on the sphere. On the other hand a recent paper [Lyons, 2020] shows that all open hemispheres are of strong negative type. On the notion of negative type see Bingham et al. [2016].

The following theorem is easy to see.

Theorem 10.1. *Let X, Y be independent random variables with distributions μ_1, μ_2, respectively, X' is an iid copy of X and Y' is an iid copy of Y.*

1. *A necessary and sufficient condition that*

$$2E\delta(X, Y) - E\delta(X, X') - E\delta(Y, Y') \geq 0 \qquad (10.1)$$

   holds for all X, Y is that $(\mathcal{X}, \delta)$ has negative type.

2. *In (10.1), a necessary and sufficient condition that*

$$2E\delta(X, Y) - E\delta(X, X') - E\delta(Y, Y') = 0$$

   holds if and only if X and Y are identically distributed ($\mu_1 = \mu_2$) is that the metric space has strong negative type.

See Section 3.7 for the proof of a special case.

Theorem 3.2 is a special case of Theorem 10.1, which shows that Euclidean spaces have strong negative type. The same holds for hyperbolic spaces [Lyons,

2014] and for all separable Hilbert spaces [Lyons, 2013]. This is very important in applications to function valued data [Horváth and Kokoszka, 2012]. Lyons [2013] proves that if $(\mathcal{X}, \delta)$ has negative type, then $(\mathcal{X}, \delta^r)$ with $0 < r < 1$ has strong negative type. Thus Theorem 3.3 follows from the observation that in Euclidean spaces $|x - y|^2$ has negative type because $\sum_{i=1}^{n} a_i = 0$ implies that $\sum_{i,j} a_i a_j |x_i - x_j|^2 = -2|\sum_{i,j} a_i x_i|^2 \leq 0$.

Important and nice examples for metric spaces that are NOT negative definite are $\mathbb{R}^d$ with l^p metric for $3 \leq d \leq \infty$ and $2 < p \leq \infty$; that is, where the distance of x and y in $\mathbb{R}^d$ is defined as

$$l_p(x, y) := \left(\sum_{i=1}^{d} |x_i - y_i|^p \right)^{1/p}.$$

For more details, see Dor [1976], Bretagnolle et al. [1966] and Koldobsky and Lonke [1999].

The l_1 metric on $\mathbb{R}^2$ is called the taxicab metric or Manhattan metric. This metric is negative definite but not strongly.

Here is a short summary of what we know about the negative definiteness of l_p spaces on $\mathbb{R}^n$:

(i) If n is arbitrary and $p = 2$, then the corresponding Euclidean space is strongly negative definite.

(ii) If $3 \leq n \leq \infty$ and $2 < p \leq \infty$, then the metric space is not negative definite.

(iii) If n is arbitrary and $p = 1$, then the l_1 metric is negative definite but not strongly negative definite.

(iv) Finally, L^p spaces have negative type when $1 \leq p \leq 2$.

The geodesic metric on the circle or on the sphere is also negative definite but not strongly negative definite. A simple function of the geodesic distance, $D(d) := d(2\pi - d)$ was shown to be strongly negative definite (see Zucker [2018] or Section 14.5.5). On functions of negative type and covariances of space-time processes, particularly those on a sphere, see Bingham et al. [2016].

Applying the so-called additive constant theorem that we discuss in defining partial distance correlation, it is easy to see that for finite metric spaces $(\mathcal{X}, \delta)$ one can always define a strictly increasing function $D(\delta)$ such that $(\mathcal{X}, D)$ is of negative type. Take e.g. $D(0) = 0$ and for $\delta > 0$ define $D(\delta) := 1 + \delta/K$ where K is a sufficiently big number. This small perturbation of the unit simplex is of negative type.

For infinite sets, however, there does not always exist a strictly monotone increasing function $D(\delta)$ such that $(\mathcal{X}, D)$ is of negative type. Example: Take two infinite sets, A and B and let $\mathcal{X}$ be their union. Define the distance of two different elements to be 1 if they are in different sets, and 2 if they are in the same set ($\delta(x, x)$ is of course always 0 and $D(0) = 0$). The function D must have the following form: $D(1) = u$, $D(2) = v$, $0 < u < v$. Define $a_i := 1$ for n elements of A and $b_i := -1$ for n elements of B. Then the sum we need to check is $n(n-1)v - n^2u$, which is positive for large enough n. Thus, no such function D exists for this example.

If we want to apply energy inference that is based on the distances of observations, then we can always apply energy distance as long as the distances are in a metric space $(\mathcal{X}, d)$ of negative type. This is the case when the sample comes from a Euclidean space or from a Hilbert space (see Székely et al. [2007] and Lyons [2013]).

Remark 10.1. The energy distance of probability distributions on strongly negative type metric spaces is also strongly negative: the barycenter map is the required mapping into Hilbert space Lyons [2013]. Since Hilbert spaces have strong negative type, so does the set of probability distributions / probability measures. Thus, if we start with a strongly negative definite metric space, then this property is inherited by the metric space of probability distributions with the energy distance as the metric. This can be repeated as many times as we want on the next generation of probability measures on the present generation of metric spaces. This is important for Bayesian inference.

Remark 10.2 (Research Problem). Having functional data in mind, it seems relevant to study energy tests of normality for Hilbert space valued variables, like L_2-valued, random variables. The first challenge is the definition itself of normal distributions for infinite dimensional Hilbert spaces. The non-triviality of this definition is indicated by the fact that $E|Z|_d \to \infty$ as $d \to \infty$. We can overcome this difficulty as in e.g. Chapter 6 of Grenander [2008]. The next challenge is to show that the energy distance defined for Euclidean spaces remains a distance for all separable Hilbert spaces. This problem was solved in Lyons [2013]; see also this chapter. Even with these important pillars we cannot claim that all other details for testing normality in Hilbert spaces are routine exercises. Just the contrary, this seems to be a good topic for a dissertation if we aim not only a consistent test but also an applicable and efficient energy test of normality for functional data.

10.3   Banach Spaces

Now let us briefly consider the case when our observations do not come from a Hilbert space, but from a general Banach space (a complete normed vector space). These extensions are important when we want to apply energy inference to manifolds, search trees, graphs, etc. where the metric is not necessarily of negative type.

Notice that every metric space $(\mathcal{X}, d)$ is isometric to a subspace of a Banach space, namely the Banach space of bounded continuous functions $C_b(\mathcal{X})$ where the norm is the supremum norm. To see this consider the isometry $f : \mathcal{X} \to C_b(\mathcal{X})$ defined by

$$f(x)(y) = d(x, y) - d(x_0, y),$$

where x_0 is an arbitrary fixed element of $\mathcal{X}$. (The function $f(x)(y)$ is bounded by $d(x, x_0)$.) Versions of this idea (Kuratowski embedding) are due to Fréchet, Kuratowski, Banach and Mazur.

Thus in order to have energy inference in general metric spaces we need to extend energy distance from Hilbert spaces to all Banach spaces. This theory is beyond the scope of this book but we can mention some recent work in this direction.

One approach could be to abandon the triangle inequality from energy distance and work with a divergence. Pan et al. [2018] proposed the "Ball Divergence" defined as follows. Suppose $(V, \| \cdot \|)$ is a Banach space, where the norm $\| \cdot \|$ induces a metric $\delta(x, y) = \|x - y\|$ for two points $x, y \in V$. Let μ and ν be two Borel measures on the separable Banach space V. If $B(x, r)$ denotes the closed ball with center x and radius r, then the ball divergence of μ, ν is defined as follows

$$D(\mu, \nu) := \int \int_{V \times V} [\mu - \nu]^2 \left(B(x, \delta(x, y)) \right) \left(\mu(dx)\mu(dy) + \nu(dx)\nu(dy) \right).$$

The empirical ball divergence is applied for a two sample test of the hypothesis $H : \mu = \nu$ and a related measure called "ball covariance" was introduced in Pan et al. [2020] as a measure of dependence in Banach spaces.

On measuring association in general topological spaces see Deb et al. [2020].

Dubey and Müller [2019] show that Fréchet mean and variance for metric space-valued random variables can be applied to data objects that lie in abstract spaces.

10.4   Earth Mover's Distance

In 1781 the great French geometer, Gaspard Monge, asked how to minimize the transportation cost of a given mass, such as a pile of sand of a given volume to given holes of the same volume [Monge, 1781]. If the transportation cost is proportional to the distance, then this is a nice geometry problem.

10.4.1 Wasserstein Distance

For general transportation cost functions this minimization was considered by the Russian Leonid Kantorovich [Kantorovich, 1942, 1948], who received a Nobel prize in economics for a related work.

The Monge–Kantorovich problem is to move one distribution of mass onto another as efficiently as possible, where Monge's original criterion for efficiency was to minimize the average distance transported.

In 1970 R. L. Dobrushin introduced a distance between probability distributions that reflects the minimizations above. He called this distance *Wasserstein distance* because of the fundamental work in Wasserstein [1969].

Suppose that the real-valued random variables X, Y have finite expectations and probability distributions (cdf's) F and G, respectively. Then their Wasserstein distance is defined as follows:

$$\mathcal{W}(F, G) := \inf E|X - Y|$$

where the infimum is taken for all random variables with marginal cumulative distribution functions F and G and varying joint distributions. One can show that this is a metric and the infimum is taken for $X := F^{-1}(U)$, $Y := G^{-1}(U)$ where U is uniformly distributed on $[0, 1]$. Thus

$$\mathcal{W}(F, G) = \int_0^1 |F^{-1}(u) - G^{-1}(u)| du = \int_{-\infty}^{\infty} |F(x) - G(x)| dx.$$

Thus for real-valued random variables $\mathcal{W}(F, G)$ is simply the L_1 distance of F and G. Let us recall that the energy distance for real-valued random variables is a constant multiple of an L_2 distance of F and G. Thus for real-valued variables the difference between Wasserstein and energy distance is roughly the difference between L_1 and L_2 distances. (The L_∞ distance is also applied frequently and is called *Kolmogorov-Smirnov distance*, which is the supremum distance between F and G, see Section 10.10.)

The definition of Wasserstein distance can easily be generalized for metric space valued random variables. In computer science this distance is known as the earth mover's distance (EMD). One can generalize it for $\alpha > 0$ as follows:

$$\mathcal{W}_\alpha(F, G) := [\inf E|X - Y|^\alpha]^{1/\alpha}.$$

The Wasserstein metric suggests the following statistical distance between two sequences $x := (x_1, x_2, \ldots, x_n)$ and $y := (y_1, y_2, \ldots, y_n)$:

$$\mathcal{W}(x, y) := \frac{1}{n} \inf_\pi \sum_{i=1}^n |x_i - y_{\pi(i)}|,$$

where the infimum is taken for all permutation π on the integers $1, 2, \ldots, n$. This is clearly an energy type function in the sense that it depends only on the distances of data. It is also a symmetric function of the data but here we do not have jackknife invariance. This means that Wasserstein inference does not work with U or V-statistics. This is a disadvantage of Wasserstein distance compared to our energy distance.

One can easily see that for real-valued data, if the ordered sample is denoted by subscripts in brackets, then

$$\mathcal{W}(x, y) := \frac{1}{n} \sum_{i=1}^{n} |x_{(i)} - y_{(i)}|.$$

If the sample sizes are different, then the Wasserstein distance is

$$\mathcal{W}(x, y) = \int_{-\infty}^{\infty} |F_n(x) - G_m(x)| dx = \frac{1}{nm} \sum_{i=1}^{nm} |x_{(i)} - y_{(i)}|,$$

where the ordered samples are the ordered elements of the x-sample with all elements repeated m times and the ordered elements of the y sample with all elements repeated n times, respectively.

One could find examples where EMD is more powerful. Also, EMD in dimension $d = 1$ is simple: just the sum of differences of the unweighted order statistics. Assuming $n \geq m$ the number of steps we need to compute energy and also EMD is $O(n \log n)$.

Energy (using the order statistics formulas) uses weighted differences between the ordered pooled sample and each sample's order statistics. Since the energy absolute(weights) are biggest near the min and the max and V shaped, we can say energy gives extra weight to the extreme observations.

It was shown by Edmonds and Karp [1972] that the algorithmic complexity of EMD can be reduced to $O(n^3)$. This is still one order of magnitude worse than the algorithmic complexity $O(n^2)$ of computing the energy distance. Probabilistic algorithms can reduce the algorithmic complexity of both the Wasserstein distance and the energy distance, but of course probabilistic algorithms compute with high probability only (not with 100% probability) and with positive error that tends to zero as $n \to \infty$. If the data are real numbers, then even ordering them takes $O(n \log n)$ operations and this is the best possible complexity we can achieve when computing energy distances of data that are real numbers. For an important special case of $\mathcal{W}(x, y)$ when the data are random in the unit square see Ajtai et al. [1984].

The Wasserstein distance is a very natural distance and whenever we can overcome the computational complexity problem of Wasserstein distance it is a good competitor of energy distance.

On Monge-Kantorovich depth, quantiles, ranks, and signs see Chernozhukov et al. [2017].

One can define Wasserstein distance for probability measures on general metric spaces (M, ρ). If (M, ρ) is a metric space for which every probability measure is a Radon measure and $\alpha \geq 1$, denote by $P_\alpha(M)$ the set of all probability measures μ on M such that

$$\int_M \rho(x, x_0)^\alpha d\mu(x) < +\infty$$

Then the α-th Wasserstein distance in $P_\alpha(M)$ is

$$\mathcal{W}_\alpha(\mu, \nu) := (\inf E[\rho(X, Y)^\alpha])^{1/\alpha}$$

where the infimum is taken for all joint distributions of (X, Y) with marginal distributions μ and ν respectively.

Remark 10.3. Fréchet [1957] introduced a metric on the space of probability distributions on $\mathbb{R}$ having first and second moments. Fréchet's distance is a special case of Wasserstein's distance with $\alpha = 2$. The bivariate distribution which minimizes

$$d^2(F, G) := \inf E|X - Y|^2$$

where the real-valued random variables X, Y have distributions F, G, respectively, is a singular distribution H with joint cdf

$$H(X, Y) = \min[F(x), G(y)].$$

In the particular case when F and G belong to a family of distributions, which is closed with respect to changes of location and scale, the Fréchet distance takes the simple form

$$d^2(F, G) = (m_F - m_G)^2 + (\sigma_F - \sigma_G)^2,$$

where m and σ denotes the expectation and the standard deviation of the corresponding cdf.

10.4.2 Energy vs. Earth Mover's Distance

Typically there are several options for the distance of two data sets or for the distance of two probability distributions defined on a metric space. Of course the question arises: which one to use for statistical inference?

If the metric of the metric space is not negative definite (in other words if the metric space cannot be embedded isometrically into a Hilbert space), then we should not use the energy distance because in this case the energy distance can be negative. We do not have such a problem with the Wasserstein distance (Earth mover's distance).

For Euclidean and Hilbert space valued data we can use either energy distance or EMD. However, for high dimension in case of the EMD we face the so-called "curse of dimensionality." If we take two iid samples from two probability measures on an Euclidean space with dimension d, then the empirical EMD converges to the theoretical one very slowly; the speed is about $n^{-1/d}$, which is very slow if d is big. In this case energy distance is preferable. The situation might change if the probability measures are "sparse" (see Niles-Weed and Rigollet [2019]).

The following is a heuristic proof for the lower bound. Suppose that μ is a probability measure that is absolutely continuous with respect to the Lebesgue measure. For any n points in $\mathbb{R}^d$, and small $\epsilon > 0$, the union of balls of radius ϵ around these points has total μ measure at most $O(n\epsilon^{-d})$. Let c be a small constant, and $\epsilon = cn^{-1/d}$. This quantity can be made smaller than $1/2$, in which case a constant fraction of the mass of μ must be moved at least a distance $cn^{-1/d}$ in order to reach any of these points. Therefore, the Wasserstein distance between μ and any measure supported on those n points is at least $n^{-1/d}$. This implies that the Wasserstein distance between the empirical measure and its population counterpart is at least $n^{-1/d}$, hence the convergence rate can be no faster than this in general.

For the upper bound, the most transparent proof uses entropy methods a la Dudley; here we do not go into details.

Concerning Wasserstein distances in Hilbert spaces see Lei [2020]. Here the rates depend a great deal on the decay conditions imposed on the moments of the underlying measure(s).

10.5   Minimum Energy Distance (MED) Estimators

Minimum energy distance (MED) estimators or Energy estimators, or simply E-estimators, minimize the energy distance between the sample empirical distribution and a parametric family of distributions, or equivalently minimize

$$\min_\theta \left[\frac{2}{n} \sum_{i=1}^n E|x_i - Y| - \frac{1}{n^2} \sum_{i=1}^n \sum_{j=1}^n |x_i - x_j| - E|Y - Y'| \right]$$

with respect to θ where the iid random variables Y, Y' have distribution F_θ. For an example, we can minimize the energy goodness-of-fit statistic for Pareto Type I distribution to estimate the tail index parameter. See Section 8.2.5; the formulation based on the transformation to an exponential model is a simple objective function to minimize.

Energy estimators are clearly special cases of M-estimators that minimize

$$\prod_{i=1}^n g(x_i; \theta)$$

with respect to the parameter θ where g is some function of an observation x_i and a parameter θ. On M-estimators see Huber [1981].

Let Q_θ be a parametric family of probability distributions and P_n the empirical distribution computed from the sample $X_1, X_2, \ldots, X_n$; that is $P_n(x) := (1/n) \sum_{i=1}^{n} I(X_i \leq x)$, where $I(\cdot)$ is the indicator function.

The energy minimum estimator of θ is

$$\theta^* := \operatorname{argmin}_\theta \mathcal{E}(P_n, Q_\theta)$$

if such a θ^* exists and is unique. This estimator can be a competitor of classical estimators like least squares, maximum likelihood, etc. and can be studied via the general theory of M-estimators that maximize (or minimize) more general functions like likelihood functions. For details on M-estimators, see Huber [1981].

Analytic solutions for energy minimization problems typically do not exist, but we can apply iterative methods like the stochastic gradient descent method to approximate θ^*. This iteration has the form

$$\theta_1 := \theta - \eta \nabla_\theta \mathcal{E}(P_n, Q_\theta),$$

where η is a parameter called *learning rate*.

The energy distance has an extra good property. It has unbiased sample gradients; that is, if Q_θ is a parametric family of distributions and P_n is the plug-in estimator of P based on n iid observations, then for the expectation we have

$$E[\nabla_\theta \mathcal{E}(P_n, Q_\theta)] = \nabla_\theta \mathcal{E}(P, Q_\theta).$$

In one dimension this follows from the fact that the energy distance is the same as the squared Cramér distance and its derivative is a linear function of the plug-in estimator. The same argument shows that the Wasserstein distance does not have unbiased sample gradients and might not converge to the true value of θ.

10.6 Energy in Hyperbolic Spaces and in Spheres

The most frequently applied spaces where our samples come from are Euclidean and Hilbert spaces. The next level of flexibility is when the *topological space* (see Section 10.11) is locally Euclidean, like in case of a sphere. A d-dimensional manifold is a topological space where each point has a neighborhood that is homeomorphic (topologically isomorphic or 1-1 continuously equivalent) to the d-dimensional Euclidean space. Curvature is one of the most important properties in manifolds. The curvature of the sphere at every point is the reciprocal of the radius of the sphere. For planes the curvature is zero at every point. If the manifold is "smooth," then we can define curvature at every point by approximating spheres. Smooth manifolds are called Riemannian. Their metric is the *Riemannian metric*. If a manifold has constant positive curvature, then it is a sphere; if the manifold has a constant negative curvature, then it is a hyperbolic sphere. For spheres and hyperbolic spaces Lyons proved the following fundamental results.

Lyons [2014] showed that all hyperbolic spaces are of strong negative type thus the energy distance always applies in these spaces. On the other hand, if we need the property of strong negative type in spheres, then we need to exclude one of diametrically opposite points; open hemispheres are of strong negative type with the geodesic distance [Lyons, 2020]. For an interesting application of this property of circles, see Section 14.5.5 on identifying periods of signals.

10.7   The Space of Positive Definite Symmetric Matrices

The space of positive definite matrices has a natural affine-invariant Riemannian metric, which is often considered the best one to use for the purposes of statistics and machine learning. This metric is related to distances in parameter spaces in statistics, especially to the Fisher-information metric. These metric spaces are known to have negative curvature and are simply connected. On the Fisher information metric and the related information geometry see Amari and Nagaoka [2007]. By Chentsov's theorem [Cencov, 2000] the Fisher information metric on statistical models is the only Riemannian metric (up to rescaling) that is invariant under sufficient statistics. The parameter space of all (multivariate) normal distributions forms a statistical manifold with hyperbolic geometry. Hyperbolic spaces have negative type; see Lyons [2014]. Thus, one might suspect that in general the space of all positive definite symmetric matrices always has negative type and thus energy inference applies for all these Riemann spaces.

Unfortunately this general conjecture is not true. A recent work [Feragen and Fuster, 2017] shows that some 3×3 positive definite symmetric matrices are not conditionally negative semidefinite. (They were working with 500 random 3×3 matrices.)

Russell Lyons (private communication), using the same or similar code, showed that in fact such points exist even for 2×2 matrices and the number of "points,' that is, 2×2 matrices used can be reduced from 500 to 28. This is actually very interesting, because for 2×2 positive definite symmetric matrices the space is the Riemannian product of the hyperbolic plane and the real line, both of which have strongly negative type. Therefore, negative type is not closed under formation of Euclidean products. In addition, the space of 2×2 positive definite symmetric matrices is homogeneous and has nonpositive curvature and dimension 3, whereas all Hadamard spaces (non-linear generalizations of Hilbert spaces) have negative type. Consequently, strong negative type is not only hard to prove, even negative type may not be so common.

10.8 Energy and Machine Learning

In the theory of machine learning, conditionally negative definite distances $\delta(x, y)$ are replaced by positive definite kernels $k(x, y)$.

Specifically our energy distance with conditionally negative definite distance δ is

$$\mathcal{E}_{n,m}(\mathbf{X}, \mathbf{Y}) = \frac{2}{nm} \sum_{i=1}^{n} \sum_{j=1}^{m} \delta(X_i, Y_j)$$

$$- \frac{1}{n^2} \sum_{i=1}^{n} \sum_{j=1}^{n} \delta(X_i, X_j) - \frac{1}{m^2} \sum_{i=1}^{m} \sum_{i=1}^{m} \delta(Y_i, Y_j). \qquad (10.2)$$

In machine learning the conditionally negative definite distance δ is replaced by the positive definite kernel k and the resulting number is called maximal mean discrepancy (MMD):

$$MMD_{n,m}(\mathbf{X}, \mathbf{Y}) = - \frac{2}{nm} \sum_{i=1}^{n} \sum_{j=1}^{m} k(X_i, Y_j)$$

$$+ \frac{1}{n^2} \sum_{i=1}^{n} \sum_{j=1}^{n} k(X_i, X_j) + \frac{1}{m^2} \sum_{i=1}^{m} \sum_{i=1}^{m} k(Y_i, Y_j). \qquad (10.3)$$

According to the "kernel trick" there is a one-to-one correspondence between a conditionally negative definite semi-metric δ and a positive definite kernel k.

Proposition 10.1 (kernel trick).

(i) *If δ is a semi-metric and z is an arbitrary fixed element of the metric space, then*

$$k(x, y) = \delta(x, z) + \delta(y, z) - 2\delta(x, y)$$

is positive definite if δ is conditionally negative definite (or of negative type). This k is called a distance kernel induced by δ and centered at z.

(ii) *If k is positive definite kernel, then*

$$\delta(x, y) := k(x, x) + k(y, y) - 2k(x, y)$$

defines a semimetric of negative type. In this case we say that k generates δ.

Proof. (i) Let $a_1, a_2, \ldots a_n$ be arbitrary real numbers, and introduce $a_0 := -(a_1 + a_2 + \cdots + a_n)$. (This a_0 plays the role of z.) Then

$$0 \geq \sum_{j,k=0}^{n} a_i a_j \delta(x_i, x_j)$$

$$= \sum_{j,k=1}^{n} a_i a_j \delta(x_i, x_j) + \sum_{i=1}^{n} a_0 a_j \delta(x_i, x_j) + \sum_{j=1}^{n} a_0 a_i \delta(x_i, x_j) + a_0^2 \delta(x_0, x_0)$$

$$= \sum_{i,j=1}^{n} a_i a_j [\delta(x_i, x_j) - \delta(x_i, x_0) - \delta(x_0, xj) + \delta(x_0, x_0)]$$

$$= - \sum_{j,k=1}^{n} a_i a_j k(x_i, x_j),$$

so k is positive definite.

(ii) If $a_1, a_2, \ldots, a_n$ are arbitrary real numbers such that $\sum_{i=1}^{n} a_i = 0$, then

$$\sum_{i,j=1}^{n} a_i a_j \delta(x_i, x_j) =$$

$$\sum_{i=1}^{n} a_i k(x_i, x_i) \sum_{j=1}^{n} a_j + \sum_{j=1}^{n} a_j k(x_j, x_j) \sum_{i=1}^{n} a_i - \sum_{i,j=1}^{n} a_i a_j k(x_i, x_j)$$

$$= - \sum_{i,j=1}^{n} a_i a_j k(x_i, x_j) \leq 0.$$

$\square$

For more details, see Berg et al. [1984], Vapnik [1995], Wahba [1990], Schölkopf and Smola [2002], Sejdinovic et al. [2013], and Shen and Vogelstein [2018].

This "kernel trick" implies the equivalence of the energy inference of statisticians and the kernel method of machine learning.

In machine learning the maximum mean discrepancy (MMD) between two probability measures μ, ν is defined as follows:

$$MMD_{\mathcal{F}}(\mu, \nu) := \sup_{f \in \mathcal{F}} \int f d(\mu - \nu)$$

where $\mathcal{F}$ is a collection of finitely integrable functions with respect to μ and ν.

Now, if $\mathcal{F} = H(K)$ where K is the kernel of a reproducing kernel Hilbert space (these kernels are known to be positive definite), then

$$MMD_{H_K}^2 = E[K(X, X')] + E[K(Y, Y')] - 2E[K(X, Y)]$$

as long as the expectations are finite. The kernel trick above shows that MMD is equivalent to energy distance with conditionally negative definite distances δ.

Interestingly we can have another equivalent form of MMD as follows.

If the characteristic functions of the probability measures μ and ν are f and g, respectively, then

$$MMD^2(\mu, \nu) = \int_{\mathbb{R}^d} |f(t) - g(t)|^2 F^{-1}\kappa(t)dt$$

if our kernel $K(x, y) = \kappa(x - y)$ and its inverse Fourier transform, $F^{-1}\kappa(t)$, has finite integral. (This condition obviously holds for Gaussian kernels but does not hold for our energy kernel $\kappa(x) = |x|$ because in this case $F^{-1}\kappa(t) = c/|t|^{d+1}$).

Interestingly (personal communication from Ming Yuan) if we maximize MMD for all kernels of the form $K(x, y) = \kappa(x - y)$ in the unit ball of $H(K)$, then the Gaussian kernel $G_s(x, y) = \exp\{-s|x - y|^2\}$ gives the (double) maximum for some scale s under the extra condition that $F^{-1}\kappa(t)$ has finite integral.

The energy approach and the machine learning approach are equivalent; in [Sejdinovic et al., 2013, Theorem 22] an appropriate unbounded non-translation-invariant kernel is given. In case of energy distance the corresponding weight function $(d\Lambda(u)/du)$ in Sejdinovic et al. [2013] is singular (not integrable), thus the energy approach does not have a direct equivalent version in machine learning. Their formula for γ_k^2 is true for integrable $d\Lambda(u)/du$ only because Bochner's theorem yields bounded translation invariant kernels. If we want to establish the equivalence from the Fourier transform viewpoint, we should be able to see the energy distance as a kind of generalized Fourier transform which yields a negative definite kernel. This negative definite kernel has an associated positive definite kernel yielding MMD.

Remark 10.4. According to a theorem of Schoenberg, δ is conditionally negative definite if and only if $\exp\{-t\delta\}$ is positive definite for all $t > 0$. It is easy to see that $\delta = \lim_{0<t\to 0}(1 - \exp\{-t\delta\})/t$.

Remark 10.5. According to a theorem of Bochner, a function is a characteristic function of a probability distribution on the n-dimensional Euclidean space if and only if it is positive definite and takes the value 1 at 0. Thus $\exp\{-|t|^\alpha\}$ is positive definite for $0 < \alpha \le 2$ because this is the characteristic function of a symmetric stable distribution. Since stable distributions are always infinitely divisible we can also conclude that $|x - y|^\alpha$ is conditionally negative definite for $0 < \alpha \le 2$. In high dimensional problems the Euclidean distance does not discriminate enough between points, thus for the purpose of energy inference like testing for independence it makes sense to replace the Euclidean distance $|x - y|$ in $\exp\{-|t|^\alpha\}$ by a "more sensitive" distance like the square root of the l_1 distance of x and y, $\sqrt{|x - y|_1}$, or the square root of other metrics of the high dimensional space.

Remark 10.6. Energy distance has many important connections to other areas of science. In one dimension the distance $|x - y|$ is a harmonic function because the second derivative of the absolute value function $|x|$ is identically zero (except at the origin). A generalization is a harmonic function $f = f(x, y, z, \dots)$ for which the Laplace operator $\Delta f = 0$. Harmonic functions are potential functions in physics and engineering, e.g. in three dimension $1/|x|$ is harmonic function, the classical Newton potential. For a Bayesian application of energy distance see Nguyen et al. [2020].

Remark 10.7. $|x - y|$ is conditionally negative definite. This was crucial in the definition of energy distance. The point is that in all metric spaces where the metric is of negative type, we can introduce the energy distance [Lyons, 2013]. This implies that energy inference works in every separable Hilbert space.

10.9   Minkowski Kernel and Gaussian Kernel

Energy inference does not need to rely on the original metric of the metric space. We have already seen that any $0 < \alpha < 2$ power of the Euclidean distance can replace the original Euclidean distance. This is especially important if the first moment of the random variables are infinite. Then we need to work with a power α that is less than 1. We can try an even more general framework to improve the power of the corresponding energy test. Change the Euclidean distance to a more general and thus more flexible Minkowski distance or l_p distance or its $0 < \alpha < p$ power. The simplest to recommend beyond the original energy distance is the Manhattan distance when $p = 1$; for simplicity we can choose $\alpha = 1/2$. According to Lyons (2013) the resulting Manhattan energy is always of strict negative type. The same refers to all Minkowski distance based energy inference where $0 < \alpha < p$. We can call these tests Minkowski kernel energy tests or MKE in short. An important special case is the Manhattan kernel energy test when $p = 1$. Minkowski kernels with $p < 2$ are especially useful in high dimensional inference where the Manhattan distance is preferable to the Euclidean one (see e.g. Agarwal et al. [2001]). If the dimension d tends to infinity as n tends to infinity, then even fractional $0 < p < 1$ is applicable with the caveat that the corresponding l_p is not a metric, but its p-th power is.

Let us quote from from Agarwal et al. [2001] mentioned above (we adapted the notation):

> ... *Many high-dimensional indexing structures and algorithms use the Euclidean distance metric as a natural extension of its traditional use in two- or three-dimensional spatial applications. ... In this paper we*

provide some surprising theoretical and experimental results in analyzing the dependency of the l_p norm on the value of p. More specifically, we show that the relative contrasts of the distances to a query point depend heavily on the L_p metric used. This provides considerable evidence that the meaningfulness of the L_k norm worsens faster within increasing dimensionality for higher values of k. Thus, for a given problem with a fixed (high) value for the dimensionality d, it may be preferable to use lower values of k. This means that the l_1 distance metric (Manhattan distance metric) is the most preferable for high dimensional applications, followed by the Euclidean metric (l_2).

The point is that for lower values of p it is easier to discriminate between the farthest distance of n points to the origin and the nearest distance to the origin using the distance metric l_p. The relative difference between the two (compared to the smaller) increases as p decreases.

In an interesting manuscript, Li and Yuan [2019] shows that for increasing the power of an energy test, under certain circumstances the Euclidean distance $|x - y|$ should be replaced by another negative definite function: $1 - \exp(-\beta|x - y|^2)$. With a suitable choice of the scale β this leads to a powerful energy test that we call Gaussian kernel energy (GKE) test. This has minimax optimal power among all distances (kernels) with finitely integrable Fourier transforms.

10.10 On Some Non-Energy Distances

For easier reference, in this section we collect a few important non-energy type distances between data. They are not functions of distances between data.

Suppose that $\mathbf{x} := \{x_1, x_2, \ldots, x_n\}$ and $\mathbf{y} := \{y_1, y_2, \ldots, y_m\}$ are two real-valued data sets, $F_n(x)$ denotes the proportion of $x_i's$ that are at most x and $G_m(y)$ denotes the proportion of $y_j's$ that are at most y. The following are (non-energy type) distances between the data sets $\mathbf{x}$ and $\mathbf{y}$.

1. The Kolmogorov-Smirnov (KS) distance of $\mathbf{x}$ and $\mathbf{y}$ is

$$KS(F, G) := \sup_x |F_n(x) - G_m(x)|.$$

This was introduced in Kolmogorov [1933]. If we take the supremum of the corresponding empirical probability measures not on half intervals $(-\infty, x]$ but on all intervals $[x - r, x + r]$, (here x is the center of the interval and $x \geq r$), then this can be generalized to all metric spaces and probability measures μ, ν. We just need to take

$$\sup_{S(x,r)} |\mu(S) - \nu(S)|$$

where $S(x,r)$ denotes a closed sphere in the metric space with center x and radius r. Unfortunately this supremum, called *discrepancy metric* is not so easy to compute.

Even the computation of the following supremum is an open problem. For any given $r \geq 0$ compute

$$\sup_x \mu_n(S)$$

where the d-dimensional sample of size $n = 2^d$ in $\mathbb{R}^d$ is $(\pm 1, \pm 1, \cdots \pm 1)$, the vertices of a cube, and μ_n is the empirical distribution of this sample. This is a nontrivial problem even when $d = 10$. For the statistical importance of this fact see Székely [2006].

2. The L_p distance of $\mathbf{x}$ and $\mathbf{y}$ for $p \geq 1$ is the $1/p$ power of

$$\int_{-\infty}^{\infty} |F_n(x) - G_m(x)|^p dx.$$

We have seen that for real-valued random variables or samples, $p = 1$ corresponds to the Wasserstein (Earth Mover's) distance and $p = 2$ corresponds to the square of the energy distance multiplied by 2. The L_∞ distance of cdf's is the KS distance.

3. Total variation distance

The total variation distance (TV) between two measures μ and ν is

$$TV(\mu, \nu) := \sup |\mu(A) - \nu(A)|,$$

where the sup is taken for all Borel sets A. It is clear that KS is at most TV. Also, $TV(\mu, \nu) = \inf P(X \neq Y)$ where the infimum is taken for all random variables (X, Y) with marginal distributions μ for X and ν for Y. The Wasserstein distance (Section 10.4) is a similar infimum of $E|X - Y|$ taken for all random variables (X, Y) with marginal distributions μ for X and ν for Y.

The total variation distance is typically too big to be useful. Take iid ± 1 random variables, both values taken by the same $1/2$ probability. Then by the central limit theorem, if $S_n := X_1 + X_2 + \cdots + X_n$ we have $S_n/\sqrt{n} \to N(0,1)$ but $TV(S_n/\sqrt{n}, Z) = 1$ for all n where Z is a standard normal random variable. At the same time both the Wasserstein distance and the KS distances go to 0 at rate $1/\sqrt{n}$. The energy distance also tends to 0.

4. The Lévy metric is

$$\inf_{\epsilon > 0} \{ F_n(x - \epsilon) - \epsilon \leq G_m(x) \leq F_n(x + \epsilon) + \epsilon, \forall x \in \mathbb{R} \}.$$

Convergence with respect to this metric is equivalent to the convergence of cdf's at all continuity points of the limit. The Lévy metric was introduced in Lévy [1925] pp. 119-200.

5. The *information distance* between data **x** and **y** can be defined as follows. Let us start with the *Kolmogorov complexity* of **x**, which is the length of the shortest computer program in a given programming language that produces **x** as output. Now suppose both sequences **x** and **y** are binary. Then their information distance is the length of the shortest computer program expressed in bits that transforms **x** to **y** or vice versa on a universal computer. This is an extension of Kolmogorov complexity to two sequences. Recall that energy distance is also an extension of a one-sequence notion to two sequences and this one-sequence notion is the Riesz energy

$$\frac{1}{n^2}\sum_{i=1}^{n}\sum_{j=1}^{n}|x_i - x_j|,$$

that is also called Gini's mean difference. For more details on Kolmogorov complexity and information distance, see Kolmogorov [1965] and Bennett et al. [1998]. On the comparison of Shannon information and Kolmogorov complexity see Grunwald and Vitányi [2010]. On algorithmic statistics see Gács et al. [2001].

6. The expected Kolmogorov complexity is Shannon's entropy. Using the notation $p_i = P(X = x_i)$ $i = 1, 2, \ldots$, the Shannon entropy is $-\sum_i p_i \log p_i$ For details see Grunwald and Vitányi [2010].

7. Kullback-Liebler divergence and f divergence

If $q_i = P(Y = y_i)$, then the Kullback-Leibler (KL) divergence (see Kullbach and Leibler [1951]) is defined as $\sum_i p_i \log(p_i/q_i)$. We can assume without loss of generality that $q_i > 0, i = 1, 2, \ldots$ and we can interpret $0 \log 0 = 0$.

The following generalization is due to Csiszár [1963].

If f is a convex function and $f(1) = 0$, then the f-divergence of **x** and **y** is $\sum_i p_i f(q_i/p_i)$. This is always non-negative by Jensen's inequality:

$$\sum_i p_i f(q_i/p_i) \geq f(\sum_i p_i(q_i/p_i)) = f(1) = 0,$$

with equality if and only if $p_i = q_i; i = 1, 2, \ldots$.

The f-divergence does not depend on **x** and **y**, only on the probabilities, so the f-divergence is not of energy type. The Kullback-Leibler divergence is also called information divergence or I-divergence. On the geometry of this divergence see Csiszár [1975, 1991].

Other special cases of f-divergence are the *Hellinger distance* where $f = (\sqrt{t} - 1)^2$, the χ^2-*distance* where $f(t) = (t - 1)^2$, and the *total variation distance* where $f(t) = (1/2)|t - 1|$.

Notice that the total variation distance is equivalent to the distance $\sum_i |p_i - q_i|$, the Hellinger distance is $\sum_i(\sqrt{p_i} - \sqrt{q_i})^2$, and finally the χ^2 distance is $\sum_i(p_i - q_i)^2/q_i$ (we suppose that $q_i > 0, i = 1, 2, \ldots$).

8. Rényi divergence [Rényi, 1961] is defined as $1/(\alpha-1)\ln\sum_i p_i^\alpha q_i^{1-\alpha}$ for $\alpha \neq$ 1. The limit as $\alpha \to 1$ is the Kullback-Leibler divergence or I-divergence.

For many other important distances see Deza and Deza [2016].

Remark 10.8. There are many distances and divergences between probability distributions. How to chose one for particular applications? Bellemare et al. [2017] state:

> *In machine learning, the Kullback-Leibler (KL) divergence is perhaps the most common way of assessing how well a probabilistic model explains observed data. Among the reasons for its popularity is that it is directly related to maximum likelihood estimation and is easily optimized. However, the KL divergence suffers from a significant limitation: it does not take into account how close two outcomes might be, but only their relative probability. This closeness can matter a great deal; in image modelling, for example, perceptual similarity is key [Rubner et al., 2000, Gao and Kleywegt, 2022]). Put another way, the KL divergence cannot reward a model that "gets it almost right."*

To address this limitation, researchers have turned to the Wasserstein metric, which does incorporate the underlying geometry between outcomes. The Wasserstein metric can be applied to distributions with non-overlapping supports, and has good out-of-sample performance [Esfahani and Kuhn, 2018]. Yet, practical applications of the Wasserstein distance, especially in deep learning, remain tentative. "...estimating the Wasserstein metric from samples yields biased gradients, and may actually lead to the wrong minimum. This precludes using stochastic gradient descent (SGD) and SGD-like methods, whose fundamental mode of operation is sample-based, when optimizing for this metric."

As a remedy in this book we propose the energy distance that respects the underlying geometry; it is homogeneous (H), sum-contractive (S), and also it has the unbiased sample gradients property (U).

In comparison KL has (U) and (S) but does not have(H) and Wasserstein has properties (H) and (S) (thus it is an "ideal" metric) but does not have property (U).

Note that even if a sequence of estimators is consistent, their approximations via SGD might not converge at all or not to the true value. For more details on the practical consequences, see Bellemare et al. [2017].

10.11   Topological Data Analysis

If we want to focus on the "closeness" of data points, the statistical analysis of complex data can be made easier if we represent the objects as points in topological spaces that describe neighborhoods of points; for example, for brain research. From a topological point of view, a coffee mug is the same as a donut shape because there is a bi-continuous transformation between them; that is, they are homeomorphic.

Many of the most common topological spaces are metric spaces. In other words, the most important topological spaces are metrizable. A metrizable space is a topological space that is homeomorphic to a metric space; there is a continuous deformation between them, a continuous mapping with a continuous inverse. According to Urysohn's metrization theorem "most" topological spaces (every Hausdorff, second-countable, regular space) is metrizable. For data science, therefore, all topological spaces that we need are metric spaces. We can take a step further. Separable metrizable spaces can be characterized as those spaces that are homeomorphic to a subspace of the Hilbert cube; that is, the countably infinite product of the unit interval. How can we prove it? Assume that the distance in our metric space is $d \leq 1$ (otherwise, use $d/(d+1)$). Choose a dense countable sequence (x_n) in our original separable metric space and define $f(x) := (d(x, x_n)/n)$. This is a point in the Hilbert cube. The metric in Hilbert cubes can be applied for energy inference. Thus from a topological point of view, our energy inference is natural, for example, when we want to test independence.

An application of topological data science is in brain research. The human brain contains some 10^{10} neurons linked by 10^{14} synaptic connections. Understanding this structure is one of the great challenges of neuroscience, but one that is hindered by a lack of appropriate mathematical tools. Today, that looks set to change thanks to the mathematical field of algebraic topology, which neurologists are gradually coming to grips with for the first time. One of the challenges is to find symmetries in topological spaces; a symmetry is anything that is invariant as the viewpoint changes. For data science most topological / metric spaces like the sphere or different shapes in the brain are manifolds that locally resemble (are homeomorphic to) Euclidean spaces or Hilbert spaces near each point. On this topic see Patrangenaru and Ellingson [2015]. Since not all metric spaces are of negative type it makes sense to find methods for testing homogeneity in metric spaces that are not of negative type. An interesting manuscript is Blumberg et al. [2018].

10.12 Exercises

Exercise 10.1. Prove that a metric space $(\mathfrak{X}, d)$ has strong negative type if and only if the space of probability measures on $(\mathfrak{X}, d)$ equipped with the energy distance D has strong negative type.

Hint: According to Schoenberg [1937] and Schoenberg [1938b] $(\mathfrak{X}, d)$ is of negative type iff there is a Hilbert space H and a map $\Phi : (\mathfrak{X}, d) \to H$ such that $d(x, x') = ||\Phi(x) - \Phi(x')||^2$. Now the barycenter map $\beta = \beta_\phi$ that maps a probability measure P with finite expectation to $\int \Phi(x) dP(x)$ is an isometry into the Hilbert space H; since Hilbert space has strong negative type, so does the set of probabilities. For more details on barycentric maps, see Proposition 3.1 of Lyons [2013] which explicitly says that if $(\mathcal{X}, d)$ has negative type as witnessed by the embedding Φ, then $(\mathcal{X}, d)$ is of strong negative type iff the barycenter map β_Φ is injective on the set of probability measures on $(\mathcal{X}, d)$ with finite first moment. The other direction of the iff is trivial because of degenerate probability distributions (that take one single value with probability 1).

Remark 10.9. Since strong negative type is inherited form $(\mathcal{X}, d)$ to the metric space of probability measures with energy distance D, we can go further and see that the space of probability measures on the space of probability measures equipped with the energy distance is also of strong negative type. These random probability measures are typical in Bayesian inference.

Exercise 10.2. Prove that Wasserstein metric spaces are not of negative type.

Hint: See Theorem 1.3 in Naor and Schechtman [2007]. It shows that Wasserstein does not have isometric embeddings into Hilbert spaces.

Exercise 10.3. There are many widely used classical distances like the Hausdorff distance between non-empty compact subsets of a set in a metric space (M, d) or the Fréchet distance between continuous curves. Do they have (strong) negative type?

Exercise 10.4. (Variogram Example)

(i) Show that variograms defined for real-valued stochastic processes or for random fields $Z(x), x \in \mathcal{X}$ as twice $Var[Z(x_1) - Z(x_2)], x_1 \in \mathcal{X}, x_2 \in \mathcal{X}$, are conditionally negative definite.

   Hint: Suppose the expected values $E(Z(x_1)) = EZ(x_2) = 0$ and thus $Var[Z(x_1) - Z(x_2)] = E[Z(x_1) - Z(x_2)]^2$ and then use the property of positive (semi)definiteness of the covariance function $E[Z(x_1)Z(x_2)]$.

(ii) Show that if $\mathcal{X}$ is the real line or a Euclidean space, then $|x_1 - x_2|$ is conditionally negative definite and the corresponding metric space (the usual Euclidean space) has strict negative type.

   Hint: Let $Z(s)$ be the standard Wiener process and compute its variogram.

(iii) Prove that on the real line (or on Euclidean spaces) $|x_1 - x_2|^\alpha$ where $0 < \alpha < 2$ is conditionally negative definite.

   Hint: Fractal Brownian motions with Hurst index H are defined as a continuous-time Gaussian processes on the real line (or in more general spaces) that start at zero, have expectation zero for all $x \in \mathcal{X}$, and have the following covariance functions:

$$E[B_H(x_1)B_H(x_2)] = \frac{1}{2}(|x_1|^{2H} + |x_2|^{2H} - |x_1 - x_2|^{2H}),$$

where H is a real number in (0, 1), called the Hurst index or Hurst parameter. The Hurst exponent describes the raggedness of the resultant motion, with a higher value leading to a smoother motion. If $H = 1/2$, then the process is in fact a classical Brownian motion or Wiener process; if $H > 1/2$, then the increments of the process are positively correlated; if $H < 1/2$, then the increments of the process are negatively correlated. Fractal Brownian motions have abrupt and dramatic reversals called "Noah effect" and trends—cycles called "Joseph effect." Based on the covariance function of fractal Brownian motion, show that $|x_1 - x_2|^{2H}$ is strictly negative definite.

(iv) Generalize (iii) for arbitrary Gaussian processes with general positive definite covariance function.

Note: The strict negative definiteness of $|x_1 - x_2|^{\alpha}$ where $\alpha = 2H$ is related to the property that the negative logarithm of the characteristic function of a symmetric stable distribution is $|x|^{\alpha}$. As a generalization, show that if $\phi(x)$ is a (real-valued) characteristic function of a symmetric infinitely divisible distribution, like the symmetric Laplace distribution, then $-\log \phi(x)$ is conditionally negative definite. (A univariate real-valued function $f(x)$ is called conditionally negative definite if $f(x_1 - x_2)$ is conditionally negative definite.)

Part II

Distance Correlation and Dependence

11

On Correlation and Other Measures of Association

CONTENTS

In his classical 1933 book A. N. Kolmogorov wrote : "The concept of mutual independence of two or more experiments holds, in a certain sense, a central position in the theory of probability."

11.1   The First Measure of Dependence: Correlation

The first measure of dependence is due to Francis Galton. He was the first who conceived the notion of correlation which he explained via an example in his 1888 paper Galton [1888]. Galton called the correlation coefficients "co-relations." The term "correlation coefficient" is due to Edgeworth [1892]. The paper came very close to the definition of correlation we use today except the standardization. For more details, see Stigler [1989, 1986]. Before Galton the word "correlation" was not an accepted statistical notion although Auguste Bravais, French physicist, already used it 1846. In the 1870s, Georg Cantor, the father of set theory, used the word "correlation" in the sense of $1 - 1$ correspondence, what today we would call bijection.

Here is a quotation from Georg Cantor's November 29, 1873 letter:

Take the totality of all whole-numbered individuals n and denote it by (n). And imagine, say, the totality of all positive real numerical quantities x and designate it by (x). The question is simply, Can (n) be correlated to (x) in a way such that to each individual of the one totality there corresponds one and only one of the other?

The product moment definition of correlation is due to Pearson who introduced the very notion of *moments* in statistics, (a concept borrowed from physics). The k-th moment of a random variable X is the expectation of its k-th power, $E(X^k)$, if it exists. If the random variables X and Y have zero first moments, and their second moments, $E(X^2) = E(Y^2) = 1$, then the correlation of X and Y is defined as the product moment $r := E(XY)$. If the expectation (mean), m is not necessarily 0 and the standard deviation σ is not necessarily 1, then first we standardize the variables $X^* := (X - m_X)/\sigma_X$, $Y^* := (Y - m_Y)/\sigma_Y$ and define the correlation of X and Y as $r = E(X^*Y^*)$.

The term standard deviation was first used in writing by Karl Pearson in 1894, following his use of it in lectures. The method of moments was also introduced by Pearson in 1894, and the product moment definition of correlation appeared in Pearson [1895].

Bivariate (even multivariate) normal distributions were defined a long time before Galton. Gauss (1823) defined them via their probability density function that is proportional to the exponential function of a quadratic form. However, before Galton, Edgeworth, and Pearson, the meaning of the coefficients in this quadratic form was not known. Now we know that in the bivariate case, if $m_X = m_Y = 0$, then this quadratic form is

$$-\frac{1}{2(1 - r^2)}\left[\frac{x^2}{\sigma_X^2} - \frac{2rxy}{\sigma_X\sigma_Y} + \frac{y^2}{\sigma_Y^2}\right],$$

thus the coefficients are functions of r, σ_X, σ_Y. In 1895-1896, Pearson derived the "est value" of r computed from the sample (X_i, Y_i), $i = 1, 2, \ldots, n$:

$$\frac{\sum(X_i - \bar{X})(Y_i - \bar{Y})}{\sum(X_i - \bar{X})^2 \sum(Y_i - \bar{Y})^2]^{1/2}}.$$

A famous equality of [Hoeffding, 1940] states:

If X and Y are real-valued random variables with finite variances, then their covariance, $Cov(X, Y) := E[(X - E(X))(Y - E(Y))]$, equals

$$Cov(X, Y) = \int_{-\infty}^{\infty}\int_{-\infty}^{\infty}[F_{X,Y}(x, y) - F_X(x)F_Y(y)]dx\,dy,$$

where F denotes cdf.

This shows that $Cov(X, Y)$ can be viewed as a signed distance from independence.

11.2   Distance Correlation

If the distribution of independent observations is normal, then r is a very good measure of dependence. In general, however, it can easily happen that $r = 0$ but there is a strong dependence between X and Y. This might mislead researchers who interpret $r = 0$ as "no dependence." As a result important dependencies will not be detected. A newly discovered remedy to this problem is the *distance correlation*. Distance correlation and related coefficients are the main topic of many chapters in this book, starting with Chapter 12.

An important application for two samples applies to testing independence of random vectors. In this case, we test whether the joint distribution of X and Y is equal to the product of their marginal distributions. Interestingly, the statistics can be expressed in a product-moment expression involving the double-centered distance matrices of the X and Y samples. The statistics based on distances are analogous to, but more general than, product-moment covariance and correlation. This suggests the names *distance covariance* (dCov) and *distance correlation* (dCor), defined in Chapter 12. However, before defining distance measures of dependence let us see a few recent results on Pearson's correlation in Section 11.4.

On some recent tests of independence, see Genest and Rémillard [2004] and Herwartz and Maxand [2020].

11.3 Other Dependence Measures

What are our axioms or requirements for a good measure of dependence? This is the topic of Chapter 22, starting with a close look at the 7 axioms of Rényi [1959]. In Chapter 22 examples of several other important coefficients are given in the context of understanding which axioms of dependence are satisfied. These examples include:

(i) Pearson's (linear) correlation
(ii) Spearman's ρ and Kendall's τ
(iii) Maximal Correlation Coefficient
(iv) The Correlation Ratio
(v) The RV Coefficient
(vi) Maximal Information Coefficient
(vii) Distance Correlation

See Section 22.3 for definitions and discussion of properties.

11.4 Representations by Uncorrelated Random Variables

Although Pearson's correlation has been around for more than a century, even today one can find interesting and challenging correlation problems to solve. Correlation = 0, i.e. uncorrelatedness, is the same as orthogonality in calculus and analysis. If the sample size is n, then we are talking about orthogonality in the n-dimensional Euclidean space; in case of population uncorrelatedness we are talking about orthogonality of random variables in the Hilbert space L_2.

In the following we list six theorems for readers to prove or to read the proof in the referenced papers if they are interested.

Let $X = (X_1, X_2, \ldots, X_n)$ be an arbitrary random vector where the coordinates X_i, $i = 1, 2, \ldots, n$ have finite variances. Then we can diagonalize the covariance matrix of X and thus we can find a linear transformation A such that $Y = AX$ becomes a random vector with uncorrelated coordinates. If the inverse A^{-1} exists, then $X = A^{-1}Y$ is a representation of X with the help of uncorrelated random variables. But here Y is a mixture of the X coordinates and in many cases we cannot interpret these mixtures; for example, if X_1 is the squared velocity and X_2 is the mass. Instead, let us consider representations that are univariate functions of the coordinates, not their mixtures. The idea that estimators of unrelated parameters should be unrelated (in some sense) is an old problem. The most natural notion of "unrelatedness" is independence. A classical theorem is that the maximum likelihood estimators of the mean and variance of a Gaussian distribution are independent.

Theorem 11.1 (Móri and Székely [2017]). *Every random vector* $X = (X_1, X_2, \ldots, X_n)$ *can be represented as functions of uncorrelated random variables* $Y_1, Y_2, \ldots, Y_n$; *that is, we can always find* $\mathbb{R} \to \mathbb{R}$ *functions* $f_1, f_2, \ldots, f_n$ *such that* $(X_1, X_2, \ldots, X_n)$ *has the same distribution as* $(f_1(Y_1), f_2(Y_2), \ldots, f_n(Y_n))$.

See Section 22.8 (1) for a proof.

The functions f_i, $i = 1, 2, \ldots, n$ cannot always be one-to-one because Móri [1992] can be reformulated as follows.

Theorem 11.2 (Móri [1992]). *A necessary and sufficient condition for random variables* X_1, X_2 *not to have the same distribution as* $f_1(Y_1)$, $f_2(Y_2)$ *where* Y_1 *and* Y_2 *are uncorrelated random variables and* f_1, f_2 *are one-to-one functions is that* X_i, $i = 1, 2$ *have the representation (equality in distribution)*

$$X_i = Z_i + c_i V_i \, I_{\{Z_i = b_i\}} \quad i = 1, 2,$$

where I *denotes indicator function,* $Z_1, Z_2, (V_1, V_2)$ *are independent,* V_1 *and* V_2 *are dependent (correlated) indicator functions,* b_i *and* c_i *are real numbers,* $c_i \neq 0$, *and* $P(Z_i = b_i) > 0$, $i = 1, 2$.

On the other hand, the following proposition shows that with very few exceptions for *all random variables* X one can find a 1–1 real function f such that X and $f(X)$ are uncorrelated.

Theorem 11.3 (Móri and Székely [2019]). *Let* X *be a square integrable random variable defined on an arbitrary probability space. Suppose the distribution of* X *is not concentrated on three or less points. Then there exists a measurable* *injective function $f : \mathbb{R} \to \mathbb{R}$ such that X and $f(X)$ are uncorrelated. This f can be chosen piecewise linear.*

Such an f cannot exist if X takes on exactly two values, because in this case uncorrelatedness is equivalent to independence. When the distribution of X is supported on exactly 3 points then a necessary and sufficient condition for f to exist is $P(X = E(X)) = 0$.

Theorem 11.4 (Móri and Székely [2017]). *Let $X_1, X_2, \ldots, X_n$ be arbitrary random variables with zero means, finite variances, and absolutely continuous distributions. Then there exist Borel sets $B_i \subset \mathbb{R}^+$, $i = 1, 2, \ldots, n$, such that if we define $f_i(t) = -t$ for $|t| \in B_i$, and $f_i(t) = t$ otherwise, then the random variables $Y_i := f_i(X_i)$ are uncorrelated, $E(Y_i) = 0$ and $\mathrm{Var}(Y_i) = \mathrm{Var}(X_i)$. Since the functions f_i are idempotent ($f_i(f_i(x)) = x$), we have that $X_i = f_i(Y_i)$ is a one-to-one piecewise linear function of uncorrelated random variables.*

Theorem 11.5. *Let (X_1, X_2) be an arbitrary bivariate Gaussian random variable with standard marginals. Then the one-to-one and piecewise linear function $f(x) = x$ for $|x| \geq c$ and $f(x) = -x$ for $|x| < c$ with a suitable constant $c = 1.539 \ldots$ makes X_1 and $f(X_2)$ uncorrelated, and $(X_1, X_2) \equiv (X_1, f(f(X_2)))$. Here the function does not depend on the correlation of X_1 and X_2.*

Can this result be generalized to n-variate Gaussian random variables? For this generalization one would need a partition of the set of positive integers into n disjoint subset N_i such that if H_k denotes the k-th Hermite polynomial, that is, $H_0(x) \equiv 1$, and $H_k(x)n(x) = (-1)^k (\frac{d}{dx})^k n(x)$, $k \geq 1$, where $n(x) = (2\pi)^{-1/2} e^{-x^2/2}$ denotes the standard normal pdf. Then

$$f_i(x) := \sum_{k \in N_i} a_{ki} H_k(x)$$

is a one-to-one function for $i = 1, 2, \ldots, n$. The explanation is given in the following theorem.

Theorem 11.6 (Móri and Székely [2017]). *Let $(X_1, X_2, \ldots, X_n)$ be arbitrary n-variate Gaussian random variable with standard marginals. Then $f_i(X_i)$, $i = 1, 2, \ldots, n$ are uncorrelated regardless of the correlation of the X variables if and only if the set of positive integers can be partitioned into n disjoint subsets N_i, $i = 1, 2, \ldots, n$, such that the Hermite expansion of $f_i(x)$ only contains terms $H_k(x)$ with indices from N_i, that is,*

$$f_i(x) = \sum_{k \in N_i} a_{ki} H_k(x).$$

In case $n = 2$, the partition $N_1 = \{1\}$, $N_2 = \{2, 3, \ldots\}$ will do (as we have seen above), but at the moment we do not know if there exist n one-to-one functions f_i with the property above for $n > 2$. This seems to be an interesting open problem.

12

Distance Correlation

CONTENTS

12.1 Introduction

In the analysis of real data, statisticians often work with random vectors that may have non-linear dependence. In this chapter, we focus on dependence coefficients *distance covariance* (dCov) and *distance correlation* (dCor) introduced in Székely, Rizzo, and Bakirov [2007] that measure all types of dependence between random vectors X and Y in arbitrary dimension. These energy dependence coefficients characterize independence between random vectors: $\mathrm{dCov}(X,Y)$ (and $\mathrm{dCor}(X,Y)$) is non-negative and equal to zero if and only if the random vectors X and Y are independent. Distance correlation is a very effective tool to detect novel associations in large data sets (see Simon and Tibshirani [2011] and Wahba [2014]).

The corresponding energy statistics have simple computing formulae, and they apply to sample sizes $n \geq 2$ (n can be much smaller than dimension).

DOI: 10.1201/9780429157158-12

To quote Newton [2009]: *Distance covariance not only provides a bona fide dependence measure, but it does so with a simplicity to satisfy Don Geman's elevator test (a method must be sufficiently simple that it can be explained to a colleague in the time it takes to go between floors on an elevator!).*

Here is a short description of how to compute the dCov statistic. (A more detailed version is found in Section 12.4 below.)

The distance covariance statistic is computed as follows. First we compute all of the pairwise distances between sample observations of the X sample, to get a distance matrix. Similarly, compute a distance matrix for the Y sample. Next, we center the entries of these distance matrices so that their row and column means are equal to zero. A very simple formula (12.12) accomplishes the centering. Now take the centered distances $\widehat{A}_{k\ell}$ and $\widehat{B}_{k\ell}$ and compute the sample distance covariance as the square root of

$$\mathcal{V}_n^2 = \frac{1}{n^2} \sum_{k,\,\ell=1}^{n} \widehat{A}_{k\ell}\widehat{B}_{k\ell}.$$

Notation: If $X \in \mathbb{R}^p$ and $Y \in \mathbb{R}^q$ are random vectors, $\mathcal{V}(X,Y)$ denotes the distance covariance coefficient and $\mathcal{R}(X,Y)$ is the distance correlation coefficient. If $\mathbf{X}$ and $\mathbf{Y}$ are data vectors or data matrices, then $\mathcal{V}_n(\mathbf{X},\mathbf{Y})$ and $\mathcal{R}_n(\mathbf{X},\mathbf{Y})$ are the dCov and dCor statistics, respectively. We also use $\mathrm{dCov}_n(\mathbf{X},\mathbf{Y})$ and $\mathrm{dCor}_n(\mathbf{X},\mathbf{Y})$ or simply dCov_n, dCor_n for the statistics.

The statistic $\mathcal{V}_n$ converges almost surely to distance covariance (dCov), $\mathcal{V}(X,Y)$, to be defined below, which is always non-negative and equals zero if and only if X and Y are independent. Once we have dCov, we can define distance variance (dVar). Then distance correlation (dCor) is computed as the normalized coefficient analogous to Pearson's correlation.

An example where one can clearly observe dependence that is non-linear is seen in Figure 12.1(b) of Example 12.1 below.

Example 12.1. The Saviotti aircraft data [Saviotti, 1996] record six characteristics of aircraft designs which appeared during the twentieth century. We consider two variables, wing span (m) and speed (km/h) for the 230 designs of the third (of three) periods. This example and the data (*aircraft*) are from Bowman and Azzalini [1997, 2021]. A scatterplot on log-log scale of the variables and contours of a nonparametric density estimate are shown in Figures 12.1(a) and 12.1(b). The nonlinear relation between speed and wing span is quite evident from the plots.

Note: The `aircraft` data is available in the *sm* package for R [Bowman and Azzalini, 2021].

Later in Chapter 13 it is shown that a test of independence based on the dCov V-statistic is consistent against general alternatives. The next example illustrates how the dCov test of independence is applied using the *energy* package for R [Rizzo and Székely, 2022]. The data in Example 12.1 is bivariate,

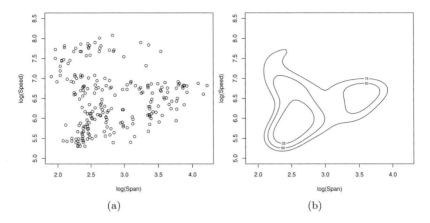

FIGURE 12.1
Scatterplot and contours of density estimate for the aircraft speed and span variables, period 3, in Example 12.1.

so it is perhaps interesting to compare the test with a test based on Pearson correlation.

Example 12.2. To test whether log(Speed) and log(Span) from Example 12.1 are independent, we apply a test based on distance correlation, as implemented in the function `dcor.test` of the *energy* package for R. This test is implemented non-parametrically as a permutation (randomization) test (see Section 13.2.1). The (Pearson) correlation test is based on the t statistic (as `cor.test` in R). The dCor test of independence of log(Speed) and log(Span) in period 3 is significant (p-value = 0.001), while the Pearson correlation test is not significant (p-value = 0.8001).

```
dcor.test(Speed3, Span3, R=1999)
##
##   dCor independence test (permutation test)
##
## data:  index 1, replicates 1999
## dCor = 0.28045, p-value = 0.001
## sample estimates:
##      dCov      dCor    dVar(X)    dVar(Y)
## 0.1218536 0.2804530 0.4872107 0.3874712

cor.test(Speed3, Span3)
##
##   Pearson's product-moment correlation
##
## data:  Speed3 and Span3
## t = 0.25349, df = 228, p-value = 0.8001
## alternative hypothesis: true correlation is not equal to 0
## 95 percent confidence interval:
##   -0.1128179  0.1458274
```

```
## sample estimates:
##        cor
## 0.01678556
```

The sample estimates are $\hat{\rho} = 0.0168$ and $\mathcal{R}_n = 0.2805$. Here we have an example of observed data where two variables are nearly uncorrelated, but dependent. We obtained essentially the same results on the correlations of ranks of the data.

Classical inference based on normal theory tests the hypothesis of multi-variate independence via a likelihood ratio statistic based on the covariance matrix of (X, Y) or their marginal ranks. These tests are not consistent against general alternatives, because like correlation measures, the statistics measure linear or monotone association. The distance covariance energy test is based on measuring the difference between the joint and marginal characteristic functions, thus it characterizes independence. For other recent consistent tests of bivariate or multivariate independence see e.g. Feuerverger [1993], Gretton and Györfi [2010, 2012].

There are many chapters in this book that are entirely or partly devoted to the dCov and dCor coefficients and their applications. The list includes Chapter 15 "Brownian Distance Covariance," Chapter 16 "U-statistics and Unbiased dCov^2," Chapter 17 "Partial Distance Correlation," Chapter 19 "The dCor t-test of Independence in High Dimension," and more.

12.2   Characteristic Function Based Covariance

As in the Introduction, we could have simply introduced definitions of the statistics at this point. Readers who prefer on a first reading to focus on applications could skim this section, and see the definitions of our dependence coefficients that follow in the next section. However, to read the proofs, the information in this section is important background.

Suppose that $X \in \mathbb{R}^p$ is a random vector with characteristic function (cf) $\hat{f}_X$, $Y \in \mathbb{R}^q$ is a random vector with cf $\hat{f}_Y$, and $\hat{f}_{X,Y}$ is the joint cf of X and Y. (Recall that the characteristic function is unique, always exists, and is bounded in magnitude by 1.) Then X and Y are independent if and only if $\hat{f}_{X,Y}(t,s) = \hat{f}_X(t)\hat{f}_Y(s)$.

Let γ be a complex valued function defined on $\mathbb{R}^p \times \mathbb{R}^q$. Define the $||\cdot||_w$-norm in the weighted L_2 space of functions on $\mathbb{R}^{p+q}$ by

$$||\gamma(t,s)||_w^2 = \int_{\mathbb{R}^{p+q}} |\gamma(t,s)|^2 w(t,s) \, dt \, ds, \tag{12.1}$$

where $w(t, s)$ is an arbitrary weight function for which the integral above exists, $w(t, s) > 0$ a.s. $t \in \mathbb{R}^p$, $s \in \mathbb{R}^q$.

Let us define a covariance dCov_w by the weighted L_2 norm as:

$$\mathrm{dCov}_w^2(X,Y) = \|f_{X,Y}(t,s) - f_X(t)f_Y(s)\|_w^2 \tag{12.2}$$

$$= \int_{\mathbb{R}^q}\int_{\mathbb{R}^p} |f_{X,Y}(t,s) - f_X(t)f_Y(s)|^2 w(t,s)\,dt\,ds,$$

for a fixed positive weight function $w(t,s)$, for which the above integral exists. The definition (12.2) immediately implies that $\mathrm{dCov}_w(X,Y) = 0$ if and only if X,Y are independent.

It is desirable that $\mathrm{dCov}_w^2(X,Y)$ be invariant to all shift and orthogonal transformations of X and Y, and scale equivariant. That is,

$$\mathrm{dCov}_w^2(a_1 + b_1 C_1 X,\ a_2 + b_2 C_2 Y) = b_1 b_2\,\mathrm{dCov}_w^2(X,Y), \tag{12.3}$$

for all vectors a_1, a_2, orthonormal matrices C_1, C_2, and scalars $b_1, b_2 > 0$. Scale equivariance of dCov_w^2 implies that the corresponding correlation, $dCor_w^2$ becomes scale invariant, which is an important property. Also, scale equivariance makes it possible to view dCov_w^2 as a scalar product (similar to Pearson's covariance).

We assume finite first moments and, for integrability of (12.2), that

$$\int_{\mathbb{R}^q}\int_{\mathbb{R}^p} \min\{1,|t|^2\}\cdot\min\{1,|s|^2\}\cdot|w(t,s)|\,dt\,ds < \infty, \tag{12.4}$$

where $|t|^2 = \langle t,t\rangle$ and $|s|^2 = \langle s,s\rangle$ are the squared Euclidean norms for $t \in \mathbb{R}^p$ and $s \in \mathbb{R}^q$, respectively, and $\langle x,t\rangle = x^\top t$ is the inner product of vectors x and t. The condition (12.4) implies that either

$$|\hat{f}_{X,Y}(t,s) - \hat{f}_X(t)\hat{f}_Y(s)|$$
$$= \left| E(e^{i\langle t,X\rangle} - 1)(e^{i\langle s,Y\rangle} - 1) - E(e^{i\langle t,X\rangle} - 1)E(e^{i\langle s,Y\rangle} - 1) \right|$$
$$\leq \mathrm{const}\cdot|t||s|,$$

(in the last step we used the conditions $E|X| < \infty$, $E|Y| < \infty$, and $E|X||Y| < \infty$)

or $|\hat{f}_{X,Y}(t,s) - \hat{f}_X(t)\hat{f}_Y(s)| \leq 2$, and therefore

$$\mathrm{dCov}_w^2(X,Y) = \int_{\mathbb{R}^q}\int_{\mathbb{R}^p} |\hat{f}_{X,Y}(t,s) - \hat{f}_X(t)\hat{f}_Y(s)|^2 w(t,s)\,dt\,ds < \infty.$$

Hence the integral (12.2) exists.

In the definitions that follow, we apply a weight function

$$w(t,s) = \frac{const}{|t|^{1+p}|s|^{1+q}}.$$

This weight function is unique if we require rigid motion invariance and scale equivariance [Székely and Rizzo, 2012].

The distance covariance statistic is derived from $\|\hat{f}^n_{X,Y}(t,s) - \hat{f}^n_X(t)\hat{f}^n_Y(s)\|^2$ (replacing characteristic functions with empirical characteristic functions of the samples).

The simple formula for the distance covariance statistic given in the introduction:

$$\mathcal{V}^2_n = \frac{1}{n^2} \sum_{k,\ell=1}^{n} \widehat{A}_{k\ell}\widehat{B}_{k\ell}$$

is derived by evaluating $\|\hat{f}^n_{X,Y}(t,s) - \hat{f}^n_X(t)\hat{f}^n_Y(s)\|^2$.

12.3 Dependence Coefficients

In this section we define the population coefficients distance covariance (dCov), distance variance (dVar), and distance correlation (dCor), and discuss some of their properties. Recall that distance covariance is a measure of the difference $\hat{f}_{X,Y} - \hat{f}_X\hat{f}_Y$.

In the following definition the constants c_p, c_q are given by

$$c_d = \frac{\pi^{(d+1)/2}}{\Gamma\left(\frac{d+1}{2}\right)}. \tag{12.5}$$

12.3.1 Definitions

Definition 12.1 (Distance Covariance). The ***distance covariance*** between random vectors $X \in \mathbb{R}^p$ and $Y \in \mathbb{R}^q$ with finite first moments is the nonnegative number $\mathcal{V}(X,Y)$ defined by

$$\mathcal{V}^2(X,Y) = \|\hat{f}_{X,Y}(t,s) - \hat{f}_X(t)\hat{f}_Y(s)\|^2$$
$$= \frac{1}{c_p c_q} \int_{\mathbb{R}^{p+q}} \frac{|\hat{f}_{X,Y}(t,s) - \hat{f}_X(t)\hat{f}_Y(s)|^2}{|t|_p^{1+p}\,|s|_q^{1+q}}\,dt\,ds. \tag{12.6}$$

By the definition of the norm $\|\cdot\|$, it is clear that $\mathcal{V}(X,Y) \geq 0$ and $\mathcal{V}(X,Y) = 0$ if and only if X and Y are independent.

The exact value of $\mathcal{V}(X,Y)$ for given X, Y can be derived without characteristic functions by applying the following Proposition.

Proposition 12.1. *If $E|X|_p < \infty$ and $E|Y|_q < \infty$, then*

$$\mathcal{V}^2(X,Y) = E\left(|X - X'|_p|Y - Y'|_q\right) + E|X - X'|_p\,E|Y - Y''|_q$$
$$- 2E\left(|X - X'|_p|Y - Y''|_q\right), \tag{12.7}$$

where (X,Y), (X',Y'), and (X'',Y'') are iid.

The proof applies Lemma 2.1 and Fubini's theorem. For another approach see Chapter 15.

Definition 12.2 (Distance Variance). ***Distance variance*** is defined as the square root of

$$\mathcal{V}^2(X) = \mathcal{V}^2(X, X) = \|\hat{f}_{X,X}(t, s) - \hat{f}_X(t)\hat{f}_X(s)\|^2.$$

Definition 12.3 (Distance Correlation Coefficient). The ***distance correlation*** (*dCor*) between random vectors X and Y with finite first moments is the nonnegative number $\mathcal{R}(X, Y)$ defined by

$$\mathcal{R}^2(X, Y) = \begin{cases} \frac{\mathcal{V}^2(X,Y)}{\sqrt{\mathcal{V}^2(X)\,\mathcal{V}^2(Y)}}, & \mathcal{V}^2(X)\mathcal{V}^2(Y) > 0; \\ 0, & \mathcal{V}^2(X)\mathcal{V}^2(Y) = 0. \end{cases} \qquad (12.8)$$

Remark 12.1. Scale invariance of distance correlation is essential. If we did not have scale invariance, then our decisions could change simply because system of measurement changed.

Remark 12.2. One might wonder why do we work with the non-integrable weight function $1/[|t|_p^{1+p}\,|s|_q^{1+q}]$ in the definition of distance covariance. The answer is very simple. If we applied an alternative, *integrable* weight function w, then for random variables X, Y with finite variance the Taylor expansion of their characteristic functions shows that the alternative distance correlation of ϵX and ϵY would tend to the square of Pearson's correlation of X and Y as $\epsilon \to 0$. Why is this a problem? Because then even if X and Y are strongly dependent, this alternative version of distance correlation of ϵX and ϵY would tend to 0 whenever their Pearson correlation is 0. This is exactly we want to avoid, especially if we insist on scale invariance of our dependence measure.

12.4   Sample Distance Covariance and Correlation

Let **X** be a data vector or data matrix of iid observations from X, and let **Y** be a data vector or data matrix of iid observations from Y.

Definition 12.4 (Double-centered distance matrix). The double-centered distance matrices are computed as in classical multidimensional scaling. Given a random sample $(x, y) = \{(x_i, y_i) : i = 1, \ldots, n\}$ from the joint distribution of random vectors X in $\mathbb{R}^p$ and Y in $\mathbb{R}^q$, compute the Euclidean distance matrix $(a_{ij}) = (|x_i - x_j|_p)$ for the X sample and $(b_{ij}) = (|y_i - y_j|_q)$ for the Y sample. The ij-th entry of $\hat{A}$ is

$$\hat{A}_{ij} = a_{ij} - \bar{a}_{i.} - \bar{a}_{.j} + \bar{a}_{..}, \qquad i, j = 1, \ldots, n,$$

where

$$\bar{a}_{i.} = \frac{1}{n}\sum_{j=1}^{n} a_{ij}, \quad \bar{a}_{.j} = \frac{1}{n}\sum_{i=1}^{n} a_{ij}, \quad \bar{a}_{..} = \frac{1}{n^2}\sum_{i,j=1}^{n} a_{ij}.$$

Similarly, the ij-th entry of $\hat{B}$ is

$$\hat{B}_{ij} = b_{ij} - \bar{b}_{i.} - \bar{b}_{.j} + \bar{b}_{..}, \quad i,j = 1,\ldots,n.$$

Definition 12.5 (Distance covariance statistic). The sample distance covariance is defined as the square root of

$$\mathcal{V}_n^2(\mathbf{X},\mathbf{Y}) = \frac{1}{n^2}\sum_{i,j=1}^{n} \hat{A}_{ij}\hat{B}_{ij}, \tag{12.9}$$

where $\hat{A}$ and $\hat{B}$ are the double-centered distance matrices of the X sample and the Y sample, respectively, and the subscript ij denotes the entry in the i-th row and j-th column.

How simple is this formula? In R [R Core Team, 2022] the $*$ operator is element-wise multiplication so if A, B are double-centered, $\mathcal{V}_n^2$ is simply the R expression mean(A*B).

Although it may not be immediately obvious that $\mathcal{V}_n^2(\mathbf{X},\mathbf{Y}) \geq 0$, this fact as well as the motivation for the definition of $\mathcal{V}_n$ will be clear from Theorem 12.4 below.

Expected Value

If X and Y are independent,

$$E[\mathcal{V}_n^2(\mathbf{X},\mathbf{Y})] = \frac{n-1}{n^2}E|X - X'||Y - Y'|.$$

For dependent X,Y, see Section 13.1.1.

Definition 12.6 (Sample distance variance). The sample *distance variance* is the square root of

$$\mathcal{V}_n^2(\mathbf{X}) = \mathcal{V}_n^2(\mathbf{X},\mathbf{X}) = \frac{1}{n^2}\sum_{i,j=1}^{n} \hat{A}_{ij}^2 . \tag{12.10}$$

The distance variance statistic is always nonnegative, and $\mathcal{V}_n^2(\mathbf{X}) = 0$ only if all of the sample observations are identical (see Székely et al. [2007]).

Remark 12.3. The statistic $\mathcal{V}_n(\mathbf{X}) = 0$ if and only if every sample observation is identical. Indeed, if $\mathcal{V}_n(\mathbf{X}) = 0$, then $\hat{A}_{kl} = 0$ for $k,l = 1,\ldots,n$. Thus $0 = \hat{A}_{kk} = -\bar{a}_{k.} - \bar{a}_{.k} + \bar{a}_{..}$ implies that $\bar{a}_{k.} = \bar{a}_{.k} = \bar{a}_{..}/2$, and

$$0 = \hat{A}_{kl} = a_{kl} - \bar{a}_{k.} - \bar{a}_{.l} + \bar{a}_{..} = a_{kl} = |X_k - X_l|_p,$$

so $X_1 = \cdots = X_n$.

Remark 12.4. Above we have defined the dCov statistic using the simple product moment formula. At first, it is hard to see how formula (12.9) is a good estimator of the coefficient $\mathcal{V}^2(X,Y)$. This statistic is actually developed by evaluating the weighted L_2 norm $\|\hat{f}_{X,Y} - \hat{f}_X \cdot \hat{f}_Y\|$, with $\hat{f}_{X,Y}, \hat{f}_X, \hat{f}_Y$ replaced by the respective empirical characteristic functions. This will become clear in the technical results that follow.

Definition 12.7 (Distance correlation statistic). The squared distance correlation is defined by

$$\mathcal{R}_n^2(\mathbf{X},\mathbf{Y}) = \begin{cases} \frac{\mathcal{V}_n^2(\mathbf{X},\mathbf{Y})}{\sqrt{\mathcal{V}_n^2(\mathbf{X})\mathcal{V}_n^2(\mathbf{Y})}}, & \mathcal{V}_n^2(\mathbf{X})\mathcal{V}_n^2(\mathbf{Y}) > 0; \\ 0, & \mathcal{V}_n^2(\mathbf{X})\mathcal{V}_n^2(\mathbf{Y}) = 0. \end{cases} \tag{12.11}$$

Distance correlation satisfies

1. $0 \le \mathcal{R}_n(\mathbf{X},\mathbf{Y}) \le 1$.

2. If $\mathcal{R}_n(\mathbf{X},\mathbf{Y}) = 1$, then there exists a vector a, a non-zero real number b and an orthogonal matrix C such that $\mathbf{Y} = a + b\mathbf{X}C$, for the data matrices $\mathbf{X}$ and $\mathbf{Y}$.

A more complete list of properties of the distance dependence coefficients and statistics can be found in Section 12.5.

12.4.1 Derivation of $\mathcal{V}_n^2$

For a random sample $(\mathbf{X},\mathbf{Y}) = \{(X_k,Y_k) : k = 1,\ldots,n\}$ of n iid random vectors (X,Y) from the joint distribution of random vectors X in $\mathbb{R}^p$ and Y in $\mathbb{R}^q$, compute the Euclidean distance matrices $(a_{k\ell}) = (|X_k - X_\ell|_p)$ and $(b_{k\ell}) = (|Y_k - Y_\ell|_q)$. Define the centered distances

$$\widehat{A}_{k\ell} = a_{k\ell} - \bar{a}_{k.} - \bar{a}_{.\ell} + \bar{a}_{..}, \qquad k,\ell = 1,\ldots,n, \tag{12.12}$$

as in Definition 12.4. Similarly, define $\widehat{B}_{k\ell} = b_{k\ell} - \bar{b}_{k.} - \bar{b}_{.\ell} + \bar{b}_{..}$, for $k,\ell = 1,\ldots,n$.

Let $\hat{f}_X^n(t)$, $\hat{f}_Y^n(s)$, and $\hat{f}_{X,Y}^n(t,s)$ denote the empirical characteristic functions of the samples $\mathbf{X}$, $\mathbf{Y}$, and $(\mathbf{X},\mathbf{Y})$, respectively. It is natural to consider a statistic based on the L_2 norm of the difference between the empirical characteristic functions; that is, to substitute the empirical characteristic functions for the characteristic functions in the definition of the norm.

To evaluate $\|\hat{f}_{X,Y}(t,s) - \hat{f}_X(t)\hat{f}_Y(s)\|^2$ the crucial observation is the following lemma.

Lemma 12.1. If $0 < \alpha < 2$, then for all x in $\mathbb{R}^d$

$$\int_{\mathbb{R}^d} \frac{1 - \cos\langle t, x\rangle}{|t|_d^{d+\alpha}}\, dt = C(d,\alpha)|x|^\alpha,$$

where

$$C(d, \alpha) = \frac{2\pi^{\frac{d}{2}} \Gamma(1 - \frac{\alpha}{2})}{\alpha 2^{\alpha} \Gamma(\frac{d+\alpha}{2})}, \tag{12.13}$$

and $\Gamma(\cdot)$ is the complete gamma function. The integrals at 0 and ∞ are meant in the principal value sense: $\lim_{\varepsilon \to 0} \int_{\mathbb{R}^d \setminus \{\varepsilon B + \varepsilon^{-1} B^c\}}$, where B is the unit ball (centered at 0) in $\mathbb{R}^d$ and B^c is the complement of B.)

Lemma 12.1 is proved in Székely and Rizzo [2005a]; see Section 2.3 for a copy of the proof.

For finiteness of $\|\hat{f}_{X,Y}(t, s) - \hat{f}_X(t)\hat{f}_Y(s)\|^2$ it is sufficient that $E|X|_p < \infty$ and $E|Y|_q < \infty$. See Section 12.7 for details.

12.4.2   Equivalent Definitions for $\mathcal{V}_n^2$

A key result in Székely et al. [2007] is that the following definitions are equivalent for sample distance covariance:

$$\mathcal{V}_n^2(\mathbf{X}, \mathbf{Y}) = \|\hat{f}_{X,Y}^n(t, s) - \hat{f}_X^n(t)\hat{f}_Y^n(s)\|^2 \tag{12.14}$$

$$= \frac{1}{n^2} \sum_{k,\ell=1}^{n} a_{k,\ell} b_{k,\ell} + \bar{a}_{..} \bar{b}_{..} - \frac{2}{n^3} \sum_{k=1}^{n} \sum_{\ell,m=1}^{n} a_{k,\ell} b_{k,m} \tag{12.15}$$

$$= \frac{1}{n^2} \sum_{k,\ell=1}^{n} \hat{A}_{k\ell} \hat{B}_{k\ell}, \tag{12.16}$$

in the notation of Definition 12.4. This result is proved in two steps. The first step in the definition of $\mathcal{V}_n^2$ is to evaluate $\|\hat{f}_{X,Y}^n(t, s) - \hat{f}_X^n(t)\hat{f}_Y^n(s)\|^2$, which is derived in the proof of Theorem 12.1 as

$$\|\hat{f}_{X,Y}^n(t, s) - \hat{f}_X^n(t)\hat{f}_Y^n(s)\|^2 = S_1 + S_2 - 2S_3,$$

equal to the second definition (12.15) above. The second step is to show the algebraic equivalence of (12.15) and (12.16).

Remark 12.5. Another equivalent formula is

$$\mathcal{V}_n^2(\mathbf{X}, \mathbf{Y}) = \frac{\sum_{i \neq j} a_{ij} b_{ij}}{n^2} - \frac{2 \sum_{i=1}^{n} a_{i.} b_{i.}}{n^3} + \frac{a_{..} b_{..}}{n^4}.$$

Refer to equation (12.40) in the proof of Theorem 12.1 and see also Section 16.7, equation (16.19) for details and the corresponding U-statistic.

12.4.3   Theorem on dCov Statistic Formula

Theorem 12.1. *If $(\mathbf{X}, \mathbf{Y})$ is a random sample from the joint distribution of (X, Y), then*

$$\|\hat{f}_{X,Y}^n(t, s) - \hat{f}_X^n(t)\hat{f}_Y^n(s)\|^2 = S_1 + S_2 - 2S_3, \tag{12.17}$$

where

$$S_1 = \frac{1}{n^2} \sum_{k,\ell=1}^{n} |X_k - X_\ell|_p |Y_k - Y_\ell|_q, \tag{12.18}$$

$$S_2 = \frac{1}{n^2} \sum_{k,\ell=1}^{n} |X_k - X_\ell|_p \frac{1}{n^2} \sum_{k,\ell=1}^{n} |Y_k - Y_\ell|_q, \tag{12.19}$$

$$S_3 = \frac{1}{n^3} \sum_{k=1}^{n} \sum_{\ell,m=1}^{n} |X_k - X_\ell|_p |Y_k - Y_m|_q, \tag{12.20}$$

and

$$\mathcal{V}_n^2(\mathbf{X}, \mathbf{Y}) = S_1 + S_2 - 2S_3, \tag{12.21}$$

where $\mathcal{V}_n^2(\mathbf{X}, \mathbf{Y})$ is given by (12.9).

The proof of Theorem 12.1 in Section 12.7 is from Székely et al. [2007].

As a corollary we have that $\mathcal{V}_n^2(\mathbf{X}, \mathbf{Y}) \geq 0$, $\mathcal{V}_n^2(\mathbf{X}) \geq 0$. Hence we know that the square root of $\mathcal{V}_n^2(\mathbf{X})\mathcal{V}_n^2(\mathbf{Y})$ is real and the distance correlation is always defined and non-negative (assuming $E|X| < \infty, E|Y| < \infty$).

To summarize our formulas for dCov, dVar, and dCor statistics:

$$\mathcal{V}_n^2(\mathbf{X}, \mathbf{Y}) = \frac{1}{n^2} \sum_{k,\ell=1}^{n} \widehat{A}_{k\ell} \widehat{B}_{k\ell} \ .$$

$$\mathcal{V}_n^2(\mathbf{X}) = \mathcal{V}_n^2(\mathbf{X}, \mathbf{X}) = \frac{1}{n^2} \sum_{k,\ell=1}^{n} \widehat{A}_{k\ell}^2 \ .$$

$$\mathcal{R}_n^2(\mathbf{X}, \mathbf{Y}) = \begin{cases} \frac{\mathcal{V}_n^2(\mathbf{X},\mathbf{Y})}{\sqrt{\mathcal{V}_n^2(\mathbf{X})\mathcal{V}_n^2(\mathbf{Y})}}, & \mathcal{V}_n^2(\mathbf{X})\mathcal{V}_n^2(\mathbf{Y}) > 0; \\ 0, & \mathcal{V}_n^2(\mathbf{X})\mathcal{V}_n^2(\mathbf{Y}) = 0, \end{cases}$$

where $\widehat{A}_{k,\ell}$ and $\widehat{B}_{k,\ell}$ are the double-centered distance matrices in Definition 12.4. We also have

$$\mathcal{V}_n^2(\mathbf{X}, \mathbf{Y}) = \frac{1}{n^2} \sum_{k,\ell=1}^{n} a_{k,\ell} b_{k,\ell} + \bar{a}_{..} \bar{b}_{..} - \frac{2}{n^3} \sum_{k=1}^{n} \sum_{\ell,m=1}^{n} a_{k,\ell} b_{k,m}, \tag{12.22}$$

$$\mathcal{V}_n^2(\mathbf{X}, \mathbf{Y}) = \frac{\sum_{i \neq j} a_{ij} b_{ij}}{n^2} - \frac{2 \sum_{i=1}^{n} a_{i.} b_{i.}}{n^3} + \frac{a_{..} b_{..}}{n^4}, \tag{12.23}$$

and the corresponding formulas for $\mathcal{V}_n(\mathbf{X})$.

12.5 Properties

The definitions of $\mathcal{V}_n^2$ and $\mathcal{R}_n^2$ suggest that our distance dependence measures are analogous in at least some respects to the corresponding product moment covariance and linear correlation. Certain properties of correlation and variance may also hold for $dCor$ and $dVar$. The key property that dCov and dCor have that Cov and Cor do not have is that $dCov(X, Y) = 0$ if and only if X and Y are independent, and $dCor(X, Y) = 0$ similarly characterizes independence of X and Y. In this section we summarize several properties of the population coefficients and the statistics. Expected value, convergence and asymptotic properties are covered in Chapter 13.

The following Theorem summarizes some key properties of the coefficients.

Theorem 12.2 (Properties of dCov and dVar). *For random vectors $X \in \mathbb{R}^p$ and $Y \in \mathbb{R}^q$ such that $E(|X|_p + |Y|_q) < \infty$, the following properties hold.*

(i) $\mathcal{V}(X, Y) \geq 0$, *and* $\mathcal{V}(X, Y) = 0$ *if and only if X and Y are independent.*

(ii) $\mathcal{V}(a_1 + b_1 C_1 X, \, a_2 + b_2 C_2 Y) = \sqrt{|b_1 b_2|}\, \mathcal{V}(X, Y)$, *for all constant vectors $a_1 \in \mathbb{R}^p$, $a_2 \in \mathbb{R}^q$, scalars b_1, b_2 and orthonormal matrices C_1, C_2 in $\mathbb{R}^p$ and $\mathbb{R}^q$, respectively.*

(iii) *If the random vector (X_1, Y_1) is independent of the random vector (X_2, Y_2), then*

$$\mathcal{V}(X_1 + X_2, \, Y_1 + Y_2) \leq \mathcal{V}(X_1, Y_1) + \mathcal{V}(X_2, Y_2).$$

Equality holds if and only if X_1 and Y_1 are both constants, or X_2 and Y_2 are both constants, or X_1, X_2, Y_1, Y_2 are mutually independent.

(iv) $\mathcal{V}(X) = 0$ *implies that $X = E[X]$, almost surely.*

(v) $\mathcal{V}(a + bCX) = |b|\mathcal{V}(X)$, *for all constant vectors a in $\mathbb{R}^p$, scalars b, and $p \times p$ orthonormal matrices C.*

(vi) *If X and Y are independent, then $\mathcal{V}(X + Y) \leq \mathcal{V}(X) + \mathcal{V}(Y)$. Equality holds if and only if one of the random vectors X or Y is constant.*

(vii) $\mathcal{V}_n(\mathbf{X}, \mathbf{Y}) \geq 0$.

(viii) $\mathcal{V}_n(\mathbf{X}) = 0$ *if and only if every sample observation is identical.*

Proofs of statements (ii) and (iii) are given in Section 12.7. See also chapter exercises, or Székely et al. [2007] for other proofs.

Theorem 12.3 (Properties of dCor). *If $E(|X|_p + |Y|_q) < \infty$, then*

(i) $0 \leq \mathcal{R} \leq 1$, *and* $\mathcal{R}(X, Y) = 0$ *if and only if X and Y are independent.*

(ii) $0 \leq \mathcal{R}_n \leq 1$.

(iii) $\mathcal{R}(X, Y) = 1$ *implies that the dimensions of the linear subspaces spanned by X and Y respectively are almost surely equal, and if we assume that these subspaces are equal, then in this subspace*

$$Y = a + bCX$$

for some vector a, non-zero real number b and orthonormal matrix C.

(iv) *If $\mathcal{R}_n(\mathbf{X}, \mathbf{Y}) = 1$, then there exists a vector a, a non-zero real number b and an orthonormal matrix C such that $\mathbf{Y} = a + b\mathbf{X}C$.*

Property (iii) shows that for uncorrelated (nondegenerate) random variables X, Y, the distance correlation cannot be 1. The fact that for uncorrelated random variables $dCor < 1$ is a very good property of dCor compared to some other measures of dependence. See Chapter 22 for discussion of several other measures of dependence and their properties. See Section 18.2 for a result on the best upper bound in a special case.

Proof. In (i), $\mathcal{R}(X,Y)$ exists whenever X and Y have finite first moments, and X and Y are independent if and only if the numerator

$$\mathcal{V}^2(X,Y) = \|\hat{f}_{X,Y}(t,s) - \hat{f}_X(t)\hat{f}_Y(s)\|^2$$

of $\mathcal{R}^2(X,Y)$ is zero. Let $U = e^{i\langle t,X\rangle} - \hat{f}_X(t)$ and $V = e^{i\langle s,Y\rangle} - \hat{f}_Y(s)$. Then

$$|\hat{f}_{X,Y}(t,s) - \hat{f}_X(t)\hat{f}_Y(s)|^2 = |E[UV]|^2 \le (E[|U||V|])^2 \le E[|U|^2|V|^2]$$
$$= (1-|\hat{f}_X(t)|^2)(1-|\hat{f}_Y(s)|^2).$$

Thus

$$\int_{\mathbb{R}^{p+q}} |\hat{f}_{X,Y}(t,s) - \hat{f}_X(t)\hat{f}_Y(s)|^2 \, d\omega \le \int_{\mathbb{R}^{p+q}} |(1-|\hat{f}_X(t)|^2)(1-|\hat{f}_Y(s)|^2)|^2 \, d\omega,$$

hence $0 \le \mathcal{R}(X,Y) \le 1$, and (ii) follows by a similar argument.

(iv): If $\mathcal{R}_n(\mathbf{X},\mathbf{Y}) = 1$, then the arguments below show that X and Y are similar almost surely, thus the dimensions of the linear subspaces spanned by $\mathbf{X}$ and $\mathbf{Y}$ respectively are almost surely equal. (Here *similar* means that $\mathbf{Y}$ and $\varepsilon\mathbf{X}$ are isometric for some $\varepsilon \ne 0$.) For simplicity we can suppose that $\mathbf{X}$ and $\mathbf{Y}$ are in the same Euclidean space and both span $\mathbb{R}^p$. From the Cauchy-Bunyakovski inequality it is easy to see that $\mathcal{R}_n(\mathbf{X},\mathbf{Y}) = 1$ if and only if $A_{kl} = \varepsilon B_{kl}$, for some factor ε. Suppose that $|\varepsilon| = 1$. Then

$$|X_k - X_l|_p = |Y_k - Y_l|_q + d_k + d_l,$$

for all k,l, for some constants d_k, d_l. Then with $k = l$ we obtain $d_k = 0$, for all k. Now, one can apply a geometric argument. The two samples are isometric, so $\mathbf{Y}$ can be obtained from $\mathbf{X}$ through operations of shift, rotation and reflection, and hence $\mathbf{Y} = a + b\mathbf{X}C$ for some vector a, $b = \varepsilon$ and orthonormal matrix C. If $|\varepsilon| \ne 1$ and $\varepsilon \ne 0$, apply the geometric argument to $\varepsilon\mathbf{X}$ and $\mathbf{Y}$ and it follows that $\mathbf{Y} = a + b\mathbf{X}C$ where $b = \varepsilon$. $\square$

Additional results and properties include:

1. Distance covariance and distance correlation satisfy a **uniqueness property**. Rigid motion invariance and scale equivariance properties imply that the weight function in

$$\|\hat{f}_{X,Y}(t,s) - \hat{f}_X(t)\hat{f}_Y(s)\|^2$$
$$= \int_{\mathbb{R}^{p+q}} |\hat{f}_{X,Y}(t,s) - \hat{f}_X(t)\hat{f}_Y(s)|^2 w(t,s) \, dt \, ds$$

must be $w(t,s) = \text{const}(|t|_p^{1+p}|s|_q^{1+q})^{-1}$ a.s. t,s. See Székely and Rizzo [2012] for details and proof.

2. Distance covariance is NOT covariance of distances, but (applying (12.7)) it can be expressed in terms of Pearson's covariance of distances as:

$$\mathcal{V}^2(X,Y) = \text{Cov}(|X-X'|_p, |Y-Y'|_q) - 2\,\text{Cov}(|X-X'|_p, |Y-Y''|_q).$$

It is interesting to note that $\text{Cov}(|X-X'|p, |Y-Y'|q) = 0$ does not imply independence of X and Y. Indeed, there is a simple 2- dimensional random variable (X,Y) such that X and Y are not independent, but $|X-X'|$ and $|Y-Y'|$ are uncorrelated. See Exercise 12.5.

Edelmann, Richards, and Vogel [2020] derive the population distance variance for several commonly applied distributions. We collect some of those results here for convenient reference.

Gaussian(μ, σ) $\qquad\qquad$ $\mathcal{V}^2(X) = 4\left[\frac{1-\sqrt{3}}{\pi} + \frac{1}{3}\right]\sigma^2(X).$ $\qquad$ (12.24)

Continuous Uniform(a,b) $\qquad$ $\mathcal{V}^2(X) = \frac{2[b-a]^2}{45} = \frac{24}{45}\sigma^2(X).$ $\qquad$ (12.25)

Exponential(rate λ) $\qquad\qquad$ $\mathcal{V}^2(X) = (3\lambda^2)^{-1} = \frac{\sigma^2(X)}{3}.$ $\qquad$ (12.26)

Edelmann et al. [2020] also prove that

$$\mathcal{V}(X) \le E|X-X'|,$$

and in the one-dimensional case

$$\mathcal{V}(X) \le \sigma(X),$$

where $\sigma(X)$ is the standard deviation of X. Thus, $\mathcal{V}$ is the smallest of the three measures of spread.

12.6 Distance Correlation for Gaussian Variables

If (X,Y) are jointly bivariate normal, there is a deterministic relation between $\mathcal{R}^2(X,Y)$ and the linear correlation $\rho(X,Y)$.

Theorem 12.4 (Székely et al. [2007], Theorem 7). *If X and Y are standard normal, with correlation $\rho = \rho(X,Y)$, then*

(i) $\mathcal{R}(X,Y) \le |\rho|,$

(ii) $\mathcal{R}^2(X,Y) = \dfrac{\rho \arcsin \rho + \sqrt{1-\rho^2} - \rho \arcsin(\rho/2) - \sqrt{4-\rho^2} + 1}{1 + \pi/3 - \sqrt{3}},$

(iii) $\displaystyle\inf_{\rho\neq 0} \frac{\mathcal{R}(X,Y)}{|\rho|} = \lim_{\rho\to 0} \frac{\mathcal{R}(X,Y)}{|\rho|} = \frac{1}{2(1+\pi/3-\sqrt{3})^{1/2}} \approx 0.89066.$

See Section 12.7.4 for the proof.

The relation between $\mathcal{R}$ and ρ derived in Theorem 12.4 is shown by the plot of $\mathcal{R}^2$ vs ρ^2 in Figure 12.2.

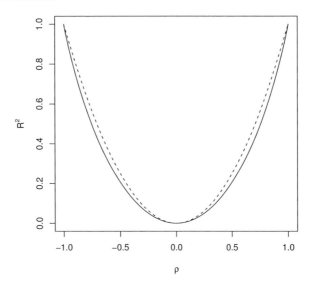

FIGURE 12.2
Dependence coefficient $\mathcal{R}^2$ (solid line) and correlation ρ^2 (dashed line) in the bivariate normal case.

12.7 Proofs

12.7.1 Finiteness of $\|\hat{f}_{X,Y}(t,s) - \hat{f}_X(t)\hat{f}_Y(s)\|^2$

For finiteness of $\|\hat{f}_{X,Y}(t,s) - \hat{f}_X(t)\hat{f}_Y(s)\|^2$ it is sufficient that $E|X|_p < \infty$ and $E|Y|_q < \infty$. By the Cauchy-Bunyakovsky inequality

$$|\hat{f}_{X,Y}(t,s) - \hat{f}_X(t)\hat{f}_Y(s)|^2 = [E(e^{i\langle t,X\rangle} - \hat{f}_X(t))(e^{i\langle s,Y\rangle} - \hat{f}_Y(s))]^2$$
$$\leq E[e^{i\langle t,X\rangle} - \hat{f}_X(t)]^2 E[e^{i\langle s,Y\rangle} - \hat{f}_Y(s)]^2 = (1 - |\hat{f}_X(t)|^2)(1 - |\hat{f}_Y(s)|^2).$$

If $E(|X|_p + |Y|_q) < \infty$, then by Lemma 12.1 and by Fubini's theorem it follows that

$$\int_{\mathbb{R}^{p+q}} |\hat{f}_{X,Y}(t,s) - \hat{f}_X(t)\hat{f}_Y(s)|^2 \, d\omega \leq \int_{\mathbb{R}^p} \frac{1 - |\hat{f}_X(t)|^2}{c_p|t|_p^{1+p}} \, dt \int_{\mathbb{R}^q} \frac{1 - |\hat{f}_Y(s)|^2}{c_q|s|_q^{1+q}} \, ds$$

$$\tag{12.27}$$

$$= E\left[\int_{\mathbb{R}^p} \frac{1 - \cos\langle t, X - X'\rangle}{c_p|t|_p^{1+p}} \, dt\right] \cdot E\left[\int_{\mathbb{R}^q} \frac{1 - \cos\langle s, Y - Y'\rangle}{c_q|s|_q^{1+q}} \, ds\right]$$

$$= E|X - X'|_p E|Y - Y'|_q < \infty.$$

$\square$

12.7.2 Proof of Theorem 12.1

Proof. Lemma 12.1 implies that there exist constants c_p and c_q such that for all X in $\mathbb{R}^p$, Y in $\mathbb{R}^q$,

$$\int_{\mathbb{R}^p} \frac{1 - \exp\{i\langle t, X\rangle\}}{|t|_p^{1+p}} \, dt = c_p |X|_p, \tag{12.28}$$

$$\int_{\mathbb{R}^q} \frac{1 - \exp\{i\langle s, Y\rangle\}}{|s|_q^{1+q}} \, ds = c_q |Y|_q, \tag{12.29}$$

$$\int_{\mathbb{R}^p} \int_{\mathbb{R}^q} \frac{1 - \exp\{i\langle t, X\rangle + i\langle s, Y\rangle\}}{|t|_p^{1+p}|s|_q^{1+q}} \, dt \, ds = c_p c_q |X|_p |Y|_q, \tag{12.30}$$

where the integrals are understood in the principal value sense. For simplicity, consider the case $p = q = 1$. The distance between the empirical characteristic functions in the weighted norm $w(t, s) = \pi^{-2} t^{-2} s^{-2}$ involves $|\hat{f}_{X,Y}^n(t, s)|^2$, $|\hat{f}_X^n(t)\hat{f}_Y^n(s)|^2$ and $\overline{\hat{f}_{X,Y}^n(t, s)} \hat{f}_X^n(t)\hat{f}_Y^n(s)$. For the first we have

$$\hat{f}_{X,Y}^n(t, s)\overline{\hat{f}_{X,Y}^n(t, s)} = \frac{1}{n^2} \sum_{k, l=1}^n \cos(X_k - X_l)t \, \cos(Y_k - Y_l)s + V_1,$$

where V_1 represents terms that vanish when integral $\|\hat{f}_{X,Y}^n(t, s) - \hat{f}_X^n(t)\hat{f}_Y^n(s)\|^2$ is evaluated. The second expression is

$$\hat{f}_X^n(t)\hat{f}_Y^n(s)\overline{\hat{f}_X^n(t)\hat{f}_Y^n(s)} = \frac{1}{n^2} \sum_{k, l=1}^n \cos(X_k - X_l)t \, \frac{1}{n^2} \sum_{k, l=1}^n \cos(Y_k - Y_l)s + V_2,$$

and the third is

$$\hat{f}_{X,Y}^n(t, s)\overline{\hat{f}_X^n(t)\hat{f}_Y^n(s)} = \frac{1}{n^3} \sum_{k, l, m=1}^n \cos(X_k - X_l)t \, \cos(Y_k - Y_m)s + V_3,$$

where V_2 and V_3 represent terms that vanish when the integral is evaluated. To evaluate the integral $\|\hat{f}_{X,Y}^n(t, s) - \hat{f}_X^n(t)\hat{f}_Y^n(s)\|^2$, apply Lemma 12.1, statements (12.28), (12.29), and (12.30) using

$$\cos u \, \cos v = 1 - (1 - \cos u) - (1 - \cos v) + (1 - \cos u)(1 - \cos v).$$

After cancellation in the numerator of the integrand there remain to evaluate integrals of the type

$$\int_{\mathbb{R}^2} (1 - \cos(X_k - X_l)t)(1 - \cos(Y_k - Y_l)s) \frac{dt \, ds}{t^2 \, s^2}$$

$$= \int_{\mathbb{R}} (1 - \cos(X_k - X_l)t) \frac{dt}{t^2} \times \int_{\mathbb{R}} (1 - \cos(Y_k - Y_l)s) \frac{ds}{s^2}$$

$$= c_1^2 |X_k - X_l||Y_k - Y_l|.$$

For random vectors X in $\mathbb{R}^p$ and Y in $\mathbb{R}^q$, the same steps are applied using $w(t,s) = \{c_p\, c_q |t|_p^{1+p} |s|_q^{1+q}\}^{-1}$. Thus

$$\|\hat{f}^n_{X,Y}(t,s) - \hat{f}^n_X(t)\hat{f}^n_Y(s)\|^2 = S_1 + S_2 - 2S_3, \tag{12.31}$$

where

$$S_1 = \frac{1}{n^2} \sum_{k,l=1}^{n} |X_k - X_l|_p |Y_k - Y_l|_q, \tag{12.32}$$

$$S_2 = \frac{1}{n^2} \sum_{k,l=1}^{n} |X_k - X_l|_p \, \frac{1}{n^2} \sum_{k,l=1}^{n} |Y_k - Y_l|_q, \tag{12.33}$$

$$S_3 = \frac{1}{n^3} \sum_{k=1}^{n} \sum_{l,m=1}^{n} |X_k - X_l|_p |Y_k - Y_m|_q. \tag{12.34}$$

To complete the proof we need to verify the algebraic identity

$$\mathcal{V}_n^2(\mathbf{X},\mathbf{Y}) = S_1 + S_2 - 2S_3. \tag{12.35}$$

Proof of statement (12.35)

By definition (12.9)

$$n^2 \mathcal{V}_n^2 = \sum_{k,l=1}^{n} \left\{ \begin{matrix} a_{kl}b_{kl} & -a_{kl}\bar{b}_{k.} & -a_{kl}\bar{b}_{.l} & +a_{kl}\bar{b}_{..} \\ -\bar{a}_{k.}b_{kl} & +\bar{a}_{k.}\bar{b}_{k.} & +\bar{a}_{k.}\bar{b}_{.l} & -\bar{a}_{k.}\bar{b}_{..} \\ -\bar{a}_{.l}b_{kl} & +\bar{a}_{.l}\bar{b}_{k.} & +\bar{a}_{.l}\bar{b}_{.l} & -\bar{a}_{.l}\bar{b}_{..} \\ +\bar{a}_{..}b_{kl} & -\bar{a}_{..}\bar{b}_{k.} & -\bar{a}_{..}\bar{b}_{.l} & +\bar{a}_{..}\bar{b}_{..} \end{matrix} \right\} \tag{12.36}$$

$$\begin{aligned}
= &\sum_{k,l} a_{kl}b_{kl} &&- \sum_k a_{k.}\bar{b}_{k.} &&- \sum_l a_{.l}\bar{b}_{.l} &&+ a_{..}\bar{b}_{..} \\
&- \sum_k \bar{a}_{k.}b_{k.} &&+ n\sum_k \bar{a}_{k.}\bar{b}_{k.} &&+ \sum_{k,l} \bar{a}_{k.}\bar{b}_{.l} &&- n\sum_k \bar{a}_{k.}\bar{b}_{..} \\
&- \sum_l \bar{a}_{.l}b_{.l} &&+ \sum_{k,l} \bar{a}_{.l}\bar{b}_{k.} &&+ n\sum_l \bar{a}_{.l}\bar{b}_{.l} &&- n\sum_l \bar{a}_{.l}\bar{b}_{..} \\
&+ \bar{a}_{..}\bar{b}_{..} &&- n\sum_k \bar{a}_{..}\bar{b}_{k.} &&- n\sum_l \bar{a}_{..}\bar{b}_{.l} &&+ n^2\bar{a}_{..}\bar{b}_{..},
\end{aligned}$$

where $a_{k.} = n\bar{a}_{k.}$, $a_{.l} = n\bar{a}_{.l}$, $b_{k.} = n\bar{b}_{k.}$, and $b_{.l} = n\bar{b}_{.l}$. Applying the identities

$$S_1 = \frac{1}{n^2} \sum_{k,l=1}^{n} a_{kl}b_{kl}, \tag{12.37}$$

$$S_2 = \frac{1}{n^2} \sum_{k,l=1}^{n} a_{kl} \, \frac{1}{n^2} \sum_{k,l=1}^{n} b_{kl} = \bar{a}_{..}\bar{b}_{..}, \tag{12.38}$$

$$n^2 S_2 = n^2 \bar{a}_{..}\bar{b}_{..} = \frac{a_{..}}{n} \sum_{l=1}^{n} \frac{b_{.l}}{n} = \sum_{k,l=1}^{n} \bar{a}_{k.}\bar{b}_{.l}, \tag{12.39}$$

$$S_3 = \frac{1}{n^3} \sum_{k=1}^{n} \sum_{l,m=1}^{n} a_{kl} b_{km} = \frac{1}{n^3} \sum_{k=1}^{n} a_{k.} b_{k.} = \frac{1}{n} \sum_{k=1}^{n} \bar{a}_{k.} \bar{b}_{k.}, \qquad (12.40)$$

to (12.36) we obtain

$$
n^2 \mathcal{V}_n^2 = \begin{array}{cccc}
n^2 S_1 & -n^2 S_3 & -n^2 S_3 & +n^2 S_2 \\
-n^2 S_3 & +n^2 S_3 & +n^2 S_2 & -n^2 S_2 \\
-n^2 S_3 & +n^2 S_2 & +n^2 S_3 & -n^2 S_2 \\
+n^2 S_2 & -n^2 S_2 & -n^2 S_2 & +n^2 S_2
\end{array} = n^2 (S_1 + S_2 - 2 S_3).
$$

Then (12.31) and (12.35) imply that $\mathcal{V}_n^2(\mathbf{X}, \mathbf{Y}) = \|\hat{f}_{X,Y}^n(t,s) - \hat{f}_X^n(t) \hat{f}_Y^n(s)\|^2$. $\qquad \square$

12.7.3   Proof of Theorem 12.2

Proof. Starting with the left side of the inequality (iii),

$$\mathcal{V}(X_1 + X_2, Y_1 + Y_2) = \|\hat{f}_{X_1+X_2,Y_1+Y_2}(t,s) - \hat{f}_{X_1+X_2}(t) \hat{f}_{Y_1+Y_2}(s)\|$$

$$= \|\hat{f}_{X_1,Y_1}(t,s) \hat{f}_{X_2,Y_2}(t,s) - \hat{f}_{X_1}(t) \hat{f}_{X_2}(t) \hat{f}_{Y_1}(s) \hat{f}_{Y_2}(s)\|$$

$$\leq \|\hat{f}_{X_1,Y_1}(t,s) \left(\hat{f}_{X_2,Y_2}(t,s) - \hat{f}_{X_2}(t) \hat{f}_{Y_2}(s) \right) \| \qquad (12.41)$$

$$+ \|\hat{f}_{X_2}(t) \hat{f}_{Y_2}(s) \left(\hat{f}_{X_1,Y_1}(t,s) - \hat{f}_{X_1}(t) \hat{f}_{Y_1}(s) \right) \|$$

$$\leq \|\hat{f}_{X_2,Y_2}(t,s) - \hat{f}_{X_2}(t) \hat{f}_{Y_2}(s)\| + \|\hat{f}_{X_1,Y_1}(t,s) - \hat{f}_{X_1}(t) \hat{f}_{Y_1}(s)\| \qquad (12.42)$$

$$= \mathcal{V}(X_1, Y_1) + \mathcal{V}(X_2, Y_2).$$

It is clear that if (a) X_1 and Y_1 are both constants, (b) X_2 and Y_2 are both constants, or (c) X_1, X_2, Y_1, Y_2 are mutually independent, then we have equality in (iii). Now suppose that we have equality in (iii), and thus we have equality above at (12.41) and (12.42), but neither (a) nor (b) hold. Then the only way we can have equality at (12.42) is if X_1, Y_1 are independent and also X_2, Y_2 are independent. But our hypothesis assumes that (X_1, Y_1) and (X_2, Y_2) are independent, hence (c) must hold.

Finally (vi) follows from (iii). In this special case $X_1 = Y_1 = X$ and $X_2 = Y_2 = Y$. Now (a) means that X is constant, (b) means that Y is constant, and (c) means that both of them are constants, because this is the only case when a random variable can be independent of itself. $\qquad \square$

See the exercises or Székely et al. [2007], Székely and Rizzo [2009a] for proofs of other properties in Theorem 12.2.

12.7.4 Proof of Theorem 12.4

Let X and Y have standard normal distributions with $\text{Cov}(X, Y) = \rho(X, Y) = \rho$. Introduce the function

$$F(\rho) = \int_{-\infty}^{\infty} \int_{-\infty}^{\infty} |\hat{f}_{X,Y}(t, s) - \hat{f}_X(t)\hat{f}_Y(s)|^2 \frac{dt \, ds}{t^2 \, s^2}.$$

Then $\mathcal{V}^2(X, Y) = F(\rho) / c_1^2 = F(\rho) / \pi^2$, and

$$\mathcal{R}^2(X, Y) = \frac{\mathcal{V}^2(X, Y)}{\sqrt{\mathcal{V}^2(X, X)\mathcal{V}^2(Y, Y)}} = \frac{F(\rho)}{F(1)}. \tag{12.43}$$

Proof. (i) If X and Y are standard normal with correlation ρ, then

$$F(\rho) = \int_{-\infty}^{\infty} \int_{-\infty}^{\infty} \left| e^{-(t^2+s^2)/2 - \rho ts} - e^{-t^2/2}e^{-s^2/2} \right|^2 \frac{dt \, ds}{t^2 \, s^2}$$

$$= \int_{\mathbb{R}^2} e^{-t^2-s^2}(1 - 2e^{-\rho ts} + e^{-2\rho ts}) \frac{dt \, ds}{t^2 \, s^2}$$

$$= \int_{\mathbb{R}^2} e^{-t^2-s^2} \sum_{n=2}^{\infty} \frac{2^n - 2}{n!}(-\rho ts)^n \frac{dt \, ds}{t^2 \, s^2}$$

$$= \int_{\mathbb{R}^2} e^{-t^2-s^2} \sum_{k=1}^{\infty} \frac{2^{2k} - 2}{(2k)!}(-\rho ts)^{2k} \frac{dt \, ds}{t^2 \, s^2}$$

$$= \rho^2 \left[\sum_{k=1}^{\infty} \frac{2^{2k} - 2}{(2k)!} \rho^{2(k-1)} \int_{\mathbb{R}^2} e^{-t^2-s^2}(ts)^{2(k-1)} dt \, ds \right].$$

Thus $F(\rho) = \rho^2 G(\rho)$, where $G(\rho)$ is a sum with all nonnegative terms. The function $G(\rho)$ is clearly non-decreasing in ρ and $G(\rho) \leq G(1)$. Therefore

$$\mathcal{R}^2(X, Y) = \frac{F(\rho)}{F(1)} = \rho^2 \frac{G(\rho)}{G(1)} \leq \rho^2,$$

or equivalently, $\mathcal{R}(X, Y) \leq |\rho|$.

(ii) Note that $F(0) = F'(0) = 0$ so $F(\rho) = \int_0^\rho \int_0^x F''(z)dz \, dx$. The second derivative of F is

$$F''(z) = \frac{d^2}{dz^2} \int_{\mathbb{R}^2} e^{-t^2-s^2}\left(1 - 2e^{-zts} + e^{-2zts}\right) \frac{dt \, ds}{t^2 \, s^2} = 4V(z) - 2V\left(\frac{z}{2}\right),$$

where

$$V(z) = \int_{\mathbb{R}^2} e^{-t^2-s^2-2zts} dt \, ds = \frac{\pi}{\sqrt{1-z^2}}.$$

Here we applied a change of variables, used the fact that the eigenvalues of the quadratic form $t^2 + s^2 + 2zts$ are $1 \pm z$, and $\int_{-\infty}^{\infty} e^{-t^2\lambda}dt = (\pi/\lambda)^{1/2}$. Then

$$F(\rho) = \int_0^\rho \int_0^x \left(\frac{4\pi}{\sqrt{1-z^2}} - \frac{2\pi}{\sqrt{1-z^2/4}} \right) dz\, dx$$

$$= 4\pi \int_0^\rho (\arcsin(x) - \arcsin(x/2))dx$$

$$= 4\pi \left(\rho \arcsin \rho + \sqrt{1-\rho^2} - \rho \arcsin(\rho/2) - \sqrt{4-\rho^2} + 1 \right),$$

and (12.43) imply (ii).

(iii) In the proof of (i) we have that $\mathcal{R}/|\rho|$ is a nondecreasing function of $|\rho|$, and $\lim_{|\rho|\to 0} \mathcal{R}(X,Y)/|\rho| = (1 + \pi/3 - \sqrt{3})^{-1/2}/2$ follows from (ii). □

12.8 Exercises

Exercise 12.1. Show that $dCor(X,Y)$ is scale invariant.

Exercise 12.2. Show that the double-centered distance matrix $\hat{A}$ of Definition 12.4 has the property that each of the row and column means of $\hat{A}$ are equal to zero.

Exercise 12.3. Prove that $\mathcal{V}_n(\mathbf{X}) = 0$ implies that all sample observations of $\mathbf{X}$ are identical.

Exercise 12.4. Prove that if $\mathbf{X}$ is an $n \times p$ data matrix, then $dVar_n(a + b\mathbf{X}C) = |b|\, dVar_n(\mathbf{X})$, for all constant vectors a in $\mathbb{R}^p$, scalars b, and $p \times p$ orthonormal matrices C.

Exercise 12.5. Is it true that $Cov(|X - X'|, |Y - Y'|) = 0$ implies that X and Y are independent? (Recall that X, X' are iid; Y, Y' are iid.)

Hint: It is not hard to show that the following example is a counterexample. Define the two-dimensional pdf

$$p(x,y) := (1/4 - q(x)q(y))I_{[-1,1]^2}(x,y)$$

with

$$q(x) := -(c/2)I_{[-1,0]} + (1/2)I_{(0,c)}$$

where $c := \sqrt{2} - 1$.

Exercise 12.6. Does the independence of $X - X'$ and $Y - Y'$ imply the independence of X and Y?

Exercise 12.7. Prove Proposition 12.1. Hint: Apply Lemma 12.1 and Fubini's theorem.

Exercise 12.8. Derive formula (12.24) for $\mathcal{V}^2(X)$ if X has a Gaussian(μ, σ) distribution. Hint: See the proof of Theorem 12.4.

Exercise 12.9. Derive formula (12.25) for $\mathcal{V}^2(X)$ if X has the continuous uniform distribution on $[a, b]$.

Exercise 12.10. Derive formula (12.26) for $\mathcal{V}^2(X)$ if X has an exponential distribution with rate λ.

13

Testing Independence

CONTENTS

One of the most important applications of distance covariance and distance correlation is the test of independence. Suppose that we have observed random samples from unspecified distributions $X \in \mathbb{R}^p$ and $Y \in \mathbb{R}^q$ and we want to test $H_0 : X, Y$ are independent vs $H_1 : X, Y$ are dependent. The observed samples are data vectors or data matrices $\mathbf{X}$ and $\mathbf{Y}$.

13.1   The Sampling Distribution of $n\mathcal{V}_n^2$

The dCov test of independence is based on the test statistic $n\mathcal{V}_n^2(\mathbf{X}, \mathbf{Y})$. A dCor test is based on the normalized statistic. Both tests are consistent against all dependent alternatives. Both tests reject the null hypothesis for large values of the test statistic; that is, the upper tail is significant.

DOI: 10.1201/9780429157158-13

The asymptotic properties of the statistics and limit of the sampling distribution of $n\mathcal{V}_n^2(\mathbf{X}, \mathbf{Y})$ as $n \to \infty$ follow in the next sections.

First, let us see an example with simulated data that shows the form of the sampling distribution of the test statistic $n\mathcal{V}_n^2(\mathbf{X}, \mathbf{Y})$ under each hypothesis.

Example 13.1. An example illustrating the shape of sampling distribution of $n\mathcal{V}_n^2$ under independence is shown in Figure 13.1(a). The simulated data are generated from independent standard lognormal distributions. A probability histogram of $n\mathcal{V}_n^2$ on 1000 generated samples size 100 is shown. The rejection region is in the upper tail, where the approximate 95th percentile is close to 6. Shown in Figure 13.1(b) is the sampling distribution for data generated from dependent lognormal samples. The curve is a kernel density estimate using the R `density` function.

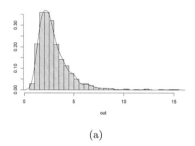 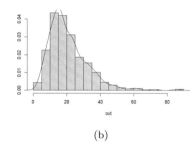

(a)   (b)

FIGURE 13.1
Sampling distribution of $n\,\mathrm{dCov}_n^2$ for (a) independent lognormal samples and (b) dependent lognormal samples.

We can clearly see that the test statistic in (b) is usually significant if we would consider $n\mathcal{V}_n^2 > 6$ to be significant.

By Corollary 13.1 we have that under H_0, $E[n\mathcal{V}_n^2(X, Y)]$ converges to the constant $E|X - X'||Y - Y'|$ as n increases. The sampling distribution under H_0 depends on the sampled distributions X and Y.

Although in this example we simulated the data from fully specified distributions, and could estimate a critical value, that is not the general problem that we need to solve. We need a nonparametric test for arbitrary observed samples $\mathbf{X}$ and $\mathbf{Y}$. To estimate the sampling distribution we apply a permutation or randomization test.

This example may help to visualize the meaning of Corollary 13.3. Although the limit distribution Q in Corollary 13.3 depends on the underlying distributions of X and Y, under independence the general shape of the distribution Q resembles the density estimate shown in Figure 13.1(a).

13.1.1 Expected Value and Bias of Distance Covariance

We derive the expected value of the V-statistic for (squared) distance covariance, $\mathcal{V}_n^2(X, Y)$ for X and Y in arbitrary dimensions, under independence or dependence. In the following, $|\cdot|$ denotes the Euclidean norm.

Proposition 13.1. *For random samples* (X_i, Y_i), $i = 1, \ldots, n$ *from the joint distribution* (X, Y) *in* $\mathbb{R}^p \times \mathbb{R}^q$,

$$E[\mathcal{V}_n^2(\mathbf{X}, \mathbf{Y})] = \frac{(n-1)(n-2)^2}{n^3} \mathcal{V}^2(X, Y) \tag{13.1}$$
$$+ \frac{2(n-1)^2}{n^3} \gamma - \frac{(n-1)(n-2)}{n^3} \alpha\beta,$$

where $\gamma = E|X - X'||Y - Y'|$, $\alpha = E|X - X'|$, *and* $\beta = E|Y - Y'|$. *If* X *and* Y *are independent,*

$$E[\mathcal{V}_n^2(\mathbf{X}, \mathbf{Y})] = \frac{n-1}{n^2} \gamma. \tag{13.2}$$

Corollary 13.1. $\mathcal{V}_n^2(\mathbf{X}, \mathbf{Y})$ *is asymptotically unbiased for* $\mathcal{V}^2(X, Y)$ *under independence or dependence of* (X, Y). *Under independence, the bias in* $\mathcal{V}_n^2$ *with respect to* $\mathcal{V}^2$ *is*

$$E[\mathcal{V}_n^2(\mathbf{X}, \mathbf{Y})] = \frac{n-1}{n^2} E|X - X'||Y - Y'|.$$

Corollary 13.1 shows that under independence, the expected value of the distance covariance test statistic is $E[n\mathcal{V}_n^2(\mathbf{X}, \mathbf{Y})] \to E|X - X'||Y - Y'|$ as n increases.

Corresponding to the positive bias in $\mathcal{V}_n^2$, the distance correlation $\mathcal{R}_n^2$ is positively biased for $\mathcal{R}^2(X, Y)$. For X, Y in high dimension, distances can be quite large, so the constant $E|X - X'||Y - Y'|$ can be large, depending on the underlying distributions.

For more details on the bias of distance correlation, particularly in high dimension, see Section 19.2 and the Numerical Illustration summarized in Table 19.1 from Székely and Rizzo [2013a]. The table shows empirically how the bias of dCor is increasing with dimension. A bias corrected distance correlation is introduced, $\mathcal{R}_n^*$. One can see in Table 19.1 how the bias corrected statistic is centered on 0 in all dimensions.

In Section 14.2.4 an unbiased estimator of squared distance covariance and a bias-corrected squared distance correlation are introduced. See Chapter 16 for more details and alternate computing formulas.

13.1.2 Convergence

The following theorem is an essential property of energy dependence coefficients.

Theorem 13.1. *If $E|X|_p < \infty$ and $E|Y|_q < \infty$, then almost surely*

$$\lim_{n\to\infty} \mathcal{V}_n(\mathbf{X}, \mathbf{Y}) = \mathcal{V}(X,Y). \qquad (13.3)$$

See Section 13.4.2 for the proof given in Székely et al. [2007].

Corollary 13.2. *If $E(|X|_p + |Y|_q) < \infty$, then almost surely,*

$$\lim_{n\to\infty} \mathcal{R}_n^2(\mathbf{X}, \mathbf{Y}) = \mathcal{R}^2(X,Y).$$

13.1.3   Asymptotic Properties of $n\mathcal{V}_n^2$

Our proposed test of independence is based on the statistic $n\mathcal{V}_n^2$. If $E(|X|_p + |Y|_q) < \infty$ we prove that under independence $n\mathcal{V}_n^2$ converges in distribution to a quadratic form

$$Q \overset{D}{=} \sum_{j=1}^{\infty} \lambda_j Z_j^2, \qquad (13.4)$$

where Z_j are independent standard normal random variables, and $\{\lambda_j\}$ are nonnegative constants that depend on the distribution of (X,Y). A test of independence that rejects independence for large $n\mathcal{V}_n^2$ is statistically consistent against all alternatives with finite first moments.

Let $\zeta(\cdot)$ denote a complex-valued zero mean Gaussian random process with covariance function

$$R(u, u_0) = \left(\hat{f}_X(t - t_0) - \hat{f}_X(t)\overline{\hat{f}_X(t_0)}\right)\left(\hat{f}_Y(s - s_0) - \hat{f}_Y(s)\overline{\hat{f}_Y(s_0)}\right),$$

where $u = (t,s), u_0 = (t_0, s_0) \in \mathbb{R}^p \times \mathbb{R}^q$.

Theorem 13.2 (Weak Convergence). *If X and Y are independent and $E(|X|_p + |Y|_q) < \infty$, then*

$$n\mathcal{V}_n^2 \xrightarrow[n\to\infty]{D} \|\zeta(t,s)\|^2.$$

A distance covariance test of independence based on $n\mathcal{V}_n^2$ or alternately on $n\mathcal{V}_n^2/S_2$ is consistent. In the *energy* package [Rizzo and Székely, 2022] we have applied the test statistic $n\mathcal{V}_n^2$ in dcov.test.

Corollary 13.3. *If $E(|X|_p + |Y|_q) < \infty$, then*

(i) *If X and Y are independent, $n\mathcal{V}_n^2 \xrightarrow[n\to\infty]{D} Q$ where Q is a nonnegative quadratic form of centered Gaussian random variables (13.4). The normalized statistic $n\mathcal{V}_n^2/S_2 \xrightarrow[n\to\infty]{D} Q$ where Q is a nonnegative quadratic form of centered Gaussian random variables (13.4) and $E[Q] = 1$.*

(ii) If X and Y are dependent, then $n\mathcal{V}_n^2/S_2 \xrightarrow[n\to\infty]{P} \infty$. and $n\mathcal{V}_n^2 \xrightarrow[n\to\infty]{P} \infty$.

To summarize, under independence $n\mathcal{V}_n^2(\mathbf{X}, \mathbf{Y})$ converges in distribution to a quadratic form $Q \overset{\mathcal{D}}{=} \sum_{j=1}^{\infty} \lambda_j Z_j^2$, where Z_j are independent standard normal random variables, and $\{\lambda_j\}$ are nonnegative constants that depend on the distribution of (X, Y) [Székely et al., 2007, Theorem 5]. Under dependence of (X, Y), $n\mathcal{V}_n^2(\mathbf{X}, \mathbf{Y}) \to \infty$ as $n \to \infty$, hence a test that rejects independence for large $n\mathcal{V}_n^2$ is consistent against dependent alternatives.

13.2   Testing Independence

A test of independence can be based on $n\mathcal{V}_n^2(\mathbf{X}, \mathbf{Y})$ (a dCov test) or $\mathcal{R}_n^2(\mathbf{X}, \mathbf{Y})$ (a dCor test). Values of the test statistic in the upper tail of the sampling distribution are significant.

In general, we need a nonparametric test because the eigenvalues in quadratic form $Q = \sum_{j=1}^{\infty} \lambda_j Z_j^2$ depend on the underlying distributions of X and Y. These distributions are generally unknown (if they were fully specified, then we already know whether they are independent). Typically the only information that we have is the observed sample data.

13.2.1   Implementation as a Permutation Test

To estimate the distribution of Q and obtain a test decision, one can use resampling. We apply a permutation test. Our data consists of paired observations (X_i, Y_i), $i = 1, \ldots, n$ where $X_i \in \mathbb{R}^p, Y_i \in \mathbb{R}^q$. Denote the index set $J = (1, \ldots, n)$ and consider a random permutation $\pi(J)$ of J. Let $\mathbf{Y}^{\pi(J)}$ be the re-indexed Y sample ordered by $\pi(J)$.

Under the null hypothesis of independence, the sampling distribution of $n\mathcal{V}_n^2(\mathbf{X}, \mathbf{Y})$ is the same as the sampling distribution $n\mathcal{V}_n^2(\mathbf{X}, \mathbf{Y}^{\pi(J)})$ for any random permutation $\pi(J)$. Generate R random permutations of $J = 1, \ldots, n$ and compute the R test statistics $n\mathcal{V}_n^{2(r)} := n\mathcal{V}_n^{2(r)}(\mathbf{X}, \mathbf{Y}^{\pi_r(J)})$, $r - 1, \ldots, R$.

For simplicity, denote the set of permutation replicates as $T^* = \{T^{(1)}, \ldots, T^{(R)}\}$, and our observed test statistic on the original data T_0. The replicates T^* are iid and generated under H_0, thus we have an estimate of the sampling distribution of $n\mathcal{V}_n^2$ if the null hypothesis is true. We use the replicates to determine whether our observed statistic T_0 is significant.

Note that T_0 is computed from one of the possible permutations of J. The p-value is computed by

$$p = \frac{1 + \sum_{r=1}^{R} I(T^{(r)} > T_0)}{1 + R}$$

and we reject independence at level α if $p \leq \alpha$. For more details on permutation tests, see e.g. Efron and Tibshirani [1993] and Davison and Hinkley [1997]. For examples in R including the distance covariance test see Rizzo [2019].

Remark 13.1 (Permutation test or randomization test). For a true permutation test, one generates all possible permutations of the index set, and this test is exact. This is rarely practical or feasible unless sample size n is very small. Instead we randomly generate a suitable number of permutations, so some authors prefer to use the term "randomization test." Here the number of replicates R can be chosen such that $(R + 1)\alpha$ is an integer.

Our permutation tests of independence control the Type I error rate at the nominal significance level. For empirical results on Type I error rates see Székely et al. [2007], Table 1.

We can also implement a dCor test as a permutation (randomization) test. Notice that the denominator of $\mathcal{R}_n$ is invariant under permutations of the sample indices, so the denominator is a constant. If we generated the same permutations, we would obtain the same p-value and test decision using either the dCov test or the dCor test.

These tests are implemented in the *energy* package for R as `dcov.test` and `dcor.test`. The test implementation is designed to be computationally efficient, taking advantage of the fact that the distance matrices never need to be re-computed.

13.2.2 Rank Test

In the case of bivariate (X, Y) one can also consider a distance covariance test of independence for rank(X), rank(Y), which has the advantage that it is distribution free and invariant with respect to monotone transformations of X and Y, but usually at a cost of lower power than the dCov(X,Y) test (see Example 14.1). The rank-dCov test can be applied to continuous or discrete data, but for discrete data it is necessary to use the correct method for breaking ties. Any ties in ranks should be broken randomly, so that a sample of size n is transformed to some permutation of the integers 1:n. A table of critical values for the statistic $n\mathcal{R}_n^2$, based on Monte Carlo results, is provided in Székely and Rizzo [2009a], Table 2.

13.2.3 Categorical Data

Energy inference also applies to categorical or nominal data, if the distances between observations are 0 or 1. Euclidean distances of course are not meaningful for more than two categories. Often we find categorical data encoded with integers; if that is the case, there is no part of computing distances or distance correlation that would fail (it is impossible for the software to understand the meaning as other than numerical). Unfortunately in this case the test would be meaningless. Users of the *energy* package must take care when working with categorical data to input distances such that distances between observations taking the same value are always 0, and otherwise must equal 1 (or some constant).

For this test, we construct a distance matrix on the group or category labels such that $d(X_i, X_j) = 0$ if $X_i = X_j$ and $d(X_i, X_j) = 1$ if $X_i \neq X_j$. If the 0-1 distance matrices of two samples are A and B, we can compute e.g.

```
dcor.test(as.dist(A), as.dist(B), R=999)
```

and this will be a valid energy test. Note that the functions dcov, dcor, dcov.test and dcor.test optionally accept distance *objects* as arguments. Here "distance object" refers to the value returned by the R function dist, and as.dist converts our 0-1 distance matrix to a object of class dist.

Note: For *energy* versions 1.7.11 and later, the dCor and dCov functions can also take a factor for either or both of the data arguments, and in this case the 0-1 distance matrices will be computed automatically for unordered factors.

13.2.4 Examples

Example 13.2 (Saviotti aircraft data, cont.). Refer to Examples 12.1 and 12.2. Here the dCor test of independence of log(Speed), log(Span) was highly significant with a p-value of .001. Now let us look into the details of the test. Figure 13.2 displays a histogram of the 1999 permutation replicates of the dCor test statistic. The vertical black line is at the value of the observed statistic 0.28045. The histogram is our estimate of the sampling distribution of the test statistic under independence. We could compute a critical value at the 95th percentile, say, which is 0.155 and we reject H_0 because $T_0 = .28045 > .155$, or simply compute a *p*-value as in dcor.test.

R code to generate this example using the *energy* package is as follows. Refer to Example 12.2 for the printed value of tst.

```
set.seed(2022)
tst <- dcor.test(Speed3, Span3, R=1999)
MASS::truehist(tst$replicates, xlab="permutation replicates")
abline(v = tst$statistic, lwd=2)
```

13.2.5 Power Comparisons

Below we summarize some power comparisons of the distance covariance test of independence with three classical tests for multivariate independence. See Székely et al. [2007] for more empirical results and power comparisons.

The likelihood ratio test (LRT) of the hypothesis $H_0 : \Sigma_{12} = 0$, with μ unknown, is based on

$$\frac{\det(S)}{\det(S_{11})\det(S_{22})} = \frac{\det(S_{22} - S_{21}S_{11}^{-1}S_{12})}{\det(S_{22})}, \qquad (13.5)$$

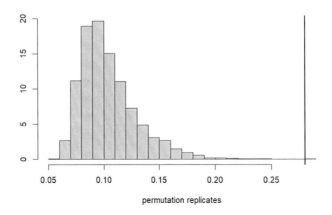

permutation replicates

FIGURE 13.2
Histogram of permutation replicates of the dCor test statistic in Example 13.2.

where $\det(\cdot)$ is the determinant, S, S_{11}, and S_{22} denote the sample covariances of $(\mathbf{X}, \mathbf{Y})$, $\mathbf{X}$, and $\mathbf{Y}$ respectively, and S_{12} is the sample covariance $\widehat{\mathrm{Cov}}(\mathbf{X}, \mathbf{Y})$. Under multivariate normality,

$$W = 2 \log \lambda = -n \log \det(I - S_{22}^{-1} S_{21} S_{11}^{-1} S_{12})$$

has the Wilks Lambda distribution: $\Lambda(q, n - 1 - p, p)$ [Wilks, 1935].

Puri and Sen [1971] (Chapter 8) proposed similar tests based on more general sample dispersion matrices $T = (T_{ij})$. The Puri-Sen tests replace S, S_{11}, S_{12} and S_{22} in (13.5) with T, T_{11}, T_{12}, and T_{22}. For example, T can be a matrix of Spearman's rank correlation statistics. For a sign test the dispersion matrix has entries $\frac{1}{n} \sum_{j=1}^{n} \mathrm{sign}(Z_{jk} - \widetilde{Z}_k)\mathrm{sign}(Z_{jm} - \widetilde{Z}_m)$, where $\widetilde{Z}_k$ is the sample median of the kth variable.

Critical values of the Wilks Lambda and Puri-Sen statistics are given by Bartlett's approximation: if n is large and $p, q > 2$, then $-(n - \frac{1}{2}(p + q + 3)) \log \det(I - S_{22}^{-1} S_{21} S_{11}^{-1} S_{12})$ has an approximate $\chi^2(pq)$ distribution [Mardia et al., 1979, Sec. 5.3.2b].

The LRT type tests measure linear or monotone association, while dCov and dCor tests measure all types of dependence. In the first example below, the data are correlated multivariate t distributions. This data has only linear correlation and we might expect that the LRT tests perform well against this type of alternative. Our results show that dCov is quite competitive with these tests, even when there is no nonlinear dependence. We also look at multivariate data that has non-linear dependence and see a dramatic difference with a clear advantage in power for dCov. These examples are part of a larger power study in Székely et al. [2007].

Each example compares the empirical power of the *dCov* test (labeled *V*) with the Wilks Lambda statistic (*W*), Puri-Sen rank correlation statistic (*S*) and Puri-Sen sign statistic (*T*). Empirical power is computed as the proportion of significant tests on 10000 random samples at significance level 0.1. For this study we used $\lfloor 200 + 5000/n \rfloor$ replicates.

Example 13.3. The marginal distributions of X and Y are $t(\nu)$ in dimensions $p = q = 5$, and $\text{Cov}(X_k, Y_l) = \rho$ for $k, l = 1, \ldots, 5$. This type of alternative has only linear dependence. The results displayed in Figure 13.3 summarize a power comparison for the case $\nu = 2$, with power estimated from 10000 test decisions for each of the sample sizes $n = 25{:}50{:}1, 55{:}100{:}5, 110{:}200{:}10$ with $\rho = 0.1$. Note: To generate correlated $t(\nu)$ data, see the *mvtnorm* package for R [Genz and Bretz, 2009].

Wilks LRT has inflated Type I error for $\nu = 1, 2, 3$, so a power comparison with W is not meaningful, particularly for $\nu = 1, 2$. The Puri-Sen rank test S performs well on this example of strictly linear dependence, with the dCov test performing almost equally well. In general, the dCov test has competitive power when compared with other tests against a wide range of alternatives. When there is also non-linear dependence we typically see that the dCov test is more powerful than competing tests, sometimes dramatically so as in the following examples.

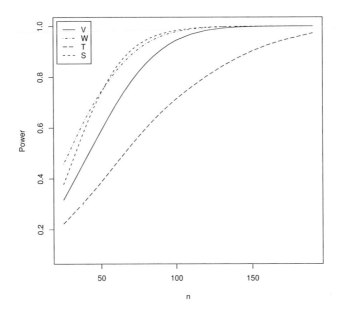

FIGURE 13.3

Example 13.3: Empirical power at 0.1 significance and sample size n, t(2) alternative.

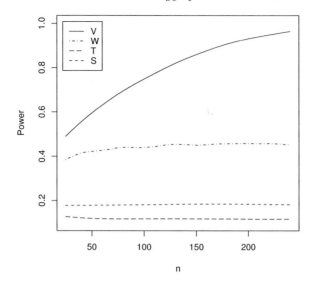

FIGURE 13.4
Example 13.4: Empirical power at 0.1 significance and sample size n against the alternative $Y = X\varepsilon$.

Example 13.4. This example has a more complicated dependence structure and there is nonlinear dependence in the data. We should expect to see the dCov test perform well, and indeed there is a dramatic advantage in power vs the LRT tests. The distribution of X is standard multivariate normal ($p = 5$), and $Y_{kj} = X_{kj}\varepsilon_{kj}$, $j = 1, \ldots, p$, where ε_{kj} are independent standard normal variables and independent of X. Comparisons are based on 10000 tests for each of the sample sizes $n = 25{:}50{:}1$, $55{:}100{:}5$, $110{:}240{:}10$. The results displayed in Figure 13.4 show that the *dCov* test is clearly superior to the LRT tests. This alternative is an example where the rank correlation and sign tests do not exhibit power increasing with sample size.

Example 13.5. Here we have another example of non-linear dependence, and again the dCov test dominates the LRT tests. The distribution of X is standard multivariate normal ($p = 5$), and $Y_{kj} = \log(X_{kj}^2)$, $j = 1, \ldots, p$. Comparisons are based on 10000 tests for each of the sample sizes $n = 25{:}50{:}1$, $55{:}100{:}5$. Simulation results are displayed in Figure 13.5. This is an example of a nonlinear relation where $n\mathcal{V}_n^2$ achieves very good power while none of the LRT type tests perform well.

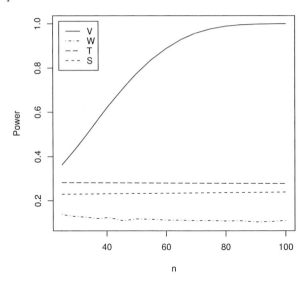

FIGURE 13.5
Example 13.5: Empirical power at 0.1 significance and sample size n against the alternative $Y = \log(X^2)$.

13.3 Mutual Independence

A population coefficient for mutual independence of d random variables, $d \geq 2$, is

$$\sum_{k=1}^{d-1} \mathcal{R}^2(X_k, [X_{k+1}, \ldots, X_d]),$$

which is non-negative and equals zero iff mutual independence holds. For example, if $d = 4$ the population coefficient is

$$\mathcal{R}^2(X_1, [X_2, X_3, X_4]) + \mathcal{R}^2(X_2, [X_3, X_4]) + \mathcal{R}^2(X_3, X_4),$$

A permutation test is easily implemented based on the corresponding sample coefficient. Recall that $\mathcal{R}_n^2$ is a biased estimator of $\mathcal{R}^2$, and the bias depends on dimension. Here we are adding statistics computed on (X, Y) in varying dimensions, so the appropriate statistic to implement the test is the bias-corrected squared dCor $\mathcal{R}_n^*$. For each term in the sum, we can permute the indices of the first variable before computing the sample $\mathcal{R}_n^*(X_k, [X_{k+1}, \ldots, X_d])$. This test is implemented in the *energy* package as `mutualIndep.test`.

It is also possible to define a population coefficient for mutual independence using a "jackknife" approach. Let $X_{(j)}$ denote the $d-1$ dimensional marginal distribution of $X \in \mathbb{R}^d$ such that the jth coordinate X_j of X is left out. We just need to test if $X_{(j)}$ is independent of X_j for $j = 1, 2, 3, \ldots, n$.

For more information on this topic see Yao et al. [2018] and Chakraborty and Zhang [2019]. A nice extension of distance covariance to more than two variables is the multivariance (see Böttcher et al. [2019]).

13.4   Proofs

13.4.1   Proof of Proposition 13.1

Proposition 13.1 asserts that:

$$E[\mathcal{V}_n^2(X,Y)] = \frac{(n-1)(n-2)^2}{n^3}\mathcal{V}^2(X,Y)$$
$$+ \frac{2(n-1)^2}{n^3}\gamma - \frac{(n-1)(n-2)}{n^3}\alpha\beta,$$

where $\gamma = E|X - X'||Y - Y'|$, $\alpha = E|X - X'|$, and $\beta = E|Y - Y'|$.

Proof. Let $(X_1, Y_1), \ldots, (X_n, Y_n)$ be a random sample from (X, Y), $X \in \mathbb{R}^p$, $Y \in \mathbb{R}^q$. The expectations involved in $E[\mathcal{V}_n]$ are linear combinations of $\gamma := E|X-X'||Y-Y'|$, $\alpha := E|X-X'|$, $\beta := E|Y-Y'|$, and $\delta := E|X-X'||Y-Y''|$.

We denote the distance matrices $A = (a_{ij})$ and $B = (b_{ij})$ for the X and Y samples, respectively. The row and column sums of A and B are denoted by

$$a_{i\cdot} = \sum_{\ell=1}^{n} a_{i\ell}, \quad a_{\cdot j} = \sum_{k=1}^{n} a_{kj}, \quad a_{\cdot\cdot} = \sum_{k,\ell=1}^{n} a_{k\ell},$$

$$b_{i\cdot} = \sum_{\ell=1}^{n} b_{i\ell}, \quad b_{\cdot j} = \sum_{k=1}^{n} b_{kj}, \quad b_{\cdot\cdot} = \sum_{k,\ell=1}^{n} b_{k\ell}.$$

Let

$$T_1 = \frac{1}{n^2}\sum_{i,j=1}^{n} a_{ij}b_{ij}; \quad T_2 = \frac{1}{n^3}\sum_{i,j,k=1}^{n} a_{ij}b_{ik} = \frac{1}{n^3}\sum_{i=1}^{n}(a_{i\cdot}b_{i\cdot});$$

$$T_3 = \frac{1}{n^2}\sum_{i,j=1}^{n} a_{ij} \cdot \frac{1}{n^2}\sum_{k,l=1}^{n} b_{kl} = \frac{1}{n^4}a_{\cdot\cdot}b_{\cdot\cdot}.$$

Then $\mathcal{V}_n^2(X,Y) = T_1 - 2T_2 + T_3$. It is clear that

$$E[T_1] = \frac{n(n-1)}{n^2}\gamma. \tag{13.6}$$

Expanding and simplifying T_2 gives

$$E[T_2] = \frac{n(n-1)}{n^3}[(n-2)\delta + \gamma]. \tag{13.7}$$

Expanding the product of the sums in T_3 we obtain

$$E[T_3] = \frac{n(n-1)}{n^4}[(n-2)(n-3)\alpha\beta + 2\gamma + 4(n-2)\delta]. \tag{13.8}$$

Now taking the expected value of $\mathcal{V}_n^2(X,Y)$, using (13.6)–(13.8), the terms with mean δ can be absorbed into $\mathcal{V}^2(X,Y)$, and after simplifying we have

$$E[\mathcal{V}_n^2(X,Y)] = \frac{(n-1)(n-2)^2}{n^3}\mathcal{V}^2(X,Y) + \frac{2(n-1)^2}{n^3}\gamma - \frac{(n-1)(n-2)}{n^3}\alpha\beta.$$

If X and Y are independent, $\mathcal{V}^2(X,Y) = 0$ and $\gamma = \alpha\beta$. Hence (13.1) becomes

$$E[\mathcal{V}_n^2(X,Y)] = \frac{n^2-n}{n^3}\gamma = \frac{n-1}{n^2}E|X-X'||Y-Y'|.$$

$\square$

13.4.2 Proof of Theorem 13.1

Proof. Define

$$\xi_n(t,s) = \frac{1}{n}\sum_{k=1}^{n} e^{i\langle t,X_k\rangle + i\langle s,Y_k\rangle} - \frac{1}{n}\sum_{k=1}^{n} e^{i\langle t,X_k\rangle}\frac{1}{n}\sum_{k=1}^{n} e^{i\langle s,Y_k\rangle},$$

so that $\mathcal{V}_n^2 = \|\xi_n(t,s)\|^2$. Then after elementary transformations

$$\xi_n(t,s) = \frac{1}{n}\sum_{k=1}^{n} u_k v_k - \frac{1}{n}\sum_{k=1}^{n} u_k \frac{1}{n}\sum_{k=1}^{n} v_k,$$

where $u_k = \exp(i\langle t,X_k\rangle) - \hat{f}_X(t)$, and $v_k = \exp(i\langle s,Y_k\rangle) - \hat{f}_Y(s)$.
For each $\delta > 0$ define the region

$$D(\delta) = \{(t,s) : \delta \le |t|_p \le 1/\delta, \ \delta \le |s|_q \le 1/\delta\} \tag{13.9}$$

and random variables

$$\mathcal{V}_{n,\delta}^2 = \int_{D(\delta)} |\xi_n(t,s)|^2 d\omega.$$

For any fixed $\delta > 0$, the weight function $w(t,s)$ is bounded on $D(\delta)$. Hence $\mathcal{V}_{n,\delta}^2$ is a combination of V-statistics of bounded random variables. For each

$\delta > 0$ by the strong law of large numbers (SLLN) for V-statistics, it follows that almost surely

$$\lim_{n\to\infty} \mathcal{V}_{n,\delta}^2 = \mathcal{V}_{\cdot,\delta}^2 = \int_{D(\delta)} |\hat{f}_{X,Y}(t,s) - \hat{f}_X(t)\hat{f}_Y(s)|^2 d\omega.$$

Clearly $\mathcal{V}_{\cdot,\delta}^2$ converges to $\mathcal{V}^2$ as δ tends to zero. Now it remains to prove that almost surely

$$\limsup_{\delta\to 0}\limsup_{n\to\infty} |\mathcal{V}_{n,\delta}^2 - \mathcal{V}_n^2| = 0. \tag{13.10}$$

For each $\delta > 0$

$$|\mathcal{V}_{n,\delta}^2 - \mathcal{V}_n^2| \le \int_{|t|_p<\delta} |\xi_n(t,s)|^2 d\omega + \int_{|t|_p>1/\delta} |\xi_n(t,s)|^2 d\omega$$
$$+ \int_{|s|_q<\delta} |\xi_n(t,s)|^2 d\omega + \int_{|s|_q>1/\delta} |\xi_n(t,s)|^2 d\omega. \tag{13.11}$$

For $z = (z_1, z_2, ..., z_p)$ in $\mathbb{R}^p$ define the function

$$G(y) = \int_{|z|<y} \frac{1-\cos z_1}{|z|^{1+p}} dz.$$

Clearly $G(y)$ is bounded by c_p and $\lim_{y\to 0} G(y) = 0$. Applying the inequality $|x+y|^2 \le 2|x|^2 + 2|y|^2$ and the Cauchy-Bunyakovsky inequality for sums, one can obtain that

$$|\xi_n(t,s)|^2 \le 2\left|\frac{1}{n}\sum_{k=1}^n u_k v_k\right|^2 + 2\left|\frac{1}{n}\sum_{k=1}^n u_k \frac{1}{n}\sum_{k=1}^n v_k\right|^2$$
$$\le \frac{4}{n}\sum_{k=1}^n |u_k|^2 \frac{1}{n}\sum_{k=1}^n |v_k|^2. \tag{13.12}$$

Hence the first summand in (13.11) satisfies

$$\int_{|t|_p<\delta} |\xi_n(t,s)|^2 d\omega \le \frac{4}{n}\sum_{k=1}^n \int_{|t|_p<\delta} \frac{|u_k|^2 dt}{c_p|t|_p^{1+p}} \frac{1}{n}\sum_{k=1}^n \int_{\mathbb{R}^q} \frac{|v_k|^2 ds}{c_q|s|_q^{1+q}}. \tag{13.13}$$

Here $|v_k|^2 = 1 + |\hat{f}_Y(s)|^2 - e^{i\langle s,Y_k\rangle}\overline{\hat{f}_Y(s)} - e^{-i\langle s,Y_k\rangle}\hat{f}_Y(s)$, thus

$$\int_{\mathbb{R}^q} \frac{|v_k|^2 ds}{c_q|s|_q^{1+q}} = (2E_Y|Y_k - Y| - E|Y - Y'|) \le 2(|Y_k| + E|Y|),$$

where the expectation E_Y is taken with respect to Y, and $Y' \overset{D}{=} Y$ is independent of Y_k. Further, after a suitable change of variables

$$\int_{|t|_p<\delta} \frac{|u_k|^2 dt}{c_p|t|_p^{1+p}} = 2E_X|X_k - X|G(|X_k - X|\delta) - E|X - X'|G(|X - X'|\delta)$$
$$\le 2E_X|X_k - X|G(|X_k - X|\delta),$$

where the expectation E_X is taken with respect to X, and $X' \overset{D}{=} X$ is independent of X_k. Therefore, from (13.13)

$$\int_{|t|_p < \delta} |\xi_n(t, s)|^2 d\omega \leq 4 \frac{2}{n} \sum_{k=1}^{n} (|Y_k| + E|Y|) \frac{2}{n} \sum_{k=1}^{n} E_X |X_k - X| G(|X_k - X|\delta).$$

By the SLLN

$$\limsup_{n \to \infty} \int_{|t|_p < \delta} |\xi_n(t, s)|^2 d\omega \leq 4 \cdot 2 \cdot 2E|Y| \cdot 2E|X_1 - X_2| G(|X_1 - X_2|\delta)$$

almost surely. Therefore, by the Lebesque bounded convergence theorem for integrals and expectations

$$\limsup_{\delta \to 0} \limsup_{n \to \infty} \int_{|t|_p < \delta} |\xi_n(t, s)|^2 d\omega = 0$$

almost surely.

Consider now the second summand in (13.11). Inequalities (13.12) imply that $|u_k|^2 \leq 4$ and $\frac{1}{n} \sum_{k=1}^{n} |u_k|^2 \leq 4$, hence

$$\int_{|t|_p > 1/\delta} \frac{|u_k|^2 dt}{c_p |t|_p^{1+p}} \leq 16 \int_{|t|_p > 1/\delta} \frac{dt}{c_p |t|_p^{1+p}} \int_{\mathbb{R}^q} \frac{1}{n} \sum_{k=1}^{n} |v_k|^2 \frac{ds}{c_q |s|_q^{1+q}}$$

$$\leq 16 \, \delta \, \frac{2}{n} \sum_{k=1}^{n} (|Y_k| + E|Y|).$$

Thus, almost surely

$$\limsup_{\delta \to 0} \limsup_{n \to \infty} \int_{|t|_p > 1/\delta} |\xi_n(t, s)|^2 d\omega = 0.$$

One can apply a similar argument to the remaining summands in (13.11) to obtain (13.10). $\square$

13.4.3 Proof of Corollary 13.3

Proof. (i) The independence of X and Y implies that ζ_n and thus ζ is a zero mean process. According to Kuo [1975] Section 1.2, the squared norm $\|\zeta\|^2$ of the zero mean Gaussian process ζ has the representation

$$\|\zeta\|^2 \overset{D}{=} \sum_{j=1}^{\infty} \lambda_j Z_j^2, \qquad (13.14)$$

where Z_j are independent standard normal random variables, and the non-negative constants $\{\lambda_j\}$ depend on the distribution of (X, Y). Hence, under independence, nV_n^2 converges in distribution to a quadratic form (13.14). It follows from (12.27) that

$$E\|\zeta\|^2 = \int_{\mathbb{R}^{p+q}} R(u, u)\, d\omega = \int_{\mathbb{R}^{p+q}} (1 - |\hat{f}_X(t)|^2)(1 - |\hat{f}_Y(s)|^2)\, d\omega$$
$$= E(|X - X'|_p |Y - Y'|_q).$$

By the SLLN for V-statistics $S_2 \xrightarrow[n\to\infty]{a.s.} E(|X - X'|_p |Y - Y'|_q)$. Therefore $nV_n^2/S_2 \xrightarrow[n\to\infty]{D} Q$, where $E[Q] = 1$ and Q is the quadratic form (13.4).

(ii) Suppose that X and Y are dependent and $E(|X|_p + |Y|_q) < \infty$. Then $V(X, Y) > 0$, Theorem 13.1 implies that $V_n^2(\mathbf{X}, \mathbf{Y}) \xrightarrow[n\to\infty]{a.s.} V^2(X, Y) > 0$, and therefore $nV_n^2(\mathbf{X}, \mathbf{Y}) \xrightarrow[n\to\infty]{P} \infty$. By the SLLN, S_2 converges to a constant and therefore $nV_n^2/S_2 \xrightarrow[n\to\infty]{P} \infty$. $\square$

13.4.4   Proof of Theorem 13.2

Theorem 13.2. Define the empirical process

$$\zeta_n(u) = \zeta_n(t, s) = \sqrt{n}\xi_n(t, s) = \sqrt{n}(\hat{f}_{X,Y}^n(t, s) - \hat{f}_X^n(t)\hat{f}_Y^n(s)).$$

Under the independence hypothesis, $E[\zeta_n(u)] = 0$, and $E[\zeta_n(u)\, \overline{\zeta_n(u_0)}] = \frac{n-1}{n} R(u, u_0)$. In particular, $E|\zeta_n(u)|^2 = \frac{n-1}{n}(1 - |\hat{f}_X(t)|^2)(1 - |\hat{f}_Y(s)|^2) \le 1$.

For each $\delta > 0$ we construct a sequence of random variables $\{Q_n(\delta)\}$ with the following properties:

1. $Q_n(\delta)$ converges in distribution to a random variable $Q(\delta)$.

2. $E|Q_n(\delta) - \zeta_n| \le \delta$.

3. $E|Q(\delta) - \zeta| \le \delta$.

Then the weak convergence of $\|\zeta_n\|^2$ to $\|\zeta\|^2$ follows from the convergence of the corresponding characteristic functions.

The sequence $Q_n(\delta)$ is defined as follows. Given $\epsilon > 0$, choose a partition $\{D_k\}_{k=1}^N$ of $D(\delta)$ (13.9) into $N = N(\epsilon)$ measurable sets with diameter at most ϵ. Define

$$Q_n(\delta) = \sum_{k=1}^{N} \int_{D_k} |\zeta_n|^2 d\omega.$$

For a fixed $M > 0$ let

$$\beta(\epsilon) = \sup_{u, u_0} E\left||\zeta_n(u)|^2 - |\zeta_n(u_0)|^2\right|,$$

where the supremum is taken over all $u = (t, s)$ and $u_0 = (t_0, s_0)$ such that $\max\{|t|, |t_0|, |s|, |s_0|\} < M$, and $|t - t_0|^2 + |s - s_0|^2 < \epsilon^2$. Then $\lim_{\epsilon \to 0} \beta(\epsilon) = 0$ for every fixed $M > 0$, and for fixed $\delta > 0$

$$E \left| \int_{D(\delta)} |\zeta_n(u)|^2 d\omega - Q_n(\delta) \right| \leq \beta(\epsilon) \int_D |\zeta_n(u)|^2 d\omega \xrightarrow[\epsilon \to 0]{} = 0.$$

On the other hand

$$\left| \int_D |\zeta_n(u)|^2 d\omega - \int_{\mathbb{R}^{p+q}} |\zeta_n(u)|^2 d\omega \right| \leq \int_{|t|<\delta} |\zeta_n(u)|^2 d\omega + \int_{|t|>1/\delta} |\zeta_n(u)|^2 d\omega$$

$$+ \int_{|s|<\delta} |\zeta_n(u)|^2 d\omega + \int_{|s|>1/\delta} |\zeta_n(u)|^2 d\omega.$$

By similar steps as in the proof of Theorem 13.1, one can derive that

$$E \left[\int_{|t|<\delta} |\zeta_n(u)|^2 d\omega + \int_{|t|>1/\delta} |\zeta_n(u)|^2) d\omega \right]$$

$$\leq \frac{n-1}{n} \left(E|X_1 - X_2|G(|X_1 - X_2|\delta) + w_p \delta \right) E|Y_1 - Y_2|\delta \xrightarrow[\delta \to 0]{} 0,$$

where w_p is a constant depending only on p, and similarly

$$E \left[\int_{|s|<\delta} |\zeta_n(u)|^2 d\omega + \int_{|s|>1/\delta} |\zeta_n(u)|^2) d\omega \right] \xrightarrow[\delta \to 0]{} 0.$$

Similar inequalities also hold for the random process $\zeta(t, s)$ with

$$Q(\delta) = \sum_{k=1}^N \int_{D_k} |\zeta(u)|^2 d\omega.$$

The weak convergence of $Q_n(\delta)$ to $Q(\delta)$ as $n \to \infty$ follows from the multivariate central limit theorem, and therefore $n\mathcal{V}_n^2 = \|\zeta_n\|^2 \xrightarrow[n \to \infty]{D} \|\zeta\|^2$. □

13.5   Exercises

Exercise 13.1. Let $\mathbf{X}$ be a sample $\{X_1, \ldots, X_n\}$, and let $A = (a_{ij})$ be the matrix of pairwise Euclidean distances $a_{ij} = |X_i - X_j|$. Suppose that π is a permutation of the integers $\{1, \ldots, n\}$. In a permutation test we re-index one of the samples under a random permutation π. For computing $\mathcal{V}_n^2$, etc., it is equivalent to re-index both the rows and columns of A by π. Are the re-indexing operation and the double-centering operation interchangeable? That is, do we get the same square matrix under (1) and (2)?

(1) Re-index A then double-center the result.

(2) Double-center A then re-index the result.

Hint: First compare (1) and (2) on a toy example empirically, to see what to prove or disprove. In R code, if p=sample(1:n) is the permutation vector, then A[p, p] is the re-indexed distance matrix.

Exercise 13.2. Let $\widehat{A}$ be a double-centered distance matrix. Prove or disprove the following statements:

1. If B is the matrix obtained by double-centering $\widehat{A}$, then $B = \widehat{A}$.

2. If c is a constant and B denotes the matrix obtained by adding c to the off-diagonal elements of $\widehat{A}$, then $\widehat{B} = \widehat{A}$.

Compare your conclusions to Lemma 17.1.

14

Applications and Extensions

CONTENTS

14.1 Applications

14.1.1 Nonlinear and Non-monotone Dependence

Suppose that one wants to test the independence of X and Y, where X and Y cannot be observed directly, but can only be measured with independent errors. Consider the following.

DOI: 10.1201/9780429157158-14

(i) Suppose that X_i can only be measured through observation of $A_i = X_i + \varepsilon_i$, where ε_i are independent of X_i, and similarly for Y_i.

(ii) One can only measure (non) random functions of X and Y, e.g. $A_i = \phi(X_i)$ and $B_i = \psi(Y_i)$.

(iii) Suppose both (i) and (ii) for certain types of random ϕ and ψ.

In all of these cases, even if (X, Y) were jointly normal, the dependence between (A, B) can be such that the correlation of A and B is almost irrelevant, but $\mathrm{dCor}(A, B)$ is obviously relevant.

Example 14.1. This example is similar to the type considered in (ii), with observed data from the NIST Statistical Reference Datasets (NIST StRD) for Nonlinear Regression. The data analyzed is *Eckerle4*, data from an NIST study of circular interference transmittance [Eckerle, 1979]. There are 35 observations, the response variable is transmittance, and the predictor variable is wavelength. A plot of the data in Figure 14.1(a) reveals that there is a nonlinear relation between wavelength and transmittance. The proposed nonlinear model is

$$y = f(x; \beta) + \varepsilon = \frac{\beta_1}{\beta_2} \exp\left\{\frac{-(x - \beta_3)^2}{2\beta_2^2}\right\} + \varepsilon,$$

where $\beta_1, \beta_2 > 0$, $\beta_3 \in \mathbb{R}$, and ε is random error. In the hypothesized model, Y depends on the density of X.

Results of the dCov test of independence of wavelength and transmittance are

```
                dCov test of independence
    data:  x and y
    nV^2 = 8.1337, p-value = 0.021
    sample estimates:
          dCor
    0.4275431
```

with $\mathcal{R}_n \doteq 0.43$, and dCov is significant (p-value $= 0.021$) based on 999 replicates. In contrast, neither Pearson correlation $\hat{\rho} = 0.0356$, (p-value $= 0.839$) nor Spearman rank correlation $\hat{\rho}_s = 0.0062$ (p-value $= 0.9718$) detects the nonlinear dependence between wavelength and transmittance, even though the relation in Figure 14.1(a) appears to be nearly deterministic.

The certified estimates (best solution found) for the parameters are reported by NIST as $\hat{\beta}_1 \doteq 1.55438$, $\hat{\beta}_2 \doteq 4.08883$, and $\hat{\beta}_3 \doteq 451.541$. The residuals of the fitted model are easiest to analyze when plotted vs the predictor variable as in Figure 14.1(b). Comparing residuals and transmittance,

```
                dCov test of independence
    data:  y and res
    nV^2 = 0.0019, p-value = 0.019
    sample estimates:
          dCor
    0.4285534
```

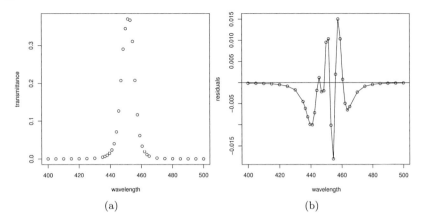

FIGURE 14.1
The Eckerle4 data (a) and plot of residuals vs predictor variable for the NIST
certified estimates (b), in Example 14.1.

we have $\mathcal{R}_n \doteq 0.43$ and the dCov test is significant (p-value $= 0.019$) based
on 999 replicates. Again the Pearson correlation is non-significant ($\hat{\rho} \doteq 0.11$,
p-value $= 0.5378$).

Although nonlinear dependence is clearly evident in both plots, note that
the methodology applies to multivariate analysis as well, for which residual
plots are much less informative.

Example 14.2. In the model specification of Example 14.1, the response
variable Y is assumed to be proportional to a normal density plus random
error. For simplicity, consider $(X, Y) = (X, \phi(X))$, where X is standard normal and $\phi(\cdot)$ is the standard normal density. Results of a Monte Carlo power
comparison of the dCov test with classical Pearson correlation and Spearman
rank tests are shown in Figure 14.2. The power estimates are computed as the
proportion of significant tests out of 10,000 at 10% significance level.

In this example, where the relation between X and Y is deterministic but
not monotone, it is clear that the dCov test is superior to product moment
correlation tests. Statistical consistency of the dCov test is evident, as its
power increases to 1 with sample size, while the power of correlation tests
against this alternative remains approximately level across sample sizes. We
also note that distance correlation applied to ranks of the data is more powerful
in this example than either correlation test, although somewhat less powerful
than the dCov test on the original (X, Y) data.

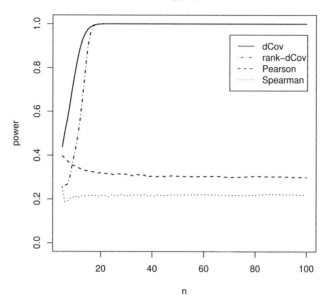

FIGURE 14.2
Example 14.2: Empirical power at 0.1 significance and sample size n.

14.1.2   Identify and Test for Nonlinearity

Example 14.3. In this example we illustrate how to isolate the nonlinear dependence between random vectors to test for nonlinearity.

Gumbel's bivariate exponential distribution [Gumbel, 1961] has density function

$$f(x, y; \theta) = [(1 + \theta x)(1 + \theta y)] \exp(-x - y - \theta xy), \qquad x, y > 0; 0 \le \theta \le 1.$$

The marginal distributions are standard exponential, so there is a strong nonlinear, but monotone dependence relation between X and Y. The conditional density is

$$f(y|x) = e^{-(1+\theta x)y}[(1 + \theta x)(1 + \theta y) - \theta], \qquad y > 0.$$

If $\theta = 0$, then $\hat{f}_{X,Y}(x, y) = \hat{f}_X(x)\hat{f}_Y(y)$ and independence holds, so $\rho = 0$. At the opposite extreme if $\theta = 1$, then $\rho = -0.40365$ (see [Kotz et al., 2000, Sec. 2.2]). Simulated data was generated using the conditional distribution function approach outlined in Johnson [1987]. Empirical power of dCov and correlation tests for the case $\theta = 0.5$ are compared in Figure 14.3(a), estimated from 10,000 test decisions each for sample sizes {10:100(10), 120:200(20), 250, 300}. This comparison reveals that the correlation test is more powerful than dCov against this alternative, which is not unexpected because $E[Y|X = x] = (1 + \theta + x\theta)/(1 + x\theta)^2$ is monotone.

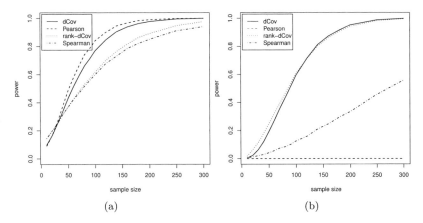

FIGURE 14.3
Power comparison of dCov and correlation tests at 10% significance level for Gumbel's bivariate exponential distribution in Example 14.3.

While we cannot split the dCor or dCov coefficient into linear and nonlinear components, we can extract correlation first and then compute dCor on the residuals. In this way one can separately analyze the linear and nonlinear components of bivariate or multivariate dependence relations.

To extract the linear component of dependence, fit a linear model $Y = X\beta + \varepsilon$ to the sample $(\mathbf{X}, \mathbf{Y})$ by ordinary least squares. It is not necessary to test whether the linear relation is significant. The residuals $\hat{\varepsilon}_i = X_i\hat{\beta} - Y_i$ are uncorrelated with the predictors $\mathbf{X}$. Apply the dCov test of independence to $(\mathbf{X}, \hat{\varepsilon})$.

Returning to the Gumbel bivariate exponential example, we have extracted the linear component and applied dCov to the residuals of a simple linear regression model. Repeating the power comparison described above on $(\mathbf{X}, \hat{\varepsilon})$ data, we obtained the power estimates shown in Figure 14.3(b). One can note that power of dCov tests is increasing to 1 with sample size, exhibiting statistical consistency against the nonlinear dependence remaining in the residuals of the linear model.

This procedure is easily applied in arbitrary dimension. One can fit a linear multiple regression model or a model with multivariate response to extract the linear component of dependence. This has important practical applications for evaluating models in higher dimension.

14.1.3 Exploratory Data Analysis

Example 14.4. This example compares dCor and Pearson correlation in exploratory data analysis. Consider the Freedman [1975] census data from

[United States Bureau of the Census, 1970] on crime rates in U. S. metropolitan areas with 1968 populations of 250,000 or more. The data set is available from the *car* package for R [Fox and Weisberg, 2019], and contains four numeric variables:

population (total 1968, in thousands)

nonwhite (percent nonwhite population, 1960)

density (population per square mile, 1968)

crime (crime rate per 100,000, 1969)

The 110 observations contain missing values. The data analyzed are the 100 cities with complete data. Pearson $\hat{\rho}$ and dCor statistics $\mathcal{R}_n$ are shown in Table 14.1. Note that there is a significant association between crime and population density measured by dCor, which is not significant when measured by $\hat{\rho}$.

TABLE 14.1
Pearson correlation and distance correlation statistics for the Freedman data of Example 14.4. Significance at $0.05, 0.01, 0.001$ for the corresponding tests is indicated by $*, **, ***$, respectively.

	Pearson			dCor		
	nonwhite	density	crime	nonwhite	density	crime
population	0.070	0.368***	0.396***	0.260*	0.615***	0.422**
nonwhite		0.002	0.294**		0.194	0.385***
density			0.112			0.250*

Analysis of this data continues in Example 14.5.

14.1.4 Identify Influential Observations

Example 14.5. (Influential observations) When $\mathcal{V}_n$ and $\mathcal{R}_n$ are computed using formula (12.9), it is straightforward to apply a jackknife procedure to identify possible influential observations or to estimate standard error of $\mathcal{V}_n$ or $\mathcal{R}_n$. A "leave-one-out" sample corresponds to $(n-1) \times (n-1)$ matrices $A_{(i)kl}$ and $B_{(i)kl}$, where the subscript (i) indicates that the i-th observation is left out. Then $\widehat{A}_{(i)kl}$ is computed from distance matrix $A = (a_{kl})$ by omitting the i-th row and the i-th column of A, and similarly $\widehat{B}_{(i)kl}$ is computed from $B = (b_{kl})$ by omitting the i-th row and the i-th column of B. Then

$$\mathcal{V}_{(i)}^2(\mathbf{X}, \mathbf{Y}) = \frac{1}{(n-1)^2} \sum_{k,l \neq i} \widehat{A}_{(i)kl} \widehat{B}_{(i)kl}, \qquad i = 1, \ldots, n$$

are the jackknife replicates of $\mathcal{V}_n^2$, obtained without re-computing matrices A and B. Similarly, $\mathcal{R}_{(i)}^2$ can be computed from the matrices A and B. A jackknife estimate of the standard error of $\mathcal{R}_n$ is thus easily obtained from the replicates $\{\mathcal{R}_{(i)}\}$ (on the jackknife see, e.g. Efron and Tibshirani Efron and Tibshirani [1993]). The jackknife replicates $\mathcal{R}_{(i)}$ can be used to identify potentially influential observations, in the sense that outliers within the sample of replicates correspond to observations X_i that increase or decrease the dependence coefficient more than other observations. These unusual replicates are not necessarily outliers in the original data.

Consider the crime data of Example 14.4. The studentized jackknife replicates $\mathcal{R}_{(i)}/\widehat{se}(\mathcal{R}_{(i)})$, $i = 1, \ldots, n$ are plotted in Figure 14.4(a). These replicates were computed on the pairs (x, y), where x is the vector (*nonwhite, density, population*) and y is *crime*. The plot suggests that Philadelphia is an unusual observation. For comparison we plot the first two principal components of the four variables in Figure 14.4(b), but Philadelphia (PHIL) does not appear to be an unusual observation in this plot or other plots (not shown), including those where log(*population*) replaces *population* in the analysis. One can see from comparing

```
                 population nonwhite density crime
   PHILADELPHIA      4829     15.7    1359   1753
```

with sample quartiles

```
        population nonwhite  density    crime
   0%       270.00    0.300    37.00   458.00
   25%      398.75    3.400   266.50  2100.25
   50%      664.00    7.300   412.00  2762.00
   75%     1167.75   14.825   773.25  3317.75
   100%   11551.00   64.300 13087.00  5441.00
```

that *crime* in Philadelphia is low while *population*, *nonwhite*, and *density* are all high relative to other cities. Recall that all Pearson correlations were positive in Example 14.4.

This example illustrates that having a single multivariate summary statistic dCor that measures dependence is a valuable tool in exploratory data analysis, and it can provide information about potential influential observations prior to model selection.

14.2 Some Extensions

14.2.1 Affine and Monotone Invariant Versions

The original distance covariance has been defined as the square root of $dCov^2(X, Y)$ rather than the squared coefficient itself. Thus $dCov(X, Y)$ has

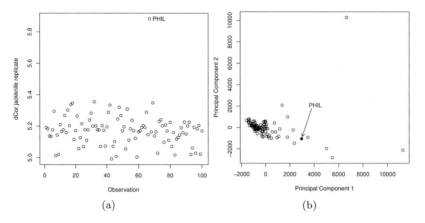

FIGURE 14.4
Jackknife replicates of dCor (a) and principal components of Freedman data (b) in Example 14.5.

the property that it is the energy distance between the joint distribution of X, Y and the product of its marginals. Under this definition, however, the distance variance, rather than the distance standard deviation, is measured in the same units as the pairwise X distances.

Alternately, one could define distance covariance to be the square of the energy distance: $dCov^2(X, Y)$. In this case, the distance standard deviation of X is measured in the same units as the X distances. Standard deviation, Gini's mean difference, and distance standard deviation are measures of dispersion: standard deviation works with deviations from the center measured by the arithmetic average of data, Gini's mean difference works with data distance without centering them, and finally distance standard deviation works with centered (doubly centered) data distances. It is easy to see that for real-valued random variables with finite variance, the distance standard deviation cannot be bigger than the standard deviation nor Gini's mean difference.

Also see Dueck et al. [2014] on affine invariant distance correlation. Here we apply the distance correlation formulae for the standardized sample. One can also apply the distance correlation formula to copula transformed random variables or to ranks of observations; this results in a version of distance correlation that is invariant with respect to monotone transformations. *The copula version of distance correlation can be considered "more equitable."* On a test of independence based on ranks of distances see Heller et al. [2013].

14.2.2 Generalization: Powers of Distances

We can broaden the applicability of distance correlation and Brownian correlation if we work with $0 < \alpha < 2$ powers of distances and with fractional

Brownian motions with Hurst parameter $0 < H = \alpha/2 < 1$, respectively [Herbin and Merzbach, 2007]. Here we do not need to suppose that X, Y have finite expectations; we just need finite α moments for some positive α. We have the same duality as before: the distance correlation computed from α powers of distances, dCor_α, equals the correlation with respect to fractional Brownian motion with Hurst parameter $0 < H = \alpha/2 < 1$.

At the endpoint $\alpha = 2$, $dCor_2$ does not characterize independence except for Gaussian distributions. The coefficient corresponding to $dCor_2$ was introduced in Escoufier [1973] and in Robert and Escoufier [1976], where it is called the RV coefficient. It does not characterize independence, but generalizes Pearson's coefficient to higher dimensions. See Example 22.3.

Definition 14.1. The distance covariance coefficient based on α power of distance is defined

$$
\begin{aligned}
\mathcal{V}_\alpha^2(X, Y) = {} & E\big(|X - X'|^\alpha |Y - Y'|^\alpha\big) + E\big(|X - X'|^\alpha\big) E\big(|Y - Y'|^\alpha\big) \\
& - 2E\big(|X - X'|^\alpha |Y - Y''|^\alpha\big). \quad (14.1)
\end{aligned}
$$

Distance variance is defined as above; that is, $dVar_\alpha(X) = \mathcal{V}_\alpha(X, X)$. The α-distance correlation $\mathcal{R}_\alpha(X, Y)$ is defined like $\mathcal{R}(X, Y)$, replacing exponent 1 on distance with exponent α:

Definition 14.2.

$$
\mathcal{R}_\alpha(X, Y) = \frac{\mathcal{V}_\alpha(X, Y)}{\sqrt{\mathcal{V}_\alpha(X, X)\mathcal{V}_\alpha(Y, Y)}},
$$

provided the denominator is positive.

For $0 < \alpha < 2$, $\mathcal{R}_\alpha$ shares all advantageous properties of $\mathcal{R}$, including that $\mathcal{R}_\alpha(X, Y) = 0$ if and only if X and Y are independent [Székely and Rizzo, 2009a].

We can go even further and work with conditionally negative definite distances δ in metric spaces. For a related interesting result see Genovese [2009].

14.2.3 Distance Correlation for Dissimilarities

As we will see in Chapter 17, distance covariance and distance correlation are also defined for dissimilarities that are non-Euclidean. The additive constant invariance is crucial for applications when instead of distance matrices we only have dissimilarity matrices (zero-diagonal, symmetric matrices). It is known [Cailliez, 1983] that if we add a big enough constant to all off-diagonal entries of a dissimilarity matrix, then we get a distance matrix, thus we can apply distance correlation methods because additive constants do not change distance correlation. In Chapter 17 (following Székely and Rizzo [2014]) an algorithm is outlined, and this approach is a strong competitor of the Mantel test for association between dissimilarity matrices. This makes dCov tests and energy statistics ready to apply to problems in e.g. community ecology, where one must often work with data in the form of non-Euclidean dissimilarity matrices. On this topic see also Omelka and Hudecová [2013].

14.2.4   An Unbiased Distance Covariance Statistic

In Székely and Rizzo [2013a] and Székely and Rizzo [2014] readers will find two closely related definitions of an unbiased distance covariance and the corresponding bias-corrected distance correlation. For both the *Distance Correlation t-test in High Dimension* [Székely and Rizzo, 2013a] and *Partial Distance Correlation* [Székely and Rizzo, 2014] it was necessary to work with an unbiased estimator. The unbiased estimator of Chapter 17 is a U-statistic. See Chapter 16 for a detailed discussion of energy U-statistics.

The U-statistic for $dCov^2(X, Y)$ can be written

$$(\widetilde{A} \cdot \widetilde{B}) := \frac{1}{n(n-3)} \sum_{i \neq j} \widetilde{A}_{i,j} \widetilde{B}_{i,j},$$

where $\widetilde{A}, \widetilde{B}$ are U-centered (not double-centered) distance matrices (refer to (16.1) and (16.2). $\mathcal{V}_n^*(\mathbf{X}, \mathbf{Y}) = (\widetilde{A} \cdot \widetilde{B})$ is an unbiased estimator of squared population distance covariance $\mathcal{V}^2(X, Y)$. The inner product notation is due to the fact that this statistic is an inner product in the Hilbert space of U-centered distance matrices [Székely and Rizzo, 2014].

The fact that the inner product statistic $(\widetilde{A} \cdot \widetilde{B})$ is an unbiased estimator of $\mathcal{V}^2(X, Y)$ follows from the result that it is a U-statistic; see Chapter 16 for details.

A bias corrected $\mathcal{R}_n^2$ is defined by normalizing the U-statistic with the bias corrected dVar statistics. Let

$$\mathcal{R}_{x,y}^* := \begin{cases} \frac{(\widetilde{A} \cdot \widetilde{B})}{|\widetilde{A}||\widetilde{B}|}, & |\widetilde{A}||\widetilde{B}| \neq 0; \\ 0, & |\widetilde{A}||\widetilde{B}| = 0. \end{cases} \qquad (14.2)$$

The bias-corrected dCor statistic is implemented in the R *energy* package by the `bcdcor` function, and the U-statistic for $\mathcal{V}^2(X, Y)$ is computed by `dcovU`. The function `dcov_Ustats` returns a vector with the bias-corrected statistic $\mathcal{R}_n^*$ and the U-statistics used to compute it.

The inner product notation for $\mathcal{V}_n^*(X, Y)$ (from Chapters 16–17) itself suggests that the bias corrected dCor statistic can be interpreted as the cosine of the angle between two U-centered matrices in the Hilbert space of these matrices.

Remark 14.1 (Why use V-statistics? Why not U-statistics?). We have defined $\mathcal{V}_n^2$ as a V-statistic. It has the property that $\mathcal{V}_n^2 \geq 0$ and $\mathcal{V}_n^2 = 0$ if and only if X, Y are independent. Thus $\mathcal{V}_n$ is a statistical distance. However, as we have seen, $\mathcal{V}_n^2$ is positively biased for $\mathcal{V}^2(X, Y)$. Why not work with

U-statistics? In fact, every energy V-statistic has a corresponding U-statistic, and the asymptotic theory is parallel. The U-statistic $\mathcal{V}_n^*$ has expected value 0 under independence, so it necessarily must take negative values with some positive probability. With the U-statistic for $\mathcal{V}^2(X,Y)$ one must work with the square rather than the square root. We gain *unbiasedness* but lose the interpretation as a statistical distance. Both the V-statistic and the U-statistic are valid and convenient to apply. It may depend on the application which is preferable.

Remark 14.2 (Why take the square root?). One could also define dCov as $\mathcal{V}_n^2$ and dCor as $\mathcal{R}_n^2$ rather than by their respective square roots. There are reasons to prefer each definition, but to be consistent with the original definitions in Székely et al. [2007], Székely and Rizzo [2009a], we define dCov as $\mathcal{V}(X,Y)$ and dCor as $\mathcal{R}(X,Y)$ (the square root). When we deal with unbiased statistics, we no longer have the non-negativity property, so we cannot take the square root and need to work with the square. Later, to develop other statistics [Székely and Rizzo, 2013a, 2014] such as partial distance covariance, unbiased statistics were needed. The names distance covariance and distance correlation were already widely used in the literature by that time under the original definitions.

14.3 Distance Correlation in Metric Spaces

14.3.1 Hilbert Spaces and General Metric Spaces

It is easy to generalize the definition of empirical distance correlation to general metric spaces. We just need to replace the Euclidean distance in our definition of distance covariance with the metric δ of the metric space $(\mathcal{X}, \delta)$. Even the population value is easy to define following Section 17.6.

The main question is *under what conditions can we claim that the population $dCor(X,Y) = 0$ implies that X, Y are independent*. In a beautiful paper, Lyons [2013] proved that a necessary and sufficient condition is that $(\mathcal{X}, \delta)$ is strong negative type. See also Jakobsen [2017] and Lyons [2018]. Energy distance in metric spaces is covered in Section 10.1.

In general, one can define an energy distance $\mathcal{E}$ for all pairs of random variables X, Y that take their values in a metric space $(\mathcal{X}, \delta)$ with distance function δ as

$$\mathcal{E}(X,Y;\delta) := 2E[\delta(X,Y)] - E[\delta(X,X')] - E[\delta(Y,Y')],$$

provided that these expectations exist. A necessary and sufficient condition that the generalization above $\mathcal{E}(X,Y;\delta)$ characterizes equality of distributions is established in Proposition 10.1.

Definition 14.3 (negative type). A metric space $(\mathfrak{X}, \delta)$ has *negative type* if for all $n \geq 1$ and all sets of n red points x_i and n green points x_i' in $\mathfrak{X}$, we have

$$2 \sum_{i,j} \delta(x_i, x_j') - \sum_{i,j} \delta(x_i, x_j) - \sum_{i,j} \delta(x_i', x_j') \geq 0.$$

Definition 14.4 (conditional negative definiteness). A metric space (c, δ) is conditionally negative definite if for all $n \geq 1$, $x_1, x_2, \ldots, x_n \in \mathfrak{X}$, and $a_1, a_2, \ldots, a_n$ real numbers with $\sum_{i=1}^{n} a_i = 0$, we have

$$\sum_{i,j} a_i a_j \delta(x_i, x_j) \leq 0.$$

The metric space $(\mathfrak{X}, \delta)$ has *strict negative type* if for every $n \geq 1$ and for all distinct points $x_1, x_2, \ldots, x_n$ equality holds in the inequality above only if $a_i = 0$ for all i.

This theorem [Lyons, 2013] then implies that in all separable Hilbert spaces $dCor(X, Y) = 0$ implies X, Y are independent. Other examples include all hyperbolic spaces [Lyons, 2014] and spheres if we exclude a point from $\mathfrak{X}$ whenever its diagonal opposite is included [Lyons, 2020]; that is, all open hemispheres are of strong negative type. It was known that spheres have negative type, but only subsets with at most one pair of antipodal points have strict negative type. These are conditions on the (angular) distances within any finite subset of points. We have shown that subsets with at most one pair of antipodal points have strong negative type, a condition on every probability distribution of points.

14.3.2   Testing Independence in Separable Metric Spaces

Distance correlation can be applied when the metric space $(\mathfrak{X}, \delta)$ is "nice," meaning that the metric is strong negative type; that is, if it can be embedded isometrically into a Hilbert space.

However, in general this is not possible. Then we need to give up something. In most cases, we apply distance correlation for testing independence of random variables. We can try to embed $(\mathfrak{X}, \delta)$ into a "nice" metric space $(\mathfrak{X}^*, \delta^*)$ that is Borel isomorphic to $(\mathfrak{X}, \delta)$. Such an embedding might dramatically change distances and thus the distance correlation, but it does not change when $dCor(X, Y) = 0$ because Borel functions of independent random variables remain independent.

In short: $dCor(X, Y) = 0$ is equivalent to $dCor(X^*, Y^*) = 0$ where $X^* = f(X), Y^* = f(Y)$ are Borel functions of X, Y from $(\mathfrak{X}, \delta)$ to $(\mathfrak{X}^*, \delta^*)$ and f is the Borel isomorphism (a 1–1 Borel measurable function) between the two metric spaces. According to Kuratowski's theorem, two standard Borel spaces are Borel isomorphic if and only if they have the same cardinality. Complete and separable metric spaces are called "standard." They are Borel isomorphic either to R, Z or a finite space. Thus we know the existence of the Borel isomorphism f. If we can construct f, then we can easily check the independence of the real-valued random variables $f(X), f(Y)$ via distance correlation and this is equivalent to testing the independence of X, Y that take values in general metric spaces. We might want to make f continuous to avoid the negative effect of minor noise. In this case, we can choose f to be a homeomorphism between our metric space and a subspace of a Hilbert cube. This f exists if and only if our metric space is separable.

Here is how to construct such an f. Assume $d \leq 1$ (otherwise, use $d/(d+1)$). Choose a dense countable sequence (x_n) from $\mathcal{X}$ which exists because the metric space is separable, and define $f(x) := (d(x, x_n)/n)$, a point in the Hilbert cube. Here we can apply distance correlation for testing independence.

On some consistent nonparametric tests of independence, see Gretton and Györfi [2010]. Chatterjee [2021] and Deb et al. [2020] discuss the possibility of assuming the dependence measure $\Delta(X, Y) = 1$ if and only if one of the variables is a Borel measurable function of the other variable.

14.3.3 Measuring Associations in Banach Spaces

Any metric space can be embedded into a Banach space [Mazur and Ulam, 1932], so it is important to see that there is a measure of dependence in Banach spaces that are not Hilbert spaces. Energy divergence of the joint distribution and the product of the marginals is a good option; for details see Pan et al. [2020].

On measuring association in general topological spaces see Deb et al. [2020].

14.4 Distance Correlation with General Kernels

Instead of a kernel function of the usual Euclidean distances between sample observations, one can apply more general kernels $k_X = k_X(x, x')$ and $k_Y = k_Y(y, y')$ for defining the generalized distance covariance:

$$
\begin{aligned}
\mathrm{dCov}^2_{n,k_X,k_Y} := {} & \frac{1}{n^2} \sum_{k.l=1}^{n} k_X(X_k, X_\ell) k_Y(Y_k, Y_\ell) \\
& + \frac{1}{n^2} \sum_{k,\ell=1}^{n} k_X(X_k, X_\ell) \times \frac{1}{n^2} \sum_{k,\ell=1}^{n} k_Y(Y_k, Y_\ell) \\
& - \frac{2}{n^3} \sum_{k=1}^{n} \sum_{\ell,m=1}^{n} k_X(X_k, X_\ell) k_Y(Y_k, Y_m).
\end{aligned}
$$

In the following we will apply this formula for ranks of observations. Introduce the empirical copula as follows:

$$C_n(u,v) = \frac{1}{n}\sum_{i=1}^n I\left(\frac{R_i}{n+1} \le u, \frac{S_i}{n+1} \le v\right),$$

where R_i and S_i are the ranks of X_i and Y_i, respectively, and I denotes the indicator function.

Deheuvels [1979] proposed the following measure of dependence:

$$B_n := n\int_0^1\int_0^1 (C_n(u,v) - uv)^2\,du\,dv.$$

Interestingly B_n is a special case of our general $dCor^2_{n,k_X,k_Y}$ formula applied to to $k_X := 1 - \frac{R_k \vee R_L}{n}$ and $k_Y := 1 - \frac{S_k \vee S_l}{n}$. This was shown in a recent article by Liu et al. [2018].

B_n is a copula parameter because it does not depend on the marginal distributions. The beauty of B_n is that its population value clearly characterizes independence because

$$B := \int_0^1\int_0^1 (C(u,v) - uv)^2 du\,dv = 0$$

if and only if X and Y are independent. For the purpose of comparison here are the definitions of Spearman's rank correlation ρ and Kendall's τ when X,Y have continuous cdf's:

$$12\int_0^1\int_0^1 (C(u,v) - uv)\,du\,dv$$

and

$$14\int_0^1\int_0^1 (C(u,v)\,dC(u,v) - 1,$$

respectively. It is clear that if they are zero then this does not characterize independence.

We can construct a copula whose parameter is B. Recall that the Farlie-Gumbel-Morgenstern copula has the form

$$C(u,v) = uv + \theta f(u)g(v).$$

If $f(u) = g(u) \doteq u(1-u)$, then the generalized distance covariance is $B = c\theta^2$. It would be better if instead of B we could introduce a parameter that corresponds to distance correlation, say, K such that for $K = 0$ we get $C(u,v) = uv$, and for $K = 1$ we get $C(u,v) = min(u,v)$ that corresponds to identical marginals. The bad news is that for the Farlie-Gumbel-Morgenstern copula, the correlation is $\theta/3$ and its maximum is $1/3$ therefore it cannot express strong correlation.

14.5 Further Reading

There is a growing literature on applications of distance correlation in computing, engineering, statistics, life and natural sciences, social sciences and humanities. A few examples include the following.

- Lee et al. [2019] test network dependence via diffusion maps and distance-based correlations.
- Gene regulatory networks and fMRI studies [Guo et al., 2014, Hua et al., 2015, Rudas et al., 2014].
- Lu et al. [2016] proposed the combination of support vector machine with distance correlation learning.
- Chemometrics [Hemmateenejad and Baumann, 2018]; glucose monitoring [Ferenci et al., 2015].
- Sferra et al. [2017] proposed distance correlation as a novel metric for phylogenetic profiling.
- Experimental design: Shamsuzzaman et al. [2015] proposed distance correlation based nearly orthogonal space-filling experimental designs.
- Graphical models [Fan et al., 2020, Wang et al., 2015].
- Multiscale graph correlation [Shen et al., 2019].
- Farjami et al. [2018] applied the genetic algorithm to find a set of optimal positions between plaintext and its hash data with maximum dCorr.

In the following sections, we describe a few more application areas.

14.5.1 Variable Selection, DCA and ICA

Distance correlation was applied for feature screening of high dimensional data by Li et al. [2012], Yousuf and Feng [2018], Liu and Li [2020], Chen et al. [2017].

In this type of research we want to select the "most influential" s random variables from a given set of p random variables that maximize their influence on a given random variable Y. Formally, we want to find

$$S_0 := \operatorname{argmax} Q(S), |S| \leq s.$$

The function Q might be the distance correlation.

Independent component analysis (ICA) based on distance correlation was proposed by [Matteson and Tsay, 2016] and implemented in the R package *EDMeasure* [Jin et al., 2018].

A recent manuscript on a topic related to "distance canonical correlation" is Cowley et al. [2017]. In this paper the authors propose a dimensionality reduction method to identify linear projections that capture interactions between two or more sets of variables. In the abstract they write

The method, distance covariance analysis (DCA), can detect both linear and nonlinear relationships, and can take dependent variables into account. On previous testbeds and a new testbed that systematically assesses the ability to detect both linear and nonlinear interactions, DCA performs better than or comparable to existing methods, while being one of the fastest methods.

On dimension reduction see also Sheng and Yin [2013, 2016].

Variable selection by partial distance correlation is explored in Example 17.7.

14.5.2   Nonparametric MANOVA Based on dCor

A MANOVA is a multivariate analysis of variance where the response variable is multivariate. The hypothesis tested using normal theory and F statistics is that the expected response does not depend on the group variable(s). The conditions for inference are that the samples are independent, the expected response is linear in the predictor, with equal variance and normally distributed errors. For a nonparametric approach, where normality or equal variance may not hold, energy methods can be applied. If instead we decompose the distances, we have a DISCO (distance components) decomposition, and methods of Chapter 9 [Rizzo and Székely, 2010] can be applied. In this case, we can test the more general hypothesis that the distribution of the response does not differ among groups.

As we have seen, in the special case when the group variable is a one-dimensional categorical variable, the DISCO test is essentially a test of independence. This suggests that dCov and dCor can also be applied for the nonparametric MANOVA type analysis.

Clearly, distance correlation is applicable for all types of data, including categorical data (factors). In addition to two-sample or multi-sample energy tests for equal distribution, one could apply the dCov or dCor test of independence between a response variable and group labels for a nonparametric k-group test of equal distribution. See Chapter 9 for details on implementing the tests in R using the *energy* package [Rizzo and Székely, 2022] and the *vegan* package [Oksanen et al., 2022]. Panda et al. [2020] applied distance correlation for an independence test of the hypothesis whether k groups are identically distributed. Their simulations suggest that the energy k-group nonparametric MANOVA typically outperforms classical MANOVA. They also extend the idea to multi-way and multi-level tests. The methods are implemented in their Python *hyppo* package available through links at `https://hyppo.neurodata.io/`.

14.5.3 Tests of Independence with Ranks

When we say that energy statistics are functions of distances of sample elements in metric spaces then one might wonder what kind of function we can work with. The first idea is to consider U-statistics or V-statistics of distances. A second idea is to consider ranks of distances and work with functions of these ranks. We shall see that energy divergence is a function of these metric ranks.

An example in Heller et al. [2013] suggests an interesting test of independence based on the ranks of distances. The test statistic is a sum of Pearson's χ^2 statistic for 2×2 contingency tables computed from the X-distances and from the Y-distances. The authors show that the implementation of this HHG test does not need n^3 operations as the naive implementation would, only $n^2 \log n$. This is still worse than the $n \log n$ complexity of the test based on distance correlation. On balance, their empirical results suggest that the HHG test is more powerful in many examples than the test based on dCor.

We can also work with distance correlations of ranks of observations. The advantage of this is that the resulting measure is invariant with respect to monotone transformations on marginal distributions.

On another rank-based measure of correlation (REVA) see Afsari et al. [2018]. On multivariate ranks see Hallin [2017] and Hallin et al. [2021].

14.5.4 Projection Correlation

Projection correlation Zhu et al. [2017] has several appealing properties. It equals zero if and only if the two random vectors are independent, it is not sensitive to the dimensions of the two random vectors, it is invariant with respect to the group of orthogonal transformations, and its estimation is free of tuning parameters and does not require moment conditions on the random vectors.

But it does not seem to be scale invariant. This is the price we need to pay for other good properties.

The sample estimate of the projection correction is n-consistent if the two random vectors are independent and root-n-consistent otherwise. Monte Carlo simulation studies by Zhu et al. [2017] suggest that the projection correlation may have higher power than the distance correlation and the ranks of distances in tests of independence, especially when the dimensions are relatively large or the moment conditions required by the distance correlation are violated. (However, moment conditions are never an issue if in Definition 12.3 we apply exponent α such that $E|X|^\alpha + E|Y|^\alpha < \infty$. In high dimension we recommend applying a test based on the unbiased $\mathcal{V}_n^*$ or the bias-corrected $\mathcal{R}_n^*$.)

14.5.5 Detection of Periodicity via Distance Correlation

Zucker [2018] introduced "Distance Correlation Periodogram" as a tool to identify periods in signals x_i measured at times t_i, $i = 1, 2, \ldots, N$. The method is especially powerful in the "hard cases" when the signals are high dimensional and t_i is unevenly sparse. The phase distance correlation for a trial period P is defined as the distance correlation based on the distance matrices $a_{i,j} = |x_i - x_j|$ and $b_{i,j} = \phi_{i,j}(P - \phi_{i,j})$, where $\phi_{i,j} := (t_i - t_j) \mod P$.

High phase distance correlation indicates that P is a good candidate for a period. The technical part of the manuscript above introduces the "linear-circular" distance covariance. Our Lemma

$$\int_{\mathbb{R}} \frac{1 - cos(tx)}{t^2} dx = \pi|x|$$

is replaced by

$$\sum_{m=-\infty, m \neq 0}^{\infty} \frac{1 - cos(my)}{m^2} = \frac{1}{2} y(2\pi - y),$$

for $y \in [0, 2\pi)$.

The right hand side explains the definition of $b_{i,j}$ above. This definition helps to simplify the empirical formula for phase distance correlation. Another natural option would have been, say, $b_{i,j} = min(\phi_{i.j}, P - \phi_{i,j})$. According to the authors, the above definition has better performance than some other natural choices.

14.5.6 dCov Goodness-of-fit Test of Dirichlet Distribution

Some distributions can be characterized by an independence condition. Lukacs's proportion-sum independence theorem [Lukacs, 1955] gives an independence characterization of the gamma distribution. This characterization leads to an independence characterization for the Dirichlet distribution.

Multivariate beta or Dirichlet distribution is often applied as a model for the prior multinomial distribution in Bayesian analysis. A random vector $X \in [0, 1]^k$ has the Dirichlet distribution with parameters $\alpha_1, \alpha_2, \ldots, \alpha_k > 0$ if it has the pdf

$$f(x_1, \ldots, x_k) = \frac{\Gamma\left(\sum_{i=1}^{k} \alpha_i\right)}{\prod_{i=1}^{k} \Gamma(\alpha_i)} \prod_{i=1}^{k} x_i^{\alpha_i - 1},$$

where $x_1 + \cdots + x_k = 1$. The univariate case with $k = 1$ is the Beta(α_1, α_2) distribution. (The R package *gtools* [Warnes et al., 2022] provides a random generator `rdirichlet` and density function `ddirichlet` for the Dirichlet distribution.)

A distance covariance test of the Dirichlet distribution ($k \geq 2$) can be based on the following characterization result. The random variable set $\{X_1, \ldots, X_k\}$ is *completely neutral* if

$$X_1, \frac{X_2}{1 - X_1}, \frac{X_3}{1 - X_1 - X_2}, \ldots, \frac{X_k}{1 - X_1 - X_2 - \cdots - X_{k-1}}$$

are jointly independent and this property holds for every permutation of the index set.

Theorem 14.1. *The random variable set $\{X_1, \ldots, X_n\}$ is completely neutral for all permutations of the index set if and only if $X_1, \ldots, X_k$ have a Dirichlet distribution.*

Then if

$$Y_1 = X_1; \quad Y_i = \frac{X_i}{(1 - X_1 - \cdots - X_{i-1})}, \quad i = 2, \ldots, k,$$

a distance covariance or distance correlation test of mutual independence of $Y_1, \ldots, Y_k$ can be applied to test the composite hypothesis that X has a Dirichlet distribution. The recommended test statistic applies the bias-corrected squared distance correlation:

$$T_n = \sum_{j=1}^{k-1} \mathcal{R}_n^*(Y_j, [Y_{j+1}, \ldots, Y_k]),$$

and the test of mutual independence can be implemented as a permutation test. On the test of mutual independence see Section 13.3. This nonparametric test is not only simpler to apply for Dirichlet than one based on the one-sample energy distance, it is more powerful (see Li [2015b]).

14.6 Exercises

Exercise 14.1. Implement the study in Example 14.3 and compare your results to the example.

Exercise 14.2. Repeat Examples 14.4 and 14.5 on the crime data.

Exercise 14.3. Repeat Examples 14.4 and 14.5 on some other available data set in any R package or other source.

15

Brownian Distance Covariance

CONTENTS

15.1   Introduction

Distance correlation, introduced in Chapter 12, is a new class of multivariate dependence coefficients applicable to random vectors of arbitrary and not necessarily equal dimension. Distance covariance and distance correlation are analogous to product-moment covariance and correlation, but generalize and extend these classical bivariate measures of dependence. Distance correlation characterizes independence: it is zero if and only if the random vectors are independent.

In this chapter the notion of covariance with respect to a *stochastic process*, which we call *Brownian distance covariance* is introduced. We show that that population distance covariance [Székely, Rizzo, and Bakirov, 2007] coincides with the covariance with respect to Brownian motion; thus both can be called *Brownian distance covariance* [Székely and Rizzo, 2009a]. In the bivariate case, Brownian covariance is the natural extension of product-moment covariance, as we obtain Pearson product-moment covariance by replacing the Brownian motion in the definition with identity.

We will see that the definition of the distance covariance coefficient, which has theoretical foundations based on characteristic functions, also has a theoretical foundation based on the new concept of covariance with respect to Brownian motion.

DOI: 10.1201/9780429157158-15

The notion of covariance of random vectors (X, Y) with respect to a stochastic process U is introduced below. This new notion $\mathrm{Cov}_U(X, Y)$ contains as distinct special cases distance covariance $\mathcal{V}^2(X, Y)$ and, for bivariate (X, Y), covariance σ_{xy}^2. The title "Brownian Distance Covariance" of this chapter refers to $\mathrm{Cov}_W(X, Y)$, where W is a Wiener process.

Brownian covariance $\mathcal{W} = \mathcal{W}(X, Y)$ is based on Brownian motion or Wiener process for random variables $X \in \mathbb{R}^p$ and $Y \in \mathbb{R}^q$ with finite first moments. An important property of Brownian covariance is that $\mathcal{W}(X, Y) = 0$ if and only if X and Y are independent.

A surprising result develops: the Brownian covariance is equal to the distance covariance. This equivalence is not only surprising, it also shows that distance covariance is a natural counterpart of product-moment covariance. For bivariate (X, Y), by considering the simplest nonrandom function, identity (id), we obtain $\mathrm{Cov}_{id}(X, Y) = \sigma_{xy}^2$. Then by considering the most fundamental random processes, Brownian motion W, we arrive at $\mathrm{Cov}_W(X, Y) = \mathcal{V}^2(X, Y)$. Brownian correlation is a standardized Brownian covariance. If Brownian motion is replaced with the identity function in Brownian correlation, we obtain the absolute value of Pearson's correlation ρ.

A further advantage of extending Pearson correlation with distance correlation is that while uncorrelatedness ($\rho = 0$) can sometimes replace independence e.g. in proving some classical laws of large numbers, uncorrelatedness is too weak to imply a central limit theorem, even for strongly stationary summands (see Bradley [1981, 1988, 2007]). On the other hand, a central limit theorem for strongly stationary sequences of summands follows from $\mathcal{R} = 0$ type conditions [Székely and Bakirov, 2008]).

See Chapter 12 for detailed definitions of the population coefficients $\mathcal{V}(X, Y) = \mathrm{dCov}(X, Y)$, $\mathcal{R}(X, Y) = \mathrm{dCor}(X, Y)$ and the corresponding statistics $\mathcal{V}_n(\mathbf{X}, \mathbf{Y}) = \mathrm{dCov}_n(\mathbf{X}, \mathbf{Y})$ and $\mathcal{R}_n(\mathbf{X}, \mathbf{Y})$.

The following notation is used in this chapter. The scalar product of vectors t and s is denoted by $\langle t, s \rangle$. The characteristic functions of X, Y, and (X, Y) are $\hat{f}_X, \hat{f}_Y, \hat{f}_{X,Y}$, respectively. For complex-valued functions $f(\cdot)$, the complex conjugate of f is denoted by $\overline{f}$ and $|f|^2 = f\,\overline{f}$. The Euclidean norm of x in $\mathbb{R}^p$ is $|x|_p$ (subscript p omitted when clear in context) and $|\cdot|$ is the complex norm if its argument is complex. A primed variable X' is an independent copy of X; that is X and X' are independent and identically distributed (iid).

15.2　Weighted L_2 Norm

For complex functions γ defined on $\mathbb{R}^p \times \mathbb{R}^q$ the $\|\cdot\|_w$-norm in the weighted L_2 space of functions on $\mathbb{R}^{p+q}$ is defined by

$$\|\gamma(t, s)\|_w^2 = \int_{\mathbb{R}^{p+q}} |\gamma(t, s)|^2 w(t, s)\, dt\, ds, \tag{15.1}$$

where $w(t, s)$ is an arbitrary positive weight function for which the integral above exists.

With a suitable choice of weight function $w(t, s)$ we define a measure of dependence

$$\mathcal{V}^2(X, Y; w) = \|\hat{f}_{X,Y}(t, s) - \hat{f}_X(t)\hat{f}_Y(s)\|_w^2$$

$$= \int_{\mathbb{R}^{p+q}} |\hat{f}_{X,Y}(t, s) - \hat{f}_X(t)\hat{f}_Y(s)|^2 w(t, s) \, dt \, ds, \qquad (15.2)$$

which is analogous to classical covariance, but with the important property that $\mathcal{V}^2(X, Y; w) = 0$ if and only if X and Y are independent. We also define

$$\mathcal{V}^2(X; w) = \mathcal{V}^2(X, X; w) = \|\hat{f}_{X,X}(t, s) - \hat{f}_X(t)\hat{f}_X(s)\|_w^2$$

$$= \int_{\mathbb{R}^{2p}} |\hat{f}_{X,X}(t, s) - \hat{f}_X(t)\hat{f}_X(s)|^2 w(t, s) \, dt \, ds,$$

and similarly define $\mathcal{V}^2(Y; w)$. Then a standardized version of $\mathcal{V}(X, Y; w)$ is

$$\mathcal{R}_w = \frac{\mathcal{V}(X, Y; w)}{\sqrt{\mathcal{V}(X; w)\mathcal{V}(Y; w)}},$$

a type of unsigned correlation.

In the definition of the norm (15.1) there are more than one potentially interesting and applicable choices of weight function w, but not every w leads to a dependence measure that has desirable statistical properties. Let us now discuss the motivation for our particular choice of weight function leading to distance covariance.

At least two conditions should be satisfied by the standardized coefficient $\mathcal{R}_w$:

(i) $\mathcal{R}_w \geq 0$ and $\mathcal{R}_w = 0$ only if independence holds.

(ii) $\mathcal{R}_w$ is scale invariant; that is, invariant with respect to transformations $(X, Y) \mapsto (\epsilon X, \epsilon Y)$, for $\epsilon > 0$.

However, if we consider an *integrable* weight function $w(t, s)$, then we cannot satisfy both (i) and (ii), because for X and Y with finite variance

$$\lim_{\epsilon \to 0} \frac{\mathcal{V}^2(\epsilon X, \epsilon Y; w)}{\sqrt{\mathcal{V}^2(\epsilon X; w)\mathcal{V}^2(\epsilon Y; w)}} = \rho^2(X, Y).$$

The above limit is obtained by considering the Taylor expansions of the underlying characteristic functions. Thus if the weight function is integrable, $\mathcal{R}_w(X, Y)$ can be zero even if X and Y are dependent. By using a suitable *non-integrable* weight function, we can obtain an $\mathcal{R}_w$ that satisfies both properties (i) and (ii) above.

A promising solution to the choice of weight function w is suggested by the following lemma, which we restate here for convenient reference.

Lemma 15.1. *If $0 < \alpha < 2$, then for all x in $\mathbb{R}^d$*

$$\int_{\mathbb{R}^d} \frac{1 - \cos\langle t, x\rangle}{|t|_d^{d+\alpha}} \, dt = C(d, \alpha)|x|_d^\alpha,$$

where

$$C(d, \alpha) = \frac{2\pi^{\frac{d}{2}} \, \Gamma(1 - \frac{\alpha}{2})}{\alpha 2^\alpha \Gamma(\frac{d+\alpha}{2})},$$

and $\Gamma(\cdot)$ is the complete gamma function. The integrals at 0 and ∞ are meant in the principal value sense: $\lim_{\varepsilon \to 0} \int_{\mathbb{R}^d \setminus \{\varepsilon B + \varepsilon^{-1} B^c\}}$, where B is the unit ball (centered at 0) in $\mathbb{R}^d$ and B^c is the complement of B.)

A proof of Lemma 15.1 is given in Székely and Rizzo [2005a]. (That proof can be found in Chapter "Preliminaries," Section 2.3.)

Lemma 15.1 suggests the weight functions

$$w(t, s; \alpha) = (C(p, \alpha)C(q, \alpha) \, |t|_p^{p+\alpha}|s|_q^{q+\alpha})^{-1}, \quad 0 < \alpha < 2. \tag{15.3}$$

The weight functions (15.3) result in coefficients $\mathcal{R}_w$ that satisfy the scale invariance property (ii) above.

In the simplest case corresponding to $\alpha = 1$ and Euclidean norm $|x|$,

$$w(t, s) = (c_p \, c_q \, |t|_p^{1+p}|s|_q^{1+q})^{-1}, \tag{15.4}$$

where

$$c_d = C(d, 1) = \frac{\pi^{\frac{1+d}{2}}}{\Gamma(\frac{1+d}{2})}. \tag{15.5}$$

(The constant $2c_d$ is the surface area of the unit sphere in $\mathbb{R}^{d+1}$.)

Distance covariance and distance correlation are a class of dependence coefficients and statistics obtained by applying a weight function of the type (15.3), $0 < \alpha < 2$. This type of weight function leads to a simple product-average form of the covariance (12.9) analogous to Pearson covariance. Other interesting weight functions could be considered (see e.g. Bakirov, Rizzo, and Székely [2006]), but only the weight functions (15.3) lead to distance covariance type statistics (12.9).

We apply weight function (15.4) with $\alpha = 1$ and the corresponding weighted L_2 norm $\| \cdot \|$, omitting the index w, and write the dependence measure (15.2) as $\mathcal{V}^2(X, Y)$. For finiteness of $\|\hat{f}_{X,Y}(t, s) - \hat{f}_X(t)\hat{f}_Y(s)\|^2$ it is sufficient that $E|X|_p < \infty$ and $E|Y|_q < \infty$. Later, in Section 15.4, we generalize Brownian covariance applying weight function (15.3) for other exponents $0 < \alpha < 2$.

With $\alpha = 1$, the distance covariance between random vectors X and Y with finite first moments is the nonnegative number $\mathcal{V}(X, Y)$ defined by

$$\begin{aligned} \mathcal{V}^2(X, Y) &= \|\hat{f}_{X,Y}(t, s) - \hat{f}_X(t)\hat{f}_Y(s)\|^2 \\ &= \frac{1}{c_p c_q} \int_{\mathbb{R}^{p+q}} \frac{|\hat{f}_{X,Y}(t, s) - \hat{f}_X(t)\hat{f}_Y(s)|^2}{|t|_p^{1+p} \, |s|_q^{1+q}} \, dt \, ds. \end{aligned} \tag{15.6}$$

15.3 Brownian Covariance

To introduce the notion of Brownian covariance, let us begin by considering the squared product-moment covariance. Recall that a primed variable X' denotes an iid copy of the unprimed symbol X. For two real-valued random variables, the square of their classical covariance is

$$E^2[(X-E(X))(Y-E(Y))] \tag{15.7}$$
$$= E[(X-E(X))(X'-E(X'))(Y-E(Y))(Y'-E(Y'))].$$

Now we generalize the squared covariance and define the square of conditional covariance, given two real-valued stochastic processes $U(\cdot)$ and $V(\cdot)$. We obtain an interesting result when U and V are independent Weiner processes.

First, to center the random variable X in the conditional covariance, we need the following definition. Let X be a real-valued random variable and $\{U(t) : t \in \mathbb{R}^1\}$ a real-valued stochastic process, independent of X. The U-centered version of X is defined by

$$X_U = U(X) - \int_{-\infty}^{\infty} U(t)dF_X(t) = U(X) - E[U(X)\,|\,U], \tag{15.8}$$

whenever the conditional expectation exists.

Note that if id is identity, we have $X_{id} = X - E[X]$. The important examples in this chapter apply Brownian motion/Weiner processes.

15.3.1 Definition of Brownian Covariance

Let W be a two-sided one-dimensional Brownian motion/Wiener process with expectation zero and covariance function

$$|s| + |t| - |s-t| = 2\min(s,t), \qquad t,s \geq 0. \tag{15.9}$$

This is twice the covariance of the standard Wiener process. Here the factor 2 simplifies the computations, so throughout the chapter, covariance function (15.9) is assumed for W.

The *Brownian covariance* or the *Wiener covariance* of two real-valued random variables X and Y with finite first moments is a nonnegative number defined by its square

$$\mathcal{W}^2(X,Y) = \mathrm{Cov}_W^2(X,Y) = E[X_W X'_W Y_{W'} Y'_{W'}], \tag{15.10}$$

where (W,W') does not depend on (X,Y,X',Y').

Note that if W in Cov_W is replaced by the (non-random) identity function id, then $\mathrm{Cov}_{id}(X,Y) = |\mathrm{Cov}(X,Y)| = |\sigma_{xy}|$, the absolute value of Pearson's product-moment covariance. While the standardized product-moment covariance, Pearson correlation (ρ), measures the degree of linear relationship between two real-valued variables, we shall see that standardized Brownian covariance measures the degree of *all kinds of possible relationships* between two real-valued random variables.

The definition of $\text{Cov}_W(X, Y)$ can be extended to random processes in higher dimension as follows. If X is an $\mathbb{R}^p$-valued random variable, and $U(s)$ is a random process (random field) defined for all $s \in \mathbb{R}^p$ and independent of X, define the U-centered version of X by

$$X_U = U(X) - E[U(X) \,|\, U],$$

whenever the conditional expectation exists.

Definition 15.1. If X is an $\mathbb{R}^p$-valued random variable, Y is an $\mathbb{R}^q$-valued random variable, $U(s)$ and $V(t)$ are arbitrary random processes (random fields) defined for all $s \in \mathbb{R}^p$, $t \in \mathbb{R}^q$, then the (U, V) covariance of (X, Y) is defined as the nonnegative number whose square is

$$\text{Cov}^2_{U,V}(X, Y) = E[X_U X'_U Y_V Y'_V], \qquad (15.11)$$

whenever the right hand side is nonnegative and finite.

Definition 15.2 (Brownian covariance). If in Definition 15.1 the random processes (U, V) are independent Brownian motions W and W' with covariance function (15.9) on $\mathbb{R}^p$ and $\mathbb{R}^q$ respectively, the **Brownian covariance** of X and Y with finite expectations is defined by

$$\mathcal{W}^2(X, Y) = \text{Cov}^2_W(X, Y) = \text{Cov}^2_{W,W'}(X, Y) \qquad (15.12)$$
$$= E[X_W X'_W Y_{W'} Y'_{W'}],$$

where (W, W') is independent of (X, X', Y, Y').

For random variables with finite expected value define the *Brownian variance* by

$$\mathcal{W}(X) = \text{Var}_W(X) = \text{Cov}_W(X, X).$$

Definition 15.3 (Brownian correlation). The Brownian correlation is defined as

$$\text{Cor}_W(X, Y) = \frac{\mathcal{W}(X, Y)}{\sqrt{\mathcal{W}(X)\mathcal{W}(Y)}}$$

whenever the denominator is not zero; otherwise $\text{Cor}_W(X, Y) = 0$.

In the following sections, we prove that $\text{Cov}_W(X, Y)$ exists for random vectors X and Y with finite first moments, and derive the Brownian covariance in this case.

15.3.2 Existence of Brownian Covariance Coefficient

In the following, the subscript on Euclidean norm $|x|_d$ for $x \in \mathbb{R}^d$ is omitted when the dimension is self-evident.

Theorem 15.1. *If X is an $\mathbb{R}^p$-valued random variable, Y is an $\mathbb{R}^q$-valued random variable, and $E(|X| + |Y|) < \infty$, then $E[X_W X'_W Y_{W'} Y'_{W'}]$ is nonnegative and finite, and*

$$\mathcal{W}^2(X, Y) = E[X_W X'_W Y_{W'} Y'_{W'}] \tag{15.13}$$
$$= E|X - X'||Y - Y'| + E|X - X'|E|Y - Y'|$$
$$- E|X - X'||Y - Y''| - E|X - X''||Y - Y'|,$$

where (X, Y), (X', Y'), and (X'', Y'') are iid.

Proof. Observe that

$$E[X_W X'_W Y_{W'} Y'_{W'}] = E\left[E\left(X_W Y_{W'} X'_W Y'_{W'} \middle| W, W'\right)\right]$$
$$= E\left[E\left(X_W Y_{W'} \middle| W, W'\right) E\left(X'_W Y'_{W'} \middle| W, W'\right)\right]$$
$$= E\left[E\left(X_W Y_{W'} \middle| W, W'\right)\right]^2,$$

and this is always nonnegative. For finiteness it is enough to prove that all factors in the definition of $\text{Cov}_W(X, Y)$ have finite second moments. Equation (15.13) relies on the special form of the covariance function (15.9) of W. The remaining details are in Section 15.5. □

See Section 15.4 for definitions and extension of results for the general case of fractional Brownian motion with Hurst parameter $0 < H < 1$ and covariance function $|t|^{2H} + |s|^{2H} - |t - s|^{2H}$.

15.3.3 The Surprising Coincidence: BCov(X,Y) = dCov(X,Y)

Theorem 15.2. *For arbitrary $X \in \mathbb{R}^p$, $Y \in \mathbb{R}^q$ with finite first moments*

$$\mathcal{W}(X, Y) = \mathcal{V}(X, Y)$$

Proof. Both $\mathcal{V}$ and $\mathcal{W}$ are nonnegative, hence it is enough to show that their squares coincide. Lemma 15.1 can be applied to evaluate $\mathcal{V}^2(X, Y)$. In the numerator of the integral we have terms like

$$E[\cos\langle X - X', t\rangle \, \cos\langle Y - Y', s\rangle],$$

where X, X' are iid, Y, Y' are iid. Now apply the identity

$$\cos u \, \cos v = 1 - (1 - \cos u) - (1 - \cos v) + (1 - \cos u)(1 - \cos v)$$

and Lemma 15.1 to simplify the integrand. After cancellation in the numerator of the integrand there remain to evaluate integrals of the type

$$
E \int_{\mathbb{R}^{p+q}} \frac{[1 - \cos\langle X - X', t\rangle][1 - \cos\langle Y - Y', s\rangle)]}{|t|^{1+p}|s|^{1+q}} \, dt ds
$$

$$
= E \left[\int_{\mathbb{R}^p} \frac{1 - \cos\langle X - X', t\rangle}{|t|^{1+p}} \, dt \times \int_{\mathbb{R}^q} \frac{1 - \cos\langle Y - Y', s\rangle}{|s|^{1+q}} \, ds \right]
$$

$$
= c_p c_q E |X - X'| E |Y - Y'|.
$$

Applying similar steps, after further simplification, we obtain

$$
\mathcal{V}^2(X, Y) = E|X - X'||Y - Y'| + E|X - X'|E|Y - Y'|
$$
$$
- E|X - X'||Y - Y''| - E|X - X''||Y - Y'|,
$$

and this is exactly equal to the expression (15.13) obtained for $\mathcal{W}(X, Y)$ in Theorem 15.1.   □

As a corollary to Theorem 15.2, the properties of Brownian covariance for random vectors X and Y with finite first moments are therefore the same properties established for distance covariance $\mathcal{V}(X, Y)$ in Chapters 12 and 13 hold for the Brownian covariance.

The surprising result that Brownian covariance equals distance covariance dCov, exactly as defined in (15.6) for $X \in \mathbb{R}^p$ and $Y \in \mathbb{R}^q$, parallels a familiar special case when $p = q = 1$. For bivariate (X, Y) we found that $\mathcal{R}(X, Y)$ is a natural counterpart of the *absolute value of* Pearson correlation. That is, if in (15.11) U and V are the simplest nonrandom function *id*, then we obtain the square of Pearson covariance σ_{xy}^2. Next if we consider the most fundamental random processes, $U = W$ and $V = W'$, we obtain the square of distance covariance, $\mathcal{V}^2(X, Y)$.

See also Bakirov and Székely [2011] on the background and application of the interesting coincidence in Theorem 15.2. For interesting discussions on Székely and Rizzo [2009a] see Bickel and Xu [2009], Cope [2009], Feuerverger [2009], Genovese [2009], Gretton et al. [2009], Kosorok [2009], Newton [2009], Rémillard [2009] and the rejoinder Székely and Rizzo [2009b].

15.4   Fractional Powers of Distances

Above we have introduced dependence measures based on Euclidean distance and on Brownian motion with Hurst index $H = 1/2$ (self-similarity index). Our definitions and results can be extended to a one-parameter family of distance dependence measures indexed by a positive exponent $0 < \alpha < 2$ on Euclidean distance, or equivalently by an index h, where $h = 2H$ for Hurst parameters $0 < H < 1$.

If $E(|X|_p^\alpha + |Y|_q^\alpha) < \infty$ define $\mathcal{V}^{(\alpha)}$ by its square

$$\mathcal{V}^{2(\alpha)}(X,Y) = \|\hat{f}_{X,Y}(t,s) - \hat{f}_X(t)\hat{f}_Y(s)\|_\alpha^2$$

$$= \frac{1}{C(p,\alpha)C(q,\alpha)} \int_{\mathbb{R}^{p+q}} \frac{|\hat{f}_{X,Y}(t,s) - \hat{f}_X(t)\hat{f}_Y(s)|^2}{|t|_p^{\alpha+p} |s|_q^{\alpha+q}} \, dt \, ds.$$

Similarly, $\mathcal{R}^{(\alpha)}$ is the square root of

$$\mathcal{R}^{2(\alpha)} = \frac{\mathcal{V}^{2(\alpha)}(X,Y)}{\sqrt{\mathcal{V}^{2(\alpha)}(X)\mathcal{V}^{2(\alpha)}(Y)}}, \qquad 0 < \mathcal{V}^{2(\alpha)}(X), \ \mathcal{V}^{2(\alpha)}(Y) < \infty,$$

and $\mathcal{R}^{(\alpha)} = 0$ if $\mathcal{V}^{2(\alpha)}(X)\mathcal{V}^{2(\alpha)}(Y) = 0$.

Now consider the Lévy fractional Brownian motion $\{W_H^d(t), \ t \in \mathbb{R}^d\}$ with Hurst index $H \in (0,1)$, which is a centered Gaussian random process with covariance function

$$E[W_H^d(t)W_H^d(s)] = |t|^{2H} + |s|^{2H} - |t-s|^{2H}, \qquad t, s \in \mathbb{R}^d.$$

See Herbin and Merzbach [2007].

In the following, (W_H, W'_{H^*}) and (X, X', Y, Y') are supposed to be independent.

Using Lemma 15.1 it can be shown that for Hurst parameters $0 < H$, $H^* \leq 1$, $h := 2H$, and $h^* := 2H^*$,

$$\text{Cov}^2_{W_H^p, W'^{\,q}_{H^*}}(X,Y) = \frac{1}{C(p,h)C(q,h^*)} \int\limits_{\mathbb{R}^p} \int\limits_{\mathbb{R}^q} \frac{|f(t,s) - f(t)g(s)|^2 dt ds}{|t|_p^{p+h} |s|_q^{q+h^*}}$$

$$= E|X - X'|_p^h |Y - Y'|_q^{h^*} + E|X - X'|_p^h E|Y - Y'|_q^{h^*}$$

$$- E|X - X'|_p^h |Y - Y''|_q^{h^*} - E|X - X''|_p^h |Y - Y'|_q^{h^*}.$$

$$\tag{15.14}$$

Here we need to suppose that $E|X|_p^h < \infty$, $E|Y|_q^{h^*} < \infty$. Observe that when $h = h^* = 1$, (15.14) is equation (15.13) of Theorem 15.1.

The corresponding statistics are defined by replacing the exponent 1 with exponent α (or h) in the distance dependence statistics (12.9), (12.10), and (12.11). That is, in the sample distance matrices replace $a_{kl} = |X_k - X_l|_p$ with $a_{kl} = |X_k - X_l|_p^\alpha$, and replace $b_{kl} = |Y_k - Y_l|_q$ with $b_{kl} = |Y_k - Y_l|_q^\alpha$, $k, l = 1, \ldots, n$.

Theorem 13.1 can be generalized for $\|\cdot\|_\alpha$ norms, so that almost sure convergence of $\mathcal{V}_n^{(\alpha)} \to \mathcal{V}^{(\alpha)}$ follows if the α-moments are finite. Similarly one can prove the weak convergence and statistical consistency for α exponents, $0 < \alpha < 2$, provided that α moments are finite.

Note that the strict inequality $0 < \alpha < 2$ is important. Although $\mathcal{V}^{(2)}$ can be defined for $\alpha = 2$, it does not characterize independence. Indeed, the case $\alpha = 2$ (squared Euclidean distance) leads to classical product-moment correlation and covariance for bivariate (X, Y). Specifically, if $p = q = 1$, then $\mathcal{R}^{(2)} = |\rho|$, $\mathcal{R}_n^{(2)} = |\hat{\rho}|$, and $\mathcal{V}_n^{(2)} = 2|\hat{\sigma}_{xy}|$, where $\hat{\sigma}_{xy}$ is the maximum likelihood estimator of Pearson covariance $\sigma_{x,y} = \sigma(X, Y)$.

15.5 Proofs of Statements

For $\mathbb{R}^d$ valued random variables $|\cdot|_d$ denotes the Euclidean norm; whenever the dimension is self-evident we suppress the index d.

15.5.1 Proof of Theorem 15.1

Proof. The proof of Theorem 15.1 starts following the theorem. To complete this proof, we need to show that all factors in the definition of $\text{Cov}_W(X, Y)$ have finite second moments. Note that $E[W^2(t)] = 2|t|$, so that $E[W^2(X)] = 2E[|X|] < \infty$. On the other hand, by the inequality $(a + b)^2 \leq 2(a^2 + b^2)$, and by Jensen's inequality, we have

$$E[X_W]^2 = E\left[W(X) - E\left(W(X)\big|W\right)\right]^2$$
$$\leq 2\left(E[W^2(X)] + E\left[E\left(W(X)\big|W\right)\right]^2\right)$$
$$\leq 2E[W^2(X)] = 4E|X| < \infty.$$

Similarly, the random variables X'_W, $Y_{W'}$, and $Y'_{W'}$ also have finite second moments. Since X and X' are iid we have

$$E(X_W X'_W)^2 = E[E(X_W X'_W)^2 |W]$$
$$= E[E[((X_W)^2 (X'_W)^2)|W] \leq E(2E[W(X)^2 2EW(X')^2|W)$$
$$= 4E^2(W(X)^2) = 4E^2(2|X|) = 16E^2|X|,$$

hence using $ab \leq (1/2)(a^2 + b^2)$ we have

$$\mathcal{W}^2(X, Y) = E[(X_W X'_W)(Y_{W'} Y'_{W'})]$$
$$\leq (1/2)[E[(X_W X'_W)^2 + E(Y_{W'} Y'_{W'})^2]$$
$$\leq 8E^2|X| + 8E^2|Y|$$
$$= 8(E^2|X| + E^2|Y|) < \infty.$$

Above we implicitly used the fact that $E[W(X)|W] = \int_{\mathbb{R}^p} W(t) dF_X(t)$ exists a.s. This can easily be proved with the help of the Borel-Cantelli lemma, using the fact that the supremum of centered Gaussian processes have small tails (see Talagrand [1988], Landau and Shepp [1970]).

Observe that

$$
\begin{aligned}
\mathcal{W}^2(X,Y) &= E[X_W X_W' Y_{W'} Y_{W'}'] \\
&= E\left[E\left(X_W X_W' Y_{W'} Y_{W'}' \,\middle|\, X, X', Y, Y'\right)\right] \\
&= E\left[E\left(X_W X_W' \,\middle|\, X, X', Y, Y'\right) E\left(Y_{W'} Y_{W'}' \,\middle|\, X, X', Y, Y'\right)\right].
\end{aligned}
$$

Here

$$
\begin{aligned}
X_W X_W' &= \left\{W(X) - \int_{\mathbb{R}^p} W(t)dF_X(t)\right\}\left\{W(X') - \int_{\mathbb{R}^p} W(t)dF_X(t)\right\} \\
&= W(X)W(X') - \int_{\mathbb{R}^p} W(X)W(t)dF_X(t) \\
&\quad - \int_{\mathbb{R}^p} W(X')W(t)dF_X(t) + \int_{\mathbb{R}^p}\int_{\mathbb{R}^p} W(t)W(s)dF_X(t)dF_X(s).
\end{aligned}
$$

By the definition of $W(\cdot)$, we have $E[W(t)W(s)] = |t| + |s| - |t-s|$, thus

$$
\begin{aligned}
E[X_W X_W' \,|\, X, X', Y, Y'] &= |X| + |X'| - |X - X'| \\
&\quad - \int_{\mathbb{R}^p} (|X| + |t| - |X - t|)dF_X(t) \\
&\quad - \int_{\mathbb{R}^p} (|X'| + |t| - |X' - t|)dF_X(t) \\
&\quad + \int_{\mathbb{R}^p}\int_{\mathbb{R}^p} (|t| + |s| - |t - s|)dF_X(t)dF_X(s).
\end{aligned}
$$

Hence

$$
\begin{aligned}
E[X_W X_W' \,|\, X, X', Y, Y'] &= |X| + |X'| - |X - X'| \\
&\quad - (|X| + E|X| - E'|X - X'|) \\
&\quad - (|X'| + E|X| - F''|X' - X''|) \\
&\quad + (E|X| + E|X'| - E|X - X'|) \\
&= E'|X - X'| + E''|X' - X''|) - |X - X'| - E|X - X'|,
\end{aligned}
$$

where E' denotes the expectation with respect to X' and E'' denotes the expectation with respect to X''. A similar argument for Y completes the proof. $\qquad\square$

15.6 Exercises

Exercise 15.1. Prove that

$$
(E[(X-E(X))(Y - E(Y))])^2
$$
$$
= E[(X - E(X))(X' - E(X'))(Y - E(Y))(Y' - E(Y'))].
$$

Exercise 15.2. If $E(|X|_p^\alpha + |Y|_q^\alpha) < \infty$ we define $\mathcal{V}^{(\alpha)}$ by its square $\mathcal{V}^2(X, Y; w)$ (15.2) with weight function $w(t, s; \alpha)$ (15.3). That is, if $X \in \mathbb{R}^p$ and $Y \in \mathbb{R}^q$,

$$
\mathcal{V}^{2(\alpha)}(X, Y) = \|\hat{f}_{X,Y}(t, s) - \hat{f}_X(t)\hat{f}_Y(s)\|_\alpha^2
$$
$$
= \frac{1}{C(p, \alpha)C(q, \alpha)} \int_{\mathbb{R}^{p+q}} \frac{|\hat{f}_{X,Y}(t, s) - \hat{f}_X(t)\hat{f}_Y(s)|^2}{|t|_p^{\alpha+p} \, |s|_q^{\alpha+q}} \, dt \, ds,
$$

where $C(p, \alpha)$ is defined in Lemma 15.1.

Lemma 15.1 holds for all exponents $0 < \alpha < 2$. Thus the Remark that follows Lemma 15.1 can be generalized for $|X|^\alpha, |Y|^\alpha$. Without using Brownian covariance, prove that

$$
\mathcal{V}^{2(\alpha)}(X, Y) = E|X - X'|^\alpha|Y - Y'|^\alpha + E|X - X'|^\alpha E|Y - Y'|^\alpha
$$
$$
- E|X - X'||Y - Y''|^\alpha - E|X - X''|^\alpha|Y - Y'|^\alpha.
$$

Exercise 15.3. Which of the properties of dCov and dCor ($\alpha = 1$) in Section 12.5 hold for arbitrary exponents $0 < \alpha < 2$? Which hold for $\alpha = 2$?

Exercise 15.4. Discuss the result if in the definition of Brownian distance covariance, Brownian motion is replaced by other stochastic processes like the Poisson process or pseudorandom processes. For the definition of a pseudorandom process see Móri and Székely [2022].

16

U-statistics and Unbiased dCov^2

CONTENTS

16.1 An Unbiased Estimator of Squared dCov

In Chapter 12 the distance covariance statistic is derived from the L_2 norm $\|\hat{f}_{X,Y}^n(t,s) - \hat{f}_X^n(t)\hat{f}_Y^n(s)\|^2$, and we obtain a simple computing formula

$$\mathcal{V}_n^2(\mathbf{X}, \mathbf{Y}) = \frac{1}{n^2} \sum_{k,\ell=1}^{n} \widehat{A}_{k\ell}\widehat{B}_{k\ell},$$

where $\widehat{A}$ and $\widehat{B}$ are double-centered distance matrices of the samples. See Definition 12.4 for the definition of $\widehat{A}$, $\widehat{B}$.

From Proposition 13.1 we see that $\mathcal{V}_n^2$ is biased with $E[\mathcal{V}_n^2] > \mathcal{V}^2(X,Y)$ (although asymptotically unbiased). In this chapter we introduce an unbiased estimator of $\mathcal{V}^2(X,Y)$ and show that it is a U-statistic. As a first step, we introduce a modified type of double-centering, called "U-centering."

Definition 16.1 (U-centered distance matrix). Let $A = (a_{ij})$ be a symmetric, real-valued $n \times n$ matrix with zero diagonal, $n > 2$. Define the U-centered matrix $\widetilde{A}$ as follows. Let the (i,j)-th entry of $\widetilde{A}$ be defined by

$$\widetilde{A}_{i,j} = \begin{cases} a_{i,j} - \frac{1}{n-2}\sum_{\ell=1}^{n} a_{i,\ell} - \frac{1}{n-2}\sum_{k=1}^{n} a_{k,j} + \frac{1}{(n-1)(n-2)}\sum_{k,\ell=1}^{n} a_{k,\ell}, & i \neq j; \\ 0, & i = j. \end{cases}$$
(16.1)

Here "U-centered" is so named because as shown below, the corresponding inner product statistic

$$(\widetilde{A} \cdot \widetilde{B}) := \frac{1}{n(n-3)} \sum_{i \neq j} \widetilde{A}_{i,j} \widetilde{B}_{i,j}$$
(16.2)

is an unbiased estimator of $\mathcal{V}^2(X,Y)$.

To prove that $E[(\widetilde{A} \cdot \widetilde{B})] = \mathcal{V}^2(X,Y)$, we derive the expected value from the following algebraically equivalent formula.

Proposition 16.1. *For $n > 3$, the inner product distance covariance statistic* (16.2) *is algebraically equal to*

$$\mathcal{U}_n = \frac{1}{n(n-3)} \sum_{i \neq j} a_{ij} b_{ij} - \frac{2}{n(n-2)(n-3)} \sum_{i=1}^{n} a_i . b_i .$$

$$+ \frac{a_{..}b_{..}}{n(n-1)(n-2)(n-3)}.$$
(16.3)

See Section 16.9 Proof (1) for details. To see the connection between $\mathcal{U}_n$ and $\mathcal{V}_n^2$ of Chapters 12–15, compare the formula $\mathcal{U}_n$ with formula (12.23).

Proposition 16.2 (Unbiasedness). *Let (x_i, y_i), $i = 1, \ldots, n$ denote a sample of observations from the joint distribution (X, Y) of random vectors X and Y. Let $A = (a_{ij})$ be the Euclidean distance matrix of the sample $x_1, \ldots, x_n$ from the distribution of X, and $B = (b_{ij})$ be the Euclidean distance matrix of the sample $y_1, \ldots, y_n$ from the distribution of Y. Then if $E(|X| + |Y|) < \infty$, for $n > 3$,*

$$(\widetilde{A} \cdot \widetilde{B}) := \frac{1}{n(n-3)} \sum_{i \neq j} \widetilde{A}_{i,j} \widetilde{B}_{i,j}$$
(16.4)

is an unbiased estimator of squared population distance covariance $\mathcal{V}^2(X,Y)$.

Proposition 16.2 follows from Proposition 16.1 and $E[\mathcal{U}_n] = \mathcal{V}^2(X,Y)]$, which is proved in Section 16.9 (2).

16.2   The Hilbert Space of U-centered Distance Matrices

For a fixed $n \geq 4$, we define a Hilbert space generated by Euclidean distance matrices of arbitrary sets (samples) of n points in a Euclidean space $\mathbb{R}^p$,

$p \geq 1$. Consider the linear span $\mathcal{S}_n$ of all $n \times n$ distance matrices of samples $\{x_1, \ldots, x_n\}$. Let $A = (a_{ij})$ be an arbitrary element in $\mathcal{S}_n$. Then A is a real-valued, symmetric matrix with zero diagonal.

Let $\mathcal{H}_n = \{\widetilde{A} : A \in \mathcal{S}_n\}$ and for each pair of elements $C = (C_{i,j})$, $D = (D_{i,j})$ in the linear span of $\mathcal{H}_n$ define their inner product

$$\mathcal{U}_n := (C \cdot D) = \frac{1}{n(n-3)} \sum_{i \neq j} C_{ij} D_{ij}. \tag{16.5}$$

If $(C \cdot C) = 0$, then $(C \cdot D) = 0$ for any $D \in \mathcal{H}_n$.

In Theorem 17.2 it is shown that every matrix $C \in \mathcal{H}_n$ is the U-centered distance matrix of a configuration of n points in a Euclidean space $\mathbb{R}^p$, where $p \leq n - 2$.

16.3   U-statistics and V-statistics

16.3.1   Definitions

We have shown that the inner product estimator $(\widetilde{A} \cdot \widetilde{B})$ is an unbiased estimator of squared distance covariance $\mathcal{V}^2(X, Y)$. In what follows, we also prove that the statistic $\mathcal{U}_n = (\widetilde{A} \cdot \widetilde{B})$ is a U-statistic.

We have seen many examples of energy statistics which are defined as V-statistics. Readers familiar with U-statistics and V-statistics may wonder why we do not work with U-statistics. Actually, we could alternately formulate energy statistics as U-statistics.

Suppose that θ is a parameter of interest of a distribution F, and $h(x_1, \ldots, x_r)$ is a symmetric function such that $E[h(X_1, \ldots, X_r)] = \theta$. Then $h(X_1, \ldots, X_r)$ is an unbiased estimator of θ. If we have observed a sample of size $n > r$, then $h(X_1, \ldots, X_r)$ is not a natural statistic for θ because it uses only r of the sample observations. In Definition 16.2 below, the U-statistic with kernel h is defined, which is a function of the full sample of n elements and unbiased for θ.

For a positive integer r, let $S(n, r)$ denote all the distinct unordered r-subsets of $\{1, 2, \ldots, n\}$. For a set $\varphi \subset \{1, \ldots, n\}$, we define the notation $x_\varphi = \{x_i \mid i \in \varphi\}$. The following defines a real-valued U-statistic of real variables.

Definition 16.2 (U-statistics of degree r). Suppose $\{x_1, \ldots, x_n\}$ is a sample from distribution F_X and θ is a parameter of interest of the distribution. Let $h : \mathbb{R}^r \to \mathbb{R}$ be a symmetric kernel function of r variables, such that $E[h(X_1, \ldots, X_r)] = \theta$. For each $n \geq r$, the associated U-statistic of degree r, $U_{n,r} : \mathbb{R}^n \to \mathbb{R}$, is the average of $h(x_\varphi)$ over all distinct subsets φ of size r:

$$U_{n,r}(x_1, \ldots, x_n) = \binom{n}{r}^{-1} \sum_{\varphi \in S(n,r)} h(x_\varphi), \tag{16.6}$$

Definition 16.3 (V-statistics of degree r). Let $\mathcal{S}(n,r)$ be the set of all ordered subsets of size r that can be selected from the integers $\{1, \ldots, n\}$, and suppose that $h(x_1, \ldots, x_r)$ is a symmetric kernel function. For each $s \in \mathcal{S}(n,r)$ let x_s denote the r sample elements indexed by s. Then $\mathcal{S}(n,r)$ has n^r members and the corresponding V-statistic of degree r is

$$V_{n,r} = \frac{1}{n^r} \sum_{s \in \mathcal{S}(n,r)} h(x_s). \tag{16.7}$$

Given a kernel function, it is easy to formulate a U-statistic or a V-statistic with that kernel. However, given an unbiased statistic for θ, how can one determine whether or not it is a U-statistic? What is the degree and kernel function? Proposition 3 in Lenth [1983] proves that the jackknife invariance property characterizes U-statistics. Below we define jackknife invariance, provide examples, and a proof of the jackknife invariance theorem for U-statistics. We apply the results to a specific problem of interest, a U-statistic for squared distance covariance. Among the applications, this U-statistic for dCov^2 is essential for the partial distance covariance introduced in Chapter 17.

Our main goal is the U-statistic for dCov^2, but when we look at the dCov$_n^2$ statistic from Chapter 12, formulated with double-centered distance matrices, it is complicated to try to find a U-statistic because the kernel function is not obvious. We will apply the jackknife invariance theorem to verify that a reformulated squared distance covariance statistic (16.2) is a U-statistic and give its kernel.

16.3.2   Examples

To warm up, let us see a much simpler example of how energy V-statistics and U-statistics are related.

Example 16.1 (Energy goodness-of-fit statistics). Suppose we have a goodness-of-fit problem. We have observed a random sample $Y_1, \ldots, Y_n$, and want to test whether the sampled distribution F_Y is equal to a hypothesized distribution $F_0 = F_X$. In Chapter 5 we introduced the energy goodness-of-fit test using the following test statistic:

$$n\mathcal{E}_n = n \left[\frac{2}{n} \sum_{i=1}^n E|y_i - X| - E|X - X'| - \frac{1}{n^2} \sum_{i,j=1}^n |y_i - y_j| \right].$$

This is $nV_n(y_1, \ldots, y_n)$, where V_n is a V-statistic with kernel

$$h(y_1, y_2) = E|y_1 - X| + E|y_2 - X| - E|X - X'| - |y_1 - y_2|.$$

The V-statistic is defined by averaging $h(y_i, y_j)$ over all n^2 pairs (y_i, y_j):

$$V_n = \frac{1}{n^2} \sum_{i,j=1}^n h(y_i, y_j) = \frac{1}{n^2} \sum_{i,j} (E|y_i - X| + E|y_j - X| - E|X - X'| - |y_i - y_j|).$$

$$\tag{16.8}$$

After simplification it is clear that $\mathcal{E}_n = V_n$, so $\mathcal{E}_n$ is a V-statistic. We proved that $V_n \geq 0$, so its square root is a distance between the distributions F_X and F_Y.

What is the corresponding U-statistic? For the U-statistic we average the same kernel function over all $i \neq j$:

$$U_n = \binom{n}{r}^{-1} \sum_{1 \leq i < j \leq n} h(y_i, y_j)$$

$$= \frac{1}{n(n-1)} \sum_{i \neq j} (E|y_i - X| + E|y_j - X| - E|X - X'| - |y_i - y_j|)$$

$$= \frac{2}{n} \sum_{i=1}^{n} E|y_i - X| - E|X - X'| - \frac{1}{n(n-1)} \sum_{i \neq j} |y_i - y_j|. \qquad (16.9)$$

It is easy to check that the U-statistic (16.9) is unbiased for the population energy distance

$$E[U_n] = 2E|X - Y| - E|X - X'| - E|Y - Y'|.$$

Example 16.2. The U-statistic with kernel

$$h(x_1, x_2, x_3) := (1/3)\text{sgn}(\text{median}(x_1, x_2, x_3) - \text{mean}(x_1, x_2, x_3))$$

can be applied to test symmetry with unknown center (sgn() denotes sign function).

The jackknife characterization theorem can be extended to a two-sample problem, in which the data is a a bivariate sample $\{(x_1, y_1), (x_2, y_2), \ldots, (x_n, y_n)\}$ for $n \geq 1$. By replacing x_i with $(x, y)_i = (x_i, y_i)$, all the jackknife arguments still hold.

Example 16.3. The following two-sample U-statistic can be applied to test association. The kernel

$$h((x_1, y_1), (x_2, y_2), x_3, y_3))$$

has degree 3 and is defined as the symmetrized version of

$$12 I(x_1 < x_2, y_1 < y_3) - 3.$$

The corresponding U-statistic is an unbiased estimator of Spearman's

$$\rho := 12P(X_1 < X_2, Y_1 < Y_3) - 3 = \text{Cor}(F(X, \infty), F(\infty, Y))$$

where F is the joint cdf of (X, Y).

Example 16.4 (Two-sample energy statistic). Consider the energy distance $\mathcal{E}(X, Y)$ between independent random variables X and Y. If $X_1, \ldots, X_n$ and $Y_1, \ldots, Y_m$ are independent random samples from the distributions F_X and F_Y, respectively, then the U-statistic for

$$\mathcal{E}(X, Y) = 2E|X - Y| - E|X - X'| - E|Y - Y'|$$

is

$$U_{n,m} = \frac{2}{mn} \sum_{i=1}^{n} \sum_{k=1}^{m} |x_i - y_k| \tag{16.10}$$

$$- \frac{2}{n(n-1)} \sum_{1 \le i < j \le n} |x_i - x_j| - \frac{2}{m(m-1)} \sum_{1 \le k < \ell \le m} |y_k - y_\ell|,$$

with the kernel function

$$h(x_1, x_2; y_1, y_2) = \tfrac{1}{2} \big(|x_1 - y_1| + |x_1 - y_2| + |x_2 - y_1| + |x_2 - y_2|$$
$$- 2|x_1 - x_2| - 2|y_1 - y_2| \big). \tag{16.11}$$

It is easy to check that

$$E[U_{n,m}] = E[h(X, X'; Y, Y')] = 2E|X - Y| - E|X - X'| - E|Y - Y'|.$$

The V-statistic (two sample energy distance) is the corresponding V-statistic with kernel (16.11):

$$V_{n,m} = \frac{1}{n^2 m^2} \sum_{i,j=1}^{n} \sum_{k,\ell=1}^{m} h(x_i, x_j; y_k, y_\ell)$$

$$= \frac{2}{mn} \sum_{i=1}^{n} \sum_{k=1}^{m} |x_i - y_k| - \frac{1}{n^2} \sum_{i,j=1}^{n} |x_i - x_j| - \frac{1}{m^2} \sum_{k,\ell=1}^{m} |y_k - y_\ell|.$$

It is proved in Székely and Rizzo [2004a] that $V_{n,m}(X, Y) \ge 0$. Note that the U-statistic can be negative.

The U-statistic $U_{n,m}$ in (16.10) is an unbiased estimator of *energy distance* $\mathcal{E}(X, Y)$ for every n, m, while $E[V_{n,m}] = \mathcal{E}(X, Y) + n^{-1}E|X - X'| + m^{-1}E|Y - Y'|$ is asymptotically unbiased for the energy distance. The test statistic is $\frac{nm}{n+m} V_{n,m}$, which has expected value $E|X - Y| = E|X - X'| = E|Y - Y'|$ when X and Y are identically distributed.

We work with the V-statistic for the problem of testing equal distributions because of the non-negativity of $V_{n,m}$, which is better to interpret as a statistical distance, but there is a parallel U-statistic theory as well. One of the obvious differences is that energy U-statistics can be negative, so we cannot take their square root.

16.4   Jackknife Invariance and U-statistics

In this section we provide a proof of a characterization theorem for U-statistics. Also see Lenth [1983] for proof of the jackknife invariance characterization.

Let $S(n, m)$ denote the set of all size m subsets of a sample $x_1, \ldots, x_n$, and let $S_i(n, m)$ denote the set of size m subsets of $x_1, \ldots, x_n$ excluding x_i.

Definition 16.4 (Jackknife statistics). For a statistic $T_n(x_1, \ldots, x_n)$, the i-th jackknife statistic $1 \leq i \leq n$, is computed as T_{n-1} on the sample $\{x_1, \ldots, x_n\} \setminus \{x_i\}$. Introduce the following notation for jackknifed U-statistics:

$$U_{n,r}^{-i}(x_1, \ldots, x_n) := U_{n-1,r}(x_1, \ldots, x_{i-1}, x_{i+1}, \ldots, x_n), \qquad (16.12)$$

where $U_{n-1,r}(x_1, \ldots, x_{i-1}, x_{i+1}, \ldots, x_n)$ is the i-th jackknife statistic removing the element x_i.

For U-statistics, we can verify the following lemma, which is the jackknife invariance property of U-statistics.

Lemma 16.1 (Jackknife invariance property). *If $U_{n,r}$ is a U-statistic of degree r, then for all $n > r$*

$$(n - r)\binom{n}{r}U_{n,r}(x_1, \ldots, x_n) = \sum_{i=1}^{n}\binom{n-1}{r}U_{n,r}^{-i}(x_1, \ldots, x_n); \qquad (16.13)$$

or equivalently,

$$U_{n,r} = \frac{1}{n}\sum_{i=1}^{n}U_{n-1,r}^{-i}.$$

Proof. In (16.13), each term $h(x_\varphi)$ is counted $(n - r)$ times on both sides. Hence the equality holds.   □

The following is a simple example of the jackknife invariance property for a kernel of degree $r = 2$.

Example 16.5 (Jackknife invariance of sample variance). The maximum likelihood estimator (MLE) of variance is a V-statistic with kernel $h(x_j, x_k) = \frac{1}{2}(x_j - x_k)^2$. It is easy to show that

$$V_{n,2} = \frac{1}{2n^2}\sum_{j,k=1}^{n}(x_j - x_k)^2 = \frac{1}{n}\left(\sum_{j=1}^{n}x_j^2 - n\bar{x}^2\right).$$

The MLE is a biased estimator of $\sigma^2 = Var(X)$. The corresponding U-statistic is

$$U_{n,2} = \frac{1}{n(n-1)}\sum_{j<k}(x_j - x_k)^2 = \frac{1}{n-1}\left(\sum_{j=1}^{n}x_j^2 - n\bar{x}^2\right), \qquad (16.14)$$

which is the unbiased estimator of σ^2. Let us verify the jackknife invariance property of the U-statistic s^2 in (16.14). In this example, $S(n, 2)$ is the set of two-element subsets of $\{1, \ldots, n\}$, and if $\varphi = \{j, k\} \in S(n, 2)$, $x_\varphi := \{x_j, x_k\}$. Then the leave-$x_i$-out jackknife statistic is

$$U_{n-1}^{-i} = \frac{1}{n-2} \left(\sum_{j \neq i} x_j^2 - (n-1) \left(\frac{\sum_{j \neq i} x_j}{n-1} \right)^2 \right)$$

$$= \frac{1}{n-2} \left(\sum_{j=1}^{n} x_j^2 - x_i^2 - (n-1) \left(\frac{\sum_{j=1}^{n} x_j - x_i}{n-1} \right)^2 \right).$$

Hence, after simplification,

$$\sum_{i=1}^{n} U_{n-1}^{-i} = \frac{1}{n-2} \left(\frac{n(n-2)}{n-1} \sum_{j=1}^{n} x_j^2 - \frac{n-2}{n-1} \left(\sum_{j=1}^{n} x_j \right)^2 \right)$$

$$= \frac{n}{n-1} \left(\sum_{j=1}^{n} x_j^2 - n \left(\frac{\sum_{j=1}^{n} x_j}{n} \right)^2 \right) = nU_n$$

$$= n \left(\frac{1}{n-1} \sum_{j=1}^{n} x_j^2 - n\bar{x}_n^2 \right) = nU_n.$$

Using mathematical induction, one can prove that the converse of Lemma 16.1 is also true. In other words, the *jackknife invariance* for all $n > r$ is a necessary and sufficient condition for a symmetric statistic U_n to define a U-statistic.

Let θ be a parameter of interest, and suppose that there exists a symmetric function h such that $E[h(X_1, \ldots, X_r)] = \theta$. Let r be the smallest sample size for which h is defined. Then

$$U_{n,r} = \binom{n}{r} \sum_{S(n,r)} h(X_{i1}, \ldots, X_{ir}) \tag{16.15}$$

is a U statistic of degree r with kernel h.

Notice that if $n = r$, then

$$U_r = \binom{r}{r} \sum_{S(r,r)} h(X_1, \ldots, X_r) = h(X_1, \ldots, X_r)$$

so that the kernel function $h(\cdot)$ is uniquely determined by U_r.

Theorem 16.1 (Jackknife invariance theorem). *Let $U_n(x_1, \ldots, x_n)$ be a symmetric statistic of a sample $x_1, \ldots, x_n$. A necessary and sufficient condition* *for* $U_n(x_1, \ldots, x_n)$ *to be a U-statistic of degree* r *is that* $U_n(x_1, \ldots, x_n)$ *has the jackknife invariance property: the identity*

$$n \cdot U_n(x_1, \ldots, x_n) = \sum_{i=1}^{n} U_{n-1}^{-i}(x_1, \ldots, x_n) \qquad (16.16)$$

holds for all $n > r$.

Proof. Induction: Suppose that U_n is a symmetric statistic, and for any $n > r$ the jackknife invariance property holds. We need to show that if $n \geq r$,

$$U_{n,r} = \binom{n}{r}^{-1} \sum_{S(n,r)} h(x_1, \ldots, x_r),$$

where h is a symmetric function such that $E[h(X_1, \ldots, X_r)] = 0$.

For $n = r + 1$, by jackknife invariance we have

$$U_{r+1,r} = \frac{1}{r+1} \sum_{i=1}^{n} U_{r,r}^{-i} = \binom{r+1}{r}^{-1} \sum_{S(r+1,r)} h(x_{i1} \ldots, x_{ir}),$$

which has the form (16.15).

Suppose that $U_{n,r}$ has the jackknife invariance property so that (16.16) is true for all $n > r$. We need to prove that if $U_{n,r}, n > r$ satisfies (16.15), then $U_{n+1,r}$ satisfies (16.15). By the jackknife invariance and the fact that $U_{n,r}$ is a U-statistic,

$$U_{n+1,r} = \frac{1}{n+1} \sum_{i=1}^{n} U_{n,r}^{-i}(x_1, \ldots, x_r)$$

$$= \frac{1}{n+1} \sum_{i=1}^{n} \binom{n}{r}^{-1} \sum_{S_i(n+1,r)} h(x_{i1}, \ldots, x_{ir}). \qquad (16.17)$$

Now $\bigcup_{i=1}^{n+1} S_i(n+1,r) = S(n+1,r)$ and each term in the sum is counted $n + 1 - r$ times, so

$$U_{n+1,r} - \frac{n+r-1}{n+1} \binom{n}{r}^{-1} \sum_{S(n+1,r)} h(x_{i1}, \ldots, x_{ir})$$

$$= \binom{n+1}{r}^{-1} \sum_{S(n+1,r)} h(x_{i1}, \ldots, x_{ir}).$$

Thus, $U_{n+1,r}$ has the same form as (16.15). Hence the statistic $U_{n+1,r}$ is a U-statistic with kernel h of degree r.

By induction, if $U_{n,r}(x_1, \ldots, x_r)$ is a symmetric statistic with the jackknife invariance property, then $U_{n,r}$ is a U-statistic with kernel U_r. Apply Lemma 16.1 to complete the proof. $\qquad \square$

16.5   The Inner Product Estimator is a U-statistic

A U-statistic is an unbiased estimator of the expected value of the parameter $\theta := E[h(X_1, X_2, \ldots, X_r)]$, where $h(\cdot)$ is the kernel function of the U-statistic, provided $E[h(X)] < \infty$. For example, if $h(x, y)$ is a symmetric kernel function and $U_{n,2}$ is the U-statistic with kernel $h(\cdot)$, then for a random sample $\{X_1, \ldots, X_n\}$ from the distribution of X, $E[U_{n,2}(X_1, \ldots, X_n)] = E[h(X, X')]$, where X, X' are iid. This follows directly from the definition, because of the linearity of the expected value; if $x_1, x_2, \ldots, x_n$ are iid, then

$$E[U_{n,2}(X)] = \binom{n}{2}^{-1} \left[E \sum_{1 \leq j < k \leq n} h(x_j, x_k) \right]$$

$$= \frac{2}{n(n-1)} \sum_{1 \leq j < k \leq n} E[h(X, X')] = E[h(X, X')].$$

A simple example is the U-statistic for Gini's mean difference with $h(x_1, x_2) = |x_1 - x_2|$, with expected value $E|X - Y|$. Similarly, if the degree r kernel is $h(x_1, \ldots, x_r)$, then

$$E[U_{n,r}(X)] = E[h(X_1, \ldots, X_r)],$$

where $X, X_1, \ldots, X_r$ are iid.

Often, however, the kernel function is complicated and the expected value of the U-statistic is not easily evaluated directly by taking the expected value of the kernel. In that case, an alternate representation of the statistic may be more tractable.

In this chapter and Chapter 17 our main application of U-statistics is the unbiased estimator (16.2) of squared distance covariance. A direct evaluation of the expected value of the unbiased distance covariance statistic follows.

Recall that we have denoted the distance matrices $A = (a_{ij})$ and $B = (b_{ij})$ for the X and Y samples, respectively. The row and column sums of A and B are denoted by

$$a_{i.} = \sum_{\ell=1}^{n} a_{i\ell}, \qquad a_{.j} = \sum_{k=1}^{n} a_{kj}, \qquad a_{..} = \sum_{k,\ell=1}^{n} a_{k\ell},$$

$$b_{i.} = \sum_{\ell=1}^{n} b_{i\ell}, \qquad b_{.j} = \sum_{k=1}^{n} b_{kj}, \qquad b_{..} = \sum_{k,\ell=1}^{n} b_{k\ell}.$$

Lemma 16.2. *For $1 \leq i \leq n$, let $a_{i.}^{-i}$, $b_{i.}^{-i}$, $a_{..}^{-i}$, and $b_{..}^{-i}$ denote the corresponding sums if observation (x_i, y_i) is removed from the sample. Let $\mathcal{U}_{n-1}^{-i}$ denote the inner product (16.2) for sample size $n-1$ when the i-th observation*

$(x, y)_i = (x_i, y_i)$ *is removed from the sample. Then we have*

$$\mathcal{U}_{n-1}^{-i} = \frac{1}{(n-1)(n-4)} \sum_{j \neq k, j \neq i, k \neq i} a_{jk} b_{jk}$$

$$- \frac{2}{(n-1)(n-3)(n-4)} \sum_{j=1, j \neq i}^{n} a_{j.}^{-i} b_{j.}^{-i}$$

$$+ \frac{a_{..}^{-i} b_{..}^{-i}}{(n-1)(n-2)(n-3)(n-4)}. \tag{16.18}$$

This formula follows immediately by evaluating the formula $\mathcal{U}_{n-1}$ on the leave-x_i-out sample.

Theorem 16.2. $\mathcal{U}_n$ *(16.3) is a U-statistic of degree 4 and its kernel function is* $\mathcal{U}_4$.

For the proof, we just need to show that $\mathcal{U}_n$, $n > 4$ is jackknife invariant. Then by Theorem 16.1 $\mathcal{U}_n$ is a U-statistic with kernel $\mathcal{U}_4$.

Proof. For $i \neq k$ we have

$$a_{i.}^{-k} = a_{i.} - a_{ik}, \qquad\qquad a_{..}^{-k} = a_{..} - 2a_{.k},$$
$$b_{i.}^{-k} = b_{i.} - b_{ik}, \qquad\qquad b_{..}^{-k} = b_{..} - 2b_{.k}.$$

For the right hand side of (16.16), we have the following:

$$\sum_{k=1}^{n} \mathcal{U}_{n-1}^{-k} = \sum_{k=1}^{n} \frac{\sum_{i \neq j, i \neq k, j \neq k} a_{ij} b_{ij}}{(n-1)(n-4)}$$

$$- \sum_{k=1}^{n} \frac{2 \sum_{i=1, i \neq k}^{n} (a_{i.} - a_{ik})(b_{i.} - b_{ik})}{(n-1)(n-3)(n-4)}$$

$$+ \sum_{k=1}^{n} \frac{(a_{..} - 2a_{.k})(b_{..} - 2b_{.k})}{(n-1)(n-2)(n-3)(n-4)}$$

$$= \frac{(n-2) \sum_{i \neq j} a_{ij} b_{ij}}{(n-1)(n-1)} - \frac{2 \left[(n-3) \sum_{i=1}^{n} a_{i.} b_{i.} + \sum_{i \neq k} a_{ik} b_{ik} \right]}{(n-1)(n-3)(n-4)}$$

$$+ \frac{(n-4)a_{..} b_{..} + 4 \sum_{k=1}^{n} a_{k.} b_{k.}}{(n-1)(n-2)(n-3)(n-4)}$$

$$= \frac{\sum_{i \neq j} a_{ij} b_{ij}}{n-3} - \frac{2}{(n-2)(n-3)} \sum_{i=1}^{n} a_{i.} b_{i.} + \frac{a_{..} b_{..}}{(n-1)(n-2)(n-3)}.$$

Comparing with (16.3), we can verify that the above equates to $n \cdot \mathcal{U}_n$, which by Theorem 16.1 indicates that $\mathcal{U}_n$ is a U-statistic. The kernel function of the corresponding U-statistic is the inner product that was defined in (16.5) with $n = 4$, or equivalently $\mathcal{U}_4$ defined in (16.3). $\qquad\square$

16.6   Asymptotic Theory

Since $\mathcal{U}_n$ is a U-statistic we can apply the relevant limit theorems to study its asymptotic behavior. What are these theorems?

For $c = 0, 1, \ldots, r$ introduce

$$h_c(x_1, x_2, \ldots, x_c, X_{c+1}, \ldots, X_r),$$

where $X_{c+1}, \ldots, X_r$ are iid. Denote the variance of $h_c(X_1, X_2, \ldots, X_c)$ by σ_c^2.

Theorem 16.3. *If $0 < \sigma_1 < \infty$, then the limit distribution of $\sqrt{n}(U_n - \theta)$ is Gaussian with expected value 0 and variance $r^2 \sigma_1^2$.*

In energy inference the typical case is that $\sigma_1 = 0$ under the null hypothesis, thus we need limit theorems for this case, too.

Definition 16.5 (Degenerate kernel). A U-statistic has degeneracy m if $\sigma_1 = \sigma_2 = \ldots, = \sigma_m = 0$ but $\sigma_{m+1} > 0$.

Example 16.6. If $h(x_1, x_2) = x_1 x_2$, then $\sigma_1 = E(X_1) Var(X_1)$ which is degenerate of order 1 when $E(X_1) = 0$. In this case, the limit distribution of U_n is not Gaussian. It is easy to see that

$$U_n = \frac{1}{n(n-1)} \left[\left(\sum_{i=1}^n X_i \right)^2 - \sum_{i=1}^n X_i^2 \right] = \frac{1}{n-1} \left[\left(\frac{1}{\sqrt{n}} \sum_{i=1}^n X_i \right)^2 - \frac{1}{n} \sum_{i=1}^n X_i^2 \right],$$

thus the limit distribution of nU_n is the same as the distribution of

$$(Z^2 - 1) Var(X_1),$$

where Z is standard Gaussian.

The above example applies the following general result for U-statistics with degeneracy of order 1.

Theorem 16.4 (Hoeffding). *If U_n is a U-statistic with symmetric kernel h of degree r, degeneracy of order 1, and expectation θ, then the limit distribution of $n(U_n - \theta)$ is the distribution of the quadratic form*

$$\frac{r(r-1)}{2} \sum_{j=1}^\infty \lambda_j (Z_j^2 - 1),$$

where $Z_1, Z_2, \ldots$ are iid standard Gaussian variables and $\lambda_1, \lambda_2, \ldots$ are eigenvalues of a Hilbert-Schmidt integral operator.

In Example 16.6 this integral operator has kernel $A(x_1, x_2) = h_2(x_1, x_2) - \theta$. The condition $E[A^2(X_1, X_2)] < \infty$ guarantees that the Hilbert-Schmidt eigenvalue equation

$$\int_{-\infty}^{\infty} A(x_x, x_2)\phi(x_1)dx_1 = \lambda\phi(x_2)$$

has countably many real eigenvalue solutions $\lambda_1, \lambda_2, \ldots$.

Proofs can be found in Hoeffding [1948], Lee [2019] and Koroljuk and Borovskich [1994].

Theory of V-statistics is parallel to theory of U-statistics. If in Theorem 16.4 V_n is the degenerate kernel V-statistic of order 1, then the limit distribution of nV_n is the distribution of the quadratic form

$$Q = \sum_{j=1}^{\infty} \lambda_j Z_j^2,$$

where $Z_1, Z_2, \ldots$ are iid standard Gaussian variables and $\lambda_1, \lambda_2, \ldots$ are eigenvalues of a Hilbert-Schmidt integral operator.

For a general (but conservative) criterion for a test decision, see Theorem 7.3.

16.7 Relation between dCov U-statistic and V-statistic

We have computing formulas

$$U_n = \frac{\sum_{i\neq j} a_{ij}b_{ij}}{n(n-3)} - \frac{2\sum_{i=1}^{n} a_i.b_i.}{n(n-2)(n-3)} + \frac{a_{..}b_{..}}{n(n-1)(n-2)(n-3)}.$$

and

$$V_n = \frac{\sum_{i\neq j} a_{ij}b_{ij}}{n^2} - \frac{2\sum_{i=1}^{n} a_i.b_i.}{n^3} + \frac{a_{..}b_{..}}{n^4}. \tag{16.19}$$

Therefore,

$$n^4 V_n = n(n-1)(n-2)(n-3)U_n + (3n-2)S_1 - 2S_2,$$

where

$$S_1 = \sum_{i,j=1}^{n} a_{ij}b_{ij}, \quad S_2 = \sum_{i=1}^{n}(a_i.b_i.).$$

Example 16.7. For a visual comparison of the sampling distributions of U_n and V_n, we simulated correlated bivariate normal data ($\mu_1 = 0, \mu_2 = 0, \sigma_1 = 1, \sigma_2 = 1, \rho = 0.2$). We computed both nU_n and nV_n for each of 1000 simulated samples of size 100. The histograms in Figure 16.1 serve as an estimate of the sampling distributions. Comparing the distributions, we note that the general shape of the distribution is similar for both, but the U-statistic takes some negative values while $nV_n > 0$. From these plots it is also clear that the square root of the U-statistic is not meaningful.

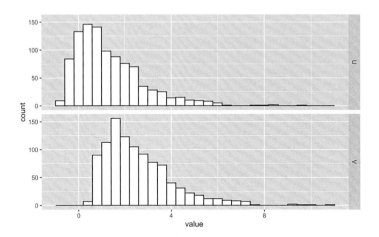

FIGURE 16.1
Histograms comparing the sampling distributions of nU_n and nV_n on simulated correlated bivariate normal data ($\rho = 0.2$).

16.7.1 Deriving the Kernel of dCov V-statistic

We can use our results for the dCov^2 U-statistic above to derive the kernel of the V-statistic in the formulation (16.19).

V-statistic for the Sum of Products

Let A and B denote the Euclidean distance matrices of the X and Y samples, respectively. If $p = (i, j, k, \ell)$ is an ordered subset of integers $\{1, \ldots, n\}$, let A_p denote the submatrix of A that contains the rows and columns of A indexed by p. Let

$$h_1(p) = \frac{1}{12} \sum_{r,s=1}^{4} A_p(r, s) B_p(r, s).$$

Then $h_1(p) = h_1(i, j, k, \ell)$ is a degree 4 kernel of a V-statistic for the mean S_1/n^2:

$$\frac{S_1}{n^2} = \frac{1}{n^2} \sum_{i,j=1}^{n} a_{ij} b_{ij} = \frac{1}{n^4} \sum_{i,j,k,\ell} h_1(i, j, k, \ell).$$

A V-statistic for the sum is

$$S_1 = \sum_{i,j=1}^{n} a_{ij}b_{ij} = \frac{n^2}{n^4} \sum_{i,j,k,\ell} h_1(i,j,k,\ell),$$

with kernel $h_{S_1}(p) = n^2 h_1(p)$.

V-statistic for the Inner Product of Row Sums

This kernel function is obtained from the identity

$$n\sum_{i=1}^{n} a_{i.}b_{i.} = \frac{1}{12} \sum_{i,j,k,\ell=1}^{n} (a_{ij}b_{ij} + a_{ji}b_{jk} + a_{ki}b_{kj} + \dots),$$

with twelve terms inside the parentheses. Each set of three products leaves out one index. Let

$$h_2(i,j,k,\ell) = \frac{1}{12} \sum_{i,j,k,\ell=1}^{n} (a_{ij}b_{ij} + a_{ji}b_{jk} + a_{ki}b_{kj} + \dots).$$

Then h_2 is a symmetric degree 4 kernel corresponding to the V-statistic for S_2/n^3:

$$\frac{S_2}{n^3} = \frac{\sum_{i=1}^{n}(a_{i.}b_{i.})}{n^3} = \frac{1}{n^4} \sum_{i,j,k,\ell=1}^{n} \frac{1}{12}(a_{ij}b_{ij} + a_{ji}b_{jk} + a_{ki}b_{kj} + \dots).$$

Hence the kernel function for the V-statistic $S_2 = \sum_{i=1}^{n} a_{i.}b_{i.}$ is $h_{S_2}(i,j,k,\ell) = n^3 h_2(i,j,k,\ell)$.

The Kernel for the V-statistic

We have that

$$n^4 V_n = n(n-1)(n-2)(n-3)U_n + (3n-2)S_1 - 2S_2,$$

and degree 4 kernel functions for the sums S_1 and S_2. To convert the kernel of the U-statistic to a kernel for a V-statistic, we introduce a symmetric 4-variate function $f : \mathbb{Z}^4 \to \{1,0\}$, such that $f(i,j,k,\ell) = 1$ if (i,j,k,ℓ) are four unique integers, and $f(i,j,k,\ell) = 0$ otherwise. An example of such a function is

$$f(p) = f(i,j,k,\ell) = \frac{\max(n^4, \Pi_{r,s=1}^{4}|p_r - p_s|)}{n^4}.$$

16.7.2 Combining Kernel Functions for V_n

Finally, we define a kernel function

$$h_U(p) = f(p)n(n-1)(n-2)(n-3)U_{4,4}(p)$$

and we obtain a degree 4 symmetric kernel for the dCov$_2$ V-statistic

$$
\begin{aligned}
h(i,j,k,\ell) &= h_U(p) + (3n-2)h_{S_1}(i,j,k,\ell) - 2h_{S_2}(i,j,k,\ell) \\
&= f(p)n(n-1)(n-2)(n-3)U_{4,4}(p) + (3n-2)h_{S_1}(i,j,k,\ell) \\
&\quad - 2h_{S_2}(i,j,k,\ell).
\end{aligned}
$$

16.8 Implementation in R

Statistics introduced in this chapter have been implemented in the *energy* package for R [Rizzo and Székely, 2022]. After loading the package with library(energy), try running the examples in the help pages using the example() function with the function name as the argument to see several examples.

- dcovU computes $\mathcal{U}_n$ for data in arbitrary dimension.
- bcdcor computes a bias-corrected dCor$^2_n(\mathbf{X},\mathbf{Y}) = \mathcal{R}^*_n(\mathbf{X},\mathbf{Y})$ from the U-statistics.
- dcovU_stats returns $\mathcal{U}_n(\mathbf{X},\mathbf{Y})$, $\mathcal{U}_n(\mathbf{X},\mathbf{X})$, $\mathcal{U}_n(\mathbf{Y},\mathbf{Y})$, and $\mathcal{R}^*_n(\mathbf{X},\mathbf{Y})$, computed from distance matrices of the samples.
- Ucenter computes the U-centered distance matrix from a distance object or data matrix, and U_center computes it from a distance or dissimilarity matrix. U_product computes the inner product formula for $\mathcal{U}_n$ from the U-centered distance matrices.
- dcov2d computes (for bivariate (X,Y) only) either the V-statistic V_n or the U-statistic U_n in Section 16.7, with type="V" (default) or type="U", respectively. This function implements an $O(n\log(n))$ computing algorithm.
- dcor2d computes the corresponding dCor$^2_n(\mathbf{X},\mathbf{Y})$ based on dcov2d statistics, type="V" (default) or type="U".

Example 16.8. We compute $\mathcal{U}_n$ and related statistics for two species of the *iris* data in $\mathbb{R}^4$ using the functions dcovU, bcdcor, and dcovU_stats from the *energy* package.

```
x <- iris[1:50, 1:4]
y <- iris[51:100, 1:4]
dcovU(x, y)
##          dCovU
```

```
## -0.002748351
bcdcor(x, y)
##      bcdcor
## -0.0271709
Dx <- as.matrix(dist(x))
Dy <- as.matrix(dist(y))
dcovU_stats(Dx, Dy)
##        dCovU        bcdcor        dVarXU        dVarYU
## -0.002748351 -0.027170902  0.065242693  0.156821104
```

We should expect that the measurements of the two species are independent random samples, so the U-statistic should fall in the lower tail. The lower tail of the distribution includes negative values of the U-statistic. If we compute the V-statistic we should see a small positive value for independent data. A significance test based on the V-statistic is below.

```
dcov(x, y)^2  #V-statistic
## [1] 0.01050803
```

```
dcov.test(x, y, R=999)
##
##   dCov independence test (permutation test)
##
## data:  index 1, replicates 999
## nV^2 = 0.5254, p-value = 0.952
## sample estimates:
##       dCov
## 0.1025087
```

The test above is a permutation test with 999 replicates. The p-value 0.952 is not significant and the independence hypothesis is not rejected.

16.9 Proofs

The proofs of Proposition 16.1 and in Székely and Rizzo [2014] Appendix A.1 as "Proof of Proposition 1." The result and proof are split into two parts in this chapter. There is a correction of a minor misprint in this version.

(1) Proof of Proposition 16.1 deriving the alternate formulation of the inner product U-statistic.

Let us adopt the notation $\widetilde{a}_{k.} := \frac{a_{k.}}{n-2}$, $\widetilde{a}_{.l} := \frac{a_{.l}}{n-2}$, and $\widetilde{a}_{..} := \frac{a_{..}}{(n-1)(n-2)}$, where $a_{k.} = \sum_{l=1}^{n} a_{kl}$, $a_{.l} = \sum_{k=1}^{n} a_{kl}$, and $a_{..} = \sum_{k,l=1}^{n} a_{kl}$. Similarly define $\widetilde{b}_{k.}, \widetilde{b}_{.l}$, and $\widetilde{b}_{..}$. Then

$$n(n-3)(\widetilde{A}\cdot\widetilde{B}) = \sum_{k\neq l} \left\{ \begin{array}{llll} a_{kl}b_{kl} & -a_{kl}\widetilde{b}_{k.} & -a_{kl}\widetilde{b}_{.l} & +a_{kl}\widetilde{b}_{..} \\ -\widetilde{a}_{k.}b_{kl} & +\widetilde{a}_{k.}\widetilde{b}_{k.} & +\widetilde{a}_{k.}\widetilde{b}_{.l} & -\widetilde{a}_{k.}\widetilde{b}_{..} \\ -\widetilde{a}_{.l}b_{kl} & +\widetilde{a}_{.l}\widetilde{b}_{k.} & +\widetilde{a}_{.l}\widetilde{b}_{.l} & -\widetilde{a}_{.l}\widetilde{b}_{..} \\ +\widetilde{a}_{..}b_{kl} & -\widetilde{a}_{..}\widetilde{b}_{k.} & -\widetilde{a}_{..}\widetilde{b}_{.l} & +\widetilde{a}_{..}\widetilde{b}_{..} \end{array} \right\}$$

$$
\begin{array}{llll}
= \sum_{k\neq l} a_{kl}b_{kl} & -\sum_{k} a_{k.}\widetilde{b}_{k.} & -\sum_{l} a_{.l}\widetilde{b}_{.l} & +a_{..}\widetilde{b}_{..} \\
-\sum_{k}\widetilde{a}_{k.}b_{kl} & +(n-1)\sum_{k}\widetilde{a}_{k.}\widetilde{b}_{k.} & +\sum_{k\neq l}\widetilde{a}_{k.}\widetilde{b}_{.l} & -(n-1)\sum_{k}\widetilde{a}_{k.}\widetilde{b}_{..} \\
-\sum_{l}\widetilde{a}_{.l}b_{.l} & +\sum_{k\neq l}\widetilde{a}_{.l}\widetilde{b}_{k.} & +(n-1)\sum_{l}\widetilde{a}_{.l}\widetilde{b}_{.l} & -(n-1)\sum_{l}\widetilde{a}_{.l}\widetilde{b}_{..} \\
+\widetilde{a}_{..}b_{..} & -(n-1)\sum_{k}\widetilde{a}_{..}\widetilde{b}_{k.} & -(n-1)\sum_{l}\widetilde{a}_{..}\widetilde{b}_{.l} & +n(n-1)\widetilde{a}_{..}\widetilde{b}_{..}.
\end{array}
$$

Let

$$T_1 = \sum_{k\neq l} a_{kl}b_{kl}, \qquad T_2 = a_{..}b_{..}, \qquad T_3 = \sum_{k} a_{k.}b_{k.}.$$

Then

$$n(n-3)(\widetilde{A}\cdot\widetilde{B}) = \left\{ \begin{array}{llll} T_1 & -\frac{T_3}{n-2} & -\frac{T_3}{n-2} & +\frac{T_2}{(n-1)(n-2)} \\ -\frac{T_3}{n-2} & +\frac{(n-1)T_3}{(n-2)^2} & +\frac{T_2-T_3}{(n-2)^2} & -\frac{T_2}{(n-2)^2} \\ -\frac{T_3}{n-2} & +\frac{T_2-T_3}{(n-2)^2} & +\frac{(n-1)T_3}{(n-2)^2} & -\frac{T_2}{(n-2)^2} \\ +\frac{T_2}{(n-1)(n-2)} & -\frac{T_2}{(n-2)^2} & -\frac{T_2}{(n-2)^2} & +\frac{nT_2}{(n-1)(n-2)^2} \end{array} \right\}$$

$$= T_1 + \frac{T_2}{(n-1)(n-2)} - \frac{2T_3}{n-2}.$$

(The last line above corrects a misprint in the original proof in Székely and Rizzo [2014]: $n(n-3)(\widetilde{A}\cdot\widetilde{B}) = T_1 + \frac{T_2}{(n-1)(n-2)} - \frac{2T_3}{n-2}$ is correct.)

Thus we have derived (16.3).

(2) Proof of Proposition 16.2 (Unbiasedness)

Proof. Proposition 16.2 asserts that $(\widetilde{A}\cdot\widetilde{B})$ is an unbiased estimator of the population coefficient $\mathcal{V}^2(X,Y)$. When the terms of $(\widetilde{A}\cdot\widetilde{B})$ are expanded, we have a linear combination of terms $a_{ij}b_{kl}$. The expected values of these terms differ according to the number of subscripts that agree. Define

$$\alpha := E[a_{kl}] = E[|X-X'|], \qquad \beta := E[b_{kl}] = E[|Y-Y'|], \qquad k\neq l,$$
$$\delta := E[a_{kl}b_{kj}] = E[|X-X'||Y-Y''|], \qquad j,k,l \text{ distinct},$$
$$\gamma := E[a_{kl}b_{kl}] = E[|X-X'||Y-Y'|], \qquad k\neq l,$$

where $(X, Y), (X', Y'), (X'', Y'')$ are iid. Due to symmetry, the expected value of each term in the expanded expression $(\widetilde{A} \cdot \widetilde{B})$ is proportional to one of $\alpha\beta$, δ, or γ, so the expected value of $(\widetilde{A} \cdot \widetilde{B})$ can be written as a linear combination of $\alpha\beta$, δ, and γ.

The population coefficient can be written as (see Székely and Rizzo [2009a], Theorem 7)

$$\mathcal{V}^2(X, Y) = E[|X - X'||Y - Y'|] + E[|X - X'|]\, E[|Y - Y'|]$$
$$- 2E[|X - X'||Y - Y''|]$$
$$= \gamma + \alpha\beta - 2\delta.$$

If

$$T_1 = \sum_{k \neq l} a_{kl} b_{kl}, \qquad T_2 = a_{..} b_{..}, \qquad T_3 = \sum_k a_{k.} b_{k.}.$$

it is easy to see that $E[T_1] = n(n-1)\gamma$, and by expanding the terms of T_2 and T_3, and combining terms that have equal expected values, one can obtain

$$E[T_2] = n(n-1)\{(n-2)(n-3)\alpha\beta + 2\gamma + 4(n-2)\delta\};$$
$$E[T_3] = n(n-1)\{(n-2)\delta + \gamma\}.$$

Then

$$E[(\widetilde{A} \cdot \widetilde{B})] = \frac{1}{n(n-3)} E\left[T_1 + \frac{T_2}{(n-1)(n-2)} - \frac{2T_3}{n-2}\right]$$
$$= \frac{1}{n(n-3)} \left\{\frac{n^3 - 5n^2 + 6n}{n-2}\gamma + n(n-3)\alpha\beta + (6n - 2n^2)\delta\right\}$$
$$= \gamma + \alpha\beta - 2\delta = \mathcal{V}^2(X, Y).$$

(The first line above corrects a misprint in the original proof: $E[(\widetilde{A} \cdot \widetilde{B})] = \frac{1}{n(n-3)} E\left[T_1 + \frac{T_2}{(n-1)(n-2)} - \frac{2T_3}{n-2}\right]$ is correct.) $\qquad\square$

16.10   Exercises

Exercise 16.1. It is obvious that $\widetilde{A} = 0$ if all sample observations are identical. More generally, show that $\widetilde{A} = 0$ if and only if the n sample observations are equally distant or at least $n-1$ of the n sample observations are identical.

Exercise 16.2. Let $\theta = \text{Cov}(X, Y)$ and consider the sample covariance statistic

$$S_{XY} = \frac{1}{n-1} \sum_{i=1}^{n} (X_i - \overline{X})(Y_i - \overline{Y}).$$

Show that S_{XY} is a U-statistic for $\sigma_{XY} = \text{Cov}(X, Y)$ What is the kernel function?

Exercise 16.3. In the proof of Proposition 16.2 (unbiasedness of the inner product estimator of $\mathcal{V}^2$) show that these formulas for expected values of T_1, T_2, T_3 are correct:

$$E[T_1] = n(n-1)\gamma$$
$$E[T_2] = n(n-1)\{(n-2)(n-3)\alpha\beta + 2\gamma + 4(n-2)\delta\};$$
$$E[T_3] = n(n-1)\{(n-2)\delta + \gamma\}.$$

17

Partial Distance Correlation

CONTENTS

17.1   Introduction

Distance covariance and distance correlation introduced in Chapter 12 characterize multivariate independence for random vectors in arbitrary, not necessarily equal dimension. This chapter covers partial distance correlation [Székely and Rizzo, 2014]. Chapter 16 covers some of the theory for the statistics introduced in this chapter.

In this chapter, we develop the partial distance covariance (pdCov) and partial distance correlation (pdCor). Among the many potential application areas of partial distance correlation are variable selection (see Example 17.7) and graphical models; see e.g. Wermuth and Cox [2013] for an example of work that motivated the question in that context.

The problem of partial distance correlation is more complex than partial correlation partly because the squared distance covariance is not an inner product in the usual linear space. For the definition of partial distance correlation we introduce a new Hilbert space where the squared distance covariance is the inner product. The partial distance correlation statistics are defined in this Hilbert space. The inner product statistic is an unbiased estimator of squared dCov. Our results show that energy dependence coefficients can also be applied to dissimilarities that are not necessarily distances.

For easy reference, we summarize some of the key definitions and properties of dCov and dCor coefficients and statistics introduced in preceding chapters. See Sections 12.3 and 12.4 for details.

The distance covariance $\mathcal{V}(X, Y)$, of two random vectors X and Y characterizes independence; that is, $\mathcal{V}(X, Y) \geq 0$ with equality to zero if and only if X and Y are independent.

Let (X, Y), (X', Y'), and (X'', Y'') be independent and identically distributed (iid), each with joint distribution (X, Y). Then

$$\mathcal{V}^2(X, Y) = E|X - X'||Y - Y'| + E|X - X'| \cdot E|Y - Y'| \qquad (17.1)$$
$$- E|X - X'||Y - Y''| - E|X - X''||Y - Y'|,$$

provided that X and Y have finite first moments. In Section 17.6 an alternate version of (17.1) is defined for X and Y taking values in a separable Hilbert space. That definition and intermediate results lead to the definition of partial distance covariance.

The distance correlation $\mathcal{R}(X, Y)$ is a standardized coefficient, $0 \leq \mathcal{R}(X, Y) \leq 1$, that also characterizes independence (Definition 12.3).

Sample dCov and dCor

The distance covariance and distance correlation statistics are functions of the double centered distance matrices of the samples. For an observed random sample $\{(x_i, y_i) : i = 1, \ldots, n\}$ from the joint distribution of random vectors X and Y, compute the Euclidean distance matrices $(a_{ij}) = (|x_i - x_j|_p)$ and $(b_{ij}) = (|y_i - y_j|_q)$. Define

$$\widehat{A}_{ij} = a_{ij} - \bar{a}_{i.} - \bar{a}_{.j} + \bar{a}_{..}, \qquad i, j = 1, \ldots, n, \qquad (17.2)$$

where $\bar{a}_{i.} = \frac{1}{n} \sum_{j=1}^{n} a_{ij}$, $\bar{a}_{.j} = \frac{1}{n} \sum_{i=1}^{n} a_{ij}$, $\bar{a}_{..} = \frac{1}{n^2} \sum_{i,j=1}^{n} a_{ij}$. Similarly, define $\widehat{B}_{ij} = b_{ij} - \bar{b}_{i.} - \bar{b}_{.j} + \bar{b}_{..}$, for $i, j = 1, \ldots, n$.

The sample distance covariance $\mathcal{V}_n(\mathbf{X}, \mathbf{Y})$ and sample distance variance $\mathcal{V}_n(\mathbf{X})$ are defined by the positive square root of

$$\mathcal{V}_n^2(\mathbf{X}, \mathbf{Y}) = \frac{1}{n^2} \sum_{i,j=1}^{n} \widehat{A}_{ij} \widehat{B}_{ij} . \qquad (17.3)$$

and

$$\mathcal{V}_n^2(\mathbf{X}) = \mathcal{V}_n^2(\mathbf{X}, \mathbf{X}) = \frac{1}{n^2} \sum_{i,j=1}^{n} \widehat{A}_{ij}^2, \qquad (17.4)$$

respectively. Sample distance correlation $\mathcal{R}_n(\mathbf{X}, \mathbf{Y})$ is the positive square root of

$$\mathcal{R}_n^2(\mathbf{X}, \mathbf{Y}) = \begin{cases} \dfrac{\mathcal{V}_n^2(\mathbf{X}, \mathbf{Y})}{\sqrt{\mathcal{V}_n^2(\mathbf{X})\mathcal{V}_n^2(\mathbf{Y})}}, & \mathcal{V}_n^2(\mathbf{X})\mathcal{V}_n^2(\mathbf{Y}) > 0; \\ 0, & \mathcal{V}_n^2(\mathbf{X})\mathcal{V}_n^2(\mathbf{Y}) = 0. \end{cases} \qquad (17.5)$$

Note that $\mathcal{V}_n(\mathbf{X}, \mathbf{Y})$ and $\mathcal{R}_n(\mathbf{X}, \mathbf{Y})$ are rigid motion invariant; they can be defined entirely in terms of the distance matrices, or equivalently in terms of the double centered distance matrices. This invariance is important, because one can also compute partial distance covariance by operating on certain transformations of the distance matrices, without reference to the original data that generated the matrices.

17.2   Hilbert Space of U-centered Distance Matrices

Why is it not straightforward to generalize distance correlation to partial distance correlation? It should be defined in a meaningful way that preserves the essential properties one would require, and allows for interpretation and inference. One approach might be to follow the definitions of the classical partial covariance and partial correlation that are based on orthogonal projections in a Euclidean space. However, there is a serious difficulty. Orthogonality in case of distance covariance and distance correlation means independence, but when we compute the orthogonal projection of a random variable onto the condition variable, the 'remainder' in the difference is typically not independent of the condition.

Alternately, the product form of sample dCov in (17.3) may suggest an inner product. Thus, we might think of working in the Hilbert space of double centered distance matrices (17.2), where the inner product is the squared distance covariance statistic (17.3). Here we have another problem: what would the projections represent? In general, the difference D of double centered distance matrices is not a double centered distance matrix of any sample. While this does not affect formal computations, we do need to be able to interpret our formulas in terms of samples to allow for meaningful inference.

An elegant solution that overcomes these difficulties while preserving the essential properties of distance covariance starts with defining an alternate type of double centering called "U-centering" (see Definition 17.1 below). The corresponding inner product is an unbiased estimator of squared population distance covariance. In the Hilbert space of "U-centered" matrices, all linear combinations, and in particular projections, are zero diagonal U-centered matrices.

Theorem 17.2 (Theorem 2 in Székely and Rizzo [2014]) connects the orthogonal projections to random samples in Euclidean space.

Methods for inference are covered below, including methods for non-Euclidean dissimilarities. The methods for dissimilarities introduced in this chapter apply to dCov and dCor as well as pdCov and pdCor.

17.2.1　U-centered Distance Matrices

The name "U-centered" refers to the fact that the corresponding inner product statistic (17.8) defined below is an unbiased estimator of squared distance covariance. In Chapter 16 we saw that (17.7) defines a U-statistic unbiased for $\mathcal{V}^2$.

Definition 17.1 (U-centered matrix). Let $A = (a_{ij})$ be a symmetric, real-valued $n \times n$ matrix with zero diagonal, $n > 2$. Define the U-centered matrix $\widetilde{A}$ as follows. Let the (i, j)-th entry of $\widetilde{A}$ be defined by

$$
\widetilde{A}_{i,j} =
\begin{cases}
a_{i,j} - \frac{1}{n-2} \sum_{\ell=1}^{n} a_{i,\ell} - \frac{1}{n-2} \sum_{k=1}^{n} a_{k,j} + \frac{1}{(n-1)(n-2)} \sum_{k,\ell=1}^{n} a_{k,\ell}, & i \neq j; \\
0, & i = j.
\end{cases}
\tag{17.6}
$$

It is obvious that $\widetilde{A} = 0$ if all sample observations are identical. More generally, $\widetilde{A} = 0$ if and only if the n sample observations are equally distant or at least $n - 1$ of the n sample observations are identical.

Proposition 17.1. *Let (x_i, y_i), $i = 1, \ldots, n$ denote a sample of observations from the joint distribution (X, Y) of random vectors X and Y. Let $A = (a_{ij})$ be the Euclidean distance matrix of the sample $x_1, \ldots, x_n$ from the distribution of X, and $B = (b_{ij})$ be the Euclidean distance matrix of the sample $y_1, \ldots, y_n$ from the distribution of Y. Then if $E(|X| + |Y|) < \infty$, for $n > 3$,*

$$
(\widetilde{A} \cdot \widetilde{B}) := \frac{1}{n(n-3)} \sum_{i \neq j} \widetilde{A}_{i,j} \widetilde{B}_{i,j}
\tag{17.7}
$$

is an unbiased estimator of squared population distance covariance $\mathcal{V}^2(X, Y)$.

Proposition 17.1 is proved in Section 16.9.

For a fixed $n \geq 4$, we define a Hilbert space generated by Euclidean distance matrices of arbitrary sets (samples) of n points in a Euclidean space $\mathbb{R}^p$, $p \geq 1$. Consider the linear span $\mathcal{S}_n$ of all $n \times n$ distance matrices of samples $\{x_1, \ldots, x_n\}$. Let $A = (a_{ij})$ be an arbitrary element in $\mathcal{S}_n$. Then A is a real-valued, symmetric matrix with zero diagonal.

Let $\mathcal{H}_n = \{\widetilde{A} : A \in \mathcal{S}_n\}$ and for each pair of elements $C = (C_{i,j})$, $D = (D_{i,j})$ in the linear span of $\mathcal{H}_n$ define their inner product

$$
\mathcal{U}_n := (C \cdot D) = \frac{1}{n(n-3)} \sum_{i \neq j} C_{ij} D_{ij}.
\tag{17.8}
$$

If $(C \cdot C) = 0$, then $(C \cdot D) = 0$ for any $D \in \mathcal{H}_n$.

In Theorem 17.2 it is shown that every matrix $C \in \mathcal{H}_n$ is the U-centered distance matrix of a configuration of n points in a Euclidean space $\mathbb{R}^p$, where $p \leq n - 2$.

By Theorem 16.2 the statistic $\mathcal{U}_n$ is a U-statistic. An alternate computing formula for $\mathcal{U}_n$ is (16.19).

17.2.2 Properties of Centered Distance Matrices

A few properties of centered distance matrices are established in the following lemma.

Lemma 17.1. *Let $\widetilde{A}$ be a U-centered distance matrix. Then*

1. *Rows and columns of $\widetilde{A}$ sum to zero.*
2. *$\widetilde{(\widetilde{A})} = \widetilde{A}$. That is, if B is the matrix obtained by U-centering an element $\widetilde{A} \in \mathcal{H}_n$, $B = \widetilde{A}$.*
3. *$\widetilde{A}$ is invariant to double centering. That is, if B is the matrix obtained by double centering the matrix $\widetilde{A}$, then $B = \widetilde{A}$.*
4. *If c is a constant and B denotes the matrix obtained by adding c to the off-diagonal elements of $\widetilde{A}$, then $\widetilde{B} = \widetilde{A}$.*

See Section 17.8 (1) for proof of Lemma 17.1.

17.2.3 Additive Constant Invariance

The last property (iv) in Lemma 17.1(iv), will be called *additive constant invariance*. This property makes it possible to apply distance correlation inference to dissimilarities. With the help of additive constant invariance, we can transform dissimilarity matrices to distance matrices that have the same distance correlation.

As Lemma 17.1(iv) is essential for our results in this chapter, it becomes clear that we cannot apply double centering as in the original definition of distance covariance, $\mathcal{V}_n^2$, in the definition of partial distance covariance. The invariance with respect to the constant c in Lemma 17.1(iv) holds for U-centered matrices, but it does not hold for double centered matrices.

Theorem 17.1. *The linear span of all $n \times n$ matrices $\mathcal{H}_n = \{\widetilde{A} : A \in \mathcal{S}_n\}$ is a Hilbert space with inner product defined by (17.8).*

Proof. Let $\mathcal{H}_n$ denote the linear span of $\mathcal{H}_n$. If $\widetilde{A}, \widetilde{B} \in \mathcal{H}_n$ and $c, d \in \mathbb{R}$, then $(c\widetilde{A} + d\widetilde{B})_{i,j} = (c\widetilde{A} + d\widetilde{B})_{i,j}$, so $(c\widetilde{A} + d\widetilde{B}) \in \mathcal{H}_n$. It is also true that $C \in \mathcal{H}_n$ implies that $-C \in \mathcal{H}_n$, and the zero element is the $n \times n$ zero matrix.

Then, since $(c\widetilde{A} + d\widetilde{B}) = \widetilde{(cA + dB)}$, for the inner product, we only need to prove that for $\widetilde{A}, \widetilde{B}, \widetilde{C} \in \mathcal{H}_n$ and real constants c, the following statements hold:

(i) $(\widetilde{A} \cdot \widetilde{A}) \geq 0$.

(ii) $(\widetilde{A} \cdot \widetilde{A}) = 0$ only if $\widetilde{A} = 0$.

(iii) $((\widetilde{cA}) \cdot \widetilde{B}) = c(\widetilde{A} \cdot \widetilde{B})$

(iv) $((\widetilde{A} + \widetilde{B}) \cdot \widetilde{C}) = (\widetilde{A} \cdot \widetilde{C}) + (\widetilde{B} \cdot \widetilde{C})$.

Statements (i) and (ii) hold because $(\widetilde{A} \cdot \widetilde{A})$ is proportional to a sum of squares. Statements (iii) and (iv) follow easily from the definition of $\mathcal{H}_n$, $\widetilde{A}$, and (17.8). $\qquad\square$

The space $\mathcal{H}_n$ is finite dimensional because it is a subspace of the space of all symmetric, zero-diagonal $n \times n$ matrices.

In the following sections $\mathcal{H}_n$ denotes the Hilbert space of Theorem 17.1 with inner product (17.8), and $|\widetilde{A}| = (\widetilde{A}, \widetilde{A})^{1/2}$ is the norm of $\widetilde{A}$.

17.3 Partial Distance Covariance and Correlation

Our definitions of population coefficients pdCov and pdCor start with definitions of dCov and dCor in Hilbert spaces. This is an important generalization of distance correlation in Euclidean spaces. However, it is perhaps easier to motivate the definitions of pdCov and pdCor by first discussing the sample coefficients. The population coefficients are then presented below in Section 17.6.

In this chapter we have modified the notation because we have three symbols for the variables rather than two. We use lower case letters rather than bold upper case to denote samples from the joint distribution of the random variables. Partial distance covariance is a coefficient of three random variables or three samples. To keep the notation compact, we denote e.g. $\text{pdCor}(X, Y; Z)$ by $\mathcal{R}^*_{x,y;z}$, and the pairwise coefficients $\mathcal{R}^*_{x,y}$, $\mathcal{R}^*_{x,z}$, $\mathcal{R}^*_{y,z}$. The sample coefficient $\text{dCor}^2_n(\mathbf{X}, \mathbf{Y}) = \mathcal{R}^*_n(\mathbf{X}, \mathbf{Y})$ will be denoted $R^*_{x,y}$ in this chapter, and sample pdCor by $R^*_{x,y;z}$.

From Theorem 17.1 a projection operator (17.9) can be defined in the Hilbert space $\mathcal{H}_n$, $n \geq 4$, and applied to define partial distance covariance and partial distance correlation for random vectors in Euclidean spaces. Let $\widetilde{A}$, $\widetilde{B}$, and $\widetilde{C}$ be elements of $\mathcal{H}_n$ corresponding to samples x, y, and z, respectively, and let

$$P_{z^\perp}(x) = \widetilde{A} - \frac{(\widetilde{A} \cdot \widetilde{C})}{(\widetilde{C} \cdot \widetilde{C})} \widetilde{C}, \qquad P_{z^\perp}(y) = \widetilde{B} - \frac{(\widetilde{B} \cdot \widetilde{C})}{(\widetilde{C} \cdot \widetilde{C})} \widetilde{C}, \qquad (17.9)$$

denote the orthogonal projection of $\widetilde{A}(x)$ onto $(\widetilde{C}(z))^\perp$ and the orthogonal projection of $\widetilde{B}(y)$ onto $(\widetilde{C}(z))^\perp$, respectively. In case $(\widetilde{C} \cdot \widetilde{C}) = 0$ the projections are defined $P_{z^\perp}(x) = \widetilde{A}$ and $P_{z^\perp}(y) = \widetilde{B}$. Clearly $P_{z^\perp}(x)$ and $P_{z^\perp}(y)$

are elements of $\mathcal{H}_n$, their dot product is defined by (17.8), and we can define sample pdCov$(X, Y; Z)$ via projections.

Definition 17.2 (Partial distance covariance). Let (x, y, z) be a random sample observed from the joint distribution of (X, Y, Z). The sample partial distance covariance (pdCov) is defined by

$$\text{pdCov}(x, y; z) = (P_{z^\perp}(x) \cdot P_{z^\perp}(y)), \tag{17.10}$$

where $P_{z^\perp}(x)$, and $P_{z^\perp}(y)$ are defined by (17.9), and

$$(P_{z^\perp}(x) \cdot P_{z^\perp}(y)) = \frac{1}{n(n-3)} \sum_{i \neq j} (P_{z^\perp}(x))_{i,j} (P_{z^\perp}(y))_{i,j}. \tag{17.11}$$

Definition 17.3 (Partial distance correlation). Let (x, y, z) be a random sample observed from the joint distribution of (X, Y, Z). Then sample partial distance correlation is defined as the cosine of the angle θ between the "vectors" $P_{z^\perp}(x)$ and $P_{z^\perp}(y)$ in the Hilbert space $\mathcal{H}_n$:

$$R^*(x, y; z) := \cos \theta = \frac{(P_{z^\perp}(x) \cdot P_{z^\perp}(y))}{|P_{z^\perp}(x)||P_{z^\perp}(y)|}, \qquad |P_{z^\perp}(x)||P_{z^\perp}(y)| \neq 0, \tag{17.12}$$

and otherwise $R^*(x, y; z) := 0$.

Energy statistics are defined for random vectors, so pdCor$(X, Y; Z)$ is a scalar coefficient defined for random vectors X, Y, and Z in arbitrary dimension.

There is a simple computing formula (17.12) for the pdCor statistic and there is a test for the hypothesis of zero pdCor based on the inner product. The statistics and tests are described in detail in Székely and Rizzo [2014] and implemented in package *energy* [Rizzo and Székely, 2022] for R (see functions pdcov, pdcor, pdcov.test.)

Example 17.1 (Test for pdCor=0). This is a simple example illustrating the permutation test implementation of the hypothesis test for pdCor $= 0$. Details about this data set are given in Example 17.7.

```
> data(Prostate, package="lasso2")
> with(Prostate,
+   energy::pdcor.test(lpsa, lcavol, lweight, R=999))


pdcor test

data:  replicates 999
pdcor = 0.44417, p-value = 0.001
sample estimates:
    pdcor
0.4441722
```

Since pdCov is defined as the inner product (17.11) of two U-centered matrices, and (unbiased squared) distance covariance (17.8) is computed as inner product, a natural question is the following. Are matrices $P_{z^\perp}(x)$ and $P_{z^\perp}(y)$ the U-centered *Euclidean* distance matrices of samples of points **U** and **V**, respectively? If so, then the sample partial distance covariance (17.11) is distance covariance of **U** and **V**, as defined by (17.8).

For every sample of points $\{x_1, \ldots, x_n\}$, $x_i \in \mathbb{R}^p$, there is a U-centered matrix $\tilde{A} = \tilde{A}(x)$ in $\mathcal{H}_n$. Conversely, given an arbitrary element H of $\mathcal{H}_n$, does there exist a configuration of points $\mathbf{U} = [u_1, \ldots, u_n]$ in some Euclidean space $\mathbb{R}^q$, for some $q \geq 1$, such that the U-centered Euclidean distance matrix of sample **U** is exactly equal to the matrix H? In the next section we prove that the answer is yes: $P_{z^\perp}(x)$, $P_{z^\perp}(y)$ of (17.11), and in general *every element* in $\mathcal{H}_n$, is the U-centered distance matrix of some sample of n points in a Euclidean space.

Simplified Computing Formula for pdCor

Let

$$R^*_{x,y} := \begin{cases} \frac{(\tilde{A} \cdot \tilde{B})}{|\tilde{A}||\tilde{B}|}, & |\tilde{A}||\tilde{B}| \neq 0; \\ 0, & |\tilde{A}||\tilde{B}| = 0, \end{cases} \tag{17.13}$$

where $\tilde{A} = \tilde{A}(x)$, $\tilde{B} = \tilde{B}(y)$ are the U-centered distance matrices of the samples x and y, and $|\tilde{A}| = (\tilde{A} \cdot \tilde{A})^{1/2}$. The statistics $R^*_{x,y}$ and $R^*_{x,y;z}$ take values in $[-1, 1]$, but they are measured in units comparable to the squared distance correlation dCor_n^2.

Proposition 17.2. *If* $(1 - (R^*_{x,z})^2)(1 - (R^*_{y,z})^2) \neq 0$, *a computing formula for* $R^*_{x,y;z}$ *in Definition (17.12) is*

$$R^*_{x,y;z} = \frac{R^*_{x,y} - R^*_{x,z} R^*_{y,z}}{\sqrt{1 - (R^*_{x,z})^2}\sqrt{1 - (R^*_{y,z})^2}}. \tag{17.14}$$

See Section 17.8 (2) for a proof.

Equation (17.14) provides a simple and familiar form of computing formula for the partial distance correlation.

Note that it is not necessary to explicitly compute projections, when (17.14) is applied. Formula (17.14) has a straightforward translation to code; see `pdcor` in the *energy* package for an implementation in R.

17.4   Representation in Euclidean Space

An $n \times n$ matrix D is called *Euclidean* if there exist points $v_1, \ldots, v_n$ in a Euclidean space such that their Euclidean distance matrix is exactly D; that is, $d_{ij}^2 = |v_i - v_j|^2 = (v_i - v_j)^T(v_i - v_j)$, $i, j = 1, \ldots, n$. Necessary and sufficient conditions that D is Euclidean are well-known results of multidimensional scaling (MDS). From Lemma 17.1, and certain results from the theory of MDS, for each element $H \in \mathcal{H}_n$ there exists a configuration of points $v_1, \ldots, v_n$ in Euclidean space such that their U-centered distance matrix is exactly equal to H.

A solution to the underlying system of equations to solve for the points v is found in Schoenberg [1935] and Young and Householder [1938]. It is a classical result that is well known in (metric) multidimensional scaling. Mardia et al. [1979] summarize the result in Theorem 14.2.1, and provide a proof. For an overview of the methodology see also Cox and Cox [2001], Gower [1966], Mardia [1978], and Torgerson [1958]. We apply the converse (b) of Theorem 14.2.1 as stated in Mardia et al. [1979], which is summarized below.

Let (d_{ij}) be a dissimilarity matrix and define $a_{ij} = -\frac{1}{2}d_{ij}^2$. Form the double centered matrix $\widehat{A} = (a_{ij} - \bar{a}_{i.} - \bar{a}_{.j} + \bar{a}_{..})$. If $\widehat{A}$ is positive semi-definite (p.s.d.) of rank p, then a configuration of points corresponding to $\widehat{D}$ can be constructed as follows. Let $\lambda_1 \geq \lambda_2 \geq \cdots \geq \lambda_p$ be the positive eigenvalues of $\widehat{A}$, with corresponding normalized eigenvectors $v_1, \ldots, v_p$, such that $v_k^T v_k = \lambda_k$, $k = 1, \ldots, p$. Then if V is the $n \times p$ matrix of eigenvectors, the rows of V are p-dimensional vectors that have interpoint distances equal to (d_{ij}), and $\widehat{A} = VV^T$ is the inner product matrix of this set of points. The solution is constrained such that the centroid of the points is the origin. There is at least one zero eigenvalue so $p \leq n - 1$.

When the matrix $\widehat{A}$ is not positive semi-definite, this leads us to the *additive constant problem*, which refers to the problem of finding a constant c such that by adding the constant to all off-diagonal entries of (d_{ij}) to obtain a dissimilarity matrix D_c, the resulting double centered matrix is p.s.d. Let $\widehat{A}_c(d_{ij}^2)$ denote the double centered matrix obtained by double centering $-\frac{1}{2}(d_{ij}^2 + c(1 - \delta^{ij}))$, where δ^{ij} is the Kronecker delta. Let $\widehat{A}_c(d_{ij})$ denote the matrix obtained by double centering $-\frac{1}{2}(d_{ij} + c(1 - \delta^{ij}))$. The smallest value of c that makes $\widehat{A}_c(d_{ij}^2)$ p.s.d. is $c^* = -2\lambda_n$, where λ_n is the smallest eigenvalue of $\widehat{A}_0(d_{ij}^2)$. Then $\widehat{A}_c(d_{ij}^2)$ is p.s.d. for every $c \geq c^*$. The number of positive eigenvalues of the p.s.d. double centered matrix $\widehat{A}_c(d_{ij}^2)$ determines the dimension required for the representation in Euclidean space.

However, we require a constant to be added to d_{ij} rather than d_{ij}^2. That is, we require a constant c, such that the dissimilarities $d_{ij}^{(c)} = d_{ij} + c(1 - \delta^{ij})$ are Euclidean. The solution by Cailliez [1983] is c^*, where c^* is the largest eigenvalue of a $2n \times 2n$ block matrix

$$\begin{bmatrix} 0 & \widehat{A}_0(d_{ij}^2) \\ I & \widehat{A}_0(d_{ij}) \end{bmatrix},$$

where 0 is the zero matrix and I is the identity matrix of size n (see Cailliez [1983] or Cox and Cox [2001], Section 2.2.8 for details). This result guarantees that there exists a constant c^* such that the adjusted dissimilarities $d_{ij}^{(c)}$ are Euclidean. In this case dimension at most $n-2$ is required [Cailliez, 1983, Theorem 1].

Finally, given an arbitrary element H of $\mathcal{H}_n$, the problem is to find a configuration of points $\mathbf{V} = [v_1, \ldots, v_n]$ such that the U-centered distance matrix of $\mathbf{V}$ is exactly equal to the element H. Thus, if $H = (h_{ij})$ we are able to find points $\mathbf{V}$ such that the Euclidean distance matrix of $\mathbf{V}$ equals $H_c = (h_{ij} + c(1 - \delta^{ij}))$, and we need $\widetilde{H}_c = H$. Now since $(c\widetilde{A} + d\widetilde{B}) = (c\widetilde{A + dB})$, we can apply Lemma 17.1 to H, and $\widetilde{H}_c = H$ follows from Lemma 17.1(ii) and Lemma 17.1(iv). Hence, by applying MDS with the additive constant theorem, and Lemma 17.1 (ii) and (iv), we obtain the configuration of points $\mathbf{V}$ such that their U-centered distances are exactly equal to the element $H \in \mathcal{H}_n$. Lemma 17.1(iv) also shows that the inner product is invariant to the constant c.

This establishes our theorem on representation in Euclidean space.

Theorem 17.2. *Let H be an arbitrary element of the Hilbert space $\mathcal{H}_n$ of U-centered distance matrices. Then there exists a sample $v_1, \ldots, v_n$ in a Euclidean space of dimension at most $n-2$, such that the U-centered distance matrix of $v_1, \ldots, v_n$ is exactly equal to H.*

Remark 17.1. The above details also serve to illustrate why a Hilbert space of double centered matrices (as applied in the original dCov$_n^2$ statistic $\mathcal{V}_n^2$ of Chapter 12) is not applicable for a meaningful definition of partial distance covariance. The diagonals of double centered distance matrices are not zero, so we cannot get an exact solution via MDS, and the inner product would depend on c. Another problem is that while $\mathcal{V}_n^2$ is always nonnegative, the inner product of projections could easily be negative.

17.5   Methods for Dissimilarities

In community ecology and other fields of application, it is often the case that only the (non-Euclidean) dissimilarity matrices are available.

Suppose that the dissimilarity matrices are symmetric with zero diagonal. An application of Theorem 17.2 provides methods for this class of non-Euclidean dissimilarities. In this case, Theorem 17.2 shows that there exist samples in Euclidean space such that their U-centered Euclidean distance matrices are equal to the dissimilarity matrices. To apply distance correlation methods to this type of problem, one only needs to obtain the Euclidean representation. Existing software implementations of classical MDS can be applied to obtain the representation derived above. For example, classical MDS based on the method outlined in Mardia [1978] is implemented in the R function cmdscale, which is in the *stats* package for R. The cmdscale function includes options to apply the additive constant of Cailliez [Cailliez, 1983] and to specify the dimension. The matrix of points $\mathbf{V}$ is returned in the component points. For an exact representation, we can specify the dimension argument equal to $n-2$.

Example 17.2. To illustrate application of Theorem 17.2 for non-Euclidean dissimilarities, we computed the Bray-Curtis dissimilarity matrix of the iris *setosa* data, a four-dimensional data set available in R. The Bray-Curtis dissimilarity defined in Cox and Cox [2001] Table 1.1 is

$$\delta_{ij} = \frac{1}{p} \frac{\sum_k |x_{ik} - x_{jk}|}{\sum_k (x_{ik} + x_{jk})}, \qquad x_i, x_j \in \mathbb{R}^p.$$

It is not a distance because it does not satisfy the triangle inequality. A Bray-Curtis method is available in the distance function of the *ecodist* or vegdist function of the *vegan* packages for R [Goslee and Urban, 2007, Oksanen et al., 2022]. We find a configuration of 50 points in $\mathbb{R}^{48}$ that have U-centered distances equal to the U-centered dissimilarities. The MDS computations are handled by the R function cmdscale. Function Ucenter, which implements U-centering, is in the R package *energy*.

```
> library(energy)
> x <- iris[1:50, 1:4]
> iris.d <- ecodist::distance(x, method="bray-curtis")
> AU <- Ucenter(iris.d)
> v <- cmdscale(as.dist(AU), k=48, add=TRUE)$points
```

The points v are a 50×48 data matrix, of 50 points in $\mathbb{R}^{48}$. Next we compare the U-centered distance matrix of the points v with the original object AU from $\mathcal{H}_n$:

```
> VU <- Ucenter(v)
> all.equal(AU, VU)
[1] TRUE
```

The last line of output shows that the points v returned by cmdscale have U-centered distance matrix VU equal to our original element AU of the Hilbert space. ◇

Note: ecodist::distance returns a class "dist" object, the same type of result as the R dist function. The *energy* function Ucenter is designed to accept either a data matrix or a class "dist" object. If the input is a distance matrix, use the U_center function of *energy*.

Example 17.2 shows that the sample distance covariance can be defined for dissimilarities via the inner product in $\mathcal{H}_n$. Alternately one can compute $\mathcal{V}_n^2(\mathbf{U}, \mathbf{V})$, where $\mathbf{U}, \mathbf{V}$ are the Euclidean representations corresponding to the two U-centered dissimilarity matrices that exist by Theorem 17.2. Using the corresponding definitions of distance variance, sample distance correlation for dissimilarities is well defined by (17.13) or $\mathcal{R}_n^2(\mathbf{U}, \mathbf{V})$. Similarly one can define pdCov and pdCor when one or more of the dissimilarity matrices of the samples is not Euclidean distance. However, as in the case of Euclidean distance, we need to define the corresponding population coefficients, and develop a test of independence. For the population definitions see Section 17.6.

17.6 Population Coefficients

17.6.1 Distance Correlation in Hilbert Spaces

The population distance covariance has been defined in terms of the joint and marginal characteristic functions of the random vectors. Here we give an equivalent definition following Lyons [2013], who generalizes distance correlation to separable Hilbert spaces. Instead of starting with the distance matrices $(a_{ij}) = (|x_i - x_j|_p)$ and $(b_{ij}) = (|y_i - y_j|_q)$, the starting point of the population definition are the bivariate distance functions $a(x, x') := |x - x'|_p$ and $b(y, y') = |y - y'|_q$, where x, x' are realizations of the random variables X and y, y' are realizations of the random variable Y.

We can also consider the random versions. Let $X \in \mathbb{R}^p$ and $Y \in \mathbb{R}^q$ be random variables with finite expectations. The random distance functions are $a(X, X') := |X - X'|_p$ and $b(Y, Y') = |Y - Y'|_q$. Here the primed random variable X denotes an independent and identically distributed (iid) copy of the variable X, and similarly Y, Y' are iid.

The population operations of double centering involves expected values with respect to the underlying population random variable. For a given random variable X with cdf F_X, we define the corresponding *double centering function with respect to X* as

$$A_X(x, x') := a(x, x') - \int_{\mathbb{R}^p} a(x, x')dF_X(x') - \int_{\mathbb{R}^p} a(x, x')dF_X(x) \quad (17.15)$$
$$+ \int_{\mathbb{R}^p} \int_{\mathbb{R}^p} a(x, x')dF_X(x')dF_X(x),$$

provided the integrals exist.

Here $A_X(x, x')$ is a real-valued function of two realizations of X, and the subscript X references the underlying random variable. Similarly for X, X' iid with cdf F_X, we define the random variable A_X as an abbreviation for

$A_X(X, X')$, which is a function of (X, X'). Similarly we define

$$B_Y(y, y') := b(y, y') - \int_{\mathbb{R}^q} b(y, y')dF_Y(y') - \int_{\mathbb{R}^q} b(y, y')dF_Y(y)$$
$$+ \int_{\mathbb{R}^q}\int_{\mathbb{R}^q} b(y, y')dF_Y(y')dF_Y(y),$$

and the random function $B_Y := B_Y(Y, Y')$.

Now for X, X' iid, and Y, Y' iid, such that X, Y have finite expectations, the population distance covariance $\mathcal{V}(X, Y)$ is defined by

$$\mathcal{V}^2(X, Y) := E[A_X B_Y]. \tag{17.16}$$

The definition (17.16) of $\mathcal{V}^2(X, Y)$ is equivalent to the original definition (12.1). However, as we will see in the next sections, (17.16) is an appropriate starting point to develop the corresponding definition of pdCov and pdCor population coefficients.

More generally, we can consider *dissimilarity functions* $a(x, x')$. A dissimilarity function is a symmetric function $a(x, x') : \mathbb{R}^p \times \mathbb{R}^p \to \mathbb{R}$ with $a(x, x) = 0$. The corresponding random dissimilarity functions $a(X, X')$ are random variables such that $a(X, X') = a(X', X)$, and $a(X, X) = 0$. Double centered dissimilarities are formally defined by the same equations as double centered distances in (17.15).

The following lemma establishes that linear combinations of double-centered dissimilarities are double-centered dissimilarities.

Lemma 17.2. *Suppose that $X \in \mathbb{R}^p$, $Y \in \mathbb{R}^q$, $a(x, x')$ is a dissimilarity on $\mathbb{R}^p \times \mathbb{R}^p$, and $b(y, y')$ is a dissimilarity on $\mathbb{R}^q \times \mathbb{R}^q$. Let $A_X(x, x')$ and $B_Y(y, y')$ denote the dissimilarity obtained by double-centering $a(x, x')$ and $b(y, y')$, respectively. Then if c_1 and c_2 are real scalars,*

$$c_1 A_X(x, x') + c_2 B_Y(y, y') = D_T(t, t'),$$

where $T = [X, Y] \in \mathbb{R}^p \times \mathbb{R}^q$, and $D_T(t, t')$ is the result of double-centering $d(t, t') = c_1 a(x, x') + c_2 b(y, y')$.

See Section 17.8 (3) for the proof.

The linear span of double-centered distance functions $A_X(x, x')$ is a subspace of the space of double-centered dissimilarity functions with the property $a(x, x') = O(|x| + |x'|)$. In this case all integrals in (17.15) and A_X are finite if X has finite expectation. The linear span of the random functions A_X for random vectors X with $E|X| < \infty$ is clearly a linear space, such that the linear extension of (17.16)

$$E[A_X B_Y] = \mathcal{V}^2(X, Y)$$

to the linear span is an inner product space or pre-Hilbert space; its completion with respect to the metric arising from its inner product

$$(A_X \cdot B_Y) := E[A_X B_Y] \tag{17.17}$$

(and norm) is a Hilbert space which we denote by $\mathcal{H}$.

17.6.2   Population pdCov and pdCor Coefficients

Definition 17.4 (Population partial distance covariance). Introduce the scalar coefficients

$$\alpha := \frac{\mathcal{V}^2(X,Z)}{\mathcal{V}^2(Z,Z)}, \qquad \beta := \frac{\mathcal{V}^2(Y,Z)}{\mathcal{V}^2(Z,Z)}.$$

If $\mathcal{V}^2(Z,Z) = 0$ define $\alpha = \beta = 0$. The double-centered projections of A_X and B_Y onto the orthogonal complement of C_Z in Hilbert space $\mathcal{H}$ are defined

$$P_{Z^\perp}(X) := A_X(X,X') - \alpha C_Z(Z,Z'), \quad P_{Z^\perp}(Y) := B_Y(Y,Y') - \beta C_Z(Z,Z'),$$

or in short $P_{Z^\perp}(X) = A_X - \alpha C_Z$ and $P_{Z^\perp}(Y) = B_Y - \beta C_Z$, where C_Z denotes double-centered with respect to the random variable Z.

The population partial distance covariance is defined by the inner product

$$(P_{Z^\perp}(X) \cdot P_{Z^\perp}(Y)) := E[(A_X - \alpha C_Z) \cdot (B_Y - \beta C_Z)].$$

Definition 17.5 (Population pdCor). Population partial distance correlation is defined

$$\mathcal{R}^*(X,Y;Z) := \frac{(P_{Z^\perp}(X) \cdot P_{Z^\perp}(Y))}{|P_{Z^\perp}(X)||P_{Z^\perp}(Y)|},$$

where $|P_{Z^\perp}(X)| = (P_{Z^\perp}(X) \cdot P_{Z^\perp}(X))^{1/2}$. If $|P_{Z^\perp}(X)||P_{Z^\perp}(Y)| = 0$ we define $\mathcal{R}^*(X,Y;Z) = 0$.

Note that if $\alpha = \beta = 0$, we have $(P_{Z^\perp}(X) \cdot P_{Z^\perp}(Y)) = E[A_X \cdot B_Y] = \mathcal{V}^2(X,Y)$, and $\mathcal{R}^*(X,Y;Z) = \mathcal{R}^*(X,Y)$.

The population coefficient of partial distance correlation can be evaluated in terms of the pairwise distance correlations using Formula (17.18) below. Theorem 17.3 establishes that (17.18) is equivalent to Definition 17.5, and therefore serves as an alternate definition of population pdCor.

Theorem 17.3 (Population pdCor). *The following definition of population partial distance correlation is equivalent to Definition 17.5.*

$$\mathcal{R}^*(X,Y;Z) = \tag{17.18}$$

$$\begin{cases} \frac{\mathcal{R}^2(X,Y) - \mathcal{R}^2(X,Z)\mathcal{R}^2(Y,Z)}{\sqrt{1-\mathcal{R}^4(X,Z)}\sqrt{1-\mathcal{R}^4(Y,Z)}}, & \mathcal{R}(X,Z) \neq 1 \text{ and } \mathcal{R}(Y,Z) \neq 1; \\ 0, & \mathcal{R}(X,Z) = 1 \text{ or } \mathcal{R}(Y,Z) = 1. \end{cases}$$

where $\mathcal{R}(X,Y)$ denotes the population distance correlation.

The proof of Theorem 17.3 is given in Section 17.8 (4).

17.6.3   On Conditional Independence

Partial distance correlation is a scalar quantity that captures dependence, while conditional inference (conditional distance correlation) is a more complex notion as it is a function of the condition. One might hope that partial distance correlation $\mathcal{R}^*(X, Y; Z) = 0$ if and only if X and Y are conditionally independent given Z, but this is not the case. Both notions capture *overlapping* aspects of dependence, but mathematically they are not equivalent. The following examples illustrate that $\mathrm{pdCor}(X, Y; Z) = 0$ is not equivalent to conditional independence of X and Y given Z.

Example 17.3. Suppose that Z_1, Z_2, Z_3 are iid standard normal variables, $X = Z_1 + Z_3$, $Y = Z_2 + Z_3$, and $Z = Z_3$. Then $\rho(X, Y) = 1/2$, $\rho(X, Z) = \rho(Y, Z) = 1/\sqrt{2}$, $\mathrm{pcor}(X, Y; Z) = 0$, and X and Y are conditionally independent of Z. One can evaluate $\mathrm{pdCor}(X, Y; Z)$ by applying (17.18) and the following result. Suppose that (X, Y) are jointly bivariate normal with correlation ρ. Then by [Székely et al., 2007, Theorem 7(ii)],

$$\mathcal{R}^2(X, Y) = \frac{\rho \arcsin \rho + \sqrt{1 - \rho^2} - \rho \arcsin(\rho/2) - \sqrt{(4 - \rho^2)} + 1}{1 + \pi/2 - \sqrt{3}}. \quad (17.19)$$

In this example $\mathcal{R}^2(X, Y) = 0.2062$, $\mathcal{R}^2(X, Z) = \mathcal{R}^2(Y, Z) = 0.4319$, and $\mathrm{pdCor}(X, Y; Z) = 0.0242$.

Example 17.4. In the other direction we can construct a trivariate normal (X, Y, Z) such that $\mathcal{R}^*(X, Y; Z) = 0$ but $\mathrm{pcor}(X, Y; Z) \neq 0$. According to Baba et al. [2004], if (X, Y, Z) are trivariate normal, then $\mathrm{pcor}(X, Y; Z) = 0$ if and only if X and Y are conditionally independent given Z. A specific numerical example is as follows. By inverting equation (17.19), given any $\mathcal{R}^2(X, Y)$, one can solve for $\rho(X, Y)$ in $[0, 1]$. When $\mathcal{R}(X, Y) = 0.04$, and $\mathcal{R}(X, Z) = \mathcal{R}(Y, Z) = 0.2$, the nonnegative solutions are $\rho(X, Y) = 0.22372287$, $\rho(X, Z) = \rho(Y, Z) = 0.49268911$. The corresponding 3×3 correlation matrix is positive definite. Let (X, Y, Z) be trivariate normal with zero means, unit variances, and correlations $\rho(X, Y) = 0.22372287$, $\rho(X, Z) = \rho(Y, Z) = 0.49268911$. Then $\mathrm{pcor}(X, Y; Z) = -0.025116547$ and therefore conditional independence of X and Y given Z does not hold, while $\mathcal{R}^*(X, Y; Z)$ is exactly zero.

In the non-Gaussian case, it is not true that zero partial correlation is equivalent to conditional independence, while partial distance correlation has the advantages that it can capture non-linear dependence, and is applicable to vector valued random variables.

Conditional distance covariance

If in the characteristic function definition of distance covariance of X and Y, (12.6), we replace the characteristic functions of X and Y with their conditional characteristic functions given the random variable Z, then we get the definition of conditional distance covariance. With the help of conditional distance covariance we can define the conditional distance correlation. This equals zero iff X and Y are independent given Z.

Wang et al. [2015] explicitly address the problem of conditional independence with the help of conditional distance correlation, introducing a conditional distance correlation coefficient for random vectors in arbitrary dimensions. Based on the sample version, they proposed a test for conditional independence, and their simulation results suggest that the test is more powerful than some other recently developed tests for conditional independence.

17.7 Empirical Results and Applications

The test for zero partial distance correlation is a test of whether the inner product $(P_{Z^\perp}(X) \cdot P_{Z^\perp}(Y))$ of the projections is zero. It is implemented in `pdcor.test` and `pdcov.test` of the *energy* package as a permutation test, where for a test of $\mathrm{pdCor}(X, Y; Z) = 0$ the sample indices of the X sample are randomized for each replicate to obtain the sampling distribution of the test statistic under the null hypothesis.

We compared the pdCov test with the partial (linear) correlation test (pcor) and the partial Mantel test. The linear partial correlation $r(x, y; z)$ measures the partial correlation between one dimensional data vectors x and y with z removed (or controlling for z). The partial correlation test is usually implemented as a t test. We applied the t-test in package *ppcor* [Kim, 2015]. The partial Mantel test is a test of the hypothesis that there is a linear association between the distances of X and Y, controlling for Z. This extension of the Mantel test [Mantel, 1967] was proposed by Smouse et al. [1986] for a partial correlation analysis on three distance matrices. The Mantel and partial Mantel tests are commonly applied in community ecology (see e.g. Legendre and Legendre [2012]), population genetics, sociology, etc. The partial Mantel test is usually implemented as a permutation (randomization) test. See Legendre [2000] for a detailed algorithm and simulation study comparing different methods of computing a partial Mantel test statistic. Based on the results reported by Legendre, we implemented the method of permutation of the raw data. The algorithm is given in detail on page 44 by Legendre [2000], and it is very similar to the algorithm we have applied for the *energy* tests. This method (permutation of raw data) for the partial Mantel test is implemented in the *ecodist* package [Goslee and Urban, 2007] and also the *vegan* package [Oksanen et al., 2022] for R. In this comparison we used the *ecodist* `mantel` function.

In each permutation test $R = 999$ replicates were generated. See Chapter 4 for details on permutation tests. Estimated power was computed from a simulation size of 10,000 tests in each case; for $n = 10$ the number of tests was 100,000. The standard error is at most 0.005 (0.0016 for $n = 10$).

Remark 17.2. Both the pdCov and partial Mantel tests are based on distances. One may ask "is distance covariance different or more general than covariance of distances?" The answer is yes; it can be shown that

$$\mathrm{dCov}^2(X, Y) = \mathrm{Cov}(|X - X'|, |Y - Y'|) - 2\,\mathrm{Cov}(|X - X'|, |Y - Y''|)$$
$$= \mathrm{Cov}(|X_1 - X_2|, |Y_1 - Y_2|) - 2\,\mathrm{Cov}(|X_1 - X_2|, |Y_1 - Y_3|),$$

where (X_i, Y_i), $i = 1, 2, 3$ are iid with joint distribution (X, Y). The dCov tests are tests of independence of X and Y $(\mathrm{dCov}^2(X, Y) = 0)$, while the Mantel test is a test of the hypothesis $\mathrm{Cov}(|X - X'|, |Y - Y'|) = 0$. An example of dependent data such that their distances are uncorrelated but $\mathrm{dCov}^2(X, Y) > 0$ is given by Lyons [2013] and in Exercise 12.5. Thus, distance covariance tests are more general than Mantel tests, in the sense that distance covariance measures all types of departures from independence.

Example 17.5. In this example, power of tests is compared for correlated trivariate normal data with standard normal marginal distributions. The pairwise correlations are $\rho(X, Y) = \rho(X, Z) = \rho(Y, Z) = 0.5$. The power comparison summarized in Figure 17.1(a) shows that pdCor has higher power than pcor or partial Mantel tests. In a separate simulation (not shown) the Type I error rates in Examples 17.5 and 17.6 were correctly controlled for pdCov and partial Mantel permutation tests, but the pcor t-test had inflated Type I error for $n \leq 30$.

Example 17.6. This example is a modification of Example 17.5 such that the variables $\log X$, Y, and Z are correlated, and the marginal distribution of X is standard lognormal, while the marginal distributions of Y and Z are each standard normal. The pairwise correlations are $\rho(\log X, Y) = \rho(\log X, Z) = \rho(Y, Z) = 0.5$. The power comparison summarized in Figure 17.1(b) shows that pdCor has higher power than pcor or partial Mantel tests. Again the pdCov test has superior power performance. The relative performance of the pcor and partial Mantel tests are reversed (partial Mantel test with lowest power) in this example compared with Example 17.5.

Example 17.7 (Variable selection). This example considers the prostate cancer data from a study by Stamey et al. [1989], and is discussed in [Hastie et al., 2009, Ch. 3] in the context of variable selection. Currently the data is available at `https://hastie.su.domains/ElemStatLearn/`. The data is from men who were about to have a radical prostatectomy. The response variable *lpsa* measures log PSA (log of the level of prostate-specific antigen). The predictor variables under consideration are eight clinical measures:

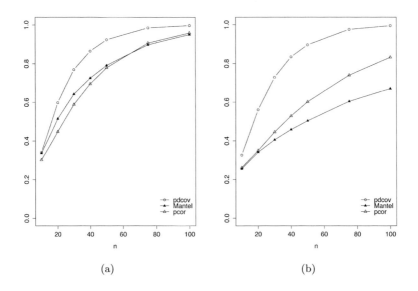

(a)   (b)

FIGURE 17.1
Power comparisons for partial distance covariance, partial Mantel test, and partial correlation test at significance level $\alpha = 0.10$. Figure (a) summarizes Example 17.5 (correlated standard normal data). Figure (b) summarizes Example 17.6 (correlated non-normal data).

lcavol	log cancer volume
lweight	log prostate weight
age	age
lbph	log of the amount of benign prostatic hyperplasia
svi	seminal vesicle invasion
lcp	log of capsular penetration
gleason	Gleason score
pgg45	percent of Gleason scores 4 or 5

Here the goal was to fit a linear model to predict the response *lpsa* given one or more of the predictor variables above. The train/test set indicator is in the last column of the data set. For comparison with the discussion and analysis in Hastie et al. [2009], we standardized each variable, and used the training set of 67 observations for variable selection.

Feature screening by distance correlation has been investigated by Li et al. [2012]. In this example we introduce a partial distance correlation criterion for variable selection. For simplicity, we implement a simple variant of forward selection. This criterion, when applied to a linear model, can help to identify possible important variables that have a strong non-linear association with the response, and thus help researchers to improve a linear model by transforming variables or improve prediction by extending to a nonlinear model.

In the initial step of pdCor forward selection, the first variable to enter the model is the variable x_j for which distance correlation $R_{x_j,y}$ with response y is largest. After the initial step, we have a model with one predictor x_j, and we compute $\text{pdCor}(y, x_k; x_j)$, for the variables $x_k \neq x_j$ not in the model, then select the variable x_k for which $\text{pdCor}(y, x_k; x_j)$ is largest. Then continue, at each step computing $\text{pdCor}(y, x_j; w)$ for every variable x_j not yet in the model, where w is the vector of predictors currently in the model. The variable to enter next is the one that maximizes $\text{pdCor}(y, x_j; w)$.

According to the pdCor criterion, the variables selected enter the model in the order: *lcavol, lweight, svi, gleason, lbph, pgg45*.

If we set a stopping rule at 5% significance for the pdCor coefficient, then we stop after adding *lbph* (or possibly after adding *pgg45*). At the step where *gleason* and *lbph* enter, the *p*-values for significance of pdCor are 0.016 and 0.001, respectively. The *p*-value for *pgg45* is approximately 0.05.

The models selected by pdCor, best subset method, and lasso [Hastie et al., 2009, Table 3.3, p. 63]) are:

pdCor:	lpsa $\sim$ lcavol + lweight + svi + gleason;
best subsets:	lpsa $\sim$ lcavol + lweight ;
lasso:	lpsa $\sim$ lcavol + lweight + svi + lbph.

The order of selection for ordinary forward stepwise selection (Cp) is *lcavol, lweight, svi, lbph, pgg45, age, lcp, gleason*.

Comparing pdCor forward selection with lasso and forward stepwise selection, we see that the results are similar, but *gleason* is not in the lasso model and enters last in the forward stepwise selection, while it is the fourth variable to enter the pdCor selected model. The raw Gleason Score is an integer from 2 to 10, which is used to measure how aggressive is the tumor, based on a prostate biopsy. Plotting the data (see Figure 17.2) we can observe that there is a strong *non-linear* relation between *gleason* and *lpsa*.

This example illustrates that partial distance correlation has practical use in variable selection and in model checking. If we were using pdCor only to check the model selected by another procedure, it would show in this case that there is some nonlinear dependence remaining between the response and the predictors excluded from the lasso model or the best subsets model. Using the pdCor selection approach, we also learn which of the remaining predictors may be important.

Finally, it is useful to recall that pdCor has more flexibility to handle predictors that are multi-dimensional. One may want groups of variables to enter or leave the model as a set. It is often the case that when dimension of the feature space is high, many of the predictor variables are highly correlated. In this case, methods such as Partial Least Squares are sometimes applied, where a small set of derived predictors (linear combinations of features) become the

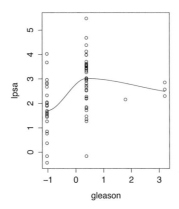

FIGURE 17.2
Scatter plot of response lpsa vs Gleason score with a loess smoother.

predictor variables. Using methods of partial distance correlation, one could evaluate the subsets (as predictor sets) without reducing the multivariate observation to a real number via linear combination. The pdCor coefficient can be computed for multivariate predictors (and for multivariate response). With the flexibility to handle multivariate response and/or multivariate predictor, pdCor offers a new method to extend the variable selection toolbox.

17.8   Proofs

(1) Properties of centered matrices

Proof of Properties (i), (ii), and (iii) are exercises. The following proof of the additive constant invariance property (iv) appears in Székely and Rizzo [2014], Appendix A.3.

Proof. In (iv) let b_{ij} be the ij-th element of B. Then $b_{ii} = 0$, and for $i \neq j$, $b_{ij} = \tilde{A}_{ij} + c$. Hence $b_{i.} = b_{.j} = (n-1)c$, $i \neq j$ and $b_{..} = n(n-1)c$. Therefore

$$\tilde{B}_{ij} = b_{ij} - \frac{2(n-1)c}{n-2} + \frac{nc}{n-2} = \tilde{A}_{ij} + c + \frac{(2-n)c}{n-2} = \tilde{A}_{ij}, \qquad i \neq j,$$

which proves (iv).   □

(2) Proof of Proposition 17.2

Proof. Here the inner product is (17.8) and the "vectors" are U-centered elements of the Hilbert space $\mathcal{H}_n$. Equation (17.14) can be derived from (17.12) using similar algebra with inner products as used to obtain the representation

$$\frac{\langle x_{z\perp}, y_{z\perp} \rangle}{\sqrt{\langle x_{z\perp}, x_{z\perp}\rangle \langle y_{z\perp}, y_{z\perp}\rangle}} = \frac{r_{xy} - r_{xz}r_{yz}}{\sqrt{1 - r_{xz}^2}\sqrt{1 - r_{yz}^2}}.$$

for the linear correlation r (see Huber [1981]). The details for pdCor are as follows. If either $|\widetilde{A}|$ or $|\widetilde{B}| = 0$, then $(\widetilde{A}, \widetilde{B}) = 0$ so by definition $R^*(x, y; z) = 0$, $R^*_{xz}R^*_{yz} = 0$, and (17.14) is also zero.

If $|\widetilde{A}||\widetilde{B}| \neq 0$ but $|\widetilde{C}| = 0$, then $R^*_{x,z} = R^*_{y,z} = 0$. In this case $P_{z\perp}(x) = \widetilde{A}$ and $P_{z\perp}(y) = \widetilde{B}$, so that

$$R^*(x, y; z) = \frac{(P_{z\perp}(x) \cdot P_{z\perp}(y))}{|P_{z\perp}(x)||P_{z\perp}(y)|} = \frac{(\widetilde{A} \cdot \widetilde{B})}{|\widetilde{A}||\widetilde{B}|} = R^*_{xy},$$

which equals expression (17.14) since $R^*_{x,z} = R^*_{y,z} = 0$.

Suppose that none of $|\widetilde{A}|, |\widetilde{B}|, |\widetilde{C}|$ are zero. Then

$$R^*(x, y; z) = (P_{z\perp}(x) \cdot P_{z\perp}(y))$$
$$= \left(\left\{ \widetilde{A} - \frac{(\widetilde{A}\cdot\widetilde{C})}{(\widetilde{C}\cdot\widetilde{C})} \cdot \widetilde{C} \right\} \cdot \left\{ \widetilde{B} - \frac{(\widetilde{B}\cdot\widetilde{C})}{(\widetilde{C}\cdot\widetilde{C})} \cdot \widetilde{C} \right\} \right)$$
$$= (\widetilde{A} \cdot \widetilde{B}) - \frac{2(\widetilde{A}\cdot\widetilde{C})(\widetilde{B}\cdot\widetilde{C})}{(\widetilde{C}\cdot\widetilde{C})} + \frac{(\widetilde{A}\cdot\widetilde{C})(\widetilde{B}\cdot\widetilde{C})(\widetilde{C}\cdot\widetilde{C})}{(\widetilde{C}\cdot\widetilde{C})^2}$$
$$= |\widetilde{A}||\widetilde{B}| \left\{ \frac{(\widetilde{A}\cdot\widetilde{B})}{|\widetilde{A}||\widetilde{B}|} - \frac{(\widetilde{A}\cdot\widetilde{C})(\widetilde{B}\cdot\widetilde{C})}{|\widetilde{A}||\widetilde{C}||\widetilde{B}||\widetilde{C}|} \right\}$$
$$= |\widetilde{A}||\widetilde{B}| \left\{ R^*_{xy} - R^*_{xz}R^*_{yz} \right\}.$$

Similarly, in the denominator of (17.12) we have

$$\sqrt{|\widetilde{A}|^2 (1 - (R^*_{xz})^2) |\widetilde{B}|^2 (1 - (R^*_{yz})^2)}$$
$$= |\widetilde{A}||\widetilde{B}|\sqrt{(1 - (R^*_{xz})^2)(1 - (R^*_{yz})^2)}.$$

Thus, if the denominator of (17.12) is not zero, the factor $|\widetilde{A}||\widetilde{B}|$ cancels from the numerator and denominator and we obtain (17.14). $\square$

(3) Proof of Lemma 17.2

Proof. It is clear that $c_1 A_X(x, x')$ is identical to the double-centered dissimilarity $c_1 a(x, x')$. It remains to show that the sum of two arbitrary elements $A_X(x, x') + B_Y(y, y')$ is a double-centered dissimilarity function. Let $T = [X, Y] \in \mathbb{R}^p \times \mathbb{R}^q$. Consider the dissimilarity function $d(t, t') := a(x, x') + b(y, y')$, where $t = [x, y]$ and $t' = [x', y']$. Then

$$
\begin{aligned}
D_X(t, t') &= a(x, x') + b(y, y') - \int [a(x, x') + b(y, y')] dF_T(t') \\
&\quad - \int [a(x, x') + b(y, y')] dF_T(t) \\
&\quad + \iint [a(x, x') + b(y, y')] dF_T(t') dF_T(t) \\
&= a(x, x') + b(y, y') - E[a(x, X') + b(y, Y')] \\
&\quad - E[a(X, x') + b(Y, y')] + E[a(X, X') + b(Y, Y')] \\
&= A_X(x, x') + B_Y(y, y').
\end{aligned}
$$

$\square$

(4) Proof of Theorem 17.3

Proof. It is straightforward to check the first three special cases.

Case (i): If Z is constant a.s., then $P_{Z^\perp}(X) = A_X$, and $P_{Z^\perp}(Y) = B_Y$. In this case both $\mathcal{R}^*(X, Y; Z)$ and (17.18) simplify to $\mathcal{R}^2(X, Y)$.

Case (ii): If X or Y is constant a.s. and Z is not a.s. constant, then we have zero in both Definition 17.5 and (17.18).

Case (iii): If none of the variables X, Y, Z are a.s. constant, but $|P_{Z^\perp}(X)| = 0$ or $|P_{Z^\perp}(Y)| = 0$, then $\mathcal{R}^*(X, Y; Z) = 0$ by definition and $\mathcal{R}^2(X, Z) = 1$ or $\mathcal{R}^2(Y, Z) = 1$. Thus (17.18) is also zero by definition.

Case (iv): In this case none of the variables X, Y, Z are a.s. constant, and $|P_{Z^\perp}(X)||P_{Z^\perp}(Y)| > 0$. Thus

$$
\begin{aligned}
E[P_{Z^\perp}(X) P_{Z^\perp}(Y)] &= (A_X - \alpha C_Z, B_Y - \beta C_Z) \quad\quad (17.20) \\
&= (A_X, B_Y) - \alpha(B_Y, C_Z) - \beta(A_X, C_Z) + \alpha\beta(C_Z, C_Z) \\
&= \mathcal{V}^2(X, Y) - \frac{\mathcal{V}^2(X,Z)\mathcal{V}^2(Y,Z)}{\mathcal{V}^2(Z,Z)} \\
&\quad - \frac{\mathcal{V}^2(X,Z)\mathcal{V}^2(Y,Z)}{\mathcal{V}^2(Z,Z)} + \frac{\mathcal{V}^2(X,Z)\mathcal{V}^2(Y,Z)\mathcal{V}^2(Z,Z)}{\mathcal{V}^2(Z,Z)\mathcal{V}^2(Z,Z)} \\
&= \mathcal{V}^2(X, Y) - \frac{\mathcal{V}^2(X,Z)\mathcal{V}^2(Y,Z)}{\mathcal{V}^2(Z,Z)}.
\end{aligned}
$$

Similarly

$$
\begin{aligned}
|P_{Z^\perp}(X)|^2 &= E[P_{Z^\perp}(X) P_{Z^\perp}(X)] = \mathcal{V}^2(X, X) - \frac{\mathcal{V}^2(X,Z)\mathcal{V}^2(X,Z)}{\mathcal{V}^2(Z,Z)} \quad (17.21) \\
&= \mathcal{V}^2(X, X) - \alpha\mathcal{V}^2(X, Z) \\
&= \mathcal{V}^2(X, X)\left(1 - \frac{(\mathcal{V}^2(X,Z))^2}{\mathcal{V}^2(X,X)\mathcal{V}^2(Z,Z)}\right) \\
&= \mathcal{V}^2(X, X)(1 - \mathcal{R}^4(X, Z)),
\end{aligned}
$$

and

$$|P_{Z^\perp}(Y)|^2 = \mathcal{V}^2(Y,Y)(1 - \mathcal{R}^4(Y,Z)). \qquad (17.22)$$

Hence, using (17.20)–(17.22)

$$
\begin{aligned}
\mathcal{R}^*(X,Y;Z) &= \frac{(P_{Z^\perp}(X)P_{Z^\perp}(Y))}{|P_{Z^\perp}(X)|\,|P_{Z^\perp}(Y)|} \\
&= \frac{\mathcal{V}^2(X,Y) - \frac{\mathcal{V}^2(X,Z)\mathcal{V}^2(Y,Z)}{\mathcal{V}^2(Z,Z)}}{\sqrt{\mathcal{V}^2(X,X)(1 - \mathcal{R}^4(X,Z))}\sqrt{\mathcal{V}^2(Y,Y)(1 - \mathcal{R}^4(Y,Z))}} \\
&= \frac{\frac{\mathcal{V}^2(X,Y)}{\mathcal{V}(X,X)\mathcal{V}(Y,Y)} - \frac{\mathcal{V}^2(X,Z)\mathcal{V}^2(Y,Z)}{\mathcal{V}(X,X)\mathcal{V}(Y,Y)\mathcal{V}^2(Z,Z)}}{\sqrt{(1 - \mathcal{R}^4(X,Z))(1 - \mathcal{R}^4(Y,Z))}} \\
&= \frac{\mathcal{R}^2(X,Y) - \mathcal{R}^2(X,Z)\mathcal{R}^2(Y,Z)}{\sqrt{(1 - \mathcal{R}^4(X,Z))(1 - \mathcal{R}^4(Y,Z))}}.
\end{aligned}
$$

Thus, in all cases Definition 17.5 and (17.18) coincide. $\qquad\square$

Proofs of two statements in this chapter appear in Chapter 16.

17.9 Exercises

Exercise 17.1. From Lemma 17.1(iv), U-centered distance matrices have an additive constant invariance property. Show that this invariance property does not hold for double-centered distance matrices.

Exercise 17.2. Suppose that $\mathbf{X}$ is a sample, $A = (a_{ij}) = |X_i - X_j|$ is the distance matrix of the sample, $\widehat{A}$ is the double-centered distance matrix and $\widetilde{A}$ is the U-centered distance matrix of the sample.

(a) State and prove necessary and sufficient conditions that $\widehat{A}$ is the zero matrix.

(b) State and prove necessary and sufficient conditions that $\widetilde{A}$ is the zero matrix.

Exercise 17.3. Prove statements (i), (ii), and (iii) of Lemma 17.1.

18

The Numerical Value of dCor

CONTENTS

18.1 Cor and dCor: How Much Can They Differ?

A natural question concerning distance correlation [Székely, Rizzo, and Bakirov, 2007, Székely and Rizzo, 2009a] for bivariate data is whether the numerical value of dCor can be compared to Pearson correlation (or it's square) in some meaningful way. For example, if Pearson's $\rho(X, Y) = .75$ should we expect that $\mathrm{dCor}(X, Y) = \mathcal{R}(X, Y)$ is close to 0.75? With a significance test we can always determine whether we can reject the null hypothesis of $\rho = 0$ or $\mathcal{R} = 0$. What can we say about the numerical values of the coefficients?

In the special case when (X, Y) are jointly distributed as bivariate normal, we have a deterministic relation between $\rho(X, Y)$ and $\mathcal{R}^2(X, Y)$:

$$\mathcal{R}^2(X, Y) = \frac{\rho \arcsin \rho + \sqrt{1 - \rho^2} - \rho \arcsin(\rho/2) - \sqrt{4 - \rho^2} + 1}{1 + \pi/3 - \sqrt{3}}.$$

Note that this is a strictly increasing, convex function of $|\rho|$, and $\mathcal{R}(X, Y) \leq |\rho|$ with equality when $\rho = 0$ or $\rho = \pm 1$. This result from Székely, Rizzo, and Bakirov [2007] is Theorem 12.4 of this book. See Figure 12.2 comparing ρ and $\mathcal{R}^2$ for the bivariate normal case.

Apart from the bivariate normal case, it is difficult to find results comparing Cor and dCor coefficients for other families of distributions.

Clearly $\mathcal{R}(X, Y) = 0$ implies $\rho(X, Y) = 0$, and $|\rho(X, Y)| = 1$ if and only if $\mathcal{R}(X, Y) = 1$. One can also show that for random variables X, Y that can each take only two possible values, $|\rho(X, Y)| = \mathcal{R}(X, Y)$. Some additional results for other bivariate parametric distributions have been derived [Dueck et al., 2017]. In the remainder of this chapter, we look at the general case.

Theorem 18.3 shows that Pearson's correlation and distance correlation can be extremely unrelated except at the boundaries (when the absolute value of correlation is 1 or when the distance correlation is 0 or 1). Indeed, the range

DOI: 10.1201/9780429157158-18

of pairs $(Cor(X, Y), dCor(X, Y))$ covers the whole open square $(-1, 1) \times (0, 1)$. Theorem 18.3 is proved by constructing a class of mixture examples (X, Y) parameterized in such a way that any given values $0 < \rho(X, Y) < 1$ and $0 < \mathcal{R}(X, Y)$ are simultaneously achieved.

Since uncorrelatedness only means the lack of linear dependence, one can naturally ask how large the distance correlation coefficient of two uncorrelated random variables can be.

For completeness, let us review the definitions of distance covariance and distance correlation. Refer to Chapter 12 for more details and examples.

Let X and Y be real-valued random variables with finite second moments. Then *the distance covariance* can be defined in the following form. Let (X, Y), (X', Y'), (X'', Y'') be iid with joint distribution (X, Y). The distance covariance (dCov) is the square root of

$$\mathcal{V}^2(X, Y) := E(|X - X'||Y - Y'|) + E(|X - X'|)E(|Y - Y'|)$$
$$- 2E(|X - X'||Y - Y''|). \qquad (18.1)$$

If the random variables X, Y have finite and positive variances, then the definition of their *distance correlation coefficient* is the following:

$$\mathcal{R}(X, Y) := \frac{\mathcal{V}(X, Y)}{\sqrt{\mathcal{V}(X, X)\mathcal{V}(Y, Y)}}. \qquad (18.2)$$

The distance correlation coefficient is nonnegative with $\mathcal{R}(X, Y) \in [0, 1]$. Both Pearson correlation and distance correlation coefficients are invariant with respect to linear transformations. $\mathcal{R}(X, Y) = 1$ if and only if $Y = aX + b$ almost surely, with real constants $a \neq 0$ and b [Székely et al., 2007]. However, there are two crucial differences between the Pearson and the distance correlation coefficients.

First, although this is not directly clear from equation (18.1), the definition of the distance correlation coefficient can be extended to variables with finite first moments only. A possible way to define the distance covariance under this less restrictive assumption is

$$\mathcal{V}^2(X, Y) := E\big(|X - X'| \cdot (|Y - Y'| - |Y - Y''| - |Y' - Y''|)\big)$$
$$+ E(|X - X'|)E(|Y - Y'|); \qquad (18.3)$$

the distance correlation coefficient is still defined by equation (18.2). The second important difference is that the lack of distance correlation defines independence; that is, $\mathcal{R}(X, Y) = 0$ if and only if X and Y are independent.

18.2 Relation Between Pearson and Distance Correlation

A natural question to ask is how large the distance correlation coefficient can be for two random variables which are *uncorrelated* (linear correlation equal to 0).

This question was investigated by Edelmann, Móri, and Székely [2021]. Theorems and examples constructed in their paper are given in this chapter along with further discussion and comments.

We will consider two different strategies for constructing extremal examples.

First, we will investigate the class of random vectors (X, Y), where X is symmetric around 0 and $Y = |X|$. While it is immediately clear that in this case $\rho(X, Y) = 0$ and $\mathcal{R}(X, Y) > 0$, we will show that the $\mathcal{R}(X, Y)$ cannot be arbitrarily close to 1.

Since uncorrelatedness only means the lack of linear dependence, one can naturally ask how large the distance correlation coefficient of two uncorrelated random variables can be. To find an extremal construction, it seems plausible to investigate the case where one of the random variables is symmetrically distributed around zero and the other one is its absolute value. Such a pair is obviously uncorrelated in the classical (Pearson) sense, and since they are equal with probability $1/2$ and the negative of each other with probability $1/2$, we may have a large distance correlation coefficient.

Below, we compute the distance correlation coefficient between a symmetrically distributed random variable and its absolute value.

Theorem 18.1. *The distance correlation coefficient of a symmetrically distributed random variable and its absolute value is less than $2^{-1/4}$ and this bound is sharp.*

Proof. Let X be a nonnegative random variable, $E(X^2) < \infty$, and let ε be independent of X with $P(\varepsilon = 1) = P(\varepsilon = -1) = 1/2$ (a random sign). Again, suppose that X, X', X'', ε, ε', and ε'' are independent; X, X', X'' are identically distributed; and ε, ε', ε'' are identically distributed. Note that here εX is symmetrically distributed with absolute value X. Clearly,

$$\mathcal{V}^2(X, X) = E\big((X - X')^2\big) + \big(E(|X - X'|)\big)^2 - 2E(|X - X'|\,|X - X''|)$$
$$= 2E(X^2) - 2\big(EX\big)^2 + \big(E(|X - X'|)\big)^2 - 2E(|X - X'|\,|X - X''|). \tag{18.4}$$

Similarly,

$$\mathcal{V}^2(\varepsilon X, \varepsilon X) = \tfrac{1}{4}\mathcal{V}^2(X, X) + E(X^2) - E(X|X - X'|), \tag{18.5}$$

which is derived by

$$
\begin{aligned}
\mathcal{V}^2&(\varepsilon X, \varepsilon X) \\
&= E\big((\varepsilon X - \varepsilon' X')^2\big) + \big(E(|\varepsilon X - \varepsilon' X'|)\big)^2 \\
&\quad - 2E(|\varepsilon X - \varepsilon' X'|\,|\varepsilon X - \varepsilon'' X''|) \\
&= E\big((X - \varepsilon' X')^2\big) + \big(E(|X - \varepsilon' X'|)\big)^2 - 2E(|X - \varepsilon' X'|\,|X - \varepsilon'' X''|) \\
&= \tfrac{1}{2} E\big((X - X')^2\big) + \tfrac{1}{2} E\big((X + X')^2\big) + \big(\tfrac{1}{2} E(|X - X'|) + EX\big)^2 \\
&\quad - \tfrac{1}{2} E(|X - X'|\,|X - X''|) - E(X|X - X'|) - EX\, E(|X - X'|) \\
&\quad - \tfrac{1}{2} E(X^2) - \tfrac{3}{2}\big(EX\big)^2 \\
&= \tfrac{3}{2} E(X^2) - \tfrac{1}{2}\big(EX\big)^2 + \tfrac{1}{4}\big(E(|X - X'|)\big)^2 \\
&\quad - \tfrac{1}{2} E(|X - X'|\,|X - X''|) - E(X|X - X'|) \\
&= \tfrac{1}{4} \mathcal{V}^2(X, X) + E(X^2) - E(X|X - X'|).
\end{aligned}
$$

Finally,

$$
\begin{aligned}
\mathcal{V}^2(X, \varepsilon X) &= E(|X - X'|\,|\varepsilon X - \varepsilon' X'|) + E(|X - X'|)E(|\varepsilon X - \varepsilon' X'|) \\
&\quad - 2E(|X - X'|\,|\varepsilon X - \varepsilon'' X''|) \\
&= E(|X - X'|\,|X - \varepsilon' X'|) + E(|X - X'|)E(|X - \varepsilon' X'|) \\
&\quad - 2E(|X - X'|\,|X - \varepsilon'' X''|) \\
&= \tfrac{1}{2} E\big((X - X')^2\big) + \tfrac{1}{2}\big(E(|X - X'|)\big)^2 - E(|X - X'|\,|X - X''|) \\
&= \tfrac{1}{2} \mathcal{V}^2(X, X).
\end{aligned}
\tag{18.6}
$$

Substituting (18.4), (18.5) and (18.6) into (18.2) we have

$$
\mathcal{R}^2(X, \varepsilon X) = \frac{\tfrac{1}{2} \mathcal{V}^2(X, X)}{\sqrt{\mathcal{V}^2(X, X)\big(\tfrac{1}{4}\mathcal{V}^2(X, X) + E(X^2) - E(X|X - X'|)\big)}}
$$

and

$$
\mathcal{R}(X, \varepsilon X) = \left(\frac{\mathcal{V}^2(X, X)}{\mathcal{V}^2(X, X) + 4E(X^2) - 4E(X|X - X'|)}\right)^{1/4}.
$$

Let X be an indicator with $E(X) = p$. Then $|X - X'|$ is an indicator as well, and

$$
\begin{aligned}
E(|X - X'|) &= 2p(1 - p), \\
E\big((X - X')^2\big) &= E(|X - X'|) = 2p(1 - p), \\
E(|X - X'|\,|X - X''|) &= p\big(E(1 - X)\big)^2 + (1 - p)\big(EX\big)^2 \\
&= p(1 - p)^2 + (1 - p)p^2 = p(1 - p), \\
E(X|X - X'|) &= pE(1 - X) = p(1 - p),
\end{aligned}
$$

hence

$$\mathcal{R}(X, \varepsilon X) = \left(\frac{(1-p)^2}{(1-p)^2 + 1}\right)^{1/4}.$$

On the other hand, we will show that

$$\mathcal{V}^2(X, X) < 4E(X^2) - 4E(X|X - X'|)$$

for arbitrary nonnegative (and not identically zero) X, hence the supremum of the distance correlation coefficient between a symmetrically distributed random variable and its absolute value is $2^{-1/4}$.

Indeed, introduce $\varphi(t) = E(|X - t|)$, then

$$E(|X - X'|) = E\big(\varphi(X)\big),$$
$$E(|X - X'||X - X''|) = E\big(\varphi(X)^2\big),$$
$$E(X|X - X'|) = E\big(X\varphi(X)\big),$$

thus

$$4E(X^2) - 4E(X|X - X'|) - \mathcal{V}^2(X, X)$$
$$= 4E(X^2) - 4E\big(X\varphi(X)\big) - 2E(X^2) + 2\big(EX\big)^2$$
$$\quad - \big(E(\varphi(X))\big)^2 + 2E\big(\varphi(X)^2\big)$$
$$= 2E\big((X - \varphi(X))^2\big) + 2E(X^2) - \big(E(\varphi(X))\big)^2$$
$$\quad - 2\big(EX - E(\varphi(X))\big)^2 + \big(2EX - E(\varphi(X))\big)^2$$
$$= \mathrm{Var}\,\big(X - \varphi(X)\big) + \big(2EX - E(\varphi(X))\big)^2,$$

and the right-hand side is strictly positive, because

$$E(\varphi(X)) = E(|X - X'|) < E(X + X') = 2EX.$$

$\square$

Is this the real supremum of the distance correlation coefficient for uncorrelated random variables? The answer is no. In Theorem 18.3 we show that there are random variables such that oct of pairs $(\rho(X, Y), \mathcal{R}(X, Y))$ cover the whole rectangle $(-1, 1) \times (0, 1)$.

The fact that $\mathcal{R}$ is defined for X and Y with finite first moments, while $\rho(X, Y)$ requires finite second moments, leads us to the conjecture that the Pearson correlation coefficient is more sensitive to dependence in the tails than the distance correlation. This conjecture motivates the construction of a specific mixture distribution showing that, up to trivial exceptions, all possible values for ρ and $\mathcal{R}$ can be simultaneously achieved.

Theorem 18.2. *For every pair (ρ, r), $-1 < \rho < 1$, $0 < r < 1$, there exist random variables X, Y with finite second moments, such that $\rho(X, Y) = \rho$, $\mathcal{R}(X, Y) = r$.*

Theorem 18.2 is the special case $\alpha = 1$ of Theorem 18.3, which is proved for general $0 < \alpha < 2$ below.

Distance covariance can be generalized by taking the α-th powers of the distances in (18.1), where $0 < \alpha < 2$ (See Section 14.2.2.) The definitions are formally the same, with distances $|X - Y|$ replaced by $|X - Y|^\alpha$. That is,

$$\mathcal{V}_\alpha^2(X, Y) := E\big(|X - X'|^\alpha |Y - Y'|^\alpha\big) + E\big(|X - X'|^\alpha\big) E\big(|Y - Y'|^\alpha\big)$$
$$- 2E\big(|X - X'|^\alpha |Y - Y''|^\alpha\big), \qquad (18.7)$$

and distance variance is $\mathcal{V}_\alpha(X, X)$. The corresponding distance correlation is

$$\mathcal{R}_\alpha(X, Y) = \frac{\mathcal{V}_\alpha(X, Y)}{\sqrt{\mathcal{V}_\alpha(X, X)\mathcal{V}_\alpha(Y, Y)}},$$

provided the denominator is positive. For $\alpha = 1$ we have the special case as in the definitions of dCov and dCor above.

Theorem 18.3. *Let $0 < \alpha < 2$. For every pair (ρ, r), $-1 < \rho < 1$, $0 < r < 1$, there exist random variables X, Y with finite moments of order 2α, such that $\rho(X, Y) = \rho$, $\mathcal{R}_\alpha(X, Y) = r$.*

Remark 18.1. In addition to the set above, the only possible values of the pair $\big(\rho(X, Y), \mathcal{R}_\alpha(X, Y)\big)$ are $(-1, 1)$, $(0, 0)$ and $(1, 1)$.

Proof. Let (U_1, U_2) have bivariate Rademacher distribution such that

$$P\big((U_1, U_2) = (1, 1)\big) = P\big((U_1, U_2) = (-1, -1)\big) = p/2,$$
$$P\big((U_1, U_2) = (-1, 1)\big) = P\big((U_1, U_2) = (1, -1)\big) = (1 - p)/2.$$

Similarly, let (V_1, V_2) have bivariate Rademacher distribution with

$$P\big((V_1, V_2) = (1, 1)\big) = P\big((V_1, V_2) = (-1, -1)\big) = q/2,$$
$$P\big((V_1, V_2) = (-1, 1)\big) = P\big((V_1, V_2) = (1, -1)\big) = (1 - q)/2.$$

Then U_1, U_2, V_1, V_2 are symmetric random variables. Let

$$0 < \varepsilon < 1, \qquad \tfrac{1}{2} < \delta < \tfrac{1}{\alpha},$$

and define $(X_\varepsilon, Y_\varepsilon)$ as the mixture of (U_1, U_2) and $\varepsilon^\delta(V_1, V_2)$ with weights ε and $1 - \varepsilon$, respectively. In other words, if Z_ε is a Bernoulli random variable with $P(Z_\varepsilon = 1) = 1 - P(Z_\varepsilon = 0) = \varepsilon$, and Z_ε is independent of U_1, U_2, V_1, V_2, then we can write

$$(X_\varepsilon, Y_\varepsilon) = Z_\varepsilon(U_1, U_2) + (1 - Z_\varepsilon)\varepsilon^\delta(V_1, V_2).$$

Elementary calculation yields

$$\mathrm{cov}(X_\varepsilon, Y_\varepsilon) = E(X_\varepsilon Y_\varepsilon) = \varepsilon(2p - 1) + (1 - \varepsilon)\varepsilon^{2\delta}(2q - 1),$$
$$\mathrm{Var}(X_\varepsilon) = \mathrm{Var}(Y_\varepsilon) = E(X_\varepsilon^2) = \varepsilon + (1 - \varepsilon)\varepsilon^{2\delta},$$

hence

$$\rho(X_\varepsilon, Y_\varepsilon) = \frac{(2p-1) + (1-\varepsilon)\varepsilon^{2\delta-1}(2q-1)}{1 + (1-\varepsilon)\varepsilon^{2\delta-1}}.$$

For fixed ε, δ and q, try to choose p in such a way that $\rho(X_\varepsilon, Y_\varepsilon) = \rho$; that is,

$$p = \tfrac{1}{2}\big[\rho + 1 + (\rho + 1 - 2q)(1-\varepsilon)\varepsilon^{2\delta-1}\big], \qquad (18.8)$$

with $0 < p < 1$. If ε is sufficiently small, this will hold.

For $\mathcal{V}_\alpha^2(X_\varepsilon, Y_\varepsilon)$, let us start with the first term on the right-hand side of (18.7):

$$E\big(|X_\varepsilon - X_\varepsilon'|^\alpha |Y_\varepsilon - Y_\varepsilon'|^\alpha\big) = \varepsilon^2 E\big(|U_1 - U_1'|^\alpha |U_2 - U_2'|^\alpha\big)$$
$$+ 2\varepsilon(1-\varepsilon) E\big(|U_1 - \varepsilon^\delta V_1'|^\alpha |U_2 - \varepsilon^\delta V_2'|^\alpha\big)$$
$$+ (1-\varepsilon)^2 \varepsilon^{2\alpha\delta} E\big(|V_1 - V_1'|^\alpha |V_2 - V_2'|^\alpha\big).$$

All three expectations on the right-hand side are clearly bounded by $2^{2\alpha} < 16$. Furthermore,

$$E\big(|U_1 - \varepsilon^\delta V_1'|^\alpha |U_2 - \varepsilon^\delta V_2'|^\alpha\big) = \frac{pq + (1-p)(1-q)}{2}\big[(1-\varepsilon^\delta)^{2\alpha} + (1+\varepsilon^\delta)^{2\alpha}\big]$$
$$+ \frac{p(1-q) + (1-p)q}{2}(1-\varepsilon^{2\delta})^\alpha.$$

Expanding the terms $(1-\varepsilon^\delta)^{2\alpha}$ and $(1+\varepsilon^\delta)^{2\alpha}$ into Maclaurin series yields

$$(1-\varepsilon^\delta)^{2\alpha} = 1 - 2\alpha\varepsilon^\delta + O(\varepsilon^{2\delta}), \quad (1+\varepsilon^\delta)^{2\alpha} = 1 + 2\alpha\varepsilon^\delta + O(\varepsilon^{2\delta}).$$

Hence $\big|E\big(|U_1 - \varepsilon^\delta V_1'|^\alpha |U_2 - \varepsilon^\delta V_2'|^\alpha\big) - 1\big| = O(\varepsilon^{2\delta})$ and thus

$$E\big(|X_\varepsilon - X_\varepsilon'|^\alpha |Y_\varepsilon - Y_\varepsilon'|^\alpha\big) = 2\varepsilon + \varepsilon^{2\alpha\delta} E\big(|V_1 - V_1'|^\alpha |V_2 - V_2'|^\alpha\big) + R_1,$$

where the remainder R_1 satisfies $|R_1| \le \kappa_1\big(\varepsilon^{1+2\alpha\delta} + \varepsilon^2\big)$ with an absolute constant κ_1 not depending on ε, α and q.

Similarly, for the second term we can write

$$E\big(|X_\varepsilon - X_\varepsilon'|^\alpha\big) = \varepsilon^2 E\big(|U_1 - U_1'|^\alpha\big) + 2\varepsilon(1-\varepsilon) E\big(|U_1 - \varepsilon^\delta V_1'|^\alpha\big)$$
$$+ (1-\varepsilon)^2 \varepsilon^{\alpha\delta} E\big(|V_1 - V_1'|^\alpha\big)$$
$$= \varepsilon^{\alpha\delta} E\big(|V_1 - V_1'|^\alpha\big) + R,$$

where $|R| \le 16\varepsilon$. Similar estimation holds for $E\big(|Y_\varepsilon - Y_\varepsilon'|^\alpha\big)$. Hence

$$E\big(|X_\varepsilon - X_\varepsilon'|^\alpha\big) E\big(|Y_\varepsilon - Y_\varepsilon'|^\alpha\big) = \varepsilon^{2\alpha\delta} E\big(|V_1 - V_1'|^\alpha\big) E\big(|V_2 - V_2'|^\alpha\big) + R_2$$

with a remainder $|R_2| \le \kappa_2\, \varepsilon^{1+\alpha\delta}$, where the constant κ_2 is absolute again.

Finally, let us turn to the third term.

$$
\begin{aligned}
E\big(|X_\varepsilon - X_\varepsilon'|^\alpha |Y_\varepsilon - Y_\varepsilon''|^\alpha\big) &= \varepsilon^3 E\big(|U_1 - U_1'|^\alpha |U_2 - U_2''|^\alpha\big) \\
&+ \varepsilon^2 (1-\varepsilon) E\big(|\varepsilon^\delta V_1 - U_1'|^\alpha |\varepsilon^\delta V_2 - U_2''|^\alpha\big) \\
&+ 2\varepsilon^2 (1-\varepsilon) E\big(|U_1 - \varepsilon^\delta V_1'|^\alpha |U_2 - U_2''|^\alpha\big) \\
&+ \varepsilon (1-\varepsilon)^2 E\big(|U_1 - \varepsilon^\delta V_1'|^\alpha |U_2 - \varepsilon^\delta V_2''|^\alpha\big) \\
&+ 2\varepsilon (1-\varepsilon)^2 \varepsilon^{\alpha\delta} E\big(|V_1 - V_1'|^\alpha |\varepsilon^\delta V_2 - U_2''|^\alpha\big) \\
&+ (1-\varepsilon)^3 \varepsilon^{2\alpha\delta} E\big(|V_1 - V_1'|^\alpha |V_2 - V_2''|^\alpha\big).
\end{aligned}
$$

Similar computations as in the case of the first and the second terms yield

$$
E\big(|X_\varepsilon - X_\varepsilon'|^\alpha |Y_\varepsilon - Y_\varepsilon''|^\alpha\big) = \varepsilon + \varepsilon^{2\alpha\delta} E\big(|V_1 - V_1'|\,|V_2 - V_2''|\big) + R_3,
$$

where $|R_3| \le \kappa_3\, \varepsilon^{1+\alpha\delta}$ with an absolute constant κ_3. Indeed, the only non-trivial term is

$$
E\big(|U_1 - \varepsilon^\delta V_1'|^\alpha |U_2 - \varepsilon^\delta V_2''|^\alpha\big) = \frac{1}{2}\left[\frac{(1-\varepsilon^\delta)^{2\alpha} + (1+\varepsilon^\delta)^{2\alpha}}{2} + (1-\varepsilon^{2\delta})^\alpha\right],
$$

which we have already shown to be $1 + O(\varepsilon^{2\delta})$.

Combining the above estimations, we find the terms of order ε cancel out yielding

$$
\begin{aligned}
\mathcal{V}_\alpha^2(X_\varepsilon, Y_\varepsilon) &= \varepsilon^{2\alpha\delta} \mathcal{V}_\alpha^2(V_1, V_2) + O\big(\varepsilon^{1+\alpha\delta}\big) \\
&= \varepsilon^{2\alpha\delta} 2^{2\alpha-2}(2q-1)^2 + O\big(\varepsilon^{1+\alpha\delta}\big),
\end{aligned}
$$

where the constant involved in the big oh notation is absolute.

In the calculations above nothing was used about the actual values of p and q. Taking $p = q = 1$ we get that

$$
\begin{aligned}
\mathcal{V}_\alpha^2(X_\varepsilon, X_\varepsilon) = \mathcal{V}_\alpha^2(Y_\varepsilon, Y_\varepsilon) &= \varepsilon^{2\alpha\delta} \mathcal{V}_\alpha^2(V_1, V_2) + O\big(\varepsilon^{1+\alpha\delta}\big) \\
&= \varepsilon^{2\alpha\delta} 2^{2\alpha-2} + O\big(\varepsilon^{1+\alpha\delta}\big).
\end{aligned}
$$

Therefore we obtain

$$
\mathcal{R}_\alpha(X_\varepsilon, Y_\varepsilon) = |2q-1| + O\big(\varepsilon^{1-\alpha\delta}\big),
$$

with an absolute constant in the big oh. This will be important if we want ε or q to vary, because then p also varies according to (18.8).

Let $0 < r_1 < r < r_2$ and $q_1 = (r_1 + 1)/2$, $q_2 = (r_2 + 1)/2$. Let ε be sufficiently small so that p_1 and p_2 given by (18.8) fall between 0 and 1, furthermore $\mathcal{R}_\alpha(X_\varepsilon, Y_\varepsilon) < r$ for $q = q_1$ and $\mathcal{R}_\alpha(X_\varepsilon, Y_\varepsilon) > r$ for $q = q_2$. Let us move q from q_1 to q_2. It is easy to see that $\mathcal{R}_\alpha(X_\varepsilon, Y_\varepsilon)$ continuously depends on q, hence there must exist a q between q_1 and q_2 for which $\mathcal{R}_\alpha(X_\varepsilon, Y_\varepsilon) = r$ (and, of course, $\rho(X_\varepsilon, Y_\varepsilon) = \rho$.) $\qquad\square$

18.3 Conjecture

Conjecture 18.1. *Let $0 < \alpha < \beta \leq 2$. Then for every pair $0 < \rho, r < 1$ there exist random variables X, Y with finite moments of order α and β, resp., such that $\mathcal{R}_\alpha(X, Y) = \rho$ and $\mathrm{dCor}_\beta(X, Y) = r$.*

Remark 18.2. The intuition behind Theorem 18.3 is that $\rho(X, Y)$ is, in a certain sense, infinitely more sensitive to dependence in the tails than $\mathcal{R}_\alpha(X, Y)$. This behavior is suggested by the fact that $\rho(X, Y)$ does not exist if either of the second moments of X or Y is not finite, whereas $\mathcal{R}_\alpha(X, Y)$ can be defined for any two random variables X and Y with finite moments of order α, $0 < \alpha < 2$.

Similarly, it is clear that any version of $\mathcal{R}_\alpha(X, Y)$ does not exist when either X or Y does not feature finite moments of order α. This leads to the conjecture above, a generalization of Theorem 18.3.

19

The dCor t-test of Independence in High Dimension

CONTENTS

19.1 Introduction

Many applications require analysis of high dimensional data. Multivariate data are represented by their characteristics or features as vectors

$$(X_1, \ldots, X_p) \in \mathbb{R}^p.$$

Here we use boldface $\mathbf{X}$ to denote the $p \times n$ data matrix with observations in rows.

Distance covariance and distance correlation tests introduced in Chapter 12 do apply to testing for independence in high dimension. Unlike tests based on covariances or correlations, where we must have $p < n$ in order to estimate a covariance matrix, sample dCov and dCor are defined for arbitrary n and

$p, q \geq 1$. However, as we will see, the V-statistic $\mathcal{V}_n^2$ is biased and leads to a biased dCor statistic such that the bias is increasing with dimensions.

In this chapter, we review the modified distance correlation statistic proposed in Székely and Rizzo [2013a]. It has the property that under independence the distribution of a transformation of the statistic converges to Student t as dimension tends to infinity, under certain conditions on the coordinates. This provides a t statistic and distance correlation t-test for independence of random vectors in arbitrarily high dimension. This test is based on an unbiased estimator of distance covariance, and the resulting t-test is unbiased for every sample size greater than three and all significance levels.

We have developed consistent tests for multivariate independence applicable in arbitrary dimension based on the corresponding sample distance covariance. Generally in statistical inference, we consider that dimensions of the random variables are fixed and investigate the effect of sample size on inference. Here we restrict attention to the situation where the dimensions of the random vectors are large relative to sample size. With the help of a modified distance correlation we obtain a distance correlation test statistic $\mathcal{T}_n$ that has an asymptotic (with respect to dimension) Student t distribution under independence. Thus we obtain a distance correlation t test for multivariate independence, applicable in high dimensions. The degrees of freedom of the t statistic are such that the statistic is approximately normal for sample size $n \geq 10$. The modified distance correlation statistic for high dimensional data has a symmetric beta distribution, which is approximately normal for moderately large n, and thus we also obtain an asymptotic (high dimension) Z-test for independence.

Remark 19.1 (Notation). *Let us emphasize before proceeding that the proposed t-test is based on an unbiased estimator of $\mathcal{V}^2$, rather than the usual V-statistic. If we label the unbiased statistic U_n^2, then under independence $E[U_n^2] = 0$, and clearly U_n^2 can take negative values. Therefore, in this chapter* **we cannot and do not take the square root of population or sample coefficients**. *Using a superscript 2 is problematic in that it may suggest that we can take the square root, so instead we adopt the notation with superscript $*$, as in $\mathcal{V}^*$, $\mathcal{R}^*$, etc.*

The modified distance correlation statistic $\mathcal{R}_n^*$ converges to the square of population distance correlation $(\mathcal{R}^2)$ stochastically. The computing formula and parameters of the t, beta, and normal limit distributions are simple (linear combinations of Euclidean distances) and the tests are straightforward to apply.

Suppose that we have random samples of observations

$$(X_i, Y_i) \in \mathbb{R}^{p+q}, \qquad i = 1, \ldots, n.$$

That is, the pair of data matrices $(\mathbf{X}, \mathbf{Y})$ are a random sample from the joint distribution of (X, Y), where X and Y are assumed to take values in $\mathbb{R}^p$ and $\mathbb{R}^q$, respectively. We assume that $E|X|^2 < \infty$ and $E|Y|^2 < \infty$.

As usual, a primed symbol denotes an independent copy of the unprimed symbol; that is, X and X' are independent and identically distributed. If the argument of $|\cdot|$ is complex, then $|\cdot|$ denotes the complex norm. In this chapter, the characteristic function of a random vector X is denoted by $\hat{f}_X$, and the joint characteristic function of random vectors X and Y is denoted by $\hat{f}_{X,Y}$. The empirical characteristic functions are denoted by $\hat{f}_X^n$, $\hat{f}_Y^n$, and $\hat{f}_{X,Y}^n$.

19.1.1   Population dCov and dCor Coefficients

The distance covariance (dCov or $\mathcal{V}$) and distance correlation (dCor or $\mathcal{R}$) coefficients are defined in Section 12.3. See Definitions 12.1 and 12.3.

For all distributions with finite first moments, recall that *distance correlation* $\mathcal{R}$ generalizes the idea of *correlation*, such that:

1. $\mathcal{R}(X, Y)$ is defined for X and Y in arbitrary dimensions.
2. $\mathcal{R}(X, Y) = 0$ if and only if X and Y are independent.
3. $0 \leq \mathcal{R}(X, Y) \leq 1$.

The same properties also hold for $\mathcal{R}^2(X, Y)$.

19.1.2   Sample dCov and dCor

The sample dCov and dCor are defined in Section 12.4 (see Definitions 13.3 and 13.4). $\mathcal{V}_n^2$ is defined by substituting the empirical characteristic functions in (12.6). The resulting statistics are given by an explicit computing formula (19.1) derived in Székely, Rizzo, and Bakirov [2007], Theorem 1. Our original definition [Székely et al., 2007, Székely and Rizzo, 2009a] of the sample distance covariance is the non-negative square root of

$$\mathrm{dCov}_n^2(\mathbf{X}, \mathbf{Y}) = \frac{1}{n^2} \sum_{i,j}^{n} A_{i,j} B_{i,j}, \tag{19.1}$$

where

$$A_{i,j} = |X_i - X_j| - \frac{1}{n}\sum_{k=1}^{n}|X_k - X_j| - \frac{1}{n}\sum_{l=1}^{n}|X_i - X_l| + \frac{1}{n^2}\sum_{k,l=1}^{n}|X_k - X_l|,$$

$$B_{i,j} = |Y_i - Y_j| - \frac{1}{n}\sum_{k=1}^{n}|Y_k - Y_j| - \frac{1}{n}\sum_{l=1}^{n}|Y_i - Y_l| + \frac{1}{n^2}\sum_{k,l=1}^{n}|Y_k - Y_l|,$$

$i, j = 1, \ldots, n$, and $|\cdot|$ denotes the Euclidean norm. Thus for $i, j = 1, \ldots, n$,

$$A_{i,j} = a_{ij} - \bar{a}_i - \bar{a}_j + \bar{a}; \qquad B_{i,j} = b_{ij} - \bar{b}_i - \bar{b}_j + \bar{b},$$

where

$$a_{ij} = |X_i - X_j|, \qquad i, j = 1, \ldots, n,$$

$$a_{i.} = \sum_{k=1}^{n} a_{ik}, \qquad a_{.j} = \sum_{k=1}^{n} a_{kj}, \qquad \bar{a}_i = \bar{a}_{i.} = \frac{1}{n} a_{i.},$$

$$a_{..} = \sum_{i,j=1}^{n} a_{ij}, \qquad \bar{a} = \frac{1}{n^2} \sum_{i,j=1}^{n} a_{ij},$$

and the corresponding notation is used for the Y sample with $b_{ij} = |Y_i - Y_j|$, etc.

The sample distance correlation is defined by the non-negative square root of

$$\mathcal{R}_n^2(\mathbf{X}, \mathbf{Y}) = \frac{\mathcal{V}_n^2(\mathbf{X}, \mathbf{Y})}{\sqrt{\mathcal{V}_n^2(\mathbf{X}, \mathbf{X}) \cdot \mathcal{V}_n^2(\mathbf{Y}, \mathbf{Y})}}, \tag{19.2}$$

and $\mathcal{R}_n^2(\mathbf{X}, \mathbf{Y}) = 0$ if the denominator in (19.2) equals 0.

We have seen that for independent random vectors $\mathbf{X}$ and $\mathbf{Y}$ with finite first moments (in fixed dimensions), $n \mathcal{V}_n^2(\mathbf{X}, \mathbf{Y})$ converges in distribution to a quadratic form of centered Gaussian random variables

$$\sum_{i=1}^{\infty} \lambda_i Z_i^2,$$

as sample size n tends to infinity [Székely et al., 2007, Theorem 5], where Z_i are iid standard normal random variables and λ_i are positive constants that depend on the distributions of $\mathbf{X}$ and $\mathbf{Y}$.

Below we prove a corresponding limit theorem as dimension tends to infinity; this limit is obtained for a related, modified version of distance correlation defined in Section 19.3.

19.2   On the Bias of the Statistics

First, let us see why a modified version of sample distance covariance and distance correlation is advantageous in high dimension. We begin by observing that although dCor characterizes independence in arbitrary dimension, the numerical value of the (original) corresponding statistic can be difficult to interpret in high dimension without a formal test.

Let $\alpha = E|X - X'|$ and $\beta = E|Y - Y'|$. It is easy to see that for any fixed n,

$$E[A_{i,j}] = \begin{cases} \frac{\alpha}{n}, & i \neq j; \\ \frac{\alpha}{n} - \alpha, & i = j; \end{cases} \qquad E[B_{i,j}] = \begin{cases} \frac{\beta}{n}, & i \neq j; \\ \frac{\beta}{n} - \beta, & i = j. \end{cases}$$

It can be shown (see Section 19.8.1) that for an important class of distributions including independent standard multivariate normal $\mathbf{X}$ and $\mathbf{Y}$, for fixed n each of the statistics

$$\frac{\mathcal{V}_n^2(\mathbf{X}, \mathbf{Y})}{\alpha\beta}, \quad \frac{\mathcal{V}_n^2(\mathbf{X}, \mathbf{X})}{\alpha^2}, \quad \frac{\mathcal{V}_n^2(\mathbf{Y}, \mathbf{Y})}{\beta^2}$$

converges to $(n-1)/n^2$ as dimensions p, q tend to infinity (with n fixed). Thus, for sample distance correlation, it follows that

$$\mathcal{R}_n^2(\mathbf{X}, \mathbf{Y}) = \frac{\mathcal{V}_n^2(\mathbf{X}, \mathbf{Y})}{\sqrt{\mathcal{V}_n^2(\mathbf{X}, \mathbf{X}) \cdot \mathcal{V}_n^2(\mathbf{Y}, \mathbf{Y})}} \xrightarrow[p,q\to\infty]{} 1, \qquad (19.3)$$

even though X and Y are independent.

Here we see that although distance correlation characterizes independence, and the dCov test of independence is valid for X, Y in arbitrary dimensions, interpretation of the size of the sample distance correlation coefficient without a formal test becomes more difficult for X and Y in high dimensions. See Example 19.1 for an illustration of how the corrected statistics and t-test address this issue.

Remark 19.2. One could address this issue, which arises from the positive bias in the V-statistic, by implementing an unbiased estimator of $\mathcal{V}^2(X, Y)$. For example, one could apply a U-statistic for $\mathrm{dCov}^2(X, Y)$. This is in fact, the idea of the modified distance covariance statistic. We apply an unbiased estimator $\mathcal{V}_n^*(\mathbf{X}, \mathbf{Y})$ (although it is a different unbiased estimator than the U-statistic in (19.5) below). See $\mathcal{R}_n^*$ in Table 22.1 for an example of how the corresponding bias-corrected $\mathrm{dCor}^2(X, Y)$ corrects this issue.

Under independence, a transformation of the modified distance correlation statistic $\mathcal{R}_n^*(\mathbf{X}, \mathbf{Y})$ converges (as $p, q \to \infty$) to a Student t distribution, providing a more interpretable sample coefficient.

Numerical Illustration

Table 19.1 illustrates the original and modified distance correlation statistics with a numerical example. We generated independent samples with iid Uniform$(0,1)$ coordinates and computed $\mathcal{R}_n^2$, modified distance correlation $\mathcal{R}_n^*$ (Section 19.3), and the corresponding t and Z statistics (Section 19.4). Each row of the table reports values for one pair of samples for dimension $p = q$ and $n = 30$. The numerical value of $\mathcal{R}_n^2$ approaches 1 as dimension increases, even under independence. Without our dCov test, numerical interpretation of original $\mathcal{R}_n^2$ is difficult. In contrast, we see that the modified statistics $\mathcal{R}_n^*$ in the table are centered close to zero and stable with respect to dimension.

The expected value of the squared distance covariance V-statistic $\mathcal{V}_n^2$ is given in Proposition 13.1.

TABLE 19.1
Numerical illustration ($n = 30$) of distance correlation $\mathcal{R}_n$ and modified distance correlation $\mathcal{R}_n^*$ statistics in high dimension. The modified statistic $\mathcal{R}_n^*$ is based on an unbiased estimator of the squared population distance covariance. Each row of the table reports statistics for one sample $(\mathbf{X}, \mathbf{Y})$, where X and Y each have iid standard uniform coordinates. The t statistic $\mathcal{T}_n$ and Z statistic introduced in Section 19.4 are also reported.

p, q	$\mathcal{R}_n$	$\mathcal{R}_n^*$	$\mathcal{T}_n$	Z
1	0.4302668	0.0248998	0.5006350	0.5004797
2	0.6443883	0.1181823	2.3921989	2.3754341
4	0.7103136	−0.0098916	−0.1988283	−0.1988186
8	0.8373288	0.0367129	0.7384188	0.7379210
16	0.8922197	−0.0675606	−1.3610616	−1.3579518
32	0.9428649	−0.0768243	−1.5487268	−1.5441497
64	0.9702281	−0.1110683	−2.2463438	−2.2324451
128	0.9864912	−0.0016547	−0.0332595	−0.0332595
256	0.9931836	0.0517415	1.0413867	1.0399917
512	0.9963244	−0.0158947	−0.3195190	−0.3194787
1024	0.9983706	0.0542560	1.0921411	1.0905325
2048	0.9991116	−0.0076502	−0.1537715	−0.1537670
4096	0.9995349	−0.0863294	−1.7417010	−1.7351986
8192	0.9997596	−0.0754827	−1.5215235	−1.5171827
16384	0.9999032	0.0277193	0.5573647	0.5571505

To motivate the definitions of the modified statistics introduced in Section 19.3, let us start by reviewing Theorem 12.1. Here we proved that the L_2 distance between the joint and product of marginal characteristic functions is

$$\|\hat{f}_{X,Y}^n(t,s) - \hat{f}_X^n(t)\hat{f}_Y^n(s)\|^2 = S_1 + S_2 - 2S_3,$$

where

$$S_1 = \frac{1}{n^2}\sum_{k,\ell=1}^{n}|X_k - X_\ell|_p|Y_k - Y_\ell|_q,$$

$$S_2 = \frac{1}{n^2}\sum_{k,\ell=1}^{n}|X_k - X_\ell|_p \frac{1}{n^2}\sum_{k,\ell=1}^{n}|Y_k - Y_\ell|_q,$$

$$S_3 = \frac{1}{n^3}\sum_{k=1}^{n}\sum_{\ell,m=1}^{n}|X_k - X_\ell|_p|Y_k - Y_m|_q,$$

and $V_n^2 := S_1 + S_2 - 2S_3$ is algebraically equivalent to (19.1).

It is obvious that using notations above, we have this equivalent formula for the V-statistic:

$$\mathcal{V}_n^2(\mathbf{X}, \mathbf{Y}) = V_n^2(\mathbf{X}, \mathbf{Y}) = \frac{1}{n^2}\sum_{i,j=1}^{n}a_{ij}b_{ij} - \frac{2}{n^3}\sum_{i=1}^{n}a_{i.}b_{i.} + \bar{a}\bar{b}. \qquad (19.4)$$

An unbiased estimator of $\mathcal{V}^2(X, Y)$ is the U-statistic

$$U_n^2(\mathbf{X}, \mathbf{Y}) = \frac{1}{n(n-3)} \sum_{i \neq j} a_{ij} b_{ij} - \frac{2}{n(n-2)(n-3)} \sum_{i=1}^{n} a_{i.} b_{i.}$$

$$+ \frac{a_{..} b_{..}}{n(n-1)(n-2)(n-3)}. \tag{19.5}$$

See (16.3) in Proposition 17.1 and the proof that U_n^2 is a U-statistic, or Huo and Székely [2016].

Comparing the V-statistic (19.4) and the U-statistic (19.5) one can see that the unbiased statistic has different divisors than n, n^2, n^3, n^4, which depend on the number of non-zero terms in the sums. Our modified statistic is another unbiased statistic, different from (19.5), and similarly, one can see that different factors than n are needed in the definitions for the unbiased statistic $\mathcal{V}_n^*(\mathbf{X}, \mathbf{Y})$ of Section 19.3.

19.3 Modified Distance Covariance Statistics

A modified version of the statistic $\mathcal{V}_n^2(\mathbf{X}, \mathbf{Y})$ that avoids (19.3) can be defined starting with corrected versions of the double-centered distance matrices $A_{i,j}$ and $B_{i,j}$. Note that in the original formulation, $E[A_{i,j}] = \alpha/n$ if $i \neq j$ and $E[A_{i,i}] = \alpha/n - \alpha$. Thus in high dimension the difference α between the diagonal and off-diagonal entries of A (and B) can be large. The modified versions $A_{i,j}^*$, $B_{i,j}^*$ of $A_{i,j}$ and $B_{i,j}$ are defined by

$$A_{i,j}^* = \begin{cases} \frac{n}{n-1}(A_{i,j} - \frac{a_{ij}}{n}), & i \neq j; \\ \frac{n}{n-1}(\bar{a}_i - \bar{a}), & i = j, \end{cases} \qquad B_{i,j}^* = \begin{cases} \frac{n}{n-1}(B_{i,j} - \frac{b_{ij}}{n}), & i \neq j; \\ \frac{n}{n-1}(\bar{b}_i - \bar{b}), & i = j. \end{cases} \tag{19.6}$$

One can easily see that $E[A_{i,j}^*] = E[B_{i,j}^*] = 0$ for all i, j.

We now define modified distance covariance and modified distance correlation statistics using the corrected terms $A_{i,j}^*$ and $B_{i,j}^*$.

Definition 19.1. The modified distance covariance statistic is

$$\mathcal{V}_n^*(\mathbf{X}, \mathbf{Y}) = \frac{1}{n(n-3)} \left\{ \sum_{i,j=1}^{n} A_{i,j}^* B_{i,j}^* - \frac{n}{n-2} \sum_{i=1}^{n} A_{i,i}^* B_{i,i}^* \right\}. \tag{19.7}$$

By Proposition 19.2 below, $\mathcal{V}_n^*(\mathbf{X}, \mathbf{Y})$ is an unbiased estimator of the squared population distance covariance.

Definition 19.2. The modified distance correlation statistic is

$$\mathcal{R}_n^*(\mathbf{X}, \mathbf{Y}) = \frac{\mathcal{V}_n^*(\mathbf{X}, \mathbf{Y})}{\sqrt{\mathcal{V}_n^*(\mathbf{X}, \mathbf{X}) \mathcal{V}_n^*(\mathbf{Y}, \mathbf{Y})}}, \tag{19.8}$$

if $V_n^*(\mathbf{X}, \mathbf{X})V_n^*(\mathbf{Y}, \mathbf{Y}) > 0$, and otherwise $\mathcal{R}_n^*(\mathbf{X}, \mathbf{Y}) = 0$.

Note: It follows from Lemma 19.3 that $\sqrt{V_n^*(\mathbf{X}, \mathbf{X})V_n^*(\mathbf{Y}, \mathbf{Y})}$ is always a real number for $n \geq 3$.

While the original $\mathcal{R}_n$ statistic is between 0 and 1, $\mathcal{R}_n^*$ can take negative values; the Cauchy-Schwartz inequality implies that $|\mathcal{R}_n^*| \leq 1$. Later we will see that $\mathcal{R}_n^*$ converges to $\mathcal{R}^2$ stochastically. In the next section we derive the limit distribution of $\mathcal{R}_n^*$.

Remark 19.3. In the following we exclude $|\mathcal{R}_n^*| = 1$, corresponding to the case when the $\mathbf{X}$ sample is a linear transformation of the $\mathbf{Y}$ sample. To be more precise, $|\mathcal{R}_n^*| = 1$ implies that the linear spaces spanned by the sample observations $\mathbf{X}$ and $\mathbf{Y}$, respectively, have the same dimension; thus we can represent the two samples in the same linear space, and in this common space the $\mathbf{Y}$ sample is a linear function of the $\mathbf{X}$ sample.

19.4   The t-test for Independence in High Dimension

The main result of Székely and Rizzo [2013a] is that as p, q tend to infinity, under the independence hypothesis,

$$\mathcal{T}_n = \sqrt{\nu - 1} \cdot \frac{\mathcal{R}_n^*}{\sqrt{1 - (\mathcal{R}_n^*)^2}}$$

converges in distribution to Student t with $\nu - 1$ degrees of freedom, where $\nu = \frac{n(n-3)}{2}$. Thus for $n \geq 10$ this limit is approximately standard normal. The t-test of independence is unbiased for every $n \geq 4$ and any significance level.

The t-test of independence rejects the null hypothesis at level α if $\mathcal{T}_n > c_\alpha$, where $c_\alpha = t_{\nu-1}^{-1}(1 - \alpha)$ is the $(1 - \alpha)$ quantile of a Student t distribution with $\nu - 1$ degrees of freedom. By Theorem 19.1(i), the test has level α. Section 19.8.4 for the proof of Theorem 19.1.

Remark 19.4. The implementation of the t-test of independence is straightforward, and all empirical results presented below were performed using R software [R Core Team, 2022]. See the *energy* package for R [Rizzo and Székely, 2022] for an implementation of our methods available under general public licence.

Example 19.1. As a simple example to illustrate an application of the t-test of independence in Theorem 19.1, we revisit the type of example summarized in Table 19.1. In the table, we saw that as dimension tends to infinity, the (uncorrected) distance correlation approaches 1. Now let us compute the corrected statistics and t-test for independent standard multivariate normal

$\mathbf{X} \in \mathbb{R}^{30}$, $\mathbf{Y} \in \mathbb{R}^{30}$, with sample size $n = 30$. The result of our t-test, coded in R, is summarized below. To generate the data:

```
set.seed(1)
n <- 30
p <- q <- 30
x <- matrix(rnorm(n*p), n, p)
y <- matrix(rnorm(n*q), n, q)
```

To compute the statistic alone use the *energy* function `dcorT`, and to run the t-test `dcorT.test`:

```
> dcorT(x, y)
[1] 0.4415111

> dcorT.test(x, y)

dcor t-test of independence for high dimension

data:  x and y
T = 0.44151, df = 404, p-value = 0.3295
sample estimates:
Bias corrected dcor
         0.0219607
```

Here the corrected statistic is $\mathcal{R}_n^* = 0.0219607$ and $\mathcal{T}_n = 0.44151$, with 404 degrees of freedom, thus we do not reject independence.

For comparison the uncorrected $\mathcal{R}_n^2$ is 0.8072597 and we can always apply the permutation test `dcov.test` or `dcor.test` for arbitrary n, p, q. With 999 permutation replicates we compute

```
dCor independence test (permutation test)

data:  index 1, replicates 999
dCor = 0.89848, p-value = 0.279
```

and we do not reject independence.

19.5   Theory and Properties

Our procedure for testing independence applies distances $|X_i - X_j|$, where $E|X_i|^2 < \infty$, so without loss of generality we can assume that $E[X_i] = 0$.

For simplicity we suppose in the main proof that random vectors X and Y have iid coordinates with finite variance. It will be clear from the proofs that

324 The Energy of Data and Distance Correlation

much weaker conditions are also sufficient because what we really need is that for partial sums of squared coordinates certain limit theorems, like Weak Law of Large Numbers and Central Limit Theorem (CLT) hold. Corollary 19.2 below deals with the case when the coordinates are exchangeable. For time series even this condition is too strong.

The following related statistics will be used in deriving our main result. Let

$$W_n(\mathbf{X}, \mathbf{Y}) = \frac{2n-1}{n^3} \sum_{i,j=1}^{n} a_{ij} b_{ij} - \bar{a}\bar{b} + \frac{2}{n^2 - 2n} \sum_{i=1}^{n} (\bar{a}_i - \bar{a})(\bar{b}_i - \bar{b}), \quad (19.9)$$

and define

$$\mathcal{U}_n(\mathbf{X}, \mathbf{Y}) \overset{def}{=} \mathcal{V}_n^2(\mathbf{X}, \mathbf{Y}) - \frac{W_n(\mathbf{X}, \mathbf{Y})}{n}. \quad (19.10)$$

The statistic W_n is related to the bias in the V-statistic V_n^2.

Lemma 19.1. *The following claims hold:*

1. *For all $\mathbf{X}$ and all $n \geq 3$, $W_n(\mathbf{X}, \mathbf{X}) \geq 0$.*

2. *For all $\mathbf{X}$ and all $n \geq 3$, $\mathcal{U}_n(\mathbf{X}, \mathbf{X}) \geq 0$.*

3. *If $E|X|^2 < \infty, E|Y|^2 < \infty$, $i = 1, 2$, then for any fixed p and q,*

$$W_n(\mathbf{X}, \mathbf{Y}) \xrightarrow[n\to\infty]{\mathcal{P}} 2E|X - X'||Y - Y'| - E|X - X'| E|Y - Y'|.$$

4. *If $E|X|^2 < \infty, E|Y|^2 < \infty$ and X and Y are independent, then for any fixed p and q,*

$$W_n(\mathbf{X}, \mathbf{Y}) \xrightarrow[n\to\infty]{\mathcal{P}} \alpha\beta.$$

Proof of Lemma 19.1 is given in Section 19.8.2.

Remark 19.5. Formula (19.10) and Lemma 19.1 (iii) imply that

$$\mathcal{V}_n^2(\mathbf{X}, \mathbf{Y}) = \mathcal{U}_n(\mathbf{X}, \mathbf{Y}) + o_P(1), \quad n \to \infty,$$

where $o_P(1)$ is a term that converges to zero in probability as $n \to \infty$. Therefore $\mathcal{U}_n(\mathbf{X}, \mathbf{Y})$ really can be viewed as a modified distance covariance and $\mathcal{U}_n(\mathbf{X}, \mathbf{X})$ can be viewed as a modified distance variance.

Let

$$\mathcal{U}_n^*(\mathbf{X}, \mathbf{Y}) := n(n-3)\mathcal{V}_n^*(\mathbf{X}, \mathbf{Y}) = \sum_{i\neq j} A_{i,j}^* B_{i,j}^* - \frac{2}{n-2} \sum_{i=1}^{n} A_{i,i}^* B_{i,i}^*.$$

Lemma 19.2. *The following identity holds:*

$$n^2 \mathcal{U}_n(\mathbf{X}, \mathbf{Y}) = \frac{(n-1)^2}{n^2} \cdot \mathcal{U}_n^*(\mathbf{X}, \mathbf{Y}). \tag{19.11}$$

Proof of Lemma 19.2 is given in Section 19.8.2.

Thus the following decomposition holds for all $n \geq 3$:

$$n^2 \mathcal{V}_n^2(\mathbf{X}, \mathbf{X}) = \left(1 - \frac{1}{n}\right)^2 \cdot \mathcal{U}_n^*(\mathbf{X}, \mathbf{X}) + n \mathcal{W}_n(\mathbf{X}, \mathbf{X}),$$

where $\mathcal{U}_n^*(\mathbf{X}, \mathbf{X}) \geq 0$, $\mathcal{W}_n(\mathbf{X}, \mathbf{X}) \geq 0$.

Recall that X' denotes an independent copy of X. In what follows, (X, Y), (X', Y'), and (X'', Y'') are independent and identically distributed.

Proposition 19.1. *If X and Y have finite second moments, then*

1.

$$\mathcal{V}^2(X, Y) = E|X - X'||Y - Y'| + E|X - X'|E|Y - Y'| \\ - 2E|X - X'||Y - Y''|.$$

2. If X and Y are independent, then

$$E[\mathcal{V}_n^2(\mathbf{X}, \mathbf{Y})] = \frac{n-1}{n^2} \left\{ E|X - X'|E|Y - Y'| \right\}.$$

3. If X and Y are independent, then $E[\mathcal{V}_n^*(\mathbf{X}, \mathbf{Y})] = 0$.

See Section 19.8.3 for proof of Proposition 19.1.

Proposition 19.2. *For all n, p, q, the modified distance covariance statistic $\mathcal{V}_n^*(\mathbf{X}, \mathbf{Y})$ is an unbiased estimator of the squared distance covariance $\mathcal{V}^2(\mathbf{X}, \mathbf{Y})$, and*

$$E[\mathcal{V}_n^2(\mathbf{X}, \mathbf{Y})] = \frac{(n-1)}{n^3} \left[(n-2)^2 \mathcal{V}^2(X, Y) + 2(n-1)\mu - (n-2)\alpha\beta \right],$$

$$E[\mathcal{W}_n(\mathbf{X}, \mathbf{Y})] = \frac{n-1}{n^2} \left[(2n-1)\mu - (n-3)\alpha\beta - 2\delta \right],$$

$$E[\mathcal{U}_n(\mathbf{X}, \mathbf{Y})] = \frac{(n-1)^2(n-3)}{n^3} \mathcal{V}^2(X, Y),$$

$$E[\mathcal{U}_n^*(\mathbf{X}, \mathbf{Y})] = \frac{n^4}{(n-1)^2} E[\mathcal{U}_n(\mathbf{X}, \mathbf{Y})] = n(n-3)\mathcal{V}^2(X, Y),$$

where $\mu = E|X - X'||Y - Y'|$, $\alpha = E|X - X'|$, $\beta = E|Y - Y'|$, and $\delta = E|X - X'||Y - Y''|$.

See Section 19.8.3 for proof of Proposition 19.2.

Lemma 19.3. *If the coordinates of X and Y are iid, $0 < E|X|^2 < \infty$, $0 < E|Y|^2 < \infty$, and X and Y are independent, then for fixed n there exist independent random variables $\Omega_{i,j}, \Psi_{i,j}$, such that*

$$(i) \qquad \mathcal{U}_n^*(\mathbf{X}, \mathbf{X}) \underset{p \to \infty}{\longrightarrow} \sum_{i \neq j} \Omega_{i,j}^2 \overset{\mathcal{D}}{=} 2\sigma_X^2 \chi_\nu^2, \tag{19.12}$$

$$(ii) \qquad \mathcal{U}_n^*(\mathbf{Y}, \mathbf{Y}) \underset{q \to \infty}{\longrightarrow} \sum_{i \neq j} \Psi_{i,j}^2 \overset{\mathcal{D}}{=} 2\sigma_Y^2 \chi_\nu^2, \tag{19.13}$$

$$(iii) \qquad \mathcal{U}_n^*(\mathbf{X}, \mathbf{Y}) \underset{p,q \to \infty}{\longrightarrow} \sum_{i \neq j} \Omega_{i,j} \Psi_{i,j}, \tag{19.14}$$

where $\nu = \frac{n(n-3)}{2}$, χ_ν^2 denotes the distribution of a chisquare random variable with ν degrees of freedom,

$$\sigma_X^2 = \frac{E\langle X, X'\rangle^2}{2E|X|^2}, \qquad \sigma_Y^2 = \frac{E\langle Y, Y'\rangle^2}{2E|Y|^2},$$

Ω_{ij} are iid Normal$(0, \sigma_X^2)$, and Ψ_{ij} are iid Normal$(0, \sigma_Y^2)$.

See Section 19.8.2 for the proof of Lemma 19.3. The variables $\Omega_{i,j}$ and $\Psi_{i,j}$ are defined in the proof by equations (19.26) and (19.27).

For the corresponding correlation coefficient we have

$$\mathcal{R}_n^* = \frac{\mathcal{U}_n^*(\mathbf{X}, \mathbf{Y})}{\sqrt{\mathcal{U}_n^*(\mathbf{X}, \mathbf{X})\mathcal{U}_n^*(\mathbf{Y}, \mathbf{Y})}} \underset{p,q \to \infty}{\longrightarrow} \frac{\sum_{i \neq j} \Omega_{i,j} \Psi_{i,j}}{\sqrt{\sum_{i \neq j} \Omega_{i,j}^2 \sum_{i \neq j} \Psi_{i,j}^2}}.$$

Define the test statistic

$$\mathcal{T}_n = \sqrt{\nu - 1} \cdot \frac{\mathcal{R}_n^*}{\sqrt{1 - (\mathcal{R}_n^*)^2}}. \tag{19.15}$$

Theorem 19.1. *If the coordinates of X and Y are iid with positive finite variance, for fixed sample size $n \geq 4$ the following hold.*
(i) Under independence of X and Y,

$$P\{\mathcal{T}_n < t\} \underset{p,q \to \infty}{\longrightarrow} P\{t_{\nu-1} < t\},$$

where $\mathcal{T}_n$ is the statistic (19.15) and $\nu = \frac{n(n-3)}{2}$.
(ii) Let $c_\alpha = t_{\nu-1}^{-1}(1 - \alpha)$ denote the $(1 - \alpha)$ quantile of a Student t distribution with $\nu - 1$ degrees of freedom. The t-test of independence at significance level α that rejects the independence hypothesis whenever $\mathcal{T}_n > c_\alpha$ is unbiased.
(iii) If X and Y are dependent, then $\rho = \rho(\Omega, \Psi) > 0$ and $P(\mathcal{T}_n > c_\alpha)$ is a monotone increasing function of ρ.

When the conditions for the t-test hold, we also obtain a Z-test of independence in high dimension:

Corollary 19.1. *Under independence of X and Y, if the coordinates of X and Y are iid with positive finite variance, then the limit distribution of $(1+\mathcal{R}_n^*)/2$ is a symmetric beta distribution with shape parameter $(\nu-1)/2$. It follows that in high dimension the large sample distribution of $\sqrt{\nu-1}\,\mathcal{R}_n^*$ is approximately standard normal.*

Corollary 19.1 follows from Theorem 19.1 and well known distributional results relating Student t, beta, and normal distributions.

Corollary 19.2. *Theorem 19.1 and Corollary 19.1 hold for random vectors with exchangeable coordinates and finite variance.*

Infinite sequences of random variables are known to be conditionally iid with respect to the sigma algebra of symmetric events [de Finetti, 1937]. If the series is finite one can refer to Kerns and Székely [2006] and Diaconis and Freedman [1980]. Further, if we assume that the variance of each of the variables is finite, then in the Corollary we can assume that conditionally with respect to the sigma algebra of symmetric events the CLT holds. The only factor that might change depending on the condition is the variance of the normal limit. However, in $\mathcal{R}_n^*$ this variance factor cancels, hence $\mathcal{R}_n^*$ (and therefore $\mathcal{T}_n$) has the same distribution

19.6 Application to Time Series

Let $\{X(t), Y(t)\}$ be a strongly stationary times series where for simplicity both $X(t)$ and $Y(t)$ are real-valued. Strong stationarity guarantees that if we take a sample of p consecutive observations from $\{X(t), Y(t)\}$, then their joint distribution and thus their dependence structure does not depend on the starting point. On the other hand strong stationarity is not enough to guarantee CLT for partial sums, not even conditionally with respect to a sigma algebra. (The extra condition of m-dependence of $\{X(t), Y(t)\}$ would be enough by a classical theorem of Hoeffding but this condition is too strong.)

In order to apply our t-test, we need the conditional validity of CLT, conditioned on a sigma algebra. Then the variance of the normal limit becomes a random variable and the possible limits are scale mixtures of normals. Many of these distributions are heavy tailed, which is important in financial applications.

Let us summarize briefly conditions for CLT that are relevant in this context.

1. If our observations are exchangeable, then de Finetti [1937] (or in the finite case Kerns and Székely [2006]) applies and if the (conditional) variances are finite, then the conditional CLT follows. See also Diaconis and Freedman [1980].

2. For strongly stationary sequences we can refer to Ibragimov's conjecture from the 1960s, and to Peligrad [1985] for a proof of a somewhat weaker claim. (For strongly stationary sequences with finite variance such that $Var(S_n)$ tends to infinity, the CLT does not always follow, not even the conditional CLT; thus in general some extra conditions are needed.)

3. Stein [1972] type of dependence, who first obtained in 1972 a bound between the distribution of a sum of an m-dependent sequence of random variables and a standard normal distribution in the Kolmogorov (uniform) metric, and hence proved not only a central limit theorem, but also bounds on the rates of convergence for the given metric.

To apply our t-test we also need iid observations at least conditionally with respect to a sigma algebra. Typically, for time series, only one realization of each series is available for analysis. A random sample of iid observations is obtained (at least conditionally iid) for analysis by application of the following Proposition.

Proposition 19.3. *Fix $p < N$ and let $T_1, T_2, \ldots, T_n$ be integers in $\{1, \ldots, N - p + 1\}$. Define X_j to be the subsequence of length p starting with $X(T_j)$, and similarly define Y_j; that is,*

$$X_j = \{X(T_j), X(T_j + 1), \ldots, X(T_j + p - 1)\}, \qquad j = 1, \ldots, n,$$
$$Y_j = \{Y(T_j), Y(T_j + 1), \ldots, Y(T_j + p - 1)\}, \qquad j = 1, \ldots, n.$$

If $X \in \mathbb{R}^d$, then $X(T_j) \in \mathbb{R}^d$ and X_j is a vector in $\mathbb{R}^{pd}$; that is, $X_j = \{X(T_j)_1, \ldots, X(T_j)_d, X(T_j + 1)_1, \ldots X(T_j + 1)_d, \ldots, X(T_j + p - 1)_d\}$. If T_j are drawn at random with equal probability from $\{1, \ldots, N - p + 1\}$, these vectors $\{X_j\}$ are exchangeable; thus, conditional with respect to the sigma algebra of symmetric events, they are iid observations. Similarly, the vectors $\{Y_j\}$ are also conditionally iid, and thus if the variances are finite we can apply the t-test of independence conditioned on the sigma algebra of symmetric events. Hence the t-test of independence is applicable unconditionally.

Corollary 19.2 and Proposition 19.3 imply that conditional with respect to a sigma algebra (to the sigma algebra of symmetric events), these vectors are iid with finite variances. Thus, we can apply the t-test of independence conditioned on this sigma algebra. The variance here is random (thus we can have heavy tailed distributions) but in the formula for t the variance cancels hence the t-test of independence is applicable unconditionally.

In summary, our method is applicable for financial time series if we suppose that the differences of the logarithms of our observations (or other suitable transformation) form a strongly stationary sequence whose partial sums conditionally with respect to a sigma algebra tend to normal.

Example 19.2. In an application of the t-test of independence using the randomization method of Proposition 19.3, we applied the test to pairs of

AR(1) (autoregressive model order 1) time series. The AR(1) data with total length $N = 2500$ was generated from the model $X_t = AX_{t-1} + e_t$, where X_t is a vector of length 2, A is a matrix of autoregressive coefficients, and $e_t \in \mathbb{R}^2$ is a noise vector with mean 0. (For simplicity we let e_t be Gaussian, but the method is applicable for non-normal error distributions as well.) The bivariate AR(1) model parameters used in this simulation are

$$A = \begin{pmatrix} .25 & 0 \\ 0 & .25 \end{pmatrix}; \qquad Cov(e) = \begin{pmatrix} 1 & r \\ r & 1 \end{pmatrix},$$

and the length of subsequences is $p = 100$. The estimates of power for varying r are shown in Figure 19.1, with error bars at ± 2 standard errors. These estimates are computed as the proportion of significant t-tests of independence in 2000 replications (significance level 10%). Figure 19.1 summarizes the simulation results for two cases: sample size $n = 25$ and $n = 50$. (The length $N = 2500$ for this example was chosen to match approximately the length of the DJIA data in Example 19.3).

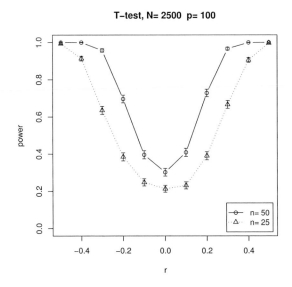

FIGURE 19.1
Proportion of significant tests of independence on dependent $AR(1)$ series of Example 19.2. The AR parameter is 0.25 and error correlation is r.

Example 19.3 (Dow Jones Industrial Index). The time series data of this example are the daily returns of the 30 stocks of the Dow Jones Industrial Index (DJIA), from from August 30, 1993 through August 29, 2003.

(The 30 stocks correspond to the composition of the index on September 1, 2003.) Pairs of stocks are tested for independence of their daily returns. The DJIA closing prices data for this example were from the R package *VaR* [Daniyarov, 2004] with the source attributed to www.nasdaq.com. Each series is identified by a stock ticker symbol; see the *VaR* package for the corresponding list of company names. (The package is archived at https://cran.r-project.org/src/contrib/Archive/VaR/.) The time series we analyze are log(returns); that is, the sequence of differences of the logarithms of prices. A plot of log(returns) (not shown) suggests that the data approximately satisfy the strong stationarity condition.

There are 2520 values of log(returns). We set $p = 100$ and $n = 50$ (see Proposition 19.3). The t statistics (T_n) here are approximately standard normal under the independence hypothesis, and a statistic greater than 1.645 is significant at the 5% level, indicating dependence. Most of the pairs of stocks had significant t statistics. The statistics for 20 pairs of stocks are shown in Table 19.2. Large positive values of T_n are significant.

Instead of a table of 435 statistics we prepared a graphical summary, using a dendrogram to summarize the 435 pairwise $\mathcal{R}^*$ statistics. This figure can be found in Székely and Rizzo [2013a] (Figure 4) along with a detailed explanation of how the clustering was obtained.

From the cluster dendrogram, one can observe, for example, that the financial stocks (AXP, C, JPM) cluster together, and the technology stocks (HPC, IBM, INTC, MSFT) also cluster together, but these two clusters are well separated and not merged until the second to last step when there are three clusters. One can also observe that five manufacturers (AA, CAT, DD, GM, IP) cluster together, as do two retail industry (HD, WMT), two drug (JNJ, MRK), and two telecommunications stocks (SBC, T).

19.7 Dependence Metrics in High Dimension

Zhu et al. [2020] show that if the dimensions $p, q \to \infty$ when n is either fixed of grows to infinity at a slower rate, $dCov_n^2(X, Y)$ is asymptotically

$$(1/\tau) \sum_{i=1}^{p} \sum_{j=1}^{q} \text{Cov}_n^2(X_i, Y_j),$$

where τ is a suitable constant that depends on the marginal distributions. Thus asymptotically $dCov_n$ is a function of Pearson's classical correlations between the coordinates, which means that surprisingly, in very high dimensions $dCov_n$ can capture the usual linear or monotone dependence only. In other words, in high dimensions all that matters is the component-wise Pearson correlation. A natural remedy is to replace these correlations by distance

TABLE 19.2
Selected pairs of stocks of the Dow Jones Industrials Index in
Example 19.3. Those pairs with the largest t-statistics indicating
that log(returns) are highly dependent, are shown on the right.
Examples of pairs of stocks in the index with non-significant T_n
statistics are shown on the left.

X	Y	T_n	p-value	X	Y	T_n	p-value
INTC	JNJ	−1.45	0.93	DD	IP	11.83	0.00
INTC	PG	−1.42	0.92	AA	HON	11.97	0.00
IBM	MCD	−0.42	0.66	DIS	GE	12.07	0.00
GM	T	−0.39	0.65	GE	JPM	13.42	0.00
EK	MO	−0.21	0.58	AXP	GE	13.51	0.00
IBM	IP	−0.00	0.50	HON	UTX	13.82	0.00
GM	JNJ	0.09	0.46	AXP	JPM	14.18	0.00
IBM	JNJ	0.19	0.42	AXP	C	14.21	0.00
MO	WMT	0.25	0.40	INTC	MSFT	15.22	0.00
DD	T	0.26	0.40	C	JPM	16.16	0.00

correlations. The suggested formula is

$$\sqrt{\binom{n}{2}} \sum_{i=1}^{p} \sum_{j=1}^{q} dCov_n^2(X_i, Y_j).$$

A similar modification is suggested for the t-test of independence. For
more details, we refer to Zhu et al. [2020]. See also Yao et al. [2018] on testing
mutual independence in high dimension.

Li et al. [2021] introduce the so-called *distance covariance matrix* whose
trace "corresponds" to distance covariance meaning that the squared sample
distance covariance is asymptotically equal to the normalized trace of the
distance covariance matrix as

$$(n, p, q) \to \infty$$

and

$$(p/n, q/n) \to (c_1, c_2) \in (0, \infty)^2.$$

Then the authors study the asymptotic behavior of the spectrum of the
distance covariance matrix. Interestingly, when the trace of this distance covariance matrix cannot effectively detect non-linear dependence, one can rely
on the largest eigenvalues of the distance covariance matrix to detect non-
linear dependence.

19.8   Proofs

19.8.1   On the Bias of Distance Covariance

In this section we show that for important special cases including standard multivariate normal X and Y, bias of the distance covariance V-statistic is such that the distance correlation of vectors X, Y can approach one as dimensions p and q tend to infinity even though X and Y are independent.

The following algebraic identity for the distance covariance statistic was established in Székely et al. [2007].

$$\mathcal{V}_n^2(\mathbf{X}, \mathbf{Y}) = \frac{1}{n^2} \sum_{i,j=1}^n A_{ij} B_{ij} = \frac{1}{n^2} \sum_{i,j=1}^n a_{ij} b_{ij} + \bar{a}\bar{b} - \frac{2}{n^3} \sum_{i,j,k=1}^n a_{ij} b_{ik}. \quad (19.16)$$

Hence,

$$\frac{\mathcal{V}_n^2(\mathbf{X}, \mathbf{Y})}{\alpha \beta} = \frac{1}{n^2} \sum_{i,j=1}^n \frac{a_{ij} b_{ij}}{\alpha \beta} + \frac{1}{n^4} \sum_{i,j=1}^n \frac{a_{ij}}{\alpha} \sum_{k,\ell=1}^n \frac{b_{k\ell}}{\beta} - \frac{2}{n^3} \sum_{i,j,k=1}^n \frac{a_{ij}}{\alpha} \frac{b_{ik}}{\beta}, \quad (19.17)$$

and

$$\frac{\mathcal{V}_n^2(\mathbf{X})}{\alpha^2} = \frac{1}{n^2} \sum_{i,j=1}^n \frac{a_{ij}^2}{\alpha^2} + \left(\frac{1}{n^2} \sum_{i,j=1}^n \frac{a_{ij}}{\alpha} \right)^2 - \frac{2}{n^3} \sum_{i,j,k=1}^n \frac{a_{ij} a_{ik}}{\alpha^2}. \quad (19.18)$$

Thus, in (19.17) and (19.18), each nonzero term a_{ij} or b_{ij}, $i \neq j$, is divided by its expected value $\alpha = E|X - X'|$ or $\beta = E|Y - Y'|$, respectively.

Suppose that $X \in \mathbb{R}^p$ and $Y \in \mathbb{R}^q$ are independent, the coordinates of X and Y are iid, and the second moments of X and Y exist. Then $X_{i1}, X_{i2}, \ldots, X_{ip}$ are iid observations from the distribution of X_{i1}, and

$$\frac{1}{p} \sum_{t=1}^p (X_{1t} - X_{2t})^2 = \frac{1}{p} \sum_{t=1}^p X_{1t}^2 - \overline{X}_1^2 + \frac{1}{p} \sum_{t=1}^p X_{2t}^2 - \overline{X}_2^2$$

$$- \frac{2}{p} \sum_{t=1}^p X_{1t} X_{2t} + 2\overline{X}_1 \overline{X}_2 + \overline{X}_1^2 - 2\overline{X}_1 \overline{X}_2 + \overline{X}_2^2$$

$$= \widehat{Var}(X_{11}) + \widehat{Var}(X_{21}) - 2\widehat{Cov}(X_{11}, X_{21}) + (\overline{X}_1 - \overline{X}_2)^2,$$

where

$$\overline{X}_i = \frac{1}{p} \sum_{t=1}^p X_{it}, \quad \widehat{Var}(X_{i1}) = \frac{1}{p} \sum_{t=1}^p X_{it}^2 - \overline{X}_i^2, \quad i = 1, 2,$$

$$\text{and } \widehat{Cov}(X_{11}, X_{21}) = \frac{1}{p} \sum_{t=1}^p (X_{1t} X_{2t}) - \overline{X}_1 \overline{X}_2.$$

Hence $\frac{1}{p}\sum_{t=1}^{p}(X_{1t} - X_{2t})^2$ converges to $2\theta^2$ as p tends to infinity, where $\theta^2 = Var(X_{11}) < \infty$. It follows that $a_{12}/\sqrt{p}$ and $\alpha/\sqrt{p}$ each converge almost surely to $\sqrt{2}\theta$ as p tends to infinity, and $\lim_{p\to\infty} a_{ij}/\alpha = 1$, $i \neq j$. Similarly for $i \neq j$, $\lim_{q\to\infty} b_{ij}/\sqrt{q} = \sqrt{2}\zeta$, where $\zeta^2 = Var(Y_{11}) < \infty$, and $\lim_{q\to\infty} b_{ij}/\beta = 1$ with probability one. Therefore

$$\lim_{p,q\to\infty} \frac{1}{n^2} \sum_{i,j=1}^{n} \frac{a_{ij}b_{ij}}{\alpha\beta} = \frac{n-1}{n},$$

$$\lim_{p,q\to\infty} \left(\frac{1}{n^4} \sum_{i,j=1}^{n} \frac{a_{ij}}{\alpha} \sum_{k,\ell=1}^{n} \frac{b_{k\ell}}{\beta} \right) = \frac{(n-1)^2}{n^2},$$

and

$$\lim_{p,q\to\infty} \frac{2}{n^3} \sum_{i,j,k=1}^{n} \frac{a_{ij}b_{ik}}{\alpha\beta} = \frac{2n(n-1)}{n^3} + \frac{2n(n-1)(n-2)}{n^3} = \frac{2(n-1)^2}{n^2}.$$

Substituting these limits in (19.17) and simplifying yields

$$\lim_{p,q\to\infty} \frac{\mathcal{V}_n^2(\mathbf{X},\mathbf{Y})}{\alpha\beta} = \frac{n-1}{n^2}.$$

By similar steps, substituting limits in (19.18) and simplifying, we obtain

$$\lim_{p\to\infty} \frac{\mathcal{V}_n^2(\mathbf{X})}{\alpha^2} = \frac{n-1}{n^2}, \qquad \lim_{q\to\infty} \frac{\mathcal{V}_n^2(\mathbf{Y})}{\beta^2} = \frac{n-1}{n^2}.$$

Hence for this class of independent random vectors each of the statistics

$$\frac{\mathcal{V}_n^2(\mathbf{X},\mathbf{Y})}{\alpha\beta}, \quad \frac{\mathcal{V}_n^2(\mathbf{X})}{\alpha^2}, \quad \frac{\mathcal{V}_n^2(\mathbf{Y})}{\beta^2}$$

converges almost surely to $(n-1)/n^2$ as dimensions p, q tend to infinity, and consequently for each fixed n the distance correlation $\mathcal{R}_n(\mathbf{X},\mathbf{Y})$ has limit one as p, q tend to infinity.

19.8.2 Proofs of Lemmas

Proof of Lemma 19.1. Proof. (1) Observe that

$$\mathcal{W}_n(\mathbf{X},\mathbf{X}) \geq \frac{1}{n^2} \sum_{i,j=1}^{n} |X_i - X_j|^2 - \bar{a}^2 \geq 0,$$

where the last inequality is the Cauchy-Bunyakovski inequality (also known as the Cauchy-Schwartz inequality). See the end of the proof of Lemma 19.3 for a proof of statement (2). Statement (3) follows from the Law of Large Numbers (LLN) for U-statistics, and (4) follows from (3) under independence. $\square$

Proof of Lemma 19.2. Lemma 19.2 establishes the identity:

$$n^2 \mathcal{U}_n(\mathbf{X}, \mathbf{Y}) = \frac{(n-1)^2}{n^2} \cdot \mathcal{U}_n^*(\mathbf{X}, \mathbf{Y}). \tag{19.19}$$

Denote $u = \frac{2}{n-2} \sum_{i=1}^n (\bar{a}_i - \bar{a})(\bar{b}_i - \bar{b})$. Then

$$
\frac{(n-1)^2}{n^2} \mathcal{U}_n^*(\mathbf{X}, \mathbf{Y}) = \sum_{i \neq j} \cdot \left(\frac{n-1}{n}\right)^2 A_{i,j}^* B_{i,j}^* - \frac{2}{n-2} \sum_{i=1}^n \left(\frac{n-1}{n}\right)^2 A_{i,i}^* B_{i,i}^*
$$

$$
= \sum_{i \neq j} \left(A_{i,j} - \frac{1}{n} a_{ij}\right)\left(B_{i,j} - \frac{1}{n} b_{ij}\right) - \frac{2}{n-2} \sum_{i=1}^n (\bar{a}_i - \bar{a})(\bar{b}_i - \bar{b})
$$

$$
= \sum_{i,j=1}^n A_{i,j} B_{i,j} - \sum_{i=1}^n A_{i,i} B_{i,i} + \frac{1}{n^2} \sum_{i,j=1}^n a_{ij} b_{ij}
$$

$$
- \frac{1}{n} \sum_{i,j=1}^n A_{i,j} b_{ij} - \frac{1}{n} \sum_{i,j=1}^n a_{ij} B_{i,j} - u. \tag{19.20}
$$

Now

$$
-A_{i,i} = 2\bar{a}_i - \bar{a} = 2(\bar{a}_i - \bar{a}) + \bar{a}, \qquad -B_{i,i} = 2\bar{b}_i - \bar{b} = 2(\bar{b}_i - \bar{b}) + \bar{b},
$$

thus

$$
\sum_{i=1}^n A_{i,i} B_{i,i} = n\bar{a}\bar{b} + 4 \sum_{i=1}^n (\bar{a}_i - \bar{a})(\bar{b}_i - \bar{b}) = n\bar{a}\bar{b} + 2(n-2)u.
$$

On the other hand, $\sum_{i,j=1}^n A_{i,j} = 0$, so

$$
\frac{1}{n} \sum_{i,j=1}^n A_{i,j} b_{ij} = \frac{1}{n} \sum_{i,j=1}^n A_{i,j} (b_{ij} - \bar{b})
$$

$$
= \frac{1}{n} \sum_{i,j=1}^n \{[a_{ij} - \bar{a}] - [\bar{a}_i - \bar{a}] - [\bar{a}_j - \bar{a}]\} (b_{ij} - \bar{b})
$$

$$
= \frac{1}{n} \sum_{i,j=1}^n (a_{ij} - \bar{a})(b_{ij} - \bar{b}) - 2 \sum_{i=1}^n (\bar{a}_i - \bar{a})(\bar{b}_i - \bar{b})
$$

$$
= \frac{1}{n} \sum_{i,j=1}^n a_{ij} b_{ij} - n\bar{a}\bar{b} - (n-2)u,
$$

and similarly,

$$
\frac{1}{n} \sum_{i,j=1}^n B_{i,j} a_{ij} = \frac{1}{n} \sum_{i,j=1}^n a_{ij} b_{ij} - n\bar{a}\bar{b} - (n-2)u.
$$

Thus the righthand side (19.20) equals

$$\sum_{i,j=1}^{n} A_{i,j} B_{i,j} - (n\bar{a}\bar{b} + 2(n-2)u) + \frac{1}{n^2} \sum_{i,j=1}^{n} a_{ij} b_{ij}$$

$$- 2 \left[\frac{1}{n} \sum_{i,j=1}^{n} a_{ij} b_{ij} - n\bar{a}\bar{b} - (n-2)u \right] - u$$

$$= \sum_{i,j=1}^{n} A_{i,j} B_{i,j} - \left[\frac{2n-1}{n^2} \sum_{i,j=1}^{n} a_{ij} b_{ij} - n\bar{a}\bar{b} - u \right]$$

$$= n^2 \mathcal{V}_n^2(\mathbf{X}, \mathbf{Y}) - n \mathcal{W}_n(\mathbf{X}, \mathbf{Y}) = n^2 \mathcal{U}_n.$$

$\square$

Notation

For the remaining proofs, we introduce the following notation.

For p-dimensional samples $\mathbf{X}_i = (X_{i1}, \ldots, X_{ip})$, $i = 1, \ldots, n$, define

$$\tau = 2E|\mathbf{X}_i|^2 = 2 \sum_{k=1}^{p} E(X_{ik}^2),$$

$$T_i(p) = \frac{|\mathbf{X}_i|^2 - E|\mathbf{X}_i|^2}{2\sqrt{\tau}} = \frac{1}{2\sqrt{\tau}} \sum_{k=1}^{p} \left[(X_{ik})^2 - E(X_{ik}^2) \right], \qquad i = 1, \ldots, n,$$

$$C_{i,j}(p) = \begin{cases} \frac{\langle \mathbf{X}_i, \mathbf{X}_j \rangle}{\sqrt{\tau}} = \frac{1}{\sqrt{\tau}} \sum_{k=1}^{p} X_{ik} X_{jk}, & i \neq j; \\ 0, & i = j, \end{cases}$$

where $\langle \mathbf{X}_i, \mathbf{X}_j \rangle$ is the dot product in $\mathbb{R}^p$. We denote the weak limits, as $p \to \infty$ of $T_i(p)$ and $C_{i,j}(p)$, by T_i and $C_{i,j}$, respectively. Note that the weak limits $C_{i,j}$ are Gaussian.

Proof of Lemma 19.3. Lemma 19.3 (i) asserts that if $E|\mathbf{X}_i|^2 < \infty$, then there exist $\Omega_{i,j}$ and $\sigma_X > 0$ such that for a fixed n

$$\mathcal{U}_n^*(\mathbf{X}, \mathbf{X}) \xrightarrow[p \to \infty]{} \sum_{i \neq j} \Omega_{i,j}^2 \overset{\mathcal{D}}{=} 2\sigma_X^2 \chi_\nu^2,$$

where $\nu = \frac{n(n-3)}{2}$ and χ_ν^2 denotes the distribution of a chisquare random variable with ν degrees of freedom.

Observe that, in distribution, the Taylor expansion of the square root implies that we have the limit

$$S_{i,j}(p) = |\mathbf{X}_i - \mathbf{X}_j| - E|\mathbf{X}_i - \mathbf{X}_j| \xrightarrow[p \to \infty]{} S_{i,j} = T_i + T_j - C_{i,j}. \qquad (19.21)$$

To see this, observe that by Taylor's Theorem we have

$$\sqrt{|X_i - X_j|^2} = \sqrt{\tau}\sqrt{1 + \frac{|\mathbf{X}_i - \mathbf{X}_j|^2}{\tau} - 1}$$

$$= \sqrt{\tau}\left(1 + \frac{1}{2}\left[\frac{|\mathbf{X}_i - \mathbf{X}_j|^2}{\tau} - 1\right] + o_p(1)\right),$$

and

$$E\sqrt{|X_i - X_j|^2} = \sqrt{\tau}E\sqrt{1 + \frac{|\mathbf{X}_i - \mathbf{X}_j|^2}{\tau} - 1}$$

$$= \sqrt{\tau}\left(1 + \frac{1}{2}\left[\frac{E|\mathbf{X}_i - \mathbf{X}_j|^2}{\tau} - 1\right] + o_p(1)\right)$$

Thus

$$\frac{|\mathbf{X}_i - \mathbf{X}_j| - E|\mathbf{X}_i - \mathbf{X}_j|}{\sqrt{\tau}} = \frac{|\mathbf{X}_i - \mathbf{X}_j|^2 - E|\mathbf{X}_i - \mathbf{X}_j|^2}{2\tau} + o_P(1).$$

Hence (19.21) is true. Using (19.21) it can be shown that (in the limit, as $p \to \infty$

$$\mathcal{U}_n^*(\mathbf{X}, \mathbf{X}) = \sum_{i \neq j}[C_{i,j} - \overline{C}_i - \overline{C}_j + \overline{C}]^2 + \lambda \sum_{i=1}^n (\overline{C}_i - \overline{C})^2, \qquad (19.22)$$

where $\lambda = -2/(n-2)$, and

$$\overline{C}_i = \frac{1}{n-1}\sum_{j=1}^n C_{i,j}, \qquad \overline{C} = \frac{1}{n^2 - n}\sum_{i,j=1}^n C_{i,j} = \frac{1}{n}\sum_{i=1}^n \overline{C}_i.$$

(The identity (19.22) is derived separately below in 19.8.2.) For finite p, replace C by $S(p)$, with corresponding subscripts and bars, throughout.

Let us prove now that (19.22) is nonnegative (this also completes the proof of Lemma 19.1). Indeed, for any constant γ

$$Q(\gamma) \overset{def}{=} \sum_{i \neq j}[(C_{i,j} - \overline{C}) - \gamma(\overline{C}_i - \overline{C}) - \gamma(\overline{C}_j - \overline{C})]^2$$

$$= \sum_{i \neq j}\left[(C_{i,j} - \overline{C})^2 + \gamma^2(\overline{C}_i - \overline{C})^2 + \gamma^2(\overline{C}_j - \overline{C})^2\right.$$

$$\left. -2\gamma(C_{i,j} - \overline{C})(\overline{C}_i - \overline{C}) - 2\gamma(C_{i,j} - \overline{C})(\overline{C}_j - \overline{C}) + 2\gamma^2(\overline{C}_i - \overline{C})(\overline{C}_j - \overline{C})\right]$$

$$= \sum_{i \neq j}(C_{i,j} - \overline{C})^2 - \left[(n-1)(4\gamma - 2\gamma^2) + 2\gamma^2\right]\sum_{i=1}^n(\overline{C}_i - \overline{C})^2.$$

So the righthand side in (19.22) equals

$$Q(1) + \lambda \sum_{i=1}^{n} (\overline{C}_i - \overline{C})^2 = \sum_{i \neq j} (C_{i,j} - \overline{C})^2 - \left[2n + \frac{2}{n-2} \right] \sum_{i=1}^{n} (\overline{C}_i - \overline{C})^2$$

$$= Q(\gamma) \geq 0, \quad \text{for } \gamma = \frac{n-1}{n-2}. \tag{19.23}$$

We have proved that $\mathcal{U}_n^*(\mathbf{X}, \mathbf{X}) \geq 0$. To complete the proof of Lemma 19.3 we need the following Cochran [1934] decomposition type lemma. Let χ_n^2 be a chi-square random variable with n degrees of freedom.

Lemma 19.4. *Let Z be a Gaussian random vector with zero mean and*

$$Q = Q_1 + Q_2, \tag{19.24}$$

where Q, Q_1, and Q_2 are non-negative quadratic forms of the coordinates of Z, and

1. $Q \overset{\mathcal{D}}{=} \chi_n^2$;

2. $Q_1 \overset{\mathcal{D}}{=} \chi_m^2$, $m \leq n$,

then Q_2 is independent of Q_1, and $Q_2 \overset{\mathcal{D}}{=} \chi_{n-m}^2$.

For a proof apply e.g. Rao Rao [1973], 3b.4, pp. 185–187: statement (i) (Fisher-Cochran Theorem) and statement (iv).

Set Q_2 equal to the righthand side of (19.23), let

$$Q_1 = \left[2n + \frac{2}{n-2} \right] \sum_{i=1}^{n} (\overline{C}_i - \overline{C})^2,$$

and

$$Q = \sum_{i \neq j} (C_{i,j} - \overline{C})^2.$$

We proved that $Q = Q_1 + Q_2$. The matrix $\|C_{i,j}\|_{i,j=1}^{n}$ is symmetric with $C_{i,i} = 0$, for all i and

$$Q = 2 \sum_{i < j} (C_{i,j} - \overline{C})^2,$$

with

$$\overline{C} = \frac{1}{n^2 - n} \sum_{i \neq j} C_{i,j} = \frac{2}{n^2 - n} \sum_{i < j} C_{i,j}.$$

Thus the quadratic form Q has rank $(n^2 - n)/2 - 1$.

From classical statistics we know that if $X_1, X_2, \ldots, X_N$ are iid Normal$(0, \sigma_X^2)$, then for $\overline{X} = \frac{1}{N} \sum_{k=1}^{N} X_k$

$$\sum_{k=1}^{N} (X_k - \overline{X})^2 \overset{\mathcal{D}}{=} \sigma_X^2 \chi_{N-1}^2.$$

Therefore, for $N = (n^2 - n)/2$, $Q \overset{D}{=} 2\sigma_X^2 \chi_{N-1}^2$, where $\sigma_X^2 = E(C_{1,2})^2$.
   Consider now the quadratic form

$$\frac{n-2}{2(n-1)^2} \cdot Q_1 = \sum_{i=1}^{n} (\overline{C}_i - \overline{C})^2, \qquad (19.25)$$

whose rank is $n-1$ because there is one single linear relationship between the vectors $\overline{C}_i - \overline{C}, i = 1, 2, \ldots, n$. Here the quadratic form (19.25) is the square of the Euclidean norm of the vector

$$(\overline{C}_1 - \overline{C}, \overline{C}_2 - \overline{C}, \ldots, \overline{C}_n - \overline{C})$$

with covariance matrix

$$\Sigma = \begin{pmatrix} d & c & \cdots & c \\ c & d & \cdots & c \\ \vdots & \vdots & \ddots & \vdots \\ c & c & \cdots & d \end{pmatrix}.$$

The moments of $\overline{C}_i$ and $\overline{C}$ are

$$E[(\overline{C}_i)^2] = \frac{\sigma_X^2}{n-1}, \qquad E[\overline{C}_i \overline{C}_j] = \frac{\sigma_X^2}{(n-1)^2},$$

$$E[\overline{C}_i \overline{C}] = \frac{1}{n} \sum_{j=1}^{n} E[\overline{C}_i \overline{C}_j] = \frac{2\sigma_X^2}{n(n-1)}, \qquad E[\overline{C}^2] = \frac{1}{n} \sum_{j=1}^{n} E[\overline{C}_i \overline{C}] = \frac{2\sigma_X^2}{n(n-1)},$$

where $\sigma_X^2 = E(C_{1,2})^2$. Therefore

$$d = E(\overline{C}_1 - \overline{C})^2 = E(\overline{C}_1)^2 - 2E(\overline{C}_1 \overline{C}) + E(\overline{C})^2 = \frac{\sigma_X^2 (n-2)}{n(n-1)},$$

$$c = E(\overline{C}_1 - \overline{C})(\overline{C}_2 - \overline{C}) = -\frac{\sigma_X^2 (n-2)}{n(n-1)^2},$$

and $d + (n-1)c = 0$. The matrix Σ has the characteristic polynomial

$$f(\lambda) = \det(\Sigma - \lambda I) = (d - \lambda + (n-1)c)(d - \lambda - c)^{n-1} = -\lambda(d - c - \lambda)^{n-1},$$

so one eigenvalue of Σ equals 0 and all other eigenvalues equal

$$d - c = \frac{\sigma_X^2 (n-2)}{(n-1)^2}.$$

Therefore

$$Q_1 \overset{D}{=} \frac{2(n-1)^2}{n-2} (d-c)\chi_{n-1}^2 = 2\sigma_X^2 \chi_{n-1}^2.$$

Applying Lemma 19.4 we obtain

$$Q_2 \overset{D}{=} 2\sigma_X^2 \chi_\nu^2$$

where $\nu = \frac{n^2-n}{2} - n = \frac{n(n-3)}{2}$. Set $m = (n-1)/(n-2)$ and

$$\Omega_{i,j} = C_{i,j} - \overline{C} - m(\overline{C}_i - \overline{C}) - m(\overline{C}_j - \overline{C}). \tag{19.26}$$

Then

$$\mathcal{U}_n^*(\mathbf{X}, \mathbf{X}) \underset{p\to\infty}{\longrightarrow} \sum_{i\neq j} \Omega_{i,j}^2 \overset{D}{=} 2\sigma_X^2 \chi_\nu^2.$$

This completes the proof of Lemma 19.3 (i).

Similarly, for the second sample, let

$$\tau_1 = 2E|\mathbf{Y}_i|^2 = 2\sum_{k=1}^{q} E(Y_{ik})^2$$

and define

$$D_{i,j}(q) = \begin{cases} \frac{\langle \mathbf{Y}_i, \mathbf{Y}_j \rangle}{\sqrt{\tau_1}} = \frac{1}{\sqrt{\tau_1}} \sum_{k=1}^{q} Y_{ik} Y_{jk}, & i \neq j; \\ 0, & i = j. \end{cases}$$

Denote the weak limits of $D_{i,j}(q)$ as $q \to \infty$ by $D_{i,j}$, $\sigma_Y^2 = E(D_{1,2})^2$, and

$$\overline{D}_i = \frac{1}{n-1} \sum_{j=1}^{n} D_{ij}, \qquad \overline{D} = \frac{1}{n} \sum_{i=1}^{n} \overline{D}_i.$$

If

$$\Psi_{i,j} = D_{i,j} - \overline{D} - m(\overline{D}_i - \overline{D}) - m(\overline{D}_j - \overline{D}), \tag{19.27}$$

then, by similar arguments as in the proof of Lemma 19.3 (i), we obtain Lemma 19.3 (ii) and (iii):

$$\mathcal{U}_n^*(\mathbf{Y}, \mathbf{Y}) \underset{q\to\infty}{\longrightarrow} \sum_{i\neq j} \Psi_{i,j}^2 \overset{D}{=} 2\sigma_Y^2 \chi_\nu^2, \qquad \mathcal{U}_n^*(\mathbf{X}, \mathbf{Y}) \underset{p,q\to\infty}{\longrightarrow} \sum_{i\neq j} \Omega_{i,j} \Psi_{i,j},$$

where $\{\Omega_{i,j}\}$ are defined by (19.26) and $\{\Psi_{i,j}\}$ are defined by (19.27). □

Proof of identity (19.22) in the proof of Lemma 19.3. We have from (19.21) that $S_{i,j} = T_i + T_j - C_{i,j}$. Notice that $S_{i,i} = 0$, $S_{i,j} = S_{j,i}$, and $E[S_{i,j}] = 0$, for all i, j. Denote

$$\overline{S}_i = \frac{1}{n-1} \sum_{j=1}^{n} S_{i,j}, \qquad \overline{S} = \frac{1}{n^2 - n} \sum_{i,j=1}^{n} S_{i,j} = \frac{1}{n} \sum_{i=1}^{n} \overline{S}_i, \qquad \overline{T} = \frac{1}{n} \sum_{i=1}^{n} T_i.$$

Using (19.21) rewrite

$$\bar{S}_i = \frac{1}{n-1} \sum_{j:j\neq i} (T_i + T_j - C_{i,j}) = \frac{(n-1)T_i + n\bar{T} - T_i}{n-1} - \bar{C}_i$$

$$= \frac{n\bar{T}}{n-1} + \frac{n-2}{n-1} T_i - \bar{C}_i,$$

$$\bar{S} = \frac{1}{n^2 - n} \sum_{i\neq j} (T_i + T_j - C_{i,j}) = \frac{1}{n^2 - n} \sum_{i\neq j} (T_i + T_j) - \bar{C}$$

$$= \frac{2n^2\bar{T} - 2n\bar{T}}{n^2 - n} - \bar{C} = 2\bar{T} - \bar{C}.$$

Since $a_{ij} = S_{ij} + \alpha$, we have

$$A_{i,j}^* = \begin{cases} a_{ij} - \frac{n}{n-1}(\bar{a}_i - \bar{a}_j + \bar{a}) & = S_{i,j} - \bar{S}_i - \bar{S}_j + \bar{S}, & i \neq j; \\ \frac{n}{n-1}(\bar{a}_i - \alpha - (\bar{a} - \alpha)) & = \bar{S}_i - \bar{S}, & i = j. \end{cases}$$

Hence, for $i \neq j$,

$$A_{i,j}^* = S_{i,j} - \bar{S}_i - \bar{S}_j + \bar{S}$$

$$= T_i + T_j - C_{i,j} - \left(\frac{n\bar{T}}{n-1} + \frac{n-2}{n-1} T_i - \bar{C}_i \right)$$

$$- \left(\frac{n\bar{T}}{n-1} + \frac{n-2}{n-1} T_j - \bar{C}_j \right) + 2\bar{T} - \bar{C}$$

$$= \frac{T_i - \bar{T}}{n-1} + \frac{T_j - \bar{T}}{n-1} - [C_{i,j} - \bar{C}_i - \bar{C}_j + \bar{C}],$$

with

$$\sum_{j:j\neq i} [C_{i,j} - \bar{C}_i - \bar{C}_j + \bar{C}] = \bar{C}_i - \bar{C},$$

and for $i = j$ we have

$$A_{i,i}^* = \bar{S}_i - \bar{S} = 2\bar{T} - \bar{C} - \left(\frac{n\bar{T}}{n-1} + \frac{n-2}{n-1} T_i - \bar{C}_i \right)$$

$$= \bar{C}_i - \bar{C} - \frac{n-2}{n-1}(T_i - \bar{T}).$$

Therefore, setting $\lambda = -2/(n-2)$ we obtain

$$\mathcal{U}_n^*(\mathbf{X}, \mathbf{X}) = \sum_{i \neq j} A^{*2}_{i,j} + \lambda \sum_{i=1}^{n} A^{*2}_{i,i}$$

$$= \sum_{i \neq j} [C_{i,j} - \overline{C}_i - \overline{C}_j + \overline{C}]^2 + \lambda \sum_{i=1}^{n} (\overline{C}_i - \overline{C})^2$$

$$- \left[\left(\frac{4}{n-1} + 2\lambda \times \frac{n-2}{n-1} \right) \sum_{i=1}^{n} (\overline{C}_i - \overline{C})(T_i - \overline{T}) \right]$$

$$+ \left(2 \times \frac{(n-2)}{(n-1)^2} + \lambda \left(\frac{n-2}{n-1} \right)^2 \right) \sum_{i=1}^{n} (T_i - \overline{T})^2$$

$$= \sum_{i \neq j} [C_{i,j} - \overline{C}_i - \overline{C}_j + \overline{C}]^2 + \lambda \sum_{i=1}^{n} (\overline{C}_i - \overline{C})^2.$$

$\square$

19.8.3 Proof of Propositions

Proof of Proposition 19.1. Proof (i): The identity

$$\mathcal{V}^2(X, Y) = E[|X - X'||Y - Y'|] + E|X - X'| \, E|Y - Y'|$$
$$- 2E[|X - X'||Y - Y''|]$$

is obtained in [Székely et al., 2007, Remark 3] by applying [Székely et al., 2007, Lemma 1] and Fubini's theorem.

Proof (ii): The result is obtained by evaluating the expected value of $\mathcal{V}_n^2(\mathbf{X}, \mathbf{Y})$ using the equivalent computing formula in identity (19.16). Under independence, we have

$$E \left[\frac{1}{n^2} \sum_{i,j=1}^{n} a_{ij} b_{ij} \right] = \frac{n-1}{n} \alpha\beta, \qquad E\left[\bar{a}\,\bar{b} \right] = \left(\frac{n-1}{n} \right)^2 \alpha\beta.$$

Similarly, evaluate the expected values of each term in (19.16) under independence and simplify. The resulting expression contains the term $\mathcal{V}^2(X, Y)$ (applying (i)), which is zero under independence, and the result follows.

Proof (iii): Again, using identity (19.16) we evaluate the expectation of $n\mathcal{U}_n(\mathbf{X}, \mathbf{Y}) = n\mathcal{V}_n^2(\mathbf{X}, \mathbf{Y}) - \mathcal{W}_n(\mathbf{X}, \mathbf{Y})$, first combining the terms of $n\mathcal{V}_n^2(\mathbf{X}, \mathbf{Y})$ and $\mathcal{W}_n(\mathbf{X}, \mathbf{Y})$ that involve $a_{ij}b_{ij}$ and $\bar{a}\bar{b}$. Under independence, the linear combination of expected values in the resulting expression sum to zero. Now, from Lemma 19.2 we have that $n^2\mathcal{U}_n(\mathbf{X}, \mathbf{Y}) = \frac{(n-1)^2}{n^2} \cdot \mathcal{U}_n^*(\mathbf{X}, \mathbf{Y})$, where $\mathcal{U}_n(\mathbf{X}, \mathbf{Y}) = \mathcal{V}_n^2(\mathbf{X}, \mathbf{Y}) - \mathcal{W}_n(\mathbf{X}, \mathbf{Y})/n$. It follows that $E[\mathcal{U}_n^*(\mathbf{X}, \mathbf{Y})] = 0$ under independence of X and Y, and therefore $E[\mathcal{V}_n^*(\mathbf{X}, \mathbf{Y})] = 0$ for independent X, Y. $\square$

Proof of Proposition 19.2. As in the proof of Proposition 19.1(ii), the expected values of $\mathcal{V}_n^2(\mathbf{X},\mathbf{Y})$ and $\mathcal{W}_n(\mathbf{X},\mathbf{Y})$ are obtained by applying identity (19.16), expanding the sums and products in the statistics, and combining the terms that have equal expected values. The expected values of $\mathcal{U}_n(\mathbf{X},\mathbf{Y})$, $\mathcal{U}_n^*(\mathbf{X},\mathbf{Y})$, and $\mathcal{V}_n^*(\mathbf{X},\mathbf{Y})$ then follow by definition. □

19.8.4   Proof of Theorem

Proof of Theorem 19.1. Introduce

$$Z_{i,j} = \frac{\Psi_{i,j}}{\sqrt{\sum_{i\neq j}\Psi_{i,j}^2}}, \qquad \sum_{i\neq j} Z_{i,j}^2 = 1.$$

Under the independence hypothesis the random variables $\{Z_{i,j}\}$ do not depend on $\{\Omega_{i,j}\}$, and for $\vartheta = \sum_{i\neq j}\Omega_{i,j}Z_{i,j}$ we have

$$\sum_{i\neq j}\Omega_{i,j}^2 - \vartheta^2 = \sum_{i\neq j}(\Omega_{i,j} - Z_{i,j}\vartheta)^2.$$

Now $Rank(\vartheta^2) = 1$ and

$$Rank\left(\sum_{i\neq j}(\Omega_{i,j} - Z_{i,j}\vartheta)^2\right) = Rank\left(\sum_{i\neq j}\Omega_{i,j}^2\right) - 1,$$

so we found one more linear relationship:

$$\sum_{i\neq j} Z_{i,j}(\Omega_{i,j} - Z_{i,j}\vartheta) = 0.$$

By Cochran's theorem, $\vartheta^2 \overset{\mathcal{D}}{=} 2\sigma_X^2\chi_1^2$, and

$$\sum_{i\neq j}(\Omega_{i,j} - Z_{i,j}\vartheta)^2 \overset{\mathcal{D}}{=} 2\sigma_X^2\chi_{\nu-1}^2,$$

which does not depend on ϑ.

For any fixed $\{Z_{i,j}\}$,

$$P\left\{\mathcal{T}_n < x \Big| \{Z_{i,j}\}_{i,j=1}^n\right\} = P\left\{\frac{\vartheta}{\sqrt{\frac{1}{\nu-1}\sum_{i\neq j}(\Omega_{i,j} - Z_{i,j}\vartheta)^2}} < x \Big| \{Z_{i,j}\}_{i,j=1}^n\right\}$$
$$= P\{t_{\nu-1} < x\},$$

where the random variable $t_{\nu-1}$ has the Student's distribution with $\nu - 1$ degrees of freedom. Therefore

$$P\{\mathcal{T}_n < x\} = E\left[P\left\{\mathcal{T}_n < x \Big| \{Z_{i,j}\}_{i,j=1}^n\right\}\right] = P\{t_{\nu-1} < x\}.$$

This proves statement (i) of Theorem 19.1.

Statement (ii) claims that the distance correlation t-test developed in Section 19.4 is unbiased. A size α hypothesis test of $H_0 : \theta \in \Theta_0 \subset \Theta$ vs $H_1 : \theta \in \Theta_1 = \Theta - \Theta_0$ is *unbiased* if its power function $\beta(\cdot)$ satisifies $\beta(\theta) \le \alpha$ if $\theta \in \Theta_0$ and $\beta(\theta) \ge \alpha$ if $\theta \in \Theta_1$.

Hence Theorem 19.1 (ii) follows by first observing that $E[\mathcal{U}_n^*(\mathbf{X},\mathbf{Y})] = 0$ under independence, and $E[\mathcal{U}_n^*(\mathbf{X},\mathbf{Y})] > 0$ under a dependent alternative (see Proposition 19.2). The test criterion is to reject the null hypothesis at level α if $\mathcal{T}_n > c_\alpha$, where $P(\mathcal{T}_n > c_\alpha) = \alpha$ under the null. If $\mathcal{T}_n > 0$ it is equivalent to reject H_0 if

$$(\mathcal{R}_n^*)^2 > \frac{c_\alpha^2}{\nu - 1 + c_\alpha^2}.$$

Since $\mathcal{V}_n^*(\mathbf{X})\mathcal{V}_n^*(\mathbf{Y})$ is the same under the null or alternative hypothesis, an equivalent criterion is to reject H_0 if

$$\mathcal{V}_n^*(\mathbf{X},\mathbf{Y}) > \sqrt{\frac{c_\alpha^2}{\nu - 1 + c_\alpha^2}\mathcal{V}_n^*(\mathbf{X})\mathcal{V}_n^*(\mathbf{Y})}, \qquad (19.28)$$

or equivalently if

$$\mathcal{U}_n^*(\mathbf{X},\mathbf{Y}) > \sqrt{\frac{c_\alpha^2}{\nu - 1 + c_\alpha^2}\mathcal{U}_n^*(\mathbf{X})\mathcal{U}_n^*(\mathbf{Y})}, \qquad (19.29)$$

where $E[\mathcal{U}_n^*(\mathbf{X},\mathbf{Y})] = n(n-3)\mathcal{V}^2(X,Y)$.

Proof of Theorem 19.1 (iii): Now following the proof of part (i) one can show that that in (19.14) of Lemma 19.3(iii)

$$\mathcal{U}_n^*(\mathbf{X},\mathbf{Y}) \underset{p,q\to\infty}{\longrightarrow} \sum_{i\ne j}\Omega_{i,j}\Psi_{i,j},$$

the random variables $(\Omega_{i,j}, \Psi_{i,j})$ are bivariate normal, have zero expected value, and under the alternative hypothesis their correlation is positive, because we have shown that $E[\mathcal{U}_n^*(\mathbf{X},\mathbf{Y})]$ is a positive constant multiple of $\mathcal{V}^2(X,Y)$.

All we need to show is that if (Ω, Ψ) are bivariate normal with zero expected value and correlation $\rho > 0$, then $P(\Omega\Psi > c)$ is a monotone increasing function of ρ. We can assume that the variances of Ω and Ψ are equal to 1. To see the monotonicity, notice that if $a^2 + b^2 = 1$, $2ab = \rho$ and X, Y are iid standard normal random variables, then for $U = aX + bY$, $V = bX + aY$ we have $\mathrm{Var}(U) = \mathrm{Var}(V) = 1$, and the covariance of U and V is $E(UV) = 2ab = \rho$. Thus (U,V) has the same distribution as (Ω, Ψ), and

$$UV = ab(X^2 + Y^2) + (a^2 + b^2)XY = \rho\frac{X^2 + Y^2}{2} + XY.$$

Thus $P(UV > c)$ is always a monotone increasing function of ρ. $\qquad\square$

19.9 Exercises

Exercise 19.1. Show that the corrected double-centered distance matrices $A^* = (A^*_{ij})$ defined in (19.6) have the property that $E[A^*_{i,j}] = 0$ for all i, j.

Exercise 19.2. Give the detailed proofs of the four statements of Proposition 19.2.

Exercise 19.3. Replicate the results of Example 19.2 using the same simulation design.

Exercise 19.4. Repeat Example 19.3 using the current 30 stocks of the Dow Jones Industrial Average and the daily returns for the most recent 10 years. (Historical stock price data is available online; in particular, adjusted closing prices can be downloaded from the Yahoo! finance site.)

Exercise 19.5. Refer to Exercise 19.4. Use hierarchical clustering to obtain a dendrogram graphical summary of the results as in Figure 4 of Székely and Rizzo [2013a]. Refer to the paper for details.

20

Computational Algorithms

CONTENTS

Several computational algorithms and strategies for energy statistics, distance covariance, distance correlation, and inference are given in this chapter. Computational algorithms for energy statistics with $O(n \log n)$ complexity exist for real-valued random variables. Randomized algorithms are discussed that reduce computing time below $O(n \log n)$. An algorithm is given for approximation of the limit distribution of some energy statistics suitable for problems of hypothesis testing or estimation.

DOI: 10.1201/9780429157158-20

20.1   Linearize Energy Distance of Univariate Samples

If X and Y are real-valued, the energy distance between two samples $\mathcal{E}_n(\mathbf{X}, \mathbf{Y})$ can be linearized, so that the computational complexity is $O(n \log n)$. We apply the following identities.

20.1.1   L-statistics Identities

An L-statistic is a linear combination of the ordered sample. Using the following identities, the distance test statistics can be linearized in the univariate case. This simplification reduces the double sums in the statistic from order n^2 to n terms, resulting in a reduction of computing time from $O(n^2)$ to $O(n \log n)$.

Suppose that $X_1, \ldots, X_n$ and $Y_1, \ldots, Y_m$ are univariate random samples, and the ordered samples are $X_{(1)}, \ldots, X_{(n)}$, $Y_{(1)}, \ldots, Y_{(m)}$. Denote by $U_1, \ldots, U_{n+m}$ the combined sample $X_1, \ldots, X_n$, $Y_1, \ldots, Y_m$, and $U_{(1)}, \ldots, U_{(n+m)}$ the ordered combined sample.

For univariate X and Y let us denote the L-statistics derived for distance matrices as follows.

$$L_n(X) = \sum_{i,j=1}^{n} |X_i - X_j|; \qquad L_n(X[i,]) = \sum_{j=1}^{n} |X_i - X_j|$$

$$L_m(Y) = \sum_{i,j=1}^{m} |Y_i - Y_j|; \qquad L_m(Y[i,]) = \sum_{j=1}^{m} |Y_i - Y_j|$$

$$L_{n,m}(X,Y) = \sum_{i=1}^{n} \sum_{j=1}^{m} |X_i - Y_j|,$$

where $L_n(X[i,])$ and $L_m(Y[i,])$ denote the row sums (column sums) of the distance matrices for the X and Y samples, respectively.

Proposition 20.1. *The following identities hold:*

$$L_n(X) = \sum_{i,j=1}^{n} |X_i - X_j| = 2 \sum_{k=1}^{n} ((2k-1) - n) X_{(k)}; \qquad (20.1)$$

$$L_m(Y) = \sum_{i,j=1}^{m} |Y_i - Y_j| = 2 \sum_{k=1}^{m} ((2k-1) - m) Y_{(k)}; \qquad (20.2)$$

$$L_{n,m}(X,Y) = \sum_{i=1}^{n} \sum_{j=1}^{m} |X_i - Y_j| \qquad (20.3)$$

$$= \sum_{k=1}^{n+m} ((2k - 1 - 2(n+m)) U_{(k)}) - \frac{1}{2} L_n(X) - \frac{1}{2} L_m(Y). \qquad (20.4)$$

For the row sums we have

$$L_n(X[i,]) = X_{(i)}(2r_i + n - 1) + \sum_{j=1}^{n} X_{(j)} - 2\sum_{r_j < r_i} X_{(j)}, \qquad (20.5)$$

where r_i is the rank of X_i, $i = 1, \ldots, n$, and a similar formula holds for $L_m(Y[i,])$.

Proof. The proof of (20.1) is simply to rearrange terms in the sum using the ordered sample, since

$$\sum_{i,j=1}^{n} |x_i - x_j| = 2\sum_{j<i}^{n}(x_{(i)} - x_{(j)})$$

$$= 2\left[(n-1)x_{(n)} + (n-2)x_{(n-1)} + (n-3)x_{(n-2)} + \cdots + (1)x_{(2)}\right]$$
$$\quad - 2\left[(n-1)x_{(1)} + (n-2)x_{(2)} + (n-3)x_{(3)} + \cdots (1)x_{(n-1)}\right]$$
$$= 2\{(1-n)x_{(1)} + (3-n)x_{(2)} + (5-n)x_{(3)} + \cdots$$
$$\quad + (n-3)x_{(n-1)} + (n-1)x_{(n)}\}$$
$$= 2\sum_{k=1}^{n}((2k-1) - n)x_{(k)}.$$

For the row and column sums:

$$L_n(X[i,]) = \sum_{j=1}^{n} |x_i - x_j| = \sum_{j=1}^{n} |x_{(i)} - x_{(j)}|$$

$$= \sum_{j<i}(x_{(i)} - x_{(j)}) + \sum_{j>i}(x_{(j)} - x_{(i)})$$

$$= (r_i - 1)x_{(i)} - (n - r_i)x_{(i)} + \sum_{j=1}^{n} x_{(j)} - 2\sum_{r_j < r_i} x_{(j)}$$

$$= x_{(i)}(2r_i + n - 1) + \sum_{j=1}^{n} x_{(j)} - 2\sum_{r_j < r_i} x_{(j)}.$$

$\square$

Identity (20.4) is derived in Rizzo [2002].

20.1.2 One-sample Energy Statistics

The energy goodness-of-fit statistic for $X \in \mathbb{R}^d$ and sample $\{y_1, \ldots, y_n\}$ is defined (see Definition 3.2) as

$$\mathcal{E}_n = \frac{2}{n}\sum_{i=1}^{n} E|y_i - X|_d + E|X - X'|_d - \frac{1}{n^2}\sum_{i,j=1}^{n} |y_i - y_j|_d.$$

If X is real-valued, then applying identity (20.1) $\mathcal{E}_n$ can be computed in $O(n \log n)$ time as

$$\mathcal{E}_n = \frac{2}{n}\sum_{i=1}^{n} E|y_i - X| + E|X - X'| - \frac{2}{n^2}\sum_{k=1}^{n}((2k-1)-n)y_{(k)}.$$

20.1.3 Energy Test for Equality of Two or More Distributions

The two sample energy test statistic is defined for $X, Y \in \mathbb{R}^d$ (see Definition 3.1) as

$$T_{n,m} = \frac{nm}{n+m}\Big(\frac{2}{nm}\sum_{i=1}^{n}\sum_{j=1}^{m}|X_i - Y_j|_d$$

$$- \frac{1}{n^2}\sum_{i=1}^{n}\sum_{j=1}^{n}|X_i - X_j|_d - \frac{1}{n^2}\sum_{i=1}^{n}\sum_{j=1}^{m}|Y_i - Y_j|_d\Big).$$

If X and Y are real-valued, then $T_{m,n}$ can be linearized by applying the L-statistics identities (20.1), (20.2) and (20.4). Then an $O(n \log n)$ computing formula for real-valued X and Y is

$$T_{n,m} = \frac{nm}{n+n}\left(\frac{2}{nm}L_{n,m}(X,Y) - \frac{1}{m^2}L_n(X) - \frac{1}{n^2}L_m(Y)\right), \qquad (20.6)$$

where

$$L_n(X) = 2\sum_{k=1}^{m}((2k-1)-m)X_{(k)}, \quad L_m(Y) = 2\sum_{k=1}^{n}((2k-1)-n)Y_{(k)},$$

$$L_{n,m}(X,Y) = \sum_{k=1}^{m+n}\left((2k-1-2(n+m))U_{(k)}\right) - \frac{1}{2}L_n(X) - \frac{1}{2}L_m(Y).$$

The multi-sample energy test statistic (based on the distance components decomposition or *disco*) introduced in Chapter 9 is a function of weighted two-sample statistics. Therefore if all of the random variables are one-dimensional, the multi-sample energy statistic for equal distributions also admits a $O(n \log n)$ computing formula by applying identities (20.1) and (20.4) to each term.

Suppose that k independent samples are drawn from the real-valued random variables $X_1, \ldots, X_k$, which have sample sizes $n_1, \ldots, n_k$, respectively and $N = \sum_{j=1}^{k} n_j$. Then the between-sample energy statistic can be written in terms of L-statistics as linear combinations of the order statistics using

$$B(X_1, \ldots, X_k) = \sum_{1 \le i < j \le k}\left\{\frac{n_i n_j}{2N}\left(2L_{n_i,n_j}(X_i, X_j) - L_{n_i}(X_i) - L_{n_j}(X_j)\right)\right\}.$$

For the distance components (disco) decomposition, the within-sample distances can be computed using

$$W(X_1,\ldots,X_k) = \sum_{j=1}^{k} \frac{n_j}{2} L_{n_j}(X_j).$$

Alternately, if S denotes the pooled sample, the total dispersion is $T(S) = \frac{N}{2}L_N(S)$ and $B(X_1,\ldots,X_k) = T(S) - W(X_1,\ldots,X_k)$.

20.2   Distance Covariance and Correlation

One can show that both the V-statistic and the U-statistic for squared distance covariance can be written as a linear combination of the following terms:

$$\sum_{i,j=1}^{n} a_{ij}b_{ij}, \quad \sum_{i=1}^{n} a_{i.}b_{i.}, \quad a_{..}b_{..},$$

where $a_{ij} = |x_i - x_j|$, $a_{i.} = \sum_{j=1}^{n} a_{ij}$, and $a_{..} = \sum_{i=1}^{n} a_{i.} = \sum_{i,j=1}^{n} a_{ij}$.

For the V-statistic (see Chapters 12, 16, and Székely and Rizzo [2013a, 2014]).

$$\begin{aligned}
\mathcal{V}_n^2 &= \frac{1}{n^2}\sum_{i,j=}^{n} a_{ij}b_{ij} + \bar{a}\bar{b} - \frac{2}{n^3}\sum_{i,j,k=1}^{n} a_{ij}b_{ik}, \\
&= \frac{1}{n^2}\sum_{i,j=}^{n} a_{ij}b_{ij} + \bar{a}\bar{b} - \frac{2}{n^3}\sum_{i=1}^{n} a_{i.}b_{i.} \quad (20.7)
\end{aligned}$$

The corresponding U statistic is

$$\begin{aligned}
\mathcal{U}_n &= \frac{1}{n(n-3)}\sum_{i\neq j} a_{ij}b_{ij} + \frac{a_{..}b_{..}}{n(n-1)(n-2)(n-3)} \\
&\quad - \frac{2}{n(n-2)(n-3)}\sum_{i=1}^{n} a_{i.}b_{i.} \quad (20.8)
\end{aligned}$$

(See Theorem 16.2.) Then a bias-corrected estimator of squared distance correlation is

$$\mathcal{R}_n^*(X,Y) = \frac{\mathcal{U}_n(X,Y)}{\sqrt{\mathcal{U}_n(X,X)\mathcal{U}_n(Y,Y)}}, \quad (20.9)$$

A few observations about computational efficiency:

- If X and Y are univariate, we can reduce the computational cost of the V-statistic (20.7) or the U-statistic (20.8) to $O(n \log n)$. The algorithm in this case requires ranks and order statistics, so it is applicable only for bivariate data. This algorithm is described in detail in Section 20.3.

- In general, including for multivariate data, the computing formula (20.7) is somewhat more computationally efficient than the product moment form $\frac{1}{n^2} \sum_{i,j} A_{ij} B_{ij}$; similarly the statistic (20.8) has faster computing time than the inner product form of unbiased squared dCov.

- We have applied a formula similar to (20.8) in the *energy* package function bcdcor [Rizzo and Székely, 2022] to compute a bias-corrected distance correlation.

- To implement a permutation test of independence, the product moment statistic (12.5) or the inner product (16.2) is more computationally efficient. The reason is that computation of the permutation replicates by these computing formulas allows re-use of all calculations to obtain the doubly centered A and B matrices.

- Although Algorithm 20.1 computes $\mathcal{U}_n$ or $\mathcal{V}_n^2$ in $O(n \log n)$ time for bivariate data, there is no obvious way implement a nonparametric test of independence efficiently like the permutation test. For the random projections approach, we have the same problem.

- In case the sample size is so large that the storage requirements of the distance matrix exceed memory limitations, then (20.7) and (20.8) allow the statistics to be computed without storing the distance matrices.

  The above observations also apply to dCor statistics and tests.

20.3   Bivariate Distance Covariance

Suppose that X and Y are real-valued. Then as derived in Huo and Székely [2016], an $O(n \log n)$ algorithm exists for computing an unbiased estimator $\mathcal{U}_n$ of squared distance covariance. The algorithm can also be applied for computing the V-statistic estimator of $\mathrm{dCov}^2(X, Y)$, with simple modification. The original algorithm was developed for computing the U-statistic (20.8) for squared distance covariance.

The $O(n \log n)$ algorithm is implemented in the functions dcov2d and dcor2d of the *energy* package.

```
> x <- iris[1:50, 1]
> y <- iris[51:100, 1]
> dcov2d(x, y)
```

```
[1] 0.00326688
> dcov2d(x, y, type="U")
```

The function computes $\mathcal{V}_n^2$ by default or $\mathcal{U}_n$ if type="U". The dcor2d function returns $\mathcal{R}_n^2$ for type="V" or $\mathcal{R}_n^*$ for type="U". The corresponding distance variances are returned by dcov2d(x,x) etc.

20.3.1 An O(n log n) Algorithm for Bivariate Data

In (20.7) and (20.8) the main computational task has two parts:

1. Compute the row sums $a_{i.}$, $b_{i.}$ of the two distance matrices, and the grand sums $a_{..}$, $b_{..}$.

2. Compute the sum of the n^2 products: $\sum_{i,j=1}^{n} a_{ij}b_{ij}$.

To compute the row sums $a_{i.}$, first sort the x_i in ascending order, and obtain the ranks $r_i = rank(x_i) = \sum_{j=1}^{n} I(x_j \leq x_i)$. As above, we denote the ordered sample $x_{(1)}, \ldots, x_{(n)}$. Then we can apply identity (20.5) to obtain

$$a_{i.} = x_{(i)}(2r_i + n - 1) + n\bar{x} - 2\sum_{r_j < r_i} x_{(j)}. \qquad (20.10)$$

One can compute the partial sums $\sum_{r_j < r_i} x_{(j)}$ recursively as a cumulative sum. Similarly,

$$b_{i.} = y_{(i)}(2s_i + n - 1) + n\bar{y} - 2\sum_{s_j < s_i} y_{(j)}, \qquad (20.11)$$

where s_i is the rank of y_i. The grand sums are simply $a_{..} = \sum_{i=1}^{n} a_{i.}$ and $b_{..} = \sum_{i=1}^{n} b_{i.}$. It is clear that the computation of row sums and grand sums can be done in $O(n \log n)$ time.

For the second task, it has been proved in Huo and Székely [2016] that one can compute the sum of products of distances in $O(n \log n)$ time.

The first step is to decompose the sum of products $a_{ij}b_{ij}$ using some algebra similar to that used to derive the L-statistics identities.

$$\sum_{i,j=1}^{n} a_{ij}b_{ij} = \sum_{i,j} |x_i - x_j||y_i - y_j| = \sum_{i,j}(x_i - x_j)(y_i - y_j)S_{ij},$$

where $S_{ij} = sign((x_i - x_j)(y_i - y_j))$. Thus, we can decompose the sum of products into four sums:

$$\sum_{i,j=1}^{n} a_{ij}b_{ij} = \sum_{i,j} x_i y_i S_{ij} - \sum_{i,j} x_i y_j S_{ij} - \sum_{i,j} x_j y_i S_{ij} + \sum_{i,j} x_j y_j S_{ij}$$

$$= \sum_{i=1}^{n} \left(x_i y_i \sum_{j=1}^{n} S_{ij} - x_i \sum_{j=1}^{n} y_j S_{ij} - y_i \sum_{j=1}^{n} x_j S_{ij} + \sum_{j=1}^{n} x_j y_j S_{ij} \right).$$

The problem is reduced to finding an $O(n \log n)$ algorithm to compute sums of the type

$$g_i(\{z_j\}) := \sum_{j=1}^{n} z_j S_{ij}. \tag{20.12}$$

Then

$$\sum_{i,j=1}^{n} a_{ij} b_{ij} = \sum_{i=1}^{n} \left(x_i y_i g_i(\{1\}) - x_i g_i(\{y_j\}) - y_i g_i(\{x_j\}) + g_i(\{x_j y_j\}) \right).$$

$$\tag{20.13}$$

To compute the sums $g_i(\{z_j\})$, we need a type of algorithm designed to reduce a search of order $O(n^2)$ to $O(n \log n)$. This type of algorithm was applied by Knight [1966] and Christensen [2005] for efficient computation of Kendall's τ statistic,

$$\tau = \frac{1}{n(n-1)} \sum_{i \neq j} S_{ij}.$$

The algorithm applies an AVL binary tree search [Adelson-Velskii and Landis, 1962, Black, 2014]. The algorithm for the AVL binary tree step is given at the end of the chapter as Algorithm 20.5.

Algorithm 20.1 (Distance Covariance of Bivariate Data). An $O(n \log n)$ algorithm to compute:

$$\mathcal{V}_n^2 = \frac{1}{n^2} \sum_{i,j=1}^{n} a_{ij} b_{ij} + \bar{a}\bar{b} - \frac{2}{n^3} \sum_{i=1}^{n} a_{i.} b_{i.} \tag{20.14}$$

or

$$\mathcal{U}_n = \frac{1}{n(n-3)} \sum_{i \neq j} a_{ij} b_{ij} + \frac{a_{..} b_{..}}{n(n-1)(n-2)(n-3)} - \frac{2}{n(n-2)(n-3)} \sum_{i=1}^{n} a_{i.} b_{i..} \tag{20.15}$$

1. Sort the X sample in increasing order, and obtain the ordered sample $x_{(i)} \leq \cdots \leq x_{(n)}$, and the corresponding ranks $r_i = \sum_{j=1}^{n} I(x_j \leq x_i)$.
2. Sort the Y sample in increasing order, and obtain the ordered sample $y_{(i)} \leq \cdots \leq y_{(n)}$, and the corresponding ranks $s_i = \sum_{j=1}^{n} I(y_j \leq y_i)$.
3. Compute (20.10) and (20.11):

$$a_{i.} = x_{(i)}(2r_i + n - 1) + n\bar{x} - 2 \sum_{r_j < r_i} x_{(j)},$$

$$b_{i.} = y_{(i)}(2s_i + n - 1) + n\bar{y} - 2 \sum_{s_j < s_i} y_{(j)},$$

4. Using results of (3) compute

$$\sum_{i=1}^{n} a_{i.}b_{i.}, \qquad a_{..} = \sum_{i=1}^{n} a_{i.}, \qquad b_{..} = \sum_{i=1}^{n} b_{i.}, \qquad \overline{ab} = \frac{a_{..}b_{..}}{n^4}.$$

5. Calculate $g_i(\{1\})$, $g_i(\{x_j\})$, $g_i(\{y_j\})$, and $g_i(\{x_jy_j\})$ following Algorithm 20.5 in Section 20.6.4.

6. Compute

$$\sum_{i,j=1}^{n} a_{ij}b_{ij} = \sum_{i \neq j} a_{ij}b_{ij}$$

$$= \sum_{i=1}^{n} (x_iy_ig_i(\{1\}) - x_ig_i(\{y_j\}) - y_ig_i(\{x_j\}) + g_i(\{x_jy_j\})).$$

7. For $\mathcal{V}_n^2$, substitute $\sum_{i,j=1}^{n} a_{ij}b_{ij}$, $\overline{ab}$, and $\sum_{i=1}^{n} a_{i.}b_{i.}$ into (20.14). The distance covariance is $\mathcal{V}_n = \sqrt{\mathcal{V}_n^2}$.
For $\mathcal{U}_n$, substitute $\sum_{i \neq j} a_{ij}b_{ij}$, $a_{..}b_{..}$, and $\sum_{i=1}^{n} a_{i.}b_{i.}$ into (20.15). This is the unbiased estimator of $\mathcal{V}^2(X,Y)$.

Example 20.1 (Benchmarks). The energy package implementation of dCov and dCor statistics is based on double centered distance matrices, and the computation is written in compiled C so it is fairly well optimized. However, in case X and Y are real-valued, benchmark results timed with `microbenchmark` [Mersmann, 2021] show that the performance advantage of the $O(n \log n)$ Algorithm 20.1 is substantial, even for small samples. In our benchmarks, the $O(n \log n)$ algorithm is faster for sample sizes about $n \geq 50$. This sample size threshold is platform and implementation dependent. For example, the Matlab and C code in the supplement to Huo and Székely [2016] does not have a performance advantage until the sample size is above 2^9 or about 500, according to their benchmarks.

20.3.2 Bias-Corrected Distance Correlation

Based on Algorithm 20.1, an $O(n \log n)$ computing algorithm for distance correlation $\mathcal{R}_n(X,Y)$ or bias-corrected squared dCor $\mathcal{R}_n^*$ follows immediately.

Algorithm 20.2 (Distance Correlation). For the original (biased) estimator of dCor(X,Y), when X and Y are both real-valued:

1. Compute $\mathcal{V}_n^2(X,Y)$, $\mathcal{V}_n^2(X) = \mathcal{V}_n^2(X,X)$ and $\mathcal{V}_n^2(Y) = \mathcal{V}_n^2(Y,Y)$ using Algorithm 20.1. Note that in the formulas $\mathcal{V}_n^2(Y), \mathcal{V}_n^2(X)$, the first sum is simply $S_1 = 2n(n-1)S^2(x)$ and $S_2 = 2n(n-1)S^2(y)$, respectively, where $S^2(x)$ is the usual unbiased estimator of sample variance.

2.

$$R_n^2(X,Y) = \begin{cases} \dfrac{\mathcal{V}_n^2(X,Y)}{\sqrt{\mathcal{V}_n^2(X)\mathcal{V}_n^2(Y)}}, & \mathcal{V}_n^2(X)\mathcal{V}_n^2(Y) > 0; \\ 0, & \mathcal{V}_n^2(X)\mathcal{V}_n^2(Y) = 0. \end{cases}$$

3. $R_n(X,Y) = (R_n^2(X,Y))^{1/2}$.

Algorithm 20.3 (Bias-corrected dCor$_n^2(X,Y)$). For the bias-corrected estimator of dCor$^2(X,Y)$, when X and Y are both real-valued:

1. Compute $\mathcal{U}_n(X,Y)$, $\mathcal{U}_n(X) = \mathcal{U}_n(X,X)$ and $\mathcal{U}_n(Y) = \mathcal{U}_n(Y,Y)$ using Algorithm 20.1. Note that in the formulas $\mathcal{U}_n^2(Y), \mathcal{U}_n^2(X)$, the first sum is simply $S_1 = 2n(n-1)S^2(x)$ and $S_2 = 2n(n-1)S^2(y)$, respectively, where $S^2(x)$ is the usual unbiased estimator of sample variance.

2.

$$R_n^*(X,Y) = \begin{cases} \dfrac{\mathcal{U}_n(X,Y)}{\sqrt{\mathcal{U}_n(X)\mathcal{U}_n(Y)}}, & \mathcal{U}_n(X)\mathcal{U}(Y) > 0; \\ 0, & \mathcal{U}_n(X)\mathcal{U}_n(Y) = 0. \end{cases}$$

Note: R_n^* may take negative values, so we do not obtain a bias-corrected estimator of $R(X,Y)$.

The bias-corrected R_n^* is approximately unbiased for the population value $R^2(X,Y)$, the squared distance correlation.

20.4 Alternate Bias-Corrected Formula

The following is an alternate computational formula for bias-corrected squared distance correlation that was introduced in Székely and Rizzo [2013a]. This formula applies for bivariate or multivariate data, and it is particularly good for an implementation that would be written entirely in a higher level language such as R, Python or Matlab (that is, not linking to a compiled C or Fortran library) or any higher level language where it is possible to vectorize the matrix operations to avoid loops. The following statistic introduced in Chapter 19 is algebraically equivalent to (20.9).

Define modified double centered distance matrices

$$A_{i,j}^* = \begin{cases} \frac{n}{n-1}(A_{i,j} - \frac{a_{ij}}{n}), & i \neq j; \\ \frac{n}{n-1}(\bar{a}_i - \bar{a}), & i = j, \end{cases} \qquad B_{i,j}^* = \begin{cases} \frac{n}{n-1}(B_{i,j} - \frac{b_{ij}}{n}), & i \neq j; \\ \frac{n}{n-1}(\bar{b}_i - \bar{b}), & i = j. \end{cases}$$

The corrected distance covariance statistic is

$$V_n^*(X,Y) = \frac{1}{n(n-1)}\left\{ \sum_{i,j=1}^n A_{i,j}^* B_{i,j}^* - \frac{n}{n-2}\sum_{i=1}^n A_{i,i}^* B_{i,i}^* \right\},$$

and the bias-corrected distance correlation statistic is

$$R_n^*(X,Y) = \frac{V_n^*(X,Y)}{\sqrt{V_n^*(X,X)V_n^*(Y,Y)}}, \tag{20.16}$$

if $V_n^*(X,X)V_n^*(Y,Y) > 0$, and otherwise $R_n^*(X,Y) = 0$. See Chapter 19 for more details. The inner product statistic introduced in Chapter 16 is also straightforward to implement in R, Python, or Matlab.

Although (20.16) is natural for application in R or Matlab, if we compare the faster compiled versions, then (20.16) is somewhat slower than the multivariate application of (20.9) or the inner product formula from Chapter 12:

$$\mathcal{R}_n^*(X,Y) := \begin{cases} \frac{(\tilde{A}\cdot\tilde{B})}{|\tilde{A}||\tilde{B}|}, & |\tilde{A}||\tilde{B}| \neq 0; \\ 0, & |\tilde{A}||\tilde{B}| = 0. \end{cases}$$

20.5 Randomized Computational Methods

For multivariate data, the $O(n \log n)$ algorithm described above does not apply. However, there are other strategies that one can use to reduce the computational complexity for distance covariance and other energy statistics. The idea of random projections onto lines to handle multivariate data was suggested by G. J. Székely. That approach replicates the $O(n \log n)$ algorithm to obtain an estimated dCov statistic.

For the approximate methods discussed in this section, we do not obtain an exact energy statistic, but an estimate of the statistic; the statistic is consistent but its sampling distribution will differ from the target statistic.

20.5.1 Random Projections

The algorithms for computing bivariate distance covariance require sorting and ranking the observations, so those algorithms do not have a direct extension to the problem of computing $\mathcal{U}_n(x,y)$ or $\mathcal{V}_n^2(x,y)$ when X or Y take values in $\mathbb{R}^d$, $d > 1$. Hence, to generalize the idea of the $O(n \log n)$ computational algorithm to random vectors in arbitrary dimension, we would like to find a way to apply the original bivariate algorithm to multivariate data. Székely suggested using random projections of X and Y onto lines; then the $O(n \log n)$ algorithm can be applied to each projected set of bivariate data. When the directions are chosen randomly, the statistics $\mathcal{U}_n$ or $\mathcal{V}_n^2$ are iid random variables and we can average them to get an estimate for the sample dCov in the multivariate case. However, a normalizing constant that depends on dimensions p and q is needed to adjust bias. Let

$$C_d = \sqrt{\pi} \frac{\Gamma((d+1)/2)}{\Gamma(d/2)}.$$

The normalizing constant is $C_p C_q$. See Huang and Huo [2017] for more details and proofs.

20.5.2   Algorithm for Squared dCov

The following is an algorithm to compute (an estimate of) unbiased squared distance covariance $\mathcal{U}_n(X, Y)$, where $X \in \mathbb{R}^p$ and $Y \in \mathbb{R}^q$. This algorithm is different than the one suggested by Huang and Huo [2017] but their main results for the unbiased distance covariance statistic still hold, and similar arguments extend the results to distance variance and bias-corrected distance correlation.

Algorithm 20.4. Here we suppose that the observed sample data are stored as an $n \times p$ data matrix X and an $n \times q$ data matrix Y, with observations in rows. The matrices preserve the indexing of the paired sample observations; (x_i, y_i) correspond to row i in both data matrices.

1. Specify the number $k > 2pq$ of random projections. (See Remark 20.1.)

2. Generate a random sample of k points on the unit sphere in $\mathbb{R}^p$. Store these points in matrix R $(p \times k)$, with each column containing one point on the unit sphere. See e.g. Rizzo [2019] Example 3.21, for an algorithm to generate the points.

3. Generate a random sample of k points on the unit sphere in $\mathbb{R}^q$, independent of R. Store these points in matrix Q $(q \times k)$, with each column containing one point on the unit sphere.

4. Compute all projections $P_x = XR^T$ and $P_y = YQ^T$, where R^T, Q^T are the transpose of matrix R and Q, respectively.

    Then P_x and P_y are $n \times k$ matrices such that each column of P_x and P_y correspond to the projected sample in a given direction.

5. For i $= 1$ to k (for each direction) do:

    (a) Let $P_x[., i]$ denote the i-th column of P_x and $P_y[., i]$ denote the i-th column of P_y.

    (b) Compute $U[i] = \mathcal{U}_n(P_x[., i], P_y[., i])$, by Algorithm 20.1.
        That is, compute the unbiased distance covariance on the i-th projections and store the result in the vector U at $U[i]$.

6. Compute the mean $\overline{U} = \frac{1}{k} \sum_{i=1}^{k} U[i]$.

7. Compute

$$C_p = \sqrt{\pi} \exp(\log(\Gamma((p+1)/2) - \log(\Gamma(p/2)));$$
$$C_q = \sqrt{\pi} \exp(\log(\Gamma((q+1)/2) - \log(\Gamma(q/2)).$$

8. Deliver $\widehat{\mathcal{U}}_n(x,y) = C_p C_q \overline{U}$, the random projection estimator of $\mathcal{U}_n(x,y)$.

Remark 20.1. The number of projections k should be large relative to dimensions. The variance of the estimator blows up if $k/(pq)$ is close to 1. Figures 20.1 and 20.2 illustrate this for small samples. The estimator is unbiased, but the variance is increasing with dimension if k and n are fixed. See Section 20.5.3.

The statistic $\widehat{\mathcal{U}}_n(x,y)$ is unbiased for $\mathcal{V}^2(X,Y)$, but it converges to $\mathcal{V}^2(X,Y)$ at a slower rate than $\mathcal{U}_n(x,y)$; the rate of convergence is $1/\sqrt{k}$, where k is the number of random directions.

Algorithm 20.4 is also valid for estimating the distance variance. If we apply the algorithm from Huang and Huo [2017] with inner product (X,X) replacing (X,Y), however, the estimator of $\mathcal{V}^2(X,X)$ will be biased. We have modified the algorithm so that it is also valid for distance variance. If we replace Y with X and q with p in Algorithm 20.4 (while generating independent matrices R and Q), then the resulting estimator $\widehat{\mathcal{U}}_n(X,X)$ is unbiased for $\mathcal{V}^2(X,X)$.

To apply a test of independence based on the statistic $\widehat{\mathcal{U}}_n$, Huang and Huo [2017] proposed estimating the asymptotic distribution by matching estimators of first and second moments to a gamma distribution. Their method requires summing all of the pairwise distances $|x_i - x_j|$ and $|y_i - y_j|$, which is $O(n^2)$ for multivariate data, although it can be done without storing the distance matrices.

20.5.3 Estimating Distance Correlation

Algorithm 20.4 can be applied three times to separately estimate $\mathcal{U}_n(x,y)$, $\mathcal{U}_n(x,x)$, and $\mathcal{U}_n(y,y)$.

There are two viable methods to estimate the squared distance correlation by random projections. Our empirical results indicate that both methods have similar performance and properties. Both methods determine an estimate for $\mathcal{R}^2(X,Y)$ that is approximately unbiased and with about the same variance. The two "good" methods of combining the estimates in the denominator of dCor are:

1. "pairwise dVar" which means to compute a product $U(x[,i]) \cdot U(y[,i])$ for each projection.

2. "unbiased dVars" where an unbiased estimate is obtained from the mean $\widehat{\mathcal{U}}_n(x,x)$ and the mean $\widehat{\mathcal{U}}_n(y,y)$ separately before multiplying them together.

For simplicity, and to be consistent with our method of computing the bias-corrected distance correlation, we implement the second method. Simply compute $\widehat{\mathcal{U}}_n(x,y)$, $\widehat{\mathcal{U}}_n(x,x)$, and $\widehat{\mathcal{U}}_n(y,y)$ separately and combine them as in Algorithm 20.3. In either approach, however, it is necessary to compute all of the distance variance estimates with independent sets of random projections in Step 3. That is, generate independent sets of projections R_1 and R_2 for $U(x,x)$ and independent sets of projections Q_1 and Q_2 for $U(y,y)$. One then has $2k$ random projections available to use for computing $\widehat{\mathcal{U}}_n(x,y)$.

Note: A method that does not work at all is to estimate distance correlation once on each projection and then average those estimates.

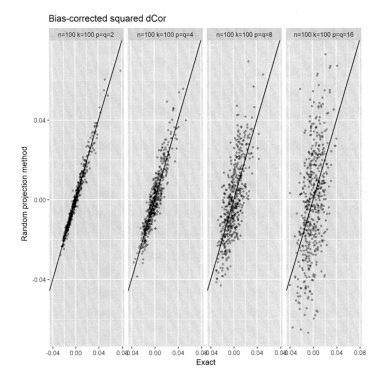

FIGURE 20.1

Comparison of the random projections estimate of bias-corrected $\mathrm{dCor}_n^2(x,y)$ and the exact calculation of the statistic for independent X and Y. The reference line corresponds to the ideal situation where the two statistics are identical. The number of projections is k and the dimensions of both samples are p.

Figure 20.1 compares the random projections estimator to the exact sample coefficient for $\mathcal{R}_n^*$ (bias-corrected) for small samples under independence. The reference line shows the ideal situation where these two statistics are identical. In the left panel where $p = q = 2$ and $k = 100$, we have $k/(pq) = 25$ and

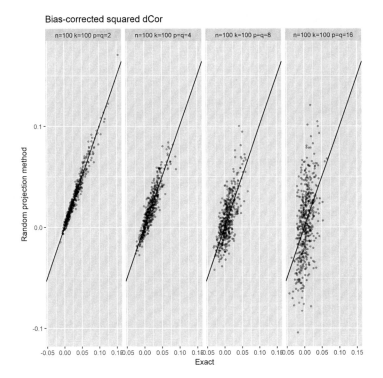

FIGURE 20.2

Comparison of the random projections estimate of bias-corrected $\mathrm{dCor}_n^2(x, y)$ and the exact calculation of the statistic for dependent data. The reference line corresponds to the ideal situation where the two statistics are identical. The number of projections is k and the dimensions of both samples are p.

the random projections estimates are nearly identical to the exact sample coefficients. In the right panel where $p = q = 16$ we have $k/(pq) < 1$ and the variance of the random projection estimates is quite large. Figure 20.2 compares the statistics for dependent data, revealing a similar pattern. With the number of projections k fixed, the variance of the random projection estimator is increasing with dimension.

The variance of both estimators decreases as $n \to \infty$, but the distance of points from the line reflects the variance in the random projection estimate of sample $\mathcal{R}_n^*$, which depends on k, the number of random projections. Increasing the sample size will shrink the spread of points along the line, but not the vertical distance of points from the line. These simulations suggest that k should be large relative to pq.

A "rule-of-thumb" guide for the number of random projections would require a more extensive study. However, the simulations summarized in Figures 20.1 and 20.2 indicate that k should be at least $2pq$ for the range of parameters in those simulations. Other simulations with fixed dimension and varying k suggest that k should be at least $\max(50, 2pq)$. A plot comparing $\mathcal{U}_n$ with $\widehat{\mathcal{U}}_n$ (not shown) reveals the same pattern.

20.6 Appendix: Binary Search Algorithm

The sum $\sum_{i,j} a_{ij} b_{ij}$ can be simplified into sums of the form

$$g_i(\{z_j\}) := \sum_{j=1}^{n} z_j S_{ij},$$

where $S_{ij} = sign((x_i - x_j)(y_i - y_j))$. We further simplify this sum as follows.

First sort the pairs (x_i, y_i) in order of increasing x_i, so that we now have $(x_{(i)}, y_{\pi(i)})$, where π is the permutation of integers $1, \ldots, n$ that was applied to the indices of sample x. For convenience, denote the re-indexed y sample by y'.

Then in each application of the sum, z_j is re-indexed by the same permutation, because z_j is a constant vector of ones, or the vector x, y, or xy.

The sum can be simplified by splitting it into sums over sets like $i < j; y'_j < y'_i$, etc. Since z are re-indexed according to the same permutation as y, we can replace the y' sample with its ranks. After simplification, we have for each i:

$$\sum_{j \neq i} z_j S_{ij} = \sum_{j=1}^{n} z_j - z_i - 2\sum_{j=1}^{i} z'_j - 2\sum_{j=1}^{i} z_j + 4\gamma_i(z), \qquad (20.17)$$

where z'_j denotes the index of z with respect to the order of y', and

$$\gamma_i(z) = \sum_{j<i; y_j<y_i} z_j.$$

Each term in (20.17) except $\gamma_i(z)$ is easily computed in n steps (assuming that the vectors were already sorted and re-indexed, which takes $O(n \log n)$ time). In terms of R functions, the permutation π is the vector order(x), and the re-indexed sample vector y' is y[order(x)].

The computation of $\gamma_i(z)$ is reduced from $O(n^2)$ to $O(n \log n)$ by the following algorithm.

20.6.1 Computation of the Partial Sums

To compute the sums $\gamma_i(z)$ in $O(n \log n)$ time, we apply a recursive binary partitioning algorithm on the set $\{1, \ldots, n\}$. The partitioning can be visualized as a binary tree.

20.6.2   The Binary Tree

Let $L = \lceil log_2 n \rceil$, so that $n \leq 2^L$. Each node or vertex in the tree corresponds to an interval of consecutive integers, $[p, q] = [p, p+1, \ldots, q]$. The top node or root of the tree is $[1, 2^L]$.

Let ℓ denote the level of the tree, counting from the terminal nodes at level 0 to the top node, $\ell = 0, 1, 2, \ldots, L$. Each level has $2^{L-\ell}$ nodes, which are a sequence of disjoint intervals of equal size whose union is $[1, 2^L]$. The left node is $[1, 2^\ell]$.

That is, on level 0 we have the singletons $[1], [2], \ldots, [2^L]$. Then

1. Level 1 nodes are $[1, 2], [3, 4], \ldots, [2^L - 1, 2^L]$.

2. Level 2 nodes are $[1, 4], [5, 8], \ldots, [2^L - 3, 2^L]$.

3. In general, for level ℓ, the nodes are $[1, 2^\ell], \ldots [2^L - 2^\ell + 1, 2^L]$.

We number the nodes on each level from left to right. The k-th interval on level ℓ is

$$I(\ell, k) = [(k - 1)2^k + 1, \; k \cdot 2^\ell],$$

$\ell = 0, \ldots, L - 1$, $k = 1, \ldots, 2^{L-\ell}$. See Figure 20.3 for the labeled tree when $n = 16 \, (L = 4)$.

20.6.3   Informal Description of the Algorithm

To compute $\gamma_i(z) = \sum_{j<i; y_j<y_i} z_j$ by the binary recursive partitioning method, we make two passes through the vector y. Assign a sum $s(\ell, k)$ to each node and initialize it to zero.

- Each integer y_j is contained in exactly one node on each level of the tree. Let us refer to the nodes containing y_j as the container nodes for y_j.

  For example, in Figure 20.3 if $y_{i-1} = 5$, then the container nodes of 5 are $[5], [5, 6], [5, 8]$, and $[1, 8]$, indexed 5, $16 + 3 = 19$, $(16 + 8) + 2 = 26$, and $(16 + 8 + 4) + 1 = 29$.

- Make one pass through the integers y, updating the sums in the container nodes of y_{i-1} of the tree. When we reach the end of the y, the sums in all containers are updated.

- Next, we make a second pass through y, to retrieve the i-th partial sum for $y_j < y_i$, $i = 2, \ldots, n$.

  For example, if $y_i = 20$, we need to combine the sums for $[1, 19] = [1, 16] \cup [17, 18] \cup [19]$, so we add the sums stored at nodes $[1, 16], [17, 18]$ and $[19]$ to get the partial sum $\gamma_i(z)$.

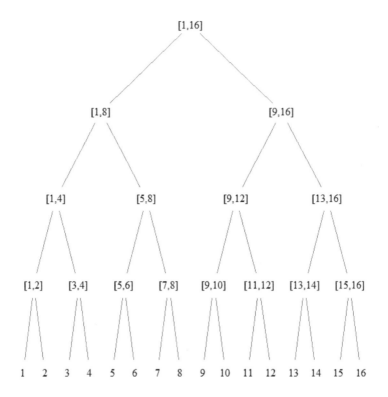

FIGURE 20.3
Binary tree for ranks of a data set size $n = 16$.

20.6.4   Algorithm

Write

$$g_i(z) = \sum_{j \neq i} z_j S_{ij} = \sum_{j=1}^{n} z_j - z_i - 2\sum_{j=1}^{i} z'_j - 2\sum_{j=1}^{i} z_j + 4\gamma_i(z), \qquad (20.18)$$

where z'_j denotes the index of z with respect to the order of y', and

$$\gamma_i(z) = \sum_{j < i; y_j < y_i} z_j.$$

The following is an $O(n \log n)$ algorithm to compute the sums $\gamma_i(z) = \sum_{j<i; y_j < y_i} z_j$, where y is a permutation of the integers $\{1, \ldots, n\}$ and z a vector of n real numbers. The closed interval notation $[j, k]$ denotes the sequence of consecutive integers $\{j, j+1, \ldots, k\}$.

Algorithm 20.5 (Binary search). Algorithm 20.1 Step 5 in $O(n \log n)$ time:

1. Set $L = \lceil \log_2 n \rceil$, and define a binary tree with the top node equal to the sequence of consecutive integers in $[1, 2^L]$.

2. Let $\ell = 0, 1, 2 \ldots, L$ denote the levels of the tree, counting from the terminal nodes at $\ell = 0$ to the top node.

3. Assign a closed interval of consecutive integers to each node, so that from left to right, the k-th node on level ℓ of the tree is the sequence of integers in the interval
$$I(\ell, k) = [(k-1)2^k + 1, \; k \cdot 2^\ell],$$
$\ell = 0, \ldots, L-1, \; k = 1, \ldots, 2^{L-\ell}$.

4. Initialize sums $s(\ell, k) = 0$ for $\ell = 0, \ldots, L-1, \; k = 1, \ldots, 2^{L-\ell}$. Each level ℓ of the tree has $2^{L-\ell}$ nodes, which are a sequence of disjoint intervals of equal size whose union is $[1, 2^L]$.

5. Convert the ragged array of sums $s(\ell, k)$ to a vector. Number the nodes from left to right on each level, from level 0 to L, and let $s(p)$ be the corresponding sums for node p.

6. First pass: Update sums in container nodes.

   (a) Set $g_1(z) = 0$.
   (b) For $i = 2$ to n:
      i. Find the list of container nodes for y_{i-1}.
      ii. Set $\text{nodes}(1) = y_{i-1}$;
      iii. For $p = 2$ to L set $\text{nodes}(p)$ equal to the index of the interval in level $p - 1$ that contains y_{i-1}.
      iv. Update the sum $s(p) \leftarrow s(p) + z_{i-1}$.

7. Second pass: Retrieve partial sums from nodes.

   (a) For $i = 2$ to n compute $\gamma_i(z)$:
      i. Find disjoint nodes P_i whose union is $[1, y_i - 1]$. This is a set of disjoint intervals $I(p) = I(\ell, k)$, at most one node from each level, such that $\bigcup_{p \in P_i} I(p) = [1, y_i - 1]$.
      ii. Compute the partial sums $\gamma_i(z) = \sum_{p \in P_i} s(p)$.

Finally, substitute $\gamma_i(z)$ into (20.18) to compute the sums $g_i(z)$ for $g_i(\{1\})$, $g_i(\{x_j\})$, $g_i(\{y_j\})$, and $g_i(\{x_j y_j\})$, then continue from Step 6 in Algorithm 20.1(a).

The exact formula for the index of the containers in each pass depends on the array indexing of the software and other programming details, which may vary according to the individual implementation. In particular, most compiled languages index the above arrays with base 0 rather than base 1.

Interested readers can refer to an implementation in R and C++ in the *energy* package, with the node lookup in the first and second passes given as separate functions in C++. The source code can be found at `github.com/mariarizzo/energy` in `/src/B-tree.cpp` and `/R/dcov2d.R` (functions `dcov2d` and `dcor2d`). An alternate implementation in Matlab and C is available in the supplementary files to Huo and Székely [2016]. See also the *Rfast* package [Papadakis et al., 2022].

21

Time Series and Distance Correlation

CONTENTS

21.1 Yule's "nonsense correlation" is Contagious

Correlation theory and distance correlation theory applies to iid observations. What can go wrong if the data are serially dependent? An example of "nonsense correlation" is discussed in "Yule's 'nonsense correlation' solved!" [Ernst et al., 2017]. This nonsense correlation is a problem for distance correlation, too. This phenomenon is caused by two factors: the observations are serially correlated and the random walks (Wiener processes) are not stationary.

Example 21.1 (Random Walks). Consider a pair of independent symmetric random walks. A simulated realization is shown in Figure 21.1. For the example shown, the sample Pearson correlation is –0.69 and sample dCor is 0.63.

A Monte Carlo experiment reveals that the sampling distribution of linear correlation, distance correlation, and bias-corrected $\mathcal{R}_n^*$ (bcdcor) all measure a spurious association. The simulation recorded the three statistics on independent pairs of random walks of length 500, on 1000 realizations. Statistics are summarized in Table 21.1 and Figure 21.2.

In Table 21.1 and Figure 21.2, notice that although the center of the (linear) correlation distribution is near zero, the sampling distribution is almost uniformly distributed on $(-1, 1)$. The statistics and histograms suggest that the nonsense correlation problem afflicts all three measures of association. In practice, the requirement that samples are iid is often overlooked, but this example shows that it is indeed critical.

DOI: 10.1201/9780429157158-21

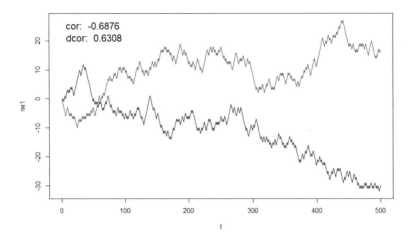

FIGURE 21.1
Two independent symmetric random walks.

21.2   Auto dCor and Testing for iid

Auto distance correlation is an application of distance correlation to time series $\{X_t : t = 0, \pm 1, \pm 2, \dots\}$. In case of auto-correlation with lag k we compute the correlation of X_t and X_{t+k}. Stationarity guarantees that this does not depend on the starting point t_0. We can do the same with distance correlation. Strong stationarity and ergodicity guarantees that the empirical distance correlation of (X_t, X_{t+k}), $t = t_0 + 1, \dots, t_0 + n$, converges to the population distance correlation and this does not depend on the starting point. An application of this new notion is testing for iid.

TABLE 21.1
Summary statistics for a Monte Carlo experiment
to measure association between independent
random walks

	correlation	dCor	bcdcor
Min.	−0.9372	0.1360	0.0135
1st Qu.	−0.3921	0.3279	0.1029
Median	−0.0237	0.4634	0.2113
Mean	−0.0013	0.4923	0.2787
3rd Qu.	0.4185	0.6511	0.4212
Max.	0.9373	0.9414	0.8861

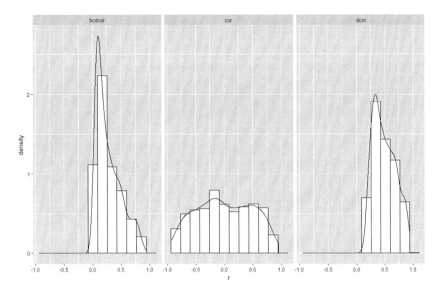

FIGURE 21.2

Nonsense correlation summarized for three measures of association: bias-corrected squared dcor (bcdcor), linear correlation (cor), and distance correlation. Histograms of the statistics with density estimates are shown for a simulation of 1000 realizations of pairs of independent symmetric walks each length 500.

For more details, see Zhou [2012] on auto-distance correlation and also the literature review on distance correlation methods for time series by Edelmann, Fokianos, and Pitsillou [2019].

In classical statistics, the typical first sentence is "Suppose the variables are independent and identically distributed" (iid). However, if we have observed say, n numbers, how can we test the independence property? For simplicity suppose that this sample comes from a time series X_t that is strongly stationary and ergodic, and the expectations $E(X_t)$ are finite. In this case, we know that the empirical auto distance correlation converges to the population distance correlation.

A test for the iid condition combines the Ljung-Box test for white noise and the auto distance correlation. The Ljung and Box [1978] statistic is

$$Q := n(n+2) \sum_{k=1}^{h} \frac{\hat{\rho}_k^2}{n-k},$$

where $\hat{\rho}_k^2$ is the empirical auto-correlation with lag k, and h is the number of lags being tested.

On auto-distance correlation and cross-distance correlation see also Zhou [2012], Fokianos and Pitsillou [2018] and Pitsillou and Fokianos [2016]. One of the ideas in these papers is to apply the Ljung-Box test modified version for testing independence. The Ljung-Box test function [Ljung and Box, 1978] is a linear combination of squared auto-correlations. If we replace all auto-correlations $\hat{\rho}_k$ with the corresponding auto distance correlations, then we get a test statistic for checking the iid property in stationary time series.

The Ljung-Box statistic applies to test the hypothesis that the population versions of these auto correlations are zero, which indicates that $\{X_t : t = 0, \pm 1, \pm 2, \dots\}$ is a white noise process (all auto correlations are zero). Now, if in the formula for Q we replace all auto correlations by the corresponding auto distance correlations, then we can test the null hypothesis of zero auto-distance correlation, which under the null hypothesis means that they are iid.

Hong [1996] proposes three classes of consistent one-sided tests for serial correlation of residuals. The tests apply a kernel-based spectral density estimator. Hong [1999] proposes tests based on the empirical characteristic function and generalized spectral density. On some other approaches for testing iid see Cho and White [2011].

Pitsillou and Fokianos [2016] describe the R package *dCovTS*, which implements their proposed statistic

$$T_n = \sum_{j=1}^{n-1} (n-j)k^2(j/p)V_X^2(j),$$

with $V_X(j)$ the auto distance covariance at lag j, and a suitable kernel $k(\cdot)$. Under pairwise independence, the empirical pairwise auto distance covariance (ADCV) is a degenerate V-statistic of order two. For a measurable, symmetric, continuous and positive semidefinite kernel $k(\cdot)$, the statistic $(n-j)V_X^2(j)$ converges to an asymptotic distribution of the same form as distance covariance for independent variables; that is, the limiting distribution is a quadratic form of iid standard normal random variables Z_k, $Q = \sum_k \lambda_k Z_k^2$. Pitsillou and Fokianos [2016] compare their proposed statistic T_n with tests based on Lyung-Box type statistics and that proposed by Hong [1999], and found T_n had better performance.

The test based on T_n is implemented for a choice of kernel functions in the R package *dCovTS* [Tsagris, Pitsillou, and Fokianos, 2021]. Several utilities for analysis of time series using auto distance covariance/correlation are also provided in the package.

21.3   Cross and Auto-dCor for Stationary Time Series

Above we explained why stationarity and ergodicity are natural assumptions for time series. Davis, Matsui, Mikosch, and Wan [2018] prove that for stationary ergodic time series if the α moment of the X and Y variables is finite for some $0 < \alpha < 2$, then the empirical α-distance covariance is a consistent estimator of the population α-distance covariance. (See Section 14.2.2.) The convergence is almost sure as the sample size tends to infinity.

Davis et al. [2018] discuss the auto-regressive process of order p, AR(p), given by the difference equation

$$X_t = \sum_{k=1}^{p} \Phi_k X_{t-k} + Z_t, \qquad t = 0, \pm 1, \pm 2, \ldots,$$

where Z_t is an iid sequence with finite κ moments for some $\kappa > 0$.

It is shown how the limit distribution of the test statistics under the null hypothesis changes when we replace the iid observation assumptions with the AR(p) assumption. The results were applied to daily stock returns of Amazon and to wind speed data.

Assuming strong mixing, Davis et al. [2018] establish the relevant asymptotic theory for the sample auto- and cross-distance correlation functions and they also apply the auto-distance correlation function (ADCF) to the residuals of an auto-regressive processes of order p as a test of goodness-of-fit. Assuming strong mixing the authors prove limit distributions for the distance covariance of stationary time series in terms of a complex valued mean-zero Gaussian process whose covariance structure is presented. Under the null that an autoregressive model is true, the limit distribution of the empirical ADCF can differ markedly from the corresponding one based on an iid sequence.

21.4   Martingale Difference dCor

Let U, V be random variables with finite second moment. Following Shao and Zhang [2014] and Park, Shao, and Yao [2015], let us call $V - E(V)$ a martingale difference with respect to U if the conditional expected value $E(V|U) = E(V)$ or equivalently if $E(V - E(V)|U) = 0$ almost surely. This property clearly holds for independent U, V and the martingale difference property implies uncorrelatedness of U, V. Thus, the martingale difference property is weaker than independence but stronger than uncorrelatedness.

Define the martingale difference divergence of $V \in \mathbb{R}$ given $U \in \mathbb{R}^q$ as the square root of

$$MDD(V|U)^2 = \frac{1}{c_q} \int_{\mathbb{R}^q} \frac{|g_{V,U}(s) - g_V \, g_U(s)|^2}{|s|_q^{1+q}} ds$$

where $g_{V,U}(s) = E(V e^{i(<s,U>)})$, $g_V = E(V)$ and $g_U(s) = E(e^{i<s,U>})$. Notice that MDD is not symmetric, which is why we called it divergence rather than distance.

The standardized version of MDD is the martingale difference correlation (MDC). It plays the same role for testing the martingale difference property as dCor played for testing independence of random variables. For the proof of theorems analogous to theorems on dCor and empirical $dCor_n$, see the papers above.

21.5   Distance Covariance for Discretized Stochastic Processes

In the original papers [Székely et al., 2007, Székely and Rizzo, 2009a], distance correlation was studied for finite-dimensional vectors. Later Lyons [2013] extended this notion to suitable metric spaces, thus also including stochastic processes X and Y. The empirical version of distance correlation requires n iid copies of (X, Y) and, in principle, knowledge of the complete sample paths. The latter assumption can be relaxed, as recent work by Dehling et al. [2020] shows; it suffices to work with discretizations of the processes X, Y. Numerical results show that the rates of convergence are excellent, depend on the smoothness of the processes, but do not require some very particular assumptions like fractional BM or OU-process.

On a test of independence for functional data, see Horváth, Hušková, and Rice [2013].

Betken, Dehling, and Kroll [2021] discuss block bootstrap for empirical distance covariance under the assumption of strictly stationary and absolutely regular sample data. They developed a test for independence of two strictly stationary and absolutely regular processes.

21.6   Energy Distance with Dependent Data: Time Shift Invariance

In Part I for energy distance we have not discussed the dependence structure of the data. Given samples $\mathbf{X}$ and $\mathbf{Y}$ we can always compute

$$\mathcal{E}_{n,m}(\mathbf{X}, \mathbf{Y}) = \frac{2}{nm} \sum_{i=1}^{n} \sum_{j=1}^{m} |X_i - Y_j|$$

$$- \frac{1}{n^2} \sum_{i=1}^{n} \sum_{j=1}^{n} |X_i - X_j| - \frac{1}{m^2} \sum_{i=1}^{m} \sum_{j=1}^{m} |Y_i - Y_j|.$$

To simply compute the statistic it does not matter if the X_i's are independent or not or if the Y_j's are independent or not.

The dependence structure becomes important if we want to claim asymptotic results, say, the empirical values $\mathcal{E}_{n,m}$ are converging in some sense. We can hope this is true if we have independent random samples; that is, the X_i's and the Y_j's are independent, $X_1, X_2, \ldots$ are iid, and $Y_1, Y_2, \ldots$ are also iid (and the underlying distances have finite expected values). In this case the strong law of large numbers is indeed true:

$$\mathcal{E}_{n,m} \to 2E|X - Y| - E|X - X'| - E|Y - Y'|$$

with probability 1 as $n, m \to \infty$ if the expectations above are finite (X and Y have the same distributions as X_i and Y_j, respectively). This law is true for iid observations according to the law of large numbers for V-statistics [Koroljuk and Borovskich, 1994], but in fact we do not need the very strong condition of being iid. What do we need?

The claim of the law of large numbers implicitly supposes that $X, X_1, X_2, \ldots$ have the same distribution (the same holds for the Y sample) and we take the expected values with respect to this common distribution. When n and m denotes time, that is, when the data comes from a time series, then it is convenient to suppose that n and m take arbitrary integer values: $n = 0, \pm 1, \pm 2, \ldots$. When the starting point does not change the picture we have *time shift invariance*. This means that the joint distribution of any finite collection of random variables

$$X_{i_1+i}, X_{i_2+i}, \ldots, X_{i_n+i}$$

does not depend on the time shift i. This property is called the *strong stationarity* of the X-data. We suppose the same invariance for the Y-data.

Under this condition $\mathcal{E}_{n,m}$ converges with probability 1 but the limit is not necessarily a constant. This is guaranteed by assuming "space-homogeneity" or using a mathematical terminology "ergodicity" which implies that the "time average," that is, the average with respect to n, m, converges with probability 1 to the "space average" meaning the expected value as $n, m \to \infty$. It is proved in Aaronson et al. [1996] that the strong law of large numbers does apply to U statistics and V-statistics when the IId condition on data is replaced by "strongly stationary and ergodic," a natural condition in time series.

We face bigger difficulties when we want to claim that the limit distribution of U-statistics or of V-statistics for strongly stationary and ergodic sequences of data can be computed the same way as if they were iid. In other words, we would like to have that the limit distribution of the V-statistic with degenerate kernel h (degree 1 degeneracy) such that $E[h^2(X_1, X_2)] < \infty$, is the same type of quadratic form as in the iid case.

As the following one sample example shows, the limit theorem of V-statistics for general stationary sequences is simply not true without any further conditions.

372 *The Energy of Data and Distance Correlation*

Example 21.2. Suppose the kernel of the V-statistic is $h(x,y) = xy$ and consider real-valued random variables X_i with finite variance. Now for stationary $X_i, i = 1, 2, \ldots$ with $EX_i = 0$, the limit distribution of the V-statistic is the same as the limit distribution of

$$\frac{1}{n}\sum_{i,j=1}^{n} X_i X_j = \left[\frac{1}{\sqrt{n}}\sum_{i=1}^{n} X_i\right]^2.$$

If the central limit theorem applies, then the limit has the same distribution as

$$\left[EX_0^2 + 2\sum_{k=1}^{\infty} E(X_0 X_k)\right] Z^2,$$

where Z is a standard normal random variable. This limit is clearly more complicated than what the eigenvalues of the operator $E[xX_0\Psi(X_0)] = \lambda\Psi(x)$ suggest because this operator has a single eigenvalue: $\lambda = E(X_0^2)$ (the corresponding eigenfunction is $\Psi(x) = x$) but the limit distribution clearly depends not only on this value, it also depends on the covariances $E(X_0 X_k)$, $k = 1, 2, \ldots$.

Even if we suppose that the stationary data are uncorrelated: $E(X_i X_j) = 0$, for general kernel h we cannot guarantee the existence of a limit distribution that depends on the eigenvalues only. The following stronger condition is sufficient. If we suppose that the X_i $i = 1, 2, \ldots$ is a martingale difference sequence, that is, if

$$E(X_i|X_0, X_1, \ldots X_{i-1}) = 0, \qquad i = 1, 2, \ldots,$$

then this implies uncorrelatedness. The following result shows that a related condition implies the desired limit distribution. Martingale difference is an important generalization of the iid property and the martingale difference property holds for some stock price models and for many other important time series where the iid property fails.

Consider the one-sample version of the data energy. This is a V-statistic with kernel

$$h(x,y) = E|x - X| + E|y - X| - E|X - X'| - |x - y|.$$

This h is positive semidefinite.

Theorem 21.1 (Leucht and Neumann [2013]). *If*

(i) $X_i, i = 0, \pm 1, \pm 2, \ldots$ *is a strictly stationary and ergodic sequence with values in $\mathbb{R}^d$,*

(ii) $h : \mathbb{R}^d \times \mathbb{R}^d \to \mathbb{R}$ *is a symmetric, continuous and positive semidefinite function, (iii)* $E(h(X_0, X_0)) < \infty$,

(iv) $E(h(x, X_i)|X_1, ..., X_{i-1}) = 0$ *a.s. for all x in the support of X_0,*

then the limit distribution of the V-statistic with kernel h can be computed the same way as if the X_i's were iid; that is, the limit distribution has the form $\sum_k \lambda_k Z_k^2$ where the Z's are iid standard normal and the λ's are eigenvalues of the integral equation $E[h(x, X_0)\Psi(x)] = \lambda\Psi(x)$.

Remark 21.1. If the expected value $E(h(x, X_i)|X_1, \ldots, X_{i-1}) = 0$, then we get $Eh(x, X_i) = 0$ for all x in the support of X_0, which is the definition of degenerate kernel. What we need for the limit distribution theorem above is more: we need (iv), which is a martingale difference type property.

22

![black bar]

Axioms of Dependence Measures

CONTENTS

22.1   Rényi's Axioms and Maximal Correlation

Our goal in this chapter is to identify a "minimalist" set of axioms that should be satisfied by an acceptable measure of dependence Δ. Our starting point is to revisit the set of seven axioms proposed by Rényi [1959].

Rényi's axioms are as follows. Let X and Y be real-valued random variables.

(A) $\Delta(X, Y)$ is defined for all random variables X and Y, neither of them being constant with probability 1.

(B) $\Delta(X, Y) = \Delta(Y, X)$ (symmetry).

(C) $0 \leq \Delta(X, Y) \leq 1$.

(D) $\Delta(X, Y) = 0$ if and only if X and Y are independent.

(E) $\Delta(X, Y) = 1$ if there is a strict dependence between X and Y; that is, either $X = g(Y)$ or $Y = f(X)$, where $g(x)$ and $f(x)$ are Borel measurable functions.

(F) If the Borel measurable functions $f(x)$ and $g(x)$ map the real axis in a one-to-one way onto itself, $\Delta(f(X), g(Y)) = \Delta(X, Y)$.

(G) If the joint distribution of X and Y is normal, then $\Delta(X, Y) = |\rho(X, Y)|$ where $\rho(X, Y)$ is the correlation coefficient of X and Y.

DOI: 10.1201/9780429157158-22   375

Most well-known dependence measures do not satisfy all of Renyi's axioms. Some of them, like symmetry, are not essential, while others like 1–1 invariance, are too strong (see Theorem 22.3 below). An important example for non-symmetric dependence measures is the correlation ratio in Example 22.7 below.

Maximal correlation is a measure of dependence that satisfies all of the above axioms [Rényi, 1959]. Maximal correlation is defined as $\sup_{f,g} \rho(f(X), g(Y))$ for all Borel-measurable functions f, g, where $\rho(X, Y)$ is Pearson correlation.

However, not even the above list of strong restrictions characterize maximal correlation, as shown by Linfoot's information-theoretical measure [Linfoot, 1957].

While Rényi's axioms clearly collect some of the most important properties of a dependence measure, not all of these axioms (A)–(G) are essential for a good measure of dependence. One might wonder which of these axioms are critically important and whether this list contains all critically important properties of dependence measures as axioms. These questions are studied in depth in "Four simple axioms of dependence measures" [Móri and Székely, 2019], on which this chapter is based.

Our goal here is to find a "minimalist" system of axioms that we can expect to be satisfied by all acceptable dependency measures Δ.

First considering Axiom (A), Δ should not be defined for all random variables (that are not constant with probability 1) because not even Pearson's correlation is defined for random variables with infinite variance. Even if we restrict (A) to random variables with finite variances, the absolute value of Pearson's correlation ρ does not satisfy Axioms (E) or (F).

As shown below, Axioms (E) and (F) can be replaced with a weaker version where 1–1 invariance is replaced by similarity invariance satisfied by $|\rho|$. There is another reason for not requiring 1–1 invariance of Δ. The 1–1 invariance condition would imply the existence of many *uncorrelated* random variables X, Y for which $\Delta(X, Y) = 1$, which is counter intuitive.

It is not surprising that there exist perfectly dependent random variables with zero Pearson's correlation. For a related statement see Kimeldorf and Sampson [1978].

In the following section it is shown that with very few exceptions, for *all random variables* X one can find a 1–1 real function f such that X and $f(X)$ are uncorrelated.

22.2   Axioms for Dependence Measures

An important property of dependence measures, namely continuity, is not there among Rényi's axioms. Here we require continuity but drop some of the

Rényi's axioms. Let us replace the axioms of symmetry and invariance with respect to all 1–1 measurable transformations, and address continuity. Our approach is closer to the spirit of the Erlangen program in geometry, where geometries are characterized via invariances. See Section 22.5 for more details.

Recall that Euclidean geometry is characterized by invariances with respect to the Euclidean group of transformations (translations, rotations, and reflections). Similarity geometry deals with geometrical objects of the same shape. We can obtain one object from another by scaling (enlarging or shrinking). Similarity transformations consist of all Euclidean transformations and all (non-zero) scaling; that is, changing the measurement units. Instead of 1–1 invariance we suppose similarity invariance only.

Definition 22.1. Let S be a non-empty set of pairs of random variables X, Y taking values in Euclidean spaces or in separable Hilbert spaces H. Suppose that X and Y are not constant with probability 1 and if $(X, Y) \in S$, then $(LX, MY) \in S$ where L, M are similarity transforms of H. Then $\Delta(X, Y) :$ $S \times S \to [0, 1]$ is called a dependence measure on S if the following four axioms hold.

(i) $\Delta(X, Y) = 0$ if and only if X and Y are independent.

(ii) $\Delta(X, Y)$ is invariant with respect to all similarity transformations of H; that is, $\Delta(LX, MY) = \Delta(X, Y)$ where L, M are similarity transformations of H.

(iii) $\Delta(X, Y) = 1$ if and only if $Y = LX$ with probability 1, where L is a similarity transformation of H. (A more symmetrical looking version is $MY = LX$ with probability 1, where M, L are similarity transformations of H.)

(iv) $\Delta(X, Y)$ is continuous; that is, if $(X_n, Y_n) \in S$, $n = 1, 2, \ldots$ such that for some positive constant K we have $E(|X_n|^2 + |Y_n|^2) \leq K$, $n = 1, 2, \ldots$ and (X_n, Y_n) converges weakly (converges in distribution) to (X, Y), then $\Delta(X_n, Y_n) \to \Delta(X, Y)$. (The condition on the boundedness of second moments can be replaced by any other condition that guarantees the convergence of expectations: $E(X_n) \to E(X)$ and $E(Y_n) \to E(Y)$); such a condition is the uniform integrability of X_n, Y_n which follows from the boundedness of second moments.)

The importance of continuity with respect to weak convergence in statistics is shown by the definition of qualitative robustness; robustness of a statistic means exactly that it is weakly continuous. We should expect a good measure of dependence to be robust. Without weak continuity we cannot expect that the empirical dependence measure converges to the population one, and this lines up with the fact that it is indeed very difficult to stably estimate coefficients like mutual information in practice. This is why we prefer continuity with respect to weak convergence to other types of continuities, such as continuity with respect to total variation distance. (Mutual information is not even continuous with respect to total variation distance, but the mutual information coefficient, whose definition will be given below, does behave stably with respect to total variation distance.)

Rényi does not assume Axiom (iv). Theorem 22.3 below explains that if he did, then no dependence measure would have satisfied all his axioms.

Remark 22.1. Functions of independent random variables are independent, thus property $\Delta(X, Y) = 0$ is invariant with respect to all 1–1 Borel measurable transformations of H. On the other hand, we do not suppose this 1–1 invariance for other values of Δ. Below we will see that such a strong condition would contradict Axiom (iv).

If $H = \mathbb{R}$, the real line, affine transformations coincide with similarities. In higher dimensions, however, affine invariance contradicts Axiom (iv); this is proved in Theorem 22.3 below. This result makes the choice of similarity invariance in our axioms even more natural.

22.3   Important Dependence Measures

Let us review examples of several important dependence measures with respect to their properties and invariances. For a recent review of dependence measures see Tjøstheim et al. [2022] and the literature review in Josse and Holmes [2014].

Example 22.1 (Pearson's ρ: similarity invariance). Let S be the set of bivariate Gaussian random variables. Pearson's correlation is

$$\rho(X, Y) = E(X - E(X))E(Y - E(Y)) = E(XY) - E(X)E(Y).$$

The absolute value of Pearson's correlation ρ satisfies Axioms (i) – (iv) on S. We know that for $X, Y \in S$, $\rho = 0$ if and only if X, Y are independent, and $|\rho| = 1$ if and only if there is a linear relationship between X and Y. On a multivariate version see Escoufier [1973] and Josse and Holmes [2014]. A major negative property of Pearson's correlation is that it does not satisfy Axiom (i) for general random variables. Local versions, for example, local Gaussian correlation [Tjøstheim and Hufthammer, 2013] have been introduced to help ρ detect non-linear relationships.

Example 22.2 (Distance correlation: similarity invariance). Distance correlation was introduced in Székely, Rizzo, and Bakirov [2007] for all random variables with finite expectations. First the distance covariance is defined as follows. If (X, Y), (X', Y') and (X'', Y'') are iid random variables with finite expectation, the distance covariance is the square root of

$$\mathrm{dCov}^2(X, Y) := E(|X - X'|\,|Y - Y'|) + E(|X - X'|)E(|Y - Y'|)$$
$$- E(|X - X'|\,|Y - Y''|) - E(|X - X''|\,|Y - Y'|).$$

See Chapters 12 and 15 for definitions and examples. Once we know the distance covariance we can define distance variance and distance correlation analogous to the variance and linear correlation. The distance correlation [Székely et al., 2007, Székely and Rizzo, 2009a, 2014, Lyons, 2013] *satisfies all of our axioms* (i)–(iv).

The coefficient $\Delta(X, Y) = \mathrm{dCor}^2(X, Y)$ can be interpreted as the cosine of the angle between elements associated with X and Y in the Hilbert space $\mathcal{H}$ of Section 17.2 defined in Székely and Rizzo [2014]. If the expectations do not exist, then we can generalize distance correlation for random variables with finite $\alpha > 0$ moments by replacing $E|X|$ with $E|X|^\alpha$. This is a very large class of random variables for applications.

Example 22.3 (The RV coefficient: similarity invariance). The ρV coefficient is defined by the population covariance matrices Σ_{XY}, Σ_{XX}, and Σ_{YY}:

$$\rho V := \frac{tr\,(\Sigma_{XY}\Sigma_{YX})}{\sqrt{tr\,(\Sigma_{XX}^2)\,tr\,(\Sigma_{YY}^2)}}.$$

The ρV coefficient is invariant to shift, rotation, and overall scaling, but (i) fails to hold, so ρV does not characterize independence. The sample coefficient RV replaces the matrices Σ_{XY} with the sample statistics $S_{XY} = \frac{1}{n-1}X'Y$.

The RV statistic can be expressed in terms of distance matrices. Let Δ_X, Δ_Y denote the Euclidean distance matrix of the X and Y samples, respectively. Then

$$RV = \frac{\langle C\Delta_X^2 C, C\Delta_Y^2 C\rangle}{\|C\Delta_X^2 C\|\|C\Delta_Y^2 C\|}, \tag{22.1}$$

where $C = \mathbf{I}_n - \frac{1}{n}\mathbf{1}\mathbf{1}^\top$, $\mathbf{I}_n$ is the $n \times n$ identity matrix and $\mathbf{1}$ is the $n \times 1$ column vector of ones. That is, the numerator of (22.1) is the inner product between the doubly centered squared Euclidean distance matrices.

Compare the formulation of the RV statistic with the squared sample distance correlation. It is equivalent to

$$\mathrm{dCor}_n^2(X, Y) = \frac{\langle C\Delta_X C, C\Delta_Y C\rangle}{\|C\Delta_X C\|\|C\Delta_Y C\|}, \tag{22.2}$$

which is equation (3.4) in Josse and Holmes [2016]. The only difference between (22.2) and the RV coefficient (22.1) is that squared distances in RV are replaced by Euclidean distances in dCor_n^2. Josse and Holmes [2016] note that "this difference implies that the dCor coefficient detects non-linear relationships whereas the RV coefficient is restricted to linear ones. Indeed, when squaring distances, many terms cancel whereas when the distances are not squared, no cancellation occurs allowing more complex associations to be detected."

Example 22.4. [Spearman's ρ and Kendall's τ: monotone invariance] If $H = \mathbb{R}$ in Definition 22.1, then the invariance of Δ with respect to all monotone transformations means that Δ is independent of the marginal distributions of X, Y. Monotone invariance implies that instead of the joint cdf $F_{X,Y}$, we can focus on the copula: $C(u; v) = F_{X,Y}(F_X^{-1}(u); F_Y^{-1}(v))$, where F_X^{-1}, F_Y^{-1} denote the inverse cdf functions. The copula can also be viewed as the joint distribution of two uniform $(0,1)$ variables. We have $C(F_X(x), F_Y(y)) = F_{X,Y}(x, y)$. Two random variables are independent if and only if $C(u, v) = uv$.

With the copula function $C(u, v)$, Spearman's ρ_S and Kendall's τ can be defined as follows:

$$\rho_S(X, Y) := 12 \int_0^1 \int_0^1 (C(u, v) - uv) du dv,$$

and

$$\tau(X, Y) := 4 \int_0^1 \int_0^1 C(u, v) dC(u, v) - 1,$$

respectively. An equivalent definition of τ is

$$\tau := P((X - X')(Y - Y') > 0) - P((X - X')(Y - Y') < 0),$$

where (X', Y') is an iid copy of (X, Y).

The absolute value of these measures satisfy (i) – (iv) for positive or negative quadrant dependent random variables. These properties mean that $C(u, v) \geq uv$ or $C(u, v) \leq uv$ for all $0 \leq u, v \leq 1$. For general random variables X, Y, (i) typically does not hold.

Example 22.5 (Distance correlation: affine and monotone invariant versions). Distance correlation applied to standardized random variables is obviously affine invariant. If we apply distance correlation to the copula $C(u; v) = F_{X,Y}(F_X^{-1}(u); F_Y^{-1}(v))$, where F_X^{-1}, F_Y^{-1} denote generalized inverse functions, then we get a monotone invariant version of distance correlation.

Example 22.6 (Maximal correlation: 1–1 invariance). Maximal correlation is defined as $\sup_{f,g} \rho(f(X), g(Y))$ where f, g are Borel-measurable functions. Maximal correlation satisfies (i), (ii), and (iii), but as we shall see in Theorem 22.3, it cannot satisfy (iv) because maximal correlation is invariant with respect to all 1–1 Borel functions of the real line.

Example 22.7 (Correlation ratio: similarity invariance). An important example for non-symmetric dependence measures is the correlation ratio, which is defined for bivariate real-valued random variables with finite and positive variance as $\Delta(X, Y) := \mathrm{Var}(E(X|Y))/\mathrm{Var}(X)$. It is not hard to prove that the maximal correlation of X, Y

$$\sup_{f,g} \rho(f(X), g(Y)),$$

where f, g are Borel-measurable functions is equal to the square root of

$$\sup_f \mathrm{Var}\{E(f(X)|Y) : \mathrm{Var}\, f(X) = 1\}.$$

For a proof see Theorem 22.1 below.

Example 22.8 (Maximal information coefficient)**.** The maximal information coefficient was introduced in Reshef et al. [2011]. For comments see Speed [2011], Simon and Tibshirani [2011]. Denote by $I(X, Y)$ the mutual information between two discrete random variables X, Y taking finitely many values (x, y):

$$I(X, Y) := \sum_{x,y} p(x, y) \log_2 \frac{p(x, y)}{p(x)p(y)}$$

where $p(x, y)$ is the probability that $(X, Y) = (x, y)$, $p(x) = P(X = x)$, and $p(y) = P(Y = y)$. Population maximal information coefficient MIC_* of a pair (X, Y) of random variables is defined as

$$MIC_*(X, Y) = \sup_G \frac{I((X, Y)|_G)}{\log \|G\|},$$

where
- G is a rectangular grid imposed on the support of (X, Y),
- $(X, Y)|_G$ denotes the discrete distribution induced by (X, Y) on the cells of G, and
- $\|G\|$ denotes the minimum of the number of rows and the number of columns of G.

Theorem 22.6 below shows that MIC has monotone invariance but does not satisfy axioms (iii) or (iv).

22.4   Invariances of Dependence Measures

In this section we first prove a theorem on Pearson's correlation under 1–1 invariance, then prove a theorem on the relationship of maximal correlation and correlation ratio. Our main result is Theorem 22.3. Finally we prove three theorems on the invariances of maximal correlation, correlation ratio, and maximal information coefficient.

Our first result shows that 1–1 invariance of Δ implies the existence of many uncorrelated random variables X, Y for which $\Delta(X, Y) = 1$. The question of whether it makes sense to have uncorrelated random variables X, Y with $\Delta(X, Y) = 1$ is one for which the answer is not trivial. For example, a uniform distribution over a circle is a highly interesting relationship that can reasonably be seen to merit a maximal score. This would push away from thinking that $\rho(X, Y) = 0$ together with $\Delta(X, Y) = 1$ is unreasonable.

Theorem 22.1. *Let X be a square integrable random variable defined on an arbitrary probability space. Suppose the distribution of X is not concentrated on at most three points. Then there exists a measurable injective function $f : \mathbb{R} \to \mathbb{R}$ such that X and $f(X)$ are uncorrelated. This f can be chosen piecewise linear.*

Such an f cannot exist if X can take exactly two values, because in that case uncorrelatedness is equivalent to independence. When the distribution of X takes exactly 3 points then a necessary and sufficient condition for f to exist is $P(X = E(X)) = 0$.

See Section 22.8 (1) for a proof of Theorem 22.1.

Theorem 22.2. *The maximal correlation of X, Y*

$$\sup_{f,g} \rho(f(X), g(Y))$$

where f, g are Borel-measurable functions is equal to the square root of the supremum of the correlation ratio, which is the square root of

$$\sup_f \text{Var}\{E(f(X)|Y) : \text{Var } f(X) = 1\}.$$

See Section 22.8 (2) for the proof of Theorem 22.2.

Theorem 22.3. *Suppose S is a set of pairs of random variables taking values in separable Hilbert space H. Suppose that X and Y are not constant with probability 1, and if $(X, Y) \in S$, then $(LX, MY) \in S$ for all affine transformations L, M of H. If the dependence measure $\Delta(X, Y)$ on S is invariant with respect to all affine transformations L, M of H where $\dim H > 1$, then Axiom (iv) cannot hold.*

If $\dim H = 1$, then affinity is the same as similarity and in this case distance correlation is affine invariant.

On the other hand, if $\Delta(X, Y)$ is invariant with respect to all 1–1 Borel measurable functions of H then even if $\dim H = 1$, Axiom (iv) cannot hold. Here we suppose that if $(X, Y) \in S$ then for all 1–1 Borel measurable functions f, g of H, $(f(X), g(Y)) \in S$.

See Section 22.8 (3) for the proof of Theorem 22.3.

For dependence measures on real-valued random variables there are further possibilities for invariances between similarities and 1–1 Borel measurable transformations. For example, the reader might want to prove that a monotone invariant Δ can satisfy our axioms. We just need to apply distance correlation to the ranks of observations (or to the copula). Are there also homeomorphism invariant versions, too? They are invariant with respect to all continuous 1–1 functions with continuous inverse. Do we have such a Δ that satisfies our axioms?

A corollary of our theorem is the following result. We will also give a short direct proof, not referring to the previous theorem.

Theorem 22.4. *The maximal correlation coefficient does not satisfy axioms (iii) and (iv).*

Proof. The first claim is trivial. To see the second claim notice that if for some real-valued random variables $X_n = Y_n$ with probability $1/n$ and with the remaining $1 - 1/n$ probability $X_n = X, Y_n = Y$ where X and Y are independent, then the maximal correlation of X_n, Y_n is 1 while in the limit they are independent. □

Theorem 22.5. *The correlation ratio satisfies Axiom (ii), but it does not satisfy any of (i), (iii), and (iv).*

See Section 22.8 (5) for the proof of Theorem 22.5.

Theorem 22.6. *The population maximal information coefficient MIC satisfies axioms (i) and (ii) but does not satisfy axioms (iii) and (iv). MIC is invariant with respect to all monotone transformations but not to all measurable 1–1 functions, hence Theorem 22.3 does not apply to MIC.*

See Section 22.8 (6) for the proof of Theorem 22.6.

Although MIC is continuous with respect to some other topologies, they do not guarantee that the empirical MIC converges to the theoretical coefficient as $n \to \infty$.

In summary, for special classes S of random variables there are many important measures of dependence, but if S contains all pairs of bounded (non-constant) random variables in an arbitrary Euclidean space or separable Hilbert space, then distance correlation seems to be the simplest one that satisfies all axioms (i)–(iv). If a dependence measure is not continuous, like the maximal correlation or maximal information coefficient, then application is problematic.

22.5   The Erlangen Program of Statistics

The Erlangen program is a method of characterizing geometries based on invariances with respect to transformation groups. It was published by Felix Klein in 1872 and was named after the University Erlangen-Nürnberg, where Klein worked. Klein defined a hierarchy of geometries: Euclidean geometry studied geometric properties that are invariant with respect to Euclidean transformations like shift and rotation. Affine geometry studies properties invariant with respect to affine transformations. This is more restrictive than projective geometry, the most general geometry considered by Klein. In projective geometry we cannot talk about circles because they are not invariant with respect to projections. In the 20-th century we could add topology that studies properties invariant with respect to topological (bijective an bi-continuous) transformations. Borel invariance is even less restrictive. Practically we just need to check if the transformation is one-to-one.

A similar scheme can be introduced in statistics. (For another type of "algebraization" see Ruzsa and Székely [1988]).

The Kolmogorov-Smirnov statistic $\sup_x |F_n(x) - G_m(x)|$ is shift invariant but not rotation invariant. Energy statistics are Euclidean. They are invariant with respect to all Euclidean transformations or isometries (distances do not change). Certain ratios of Energy statistics are similarity invariant; that is, invariant with respect to isometries and scalings. A typical example is distance correlation that is similarity invariant. Distance correlation is Thalesian in the sense that besides being invariant with respect to Euclidean transformations; that is, congruences and isometries like shifts and rotations, it is also invariant with respect to scaling. As we have seen, distance correlation is not affine invariant in dimensions bigger than 1. On affine invariant distance correlation see Dueck et al. [2014].

In certain cases we should work with other important invariances, like affine invariance when we test normality. Test statistics should be affine invariant because the class of normal distributions is closed under affine transformations.

We have seen that if the dimension of the space is more than 1, then affine invariance contradicts weak continuity. In one dimension, however, even monotone invariant versions of distance correlation exist. For a monotone invariant version we just need to apply distance correlation to the ranks of the data.

Topological data analysis works with invariances to 1–1 continuous transformations of (sparse) data. For energy inference all that matters is the metric of the space of observations. Topological data analysis provides a general framework to analyze data in a manner that is insensitive to the particular metric chosen and provides dimension reduction and robustness to noise. The strongest invariance is the invariance with respect to all 1–1 Borel functions. Maximal correlation has this strong invariance. The property of independence characterized by dCor = 0 is also Borel invariant.

As we proved, if the dimension of a Hilbert space is bigger than 1, then even affine invariance is too much: it contradicts weak continuity. In one dimension, however, all we have proved is that there is no 1–1 Borel invariant dependence measure that satisfies our four axioms. But there is a topological invariant version of distance correlation on $\mathbb{R}$ if we want to define it for random variables with continuous strictly increasing cdf's only; just apply distance correlation to $F(X)$ and $G(Y)$, where F and G are the cdf's of X and Y, respectively. In other words, for a topologically invariant version of $\mathrm{dCov}(X, Y)$, a good candidate is $\Delta(X, Y) := \mathrm{dCor}(F(X), G(Y))$. This is, of course, distance correlation applied to copulas.

Monotone invariance is typical when use rank statistics. Monotone invariance is strongly connected to non-transitivity. For example, if using $\mathrm{dCor}_n(X, Y)$ we can say that X and Y are dependent, and similarly, Y and Z are dependent (we can reject that null hypothesis that dCor is zero), then we might expect that $\mathrm{dCor}(X,Z)$ cannot be zero (if we have transitivity). However, typically we do not have transitivity. (What is true is that $\mathrm{dCor}(X,Y) = \mathrm{dCor}(Y,Z) = 1$ implies $\mathrm{dCor}(X,Z) = 1$.) This surprising non-transitivity is a kind of "good, better, worst" or "rock-paper-scissors" argument. This is typical for rank statistics. See also Arrow's paradox on social choices. It says that when voters have three or more distinct alternatives, then no ranked voting electoral system can convert the ranked preferences of individuals into a community-wide (complete and transitive) ranking while also meeting a natural set of criteria. The theorem is named after economist and Nobel Laureate Kenneth Arrow, who demonstrated the theorem in his 1950 doctoral thesis [Arrow, 2012]. See also the topic of non-transitive dice in Székely [1986].

One might want to generalize our four axioms for dependence measures and Theorem 22.1 to Banach spaces, partly because all metric spaces can be isometrically embedded into Banach spaces. The definition of *similarity* in metric spaces is the following. A 1-1 mapping of a metric space onto itself is a similarity if it is a composition of an isometry and a scaling (multiplication of all distances by the same positive constant). It is easy to prove [Lyons, 2013, Proposition 2.5] that if the distance correlation of X, Y in a metric space is equal to 1, then there is an homeomorphism f between the supports of X and Y such that $\delta(x, x') = c\delta(f(x), f(x'))$ for some $c > 0$ and $y = f(x)$ on the supports with probability one. Thus f is not only a homeomorphism, it is in fact a similarity transformation (a composition of isometry and scaling) between the supports of X and Y and $Y = f(X)$ with probability one. Every similarity f of course satisfies (iii), so similarity is an if and only if condition for the distance correlation to take the value 1.

All four axioms are easy to generalize to metric spaces. What seems to be non-trivial is to show that if the metric spaces is not strong negative type (Definition 10.4) and thus distance correlation does not satisfy axiom (i), then the axioms are not contradictory. In other words if, say, for all bounded and non-constant random variables X, Y a dependence measure $\Delta(X, Y)$ exists at all that satisfies our four axioms.

Conjecture 1. All four axioms are true in a metric space if and only if the metric space is of strong negative type (see Definition 10.4).

Conjecture 2. If our four axioms hold for Banach space valued random variables, then Δ cannot be invariant with respect to all Borel measurable 1–1 transformations. In other words Δ cannot be Borel invariant.

Conjecture 3. Suppose we replace similarity invariance by *topological invariance* (invariance with respect to all homeomorphisms) in our four axioms. We conjecture that no topologically invariant Δ can satisfy the four axioms defined for all bounded non-constant random variables in metric spaces except if the maximal correlation satisfies all axioms in the underlying (typically discrete) metric space.

Suppose we replace similarity invariance by *Borel invariance* (invariance with respect to all 1–1 Borel measurable functions) in our four axioms. By a

theorem of Kuratowski, all complete, separable metric spaces are Borel iso-
morphic to one of (1) $\mathbb{R}$ (real line), (2) $\mathbb{Z}$ (integers) or (3) a finite space. Thus
the cardinality characterizes these spaces up to Borel isomorphism, and hence
it is enough to discuss these cases when we try to find a Δ that satisfies our
four axioms for all bounded non-constant random variables. The following
theorem settles the problem.

Theorem: *In cases (2) and (3) Δ is always continuous because the topology is discrete; thus Δ = maximal correlation satisfies our four axioms. In case
of $\mathbb{R}$ we have already seen that there is no Borel invariant Δ that satisfies our
four axioms.*

Thus if the metric space is a Banach space that is Borel isomorphic with $\mathbb{R}$,
then the four axioms cannot hold when we replace similarity with 1–1 Borel
isomorphism.

In this book besides geometric type invariances we also discussed *algebraic
invariances* where we needed to check the validity of algebraic identities. Examples include additive constant invariance for unbiased distance covariance
and jackknife invariance for U-statistics.

In general, it is not true that the more invariance the better, because
very strong invariance conditions make the corresponding statistical inference
less sensitive, and sometimes, even more contradictory like non-transitivity.
Color invariance or color blindness for example is not a good thing. In physics
supreme invariance what they call supreme symmetry is supremely boring.
In this supreme invariant world all particles would exist without mass; time
would stop. In statistics the choice of the right level of invariance is crucial
for powerful inference.

22.6   Multivariate Dependence Measures

We might want to extend the axioms of dependence measures for more than
two variables. The definition is easy to extend.

Let S be a non-empty set of random variables $X_1, X_2, \ldots, X_d$ taking values in Euclidean spaces or in separable Hilbert spaces H. Suppose that $X_1, X_2, \ldots, X_d \in S$ are not constant with probability 1, with
$(L_1 X_1, L_2 X_2, \ldots, L_d X_d) \in S$ where $L_1, L_2, \ldots L_d$ are similarity transforms
of H. Then $\Delta(X_1, X_2, \ldots, X_d) : S \to [0, 1]$ is called a mutual dependence
measure on S if the following four axioms hold.

(1) $\Delta(X_1, X_2, \ldots, X_d) = 0$ if and only if $X_1, X_2, \ldots, X_d$ are mutually independent.

(2) $\Delta(X_1, X_2, \ldots, X_d)$ is invariant with respect to all similarity transformations of H; that is,

$$\Delta(L_1 X_1, L_2 X_2, \ldots, L_d X_d) = \Delta(X_1, X_2, \ldots, X_d),$$

for all similarity transformations $L_1, L_2, \ldots, L_d$ of H.

(3) $\Delta(X_1, X_2, \ldots, X_d) = 1$ if and only if $L_1 X_1 = L_2 X_2 = \ldots L_d X_d$ with probability 1, where $L_1, L_2, \ldots L_d$ are some similarity transformations of H.

(4) $\Delta(X_1, X_2, \ldots, X_d)$ is continuous; that is, if $(X_1(n), X_2(n), \ldots X_d(n)) \in S$, $n = 1, 2, \ldots$ such that for some positive constant K we have

$$E\big(|X_1(n)|^2 + |X_2(n)|^2 + \cdots + |X_d(n)|^2\big) \leq K, \qquad n = 1, 2, \ldots$$

and $(X_1(n), X_2(n), \ldots X_d(n))$ converges weakly (converges in distribution) to $(X_1, X_2, \ldots, X_d)$ then
$\Delta\big(X_1(n), X_2(n), \ldots, X_d(n)\big) \to \Delta(X_1, X_2, \ldots, X_d)$.

We can weaken some of the axioms. For example, in axioms (2) and (3) we might need to restrict similarity operators $L_1, L_2, \ldots L_d$ such that the scale is the same for all of them. This can be achieved via a "multivariance" version of distance correlation, the weighted L_2 distance of the joint characteristic function and the product of the marginals, where we apply the weight function

$$w_1 := c/[|t_1|^2 + |t_2|^2 + \cdots + |t_d|^2]^{\alpha + p_1 + p_2 + \cdots + p_d}$$

where $p_1, p_2, \ldots, p_d$ are the dimensions of X_k, $k = 1, \ldots, d$. On this weight function see Bakirov et al. [2006]. In this case we weaken some of the axioms. In axioms (2) and (3) we need to restrict similarity operators $L_1, L_2, \ldots L_d$ such that the scale is the same for all of them.

If the expectations of the random variables $X_1, X_2, \ldots, X_d$ are finite $X := (X_1, X_2, \ldots, X_d)$ and X^* denotes the random variable with the same marginal distributions as X but with independent marginals, then the weighted L_2 distance between X and X^* is the energy distance of X and X^* and thus it is finite. This distance is clearly nonnegative and equals zero if and only if $X_1, X_2, \ldots, X_d$ are mutually independent.

Now, if we standardize this energy distance as we did in Theorem 1 of Bakirov et al. [2006], then we get a standardized version $\Delta(X_1, X_2, \ldots, X_d)$ of the energy distance of X and X^*. This Δ satisfies our axioms (1) and (4) above.

For (3) we need to suppose that the scales of $L_1, L_2, \ldots, L_d$ are the same. Also, in (3) the condition of equality is that there is a measurable random set A and non-random d-dimensional vectors $a_i, b_i, i = 1, 2, \ldots, d$ such that $X_i = a_i + b_i I(A)$ (all random variables can take two values only).

We can try to get rid of these restrictions in (3) and (4) via applying another weight function, for example, the d-dimensional version of our distance covariance weight function

$$w_2 := C/[|t_1|^{p_1 + \alpha}|t_2|^{p_2 + \alpha} \ldots |t_d|^{p_d + \alpha}],$$

but so far we do not see a proof for the existence of a multivariate $\Delta(X_1, X_2, \ldots, X_d)$ with $d > 2$ that satisfies all four axioms above.

The multivariate version of Lyons' definition of distance covariance is

$$\mathcal{V}^2(X_1, X_2, X_3, \ldots) := E[A_{X_1} B_{X_2} C_{X_3} \ldots].$$

One might think that this coefficient is zero if and only if one can partition the set $\{X_1, X_2, \ldots, X_d\}$ into two nonempty subsets such that one of them is independent of the other. Unfortunately, not even this weaker version of mutual independence is true. For example, take three real-valued random variables, X, Y, Z that are not mutually independent but X is independent of (Y, Z). Then take another triplet of real-valued random variables: U, V, W that are not mutually independent but V is independent of (U, W). Now if (X_1, X_2, X_3) is a mixture of (X, Y, Z) and (U, V, W), say $(X_1, X_2, X_3) = (X, Y, Z)$ with probability $1/2$ and $(X_1, X_2, X_3) = (U, V, W)$ with probability $1/2$, then $\mathcal{V}^2(X_1, X_2, X_3) = 0$ but X_1, X_2, X_3 are not mutually independent, not even in the weaker sense above.

Although the multivariate version of the characteristic function definition of dCor satisfies Axiom (1), the scale invariance holds for identical scales of the d variables only.

Finally, here are the four axioms of mutual dependence measures for metric spaces. As we said before a mapping f from the metric space $(\mathcal{X}, \delta)$ to itself is a similarity transformation if $\delta(f(x), f(x')) = c\delta(x, x')$ for all elements $x \in \mathcal{X}$ and $x' \in \mathcal{X}$ and for some $c > 0$

Let S be a non-empty set of random variables $X_1, X_2, \ldots, X_d$ taking values in a metric space $(\mathcal{X}, \delta)$. Suppose that $X_1, X_2, \ldots, X_d$ are not constant with probability 1 and if $(X_1, X_2, \ldots, X_d) \in S$, then $(L_1 X_1, L_2 X_2, \ldots, L_d X_d) \in S$ where $L_1, L_2, \ldots L_d$ are similarity transformations of $(\mathcal{X}, \delta)$. Then $\Delta(X_1, X_2, \ldots, X_d) : S \to [0, 1]$ is called a mutual dependence measure on S if the following four axioms hold.

(I) $\Delta(X_1, X_2, \ldots, X_d) = 0$ if and only if $X_1, X_2, \ldots, X_d$ are mutually independent.

(II) $\Delta(X_1, X_2, \ldots, X_d)$ is invariant with respect to all similarity transformations of $(\mathcal{X}, \delta)$; that is, $\Delta(L_1 X_1, L_2 X_2, \ldots, L_d X_d) = \Delta(X_1, X_2, \ldots, X_d)$ where $L_1, L_2, \ldots, L_d$ are similarity transformations of $(\mathcal{X}, \delta)$.

(III) $\Delta(X_1, X_2, \ldots, X_d) = 1$ if and only if $L_1 X_1 = L_2 X_2 = \ldots L_d X_d$ with probability 1, for some similarity transformations $L_1, L_2, \ldots L_d$ of $(\mathcal{X}, \delta)$.

(IV) $\Delta(X_1, X_2, \ldots, X_d)$ is continuous; that is, if $(X_1(n), X_2(n), \ldots X_d(n)) \in S$, $n = 1, 2, \ldots$ such that $(X_1(n), X_2(n), \ldots X_d(n))$ converges weakly (converges in distribution) to $(X_1, X_2, \ldots, X_d)$, then $\Delta(X_1(n), X_2(n), \ldots, X_d(n)) \to \Delta(X_1, X_2, \ldots, X_d)$.

It is an interesting problem to characterize all metric spaces for which these axioms can hold for a given d if S contains all bounded non-constant random variables. If such a Δ does not exist, then how can the axioms be made weaker for the existence of Δ?

On this topic see Bakirov et al. [2006], Berschneider and Böttcher [2018], Böttcher [2019], Böttcher et al. [2018], Fan et al. [2017], Jin and Matteson [2018] and Pfister et al. [2018].

Yao et al. [2018] propose a distance covariance approach to testing mutual independence and for banded dependence structure for high dimensional data. The test applies pairwise distance covariance. The limiting null distribution of the test statistic is shown to be standard normal.

A distance correlation test of mutual independence is given in Section 13.3 and the test is implemented in the `mutualIndep.test` function of the *energy* package for R [Rizzo and Székely, 2022].

22.7 Maximal Distance Correlation

Maximal distance correlation is defined as follows

$$\text{maxdCor}(X,Y) = \sup_{f,g} \text{dCor}(f(X), g(Y)),$$

where f, g are Borel measurable functions for which $\text{dCor}(f(X), g(Y))$ exists.

Since functions of independent random variables remain independent, we have that for independent random variables $\text{maxdCor}(X,Y) = 0$. Thus it seems the max version of dCor is kind of superfluous. The situation is not the same as with classical correlations vs. maximal correlation. On the other hand maxdCor can help to discover lower dimensional dependencies when $\text{dCor}(X,Y)$ is very small. The supremum even with respect to all linear functions f, g can be interesting. For example, if $X := (X_1, X_2, \ldots, X_d)$ and $Y := (Y_1, Y_2, \ldots, Y_d)$ have iid coordinates and $\text{dCor}_n(X,Y)$ is very small because $(X_2, X_3, \ldots, X_d)$ is independent of Y, with the help of maxdCor we might be able to detect that $X_1 = Y_1$, say. We conjecture that as $d \to \infty$, $\text{maxdCor}(X,Y) \to 0$ in the situation above.

Maximal distance correlation is also more robust to noisy environments. But how can we compute the optimal f and g? We can apply alternating conditional expectations (ACE) [Breiman and Friedman, 1985]. They used the ACE algorithm to compute maximal correlation from correlations. We can apply the same idea to compute maximal distance correlation from distance correlation. An R implementation of ACE is available in the *acepack* package [Spector et al., 2016].

22.8 Proofs

(1) Proof of Theorem 22.1

Proof. Let X be a square integrable random variable defined on an arbitrary probability space. Suppose the distribution of X is not concentrated on at most three points.

Without loss of generality assume that $E[X] = 0$. Let Q denote the distribution of X on the Borel sets of $\mathbb{R}$. We have to find a 1–1 function f such that $\int x f(x)\, dQ = 0$.

By assumption, there exist real numbers $t_1 < t_2 < t_3$ such that each of the intervals $(-\infty, t_1], (t_1, t_2], (t_2, t_3], (t_3, +\infty)$ has positive measure (w.r.t. Q). Let δ be a suitably small positive number (the meaning of "suitably" will be made clear later). One can find $t_0 < t_1$ and $t_4 > t_3$ such that both $Q(-\infty, t_0]$ and $Q(t_4, +\infty)$ are less than δ (possibly 0).

Let the intervals $(-\infty, t_0], (t_0.t_1], (t_1, t_2], (t_2, t_3], (t_3, t_4],$ and $(t_4, +\infty)$ be denoted by $A_0, A_1, A_2, A_3, A_4,$ and A_5, respectively. Introduce

$$\mu_i = \int_{A_i} x\, dQ, \quad \sigma_i^2 = \int_{A_i} x^2 dQ, \quad 0 \le i \le 5.$$

Then $\mu_0 + \cdots + \mu_5 = 0$.

It is not hard to see that there exist real constants a_1, a_2, a_3, a_4, all different, such that

$$a_1(\mu_0 + \mu_1) + a_2\mu_2 + a_3\mu_3 + a_4(\mu_4 + \mu_5) = 0. \tag{22.3}$$

Indeed, consider the hyperplane $\mathcal{L}$ of all vectors $(a_1, a_2, a_3, a_4) \in \mathbb{R}^4$ satisfying (22.3). $\mathcal{L}$ cannot coincide with the hyperplane $\mathcal{L}_{1,2} = \{a_1 = a_2\}$, because the $\mathcal{L}_{1,2}$ is orthogonal to the vector $(1, -1, 0, 0)$, which is not parallel to $(\mu_0 + \mu_1, \mu_2, \mu_3, \mu_4 + \mu_5)$, since the latter can have at most one 0 coordinate. Thus, $\dim(\mathcal{L} \cap \mathcal{L}_{1,2}) = 2$. The same holds for $\mathcal{L}_{i,j}$, the hyperplane defined by equality $a_i = a_j$ $(i \ne j)$. Since $\mathcal{L}$ cannot be covered by six of its lower dimensional subspaces, the existence of a vector in $\mathcal{L}$ with different coordinates follows.

Let $K > \max_{1 \le i \le 4} |a_i|$. By continuity, if δ is small enough, one can find constants b_1, b_2, b_3, b_4 all different, such that $\max_{1 \le i \le 4} |b_i| < K$, and

$$-K\mu_0 + b_1\mu_1 + b_2\mu_2 + b_3\mu_3 + b_4\mu_4 + K\mu_5 = 0.$$

Finally, choose $c_0, c_1, \ldots, c_5$ in such a way that none of them are equal to 0, c_0 and c_5 are positive, and $\sum_{i=0}^{5} c_i \sigma_i^2 = 0$. This can be done, because there are at least 3 positive among the quantities σ_i^2.

Now, let $b_0 = -K$, $b_5 = K$, and $f(x) = b_i + \varepsilon c_i x$ if $x \in A_i$, $0 \le i \le 5$. Then f is injective provided ε is a sufficiently small positive number, and

$$\int_{\mathbb{R}} x f(x) \, dQ = \sum_{i=0}^{5} (b_i \mu_i + \varepsilon c_i \sigma_i^2) = 0,$$

as needed.

Such an f cannot exist if X can take on exactly two values, because in that case uncorrelatedness is equivalent to independence.

When the distribution of X is concentrated on exactly 3 points, and X is supposed to have mean 0, then such an f exists if and only if zero is not among the possible values of X. (If $E[X] = 0$ is not supposed, the necessary and sufficient condition for f to exist is $P(X = E[X]) = 0$.) Indeed, let $x_1 < x_2 < x_3$ be the possible values of X, with probabilities q_1, q_2, q_3, respectively. Then $q_1 x_1 + q_2 x_2 + q_3 x_3 = 0$, and $x_1 < 0 < x_3$. We are looking for real numbers f_1, f_2, f_3 such that $q_1 x_1 f_1 + q_2 x_2 f_2 + q_3 x_3 f_3 = 0$. If $x_2 = 0$, then it can only achieved with $f_1 = f_3$. In the complementary case $f_1 = -1$, $f_3 = 1$ and $f_2 = (q_1 x_1 - q_3 x_3)/(q_2 x_2)$ will do, because $f_2 = 1$ would imply $-q_1 x_1 + q_2 x_2 + q_3 x_3 = 0$, hence $q_1 x_1 = 0$, which is not allowed, and similarly, $f_2 = -1$ would imply $q_3 x_3 = 0$. $\qquad\square$

(2) Proof of Theorem 22.2

Proof. For the proof we can suppose without loss of generality that the Borel measurable functions f, g are such that $\operatorname{Var} f(X) = \operatorname{Var} g(Y) = 1$. Then by the Cauchy-Schwarz inequality

$$\begin{aligned}
\operatorname{Cov} f(X), g(Y)) &= E(\operatorname{Cov} f(X), g(Y)|Y) + \operatorname{Cov} E(f(X)|Y), E(g(Y)|Y)) \\
&= 0 + \operatorname{Cov} E(f(X)|Y), g(Y)) \\
&\le \sqrt{\operatorname{Var} E(f(X)|Y)} \sqrt{\operatorname{Var} g(Y)} \\
&= \sqrt{\operatorname{Var} E(f(X)|Y)}.
\end{aligned}$$

Equality holds if $g(Y) = aE(f(X)|Y) + b$ where a, b are constants, $a \ne 0$. Thus

$$\operatorname{maxcorr}^2(X, Y) = \sup_{f,g} \rho^2(f(X), g(Y))$$
$$= \sup\{\operatorname{Var} E(f(X)|Y) : \operatorname{Var} f(X) \operatorname{Var} g(Y) = 1\}.$$

$\qquad\square$

(3) Proof of Theorem 22.3

Proof. Suppose dim $H > 1$. We will show that every continuous and affine invariant dependence measure must be constant, hence violating axiom (i).

Let X, Y, X^*, and Y^* be arbitrary H-valued random variables, bounded and nonconstant. We will show that $\Delta(X,Y) = \Delta(X^*,Y^*)$. We can suppose that X^* and Y^* do not have constant coordinates at all. Then, by scale invariance, for every real number $c \neq 0$ we have $\Delta((X_1,X_2,\dots),Y) = \Delta((cX_1,X_2,\dots),Y)$, and by continuity, this remains true for $c = 0$. Similarly, we get the same if X_1 is replaced by X_1^*. Thus, X_1 can be changed to X_1^* with no effect on Δ. Gradually, all coordinates of X can be replaced by those of X^*, and then the same can be done with Y. Consequently, $\Delta(X,Y) = \Delta(X^*,Y^*)$. During these changes of coordinates we have to avoid making any of the random variables constant by changing one of their coordinates. This can be achieved if we first replace the constant coordinates by the corresponding nonconstant ones.

If $H = \mathbb{R}$, the real line, such a result cannot be true because on the real line affine transformations coincide with similarities. But if we require invariance with respect to all 1–1 Borel measurable functions, then, as a first step, we can map our scalar random variables to $\mathbb{R}^2$ with the help of a 1–1 Borel measurable function, see e.g. Gouvêa [2011]. Then the reasoning above can be applied. □

(5) Proof of Theorem 22.5

Proof. The correlation ratio does not satisfy axiom (i). Although it is zero when X and Y are independent, it can be zero in other cases, too: For example, when the conditional distribution of X given Y is symmetric.

Axiom (ii) clearly holds, because on the real line similarities coincide with linear transformations.

On the other hand, axiom (iii) does not hold because for the correlation ratio $\Delta(X,Y) = 1$ if and only if X is almost surely equal to a Borel measurable function of Y. Indeed, since $1 - \Delta(X,Y) = E[\text{Var}(X|Y)]/\text{Var}(X)$, we have $\Delta(X,Y) = 1$ if and only if $\text{Var}(X|Y) = 0$; that is, X is a Borel measurable function of Y with probability 1.

Axiom (iv) does not hold either, as shown by the following example. Let Y be a non-degenerate, bounded, integer valued random variable and X be uniformly distributed on the interval $(0,1)$ such that X is independent of Y. Define $Y_n = Y + \frac{1}{n}X$. Then $X = n\{Y_n\}$, where $\{\cdot\}$ stands for fractional part, hence $\Delta(X,Y_n) = 1$. On the other hand, (X,Y_n) tends to (X,Y) everywhere, not only in distribution, and $\Delta(X,Y) = 0$. □

(6) Proof of Theorem 22.6

MIC is introduced in Example 22.8.

Proof. It is clear that $MIC(X,Y) = 0$ if X and Y are independent. On the other hand, $MIC(X,Y) = 0$ means that $I((X,Y)|_G) = 0$ for every grid G, which implies that the discretized by G versions of X and Y are independent. Particularly we obtain that the joint distribution of (X,Y) coincides with a product measure on rectangles, hence on all bidimensional Borel sets, too.

Since monotone transformations of the coordinates map grids into grids, MIC is invariant with respect to them. Particularly, it satisfies axiom (ii), because every affine transformation on $\mathbb{R}$ is monotone.

If X is discrete, then $MIC(X,X) = 1$ if and only if there exists a partition of the real line into at least 2 parts such that X falls into every partition interval with the same probability. For example, suppose that $P(X = 0) = p \neq 1/2$, and $P(X = 1) = 1 - p$. Then $MIC(X,X) = -p \log p - (1 - p) \log(1 - p) < 1$. Thus, axiom (iii) is hurt.

Let X take the values $0, 1, 2$ with probabilities $1/4$, $1/2$, $1/4$, respectively. For computing $MIC(X,X)$ it is easy to see that there exist altogether three essentially different grids: 2×2, 2×3, and 3×3. Consequently,

$$MIC(X,X) = \max \left\{ \frac{\frac{1}{4}\log 4 + \frac{3}{4}\log\frac{4}{3}}{\log 2}, \; \frac{\frac{1}{4}\log 4 + \frac{1}{2}\log 2 + \frac{1}{4}\log 4}{\log 3} \right\}$$

$$= 0.946 \cdots < 1.$$

Let $f(x) = 3 - x$, if $1 \leq x \leq 2$, and $f(x) = x$ otherwise. This f interchanges 1 and 2, and does not move 0. It is a piecewise continuous 1–1 function, and $MIC(f(X), f(X)) = 1$, which is obtained by considering the partition $\mathbb{R} = (-\infty, 3/2) \cup [3/2, +\infty)$. Thus, MIC is not invariant with respect to measurable 1–1 transformations.

If one prefers a counterexample with continuous joint distribution, let (X,Y) be uniformly distributed over the black squares of a 4×4 checkerboard. Then the maximal information coefficient of that distribution is $1/2$, but if we permute the rows and columns of the checkerboard, we can turn it into a 2×2 checkerboard with bigger squares, and the maximal information coefficient of that distribution is 1. This permutation can be performed using piecewise linear functions. Details are left to the reader.

Finally, we show that MIC is not weakly continuous, thus axiom (iv) is not satisfied.

Let X be uniformly distributed on the interval $[0, 1]$, and define Y as the fractional part of nX. We will show that $MIC(X, Y) = 1$ for every positive integer n.

Clearly, Y is also uniformly distributed on $[0, 1]$. Let $k \geq 2$ be arbitrary, and impose an $nk \times k$ equidistant grid on the unit square. Then the distribution $(X, Y)|_G$ is discrete uniform of size nk, with discrete uniform marginals of size nk and k, respectively. Hence $I((X, Y)|_G =$

$\log(nk) + \log k - \log(nk) = \log k$, while $||G|| = k$. (In information theory, logarithm is meant on base 2, but the base does not matter here.) Thus, $MIC(X,Y) = \frac{I((X,Y)|_G)}{\log ||G||} = 1$. Now, as $n \to \infty$, the joint distribution of (X,Y) converges weakly to the uniform distribution on the unit square, which, having independent marginals, yields $MIC = 0$. $\square$

22.9   Exercises

Exercise 22.1. Which of the dependence measures discussed in the examples satisfy Axiom (i)? Justify the answers with proofs or counterexamples.

Exercise 22.2. Prove that an alternate computing formula for dCor_n^2 (Definitions 12.5–12.7) is

$$\mathrm{dCor}_n^2(X,Y) = \frac{\langle C\Delta_X C, C\Delta_Y C\rangle}{||C\Delta_X C|| \, ||C\Delta_Y C||},$$

where Δ_X, Δ_Y are the Euclidean distance matrices of the X and Y samples, and $C = \mathbf{I}_n - \frac{1}{n}\mathbf{1}\mathbf{1}^\top$. See Example 22.3.

Exercise 22.3. Prove by constructing an example that a monotone invariant Δ can satisfy Axioms (i) – (iv). Hint: One idea is to apply distance correlation to the ranks of observations (or to the copula). Are there also homeomorphism invariant versions? (They are invariant with respect to all continuous 1–1 functions with continuous inverse.)

Exercise 22.4. Prove by constructing an example that maximal correlation can be 1 for uncorrelated random variables.

23

Earth Mover's Correlation

CONTENTS

23.1 Earth Mover's Covariance

Define the Earth mover's covariance as follows:

$$eCov(X, Y) = \inf E|(X, Y) - (X^*, Y^*)|,$$

where X^* and X are identically distributed, Y^* and Y are identically distributed, and X^*, Y^* are independent. This is especially useful if the metric we want to use is not negative definite and thus the energy approach does not work.

The Earth mover's covariance, $eCov(X, Y)$, is simply the Wasserstein distance $\mathcal{W}$ between the joint distribution and the product of its marginals. The empirical version is the Wasserstein distance or minimum transportation cost between the following two mass or probability distributions:

1. $1/n$ mass at each point $(X_i, Y_i), i = 1, 2, \ldots, n$

2. $1/n^2$ mass at each point $(X_i, Y_j), i, j = 1, 2, \ldots, n$.

Because of this rephrase of the Wasserstein distance we also can call wCov the Earth mover's covariance.

Let $(\mathcal{M}, \varrho)$ be a metric space. Consider the product space $\mathcal{M} \times \mathcal{M}$ equipped with a metric d such that

$$d\big[(x_1, x_2), (x_1, x_3)\big] = \varrho(x_2, x_3),$$
$$d\big[(x_1, x_3), (x_2, x_3)\big] = \varrho(x_1, x_2),$$
$$d\big[(x_1, x_1), (x_2, x_3)\big] \geq \varrho(x_2, x_3).$$

DOI: 10.1201/9780429157158-23

For example, $d\big[(x_1, x_2), (x_3, x_4)\big] = \varrho(x_1, x_3) + \varrho(x_2, x_4)$ will do.

Introduce the Wasserstein distance of probability measures λ and μ defined on the Borel sets of the product space $\mathcal{M} \times \mathcal{M}$ as $w(\lambda, \mu) = \inf E[[(X, Y), (X', Y')]$, where the infimum is taken over all pairs of $\mathcal{M} \times \mathcal{M}$ valued random vectors (X, Y) and (X', Y') such that the distributions of (X, Y) and (X', Y') are λ and μ, respectively, but their joint distribution can otherwise be arbitrary. We also write $\mathcal{W}\big[(X, Y), (X', Y')\big]$ for the Wasserstein distance between the distributions of the random vectors (X, Y) and (X', Y').

The following inequality is of Cauchy–Bunyakovsky–Schwarz type.

Theorem 23.1.

$$\mathcal{W}^2\big[(X, Y), (X', Y')\big] \leq \mathcal{W}\big[(X, X), (X, X')\big]\, \mathcal{W}\big[(Y, Y), (Y, Y')\big], \qquad (23.1)$$

where X and X' are iid, Y and Y' are iid, and X', Y' are independent.

In fact, we can show more, namely the following.

Proposition 23.1.

$$eCov(X, Y) = \mathcal{W}\big[(X, Y), (X', Y')\big]$$
$$\leq \min\left\{\mathcal{W}\big[(X, X), (X, X')\big], \mathcal{W}\big[(Y, Y), (Y, Y')\big]\right\}.$$

Proof. Suppose that the right-hand side is equal to $\mathcal{W}\big[(Y, Y), (Y, Y')\big]$. In the sequel all random variables denoted by X with or without subscripts or superscripts will be equi-distributed with X, and the same holds for Y. Let Y_2 and Y_3 be independent, then $Ed\big[(Y_1, Y_1), (Y_2, Y_3)\big] \geq E\varrho(Y_2, Y_3) = Ed\big[(X_2, Y_2), (X_2, Y_3)\big]$, where X_2 is chosen in such a way that (X_2, Y_2) and (X, Y) are identically distributed, and Y_3 is independent of (X_2, Y_2). Then the right-hand side is greater than or equal to $\mathcal{W}\big[(X, Y), (X', Y')\big]$, while the infimum of the left-hand side as Y_1 varies is just $\mathcal{W}\big[(Y, Y), (Y, Y')\big]$. □

On the right-hand side of (23.1) $\mathcal{W}\big[(X, X), (X, X')\big]$ can be considered a kind of variance, called *the Earth mover's variance* of the distribution of X, and denoted by $eVar(X)$, while the square root of the left-hand side is a covariance-like quantity, so we can also denote it by $wCov(X, Y)$.

23.2   Earth Mover's Correlation

The *Earth mover's (Wasserstein) correlation* of the distributions of X and Y is defined as

$$eCor(X, Y) = \frac{eCov(X, Y)}{\min[eVar(X), eVar(Y)]}.$$

Remark 23.1. In fact, $eVar(Y) = E\varrho(Y, Y')$, the Gini mean difference of Y. Indeed, we have seen above that

$$Var(Y) = \inf_{Y_1} \, Ed\big[(Y_1, Y_1), (Y_2, Y_3)\big] \geq E\varrho(Y_2, Y_3),$$

and equality is attained for $Y_1 = Y_2$. Thus the Wasserstein variance is simply the Gini mean. Thus in the formula for eCor we do not divide by zero except if at lest one of X, Y is constant with probability one. In this case we do not define eCor.

Proposition 23.2. *If $Y = f(X)$ where f is a similarity transformation of our metric space, then $eCor(X, Y) = 1$.*

Proof. Because of Remark 23.1 we know that $eVar(f(X)) = ceVar(X)$. For independent X_1, X_2, X_3 we have

$$
\begin{aligned}
d[(X_1, f(X_1)), (X_2, f(X_3))] &= \varrho(X_1, X_2) + \varrho(f(X_2), f(X_3)) \\
&= \varrho(x_1, X_2) + c\varrho(X_1, X_3) \\
&\geq \min\{1, c\}[\varrho(x_1, X_2) + \varrho(X_1, X_3] \\
&= \min\{1, c\}d[(X_1, X_3), (X_2, X_3)] \\
&= \min\{1, c\}eVar(X) \\
&= \min\{eVar(X), eVar(f(X))\}.
\end{aligned}
$$

The infimum of the left hand side as X_2 and X_3 remain independent is $eCov(X, f(X))$. Thus $eCor(X, f(X)) \geq 1$. The other direction follows from Proposition 1. $\qquad\square$

The empirical eVar is the arithmetic average of the distances $|X_i - X_j|$ because the cost to transport (X_i, X_j) to the main diagonal is at least $|X_i - X_j|/n^2$ and we can achieve this via "horizontal" transportation only. This is not the case if we want to transport to n general points not necessarily on the main diagonal.

If X is a sample of size n from the uniform distribution $U[0; 1]$, we use the Euclidean metric in $\mathbb{R}^n$ and the Manhattan distance for pairs, then $eVar(X) = E|X - X'|_n$, where X and X' are independent uniform random points of the n dimensional unit cube. Particularly, for $n = 1$ we get $eVar(X) = 1/3$. For general n it is known that

$$eVar(X) = \frac{1}{\sqrt{\pi}} \int_0^\infty \left\{ 1 - \left(\frac{\sqrt{\pi}\,\mathrm{erf}(u)}{u} - \frac{1 - e^{-u^2}}{u^2} \right)^n \right\} \frac{du}{u^2},$$

where $\mathrm{erf}(u)$ is the Gauss error function at u.

We do not have a simple expression for $eVar(X)$ if n can be arbitrary. However, by the inequality $|X - X'| \geq n^{-1/2} \sum_{i=1}^n |X_i - X'_i|$ it easily follows that $eVar(X) \geq \sqrt{n}/3$. On the other hand, since the diameter of the unit cube is $\sqrt{n}$, we clearly have $eVar(X) \leq \sqrt{n}$. A somewhat better upper estimate is

$$E|X - X'| \leq \left[E|X - X'|^2\right]^{1/2}$$
$$= \left[nE(X_1 - X'_1)^2\right]^{1/2}$$
$$= [2n\,\mathrm{Var}(X_1)]^{1/2} = \sqrt{n/6}.$$

More interesting details can be found in `https://mathworld.wolfram.com/HypercubeLinePicking.html`.

23.3   Population eCor for Mutual Dependence

We can easily define the dependence measure for more than two variables. The population version of eCov for three variables is as follows:

$$eCov(X, Y, Z) = \inf E|(X, Y, Z) - (X^*, Y^*, Z^*)|.$$

Here we take the inf over all joint distributions of (X, Y, Z) with given marginals. In distribution $X = X^*$, $Y = Y^*$, $Z = Z^*$ and X^*, Y^*, Z^* are independent.

The population version of Earth mover's correlation for three random variables is

$$eCor(X, Y, Z) = \frac{eCov(X, Y, Z)}{\min[eVar(X), eVar(Y), eVar(Z)]}.$$

In general, for k random variables

$$eCov(X_1, \ldots, X_k) = \inf E|(X_1, \ldots, X_k) - (X_1^*, \ldots, X_k^*)|.$$

Thus we have a natural measure for mutual dependence of more than two random variables.

23.4   Metric Spaces

Refer to Chapter 22 where our axioms for measures of dependence are given.

A reformulation of our axioms to arbitrary metric spaces is the following.

(a) $\Delta(X, Y) = 0$ if and only if X and Y are independent.

(b) $\Delta(X, Y)$ is invariant with respect to all similarity transformations of $(\mathcal{M}, \delta)$; that is, $\Delta(f(X), g(Y)) = \Delta(X, Y)$ where f, g are similarity transformations of $(\mathcal{M}, \delta)$.

(c) $\Delta(X,Y) = 1$ if and only if $Y = f(X)$ with probability 1, where f is a similarity transformation of $(\mathcal{M}, \delta)$.

(d) $\Delta(X,Y)$ is continuous; that is, if for some positive constant K and $x_0 \in \mathcal{M}, y_0 \in \mathcal{M}$ we have $E\big(\delta^2(X_n, x_0) + \delta^2(Y_n, y_0)\big) \leq K$, $n = 1, 2, \ldots$ and (X_n, Y_n) converges weakly (i.e., converges in distribution) to (X, Y), then $\Delta(X_n, Y_n) \to \Delta(X, Y)$. (Again, the condition on the boundedness of second moments can be replaced by any other condition that guarantees the convergence of expectations: $E\delta(X_n, x_0) \to E\delta(X, x_0)$) and $E\delta(Y_n, y_0) \to E\delta(Y, y_0)$; such a condition is uniform integrability of $\delta(X_n, x_0), \delta(Y_n, y_0)$, which follows from the boundedness of second moments.)

We will also need the following weaker forms of axioms (b) and (c):

(b*) $\Delta(X,Y) = \Delta(f(X), f(Y))$ for every similarity transformation f of $(\mathcal{M}, \delta)$.

(c*) $\Delta(X,Y) = 1$ if $Y = f(X)$ with probability 1, where f is a similarity transformation of $(\mathcal{M}, \delta)$.

It is easy to see that $eCor$ as a new measure of dependence satisfies at least two of our axioms for dependence measures. Axioms (a), and (d) hold. Concerning (b) and (c) we can only prove the weaker (b*) and (c*).

Theorem 23.2. $eCor(X,Y) = eCor(f(X), f(Y))$ *for every similarity transformation f of our metric space.*

If $Y = f(X)$ where f is a similarity transformation, then $eCor(X,Y) = 1$.

Proof. From the definition it is obvious that $eCov(f(X), f(Y)) = c \cdot eCov(X,Y)$. Therefore we also have $eVar(f(X)) = c \cdot eVar(X)$, and finally $eCor(f(X), f(Y)) = eCor(X,Y)$.

For independent X_1, X_2, X_3 we have

$$
\begin{aligned}
d\big[(X_1, f(X_1)), (X_2, f(X_3))\big] &= \delta(X_1, X_2) + \delta(f(X_1), f(X_3)) \\
&= \delta(X_1, X_2) + c \cdot \delta(X_1, X_3) \\
&\geq \min\{1, c\}\big[\delta(X_1, X_2) + \delta(X_1, X_3)\big] \\
&= \min\{1, c\}\, d\big[(X_1, X_1), (X_2, X_3)\big] \\
&\geq \min\{1, c\} eVar(X) \\
&= \min\{eVar(X), eVar(f(X))\}.
\end{aligned}
$$

The infimum of the left hand side as X_2 and X_3 remain independent is $eCov(X, f(X))$. Thus $eCor(X, f(X)) \geq 1$. The other direction follows from Proposition 23.2. $\qquad\square$

Thus we proved the following result.

Theorem 23.3. *In arbitrary metric spaces $(\mathcal{M}, \rho)$ the earth mover's correlation $\Delta(X, Y) = eCor(X, Y)$ satisfies axioms (a), (b^*), (c^*), and (d).* □

It is easy to see that for an arbitrary metric space $(\mathcal{M}, \delta)$ it cannot be true that $eCor(X, Y) = 1$ always implies $Y = f(X)$ where f is a similarity. A counterexample is the following. Let $\mathcal{M}$ be the set of points of the Euclidean plane with the usual Euclidean metric. Suppose that here $eCor(X, Y) = 1$ implies $Y = f(X)$ where f is a similarity. If the random variables X and Y are supported on the x line, then we know that the similarity is $Y = aX + b$. Now define a new metric on the plane as follows: $\delta(x, y) = |x - y|$ if both x and y are on the x coordinate axis (the second coordinate is 0), otherwise for all $x \neq y$ define $\delta(x, y) = |x - y| + 1$. This does not change $eCor(X, Y) = 1$ because X and Y are supported on the x axis but $y = ax + b$ cannot be extended to the whole plane as a similarity with respect to the new metric.

Conjecture 23.1. *For Banach space valued random variables we have the iff statement in Axiom (c): $eCor(X, f(X)) = 1$ if and only if $Y = f(X)$ with probability 1, where f is a similarity transformation of the Banach space.*

Although we could not prove this conjecture it is interesting to note that by a theorem of Mazur and Ulam [1932], any bijective similarity f of any Banach space (or of any normed linear space) is affine; that is, $f(x) - f(0)$ is linear. Thus similarities in Banach spaces must have a very simple structure.

By the way, it is interesting to note that we can always embed every metric space $(\mathcal{M}, \delta)$ into the Banach space $C_b(\mathcal{M})$ of bounded continuous functions on $(\mathcal{M}, \delta)$. Just take the function $f(x)(y) := \delta(x, y) - \delta(x_0, y)$, where x_0 is an arbitrary element of $\mathcal{M}$.

Example 23.1. Let X and Y be indicators, $\Pr(X = 1) = 1 - \Pr(X = 0) = p_X$, $\Pr(Y = 1) = 1 - \Pr(Y = 0) = p_Y$, $\Pr(X = Y = 1) = p_{XY}$. Let us apply the Euclidean metric in $\mathbb{R}$ and the Manhattan distance for pairs. Then

$$eCor(X, Y) = \frac{|p_{XY} - p_X p_Y|}{\min\{p_X(1 - p_X), p_Y(1 - p_Y)\}}.$$

Indeed, in the lower bound of Remark 23.2 we have

$$
\begin{aligned}
E|X - Y| &- E|X' - Y'| \\
&= \Pr(X \neq Y) - \Pr(X' \neq Y') \\
&= \Pr(X' = Y') - \Pr(X = Y) \\
&= p_X p_Y + (1 - p_X)(1 - p_Y) - p_{XY} - (1 - p_X - p_Y + p_{XY}) \\
&= 2(p_X p_Y - p_{XY}),
\end{aligned}
$$

thus $eCov(X, Y) \geq 2|p_{XY} - p_X p_Y|$.

On the other hand, we will construct random variables X, Y, X', Y with the desired distribution in such a way that $E(|X - X'| + |Y - Y'|) = 2|p_{XY} - p_X p_Y|$.

Let U, V be independent and uniformly distributed on $[0, 1]$, and define

$$X = X' = I(U \le p_X), \quad Y' = I(V \le p_Y),$$

$$Y = I\left(U \le p_X, \ V \le \tfrac{p_{XY}}{p_X}\right) + I\left(U > p_X, \ V \le \tfrac{p_Y - p_{XY}}{1 - p_X}\right).$$

Then $\Pr(X \ne X') = 0$ and

$$\Pr(Y \ne Y') = \Pr(Y = 1, Y' = 0) + \Pr(Y = 0, Y' = 1).$$

Here

$$\Pr(Y = 1, Y' = 0) = \Pr\left(U \le p_X, \ p_Y < V \le \tfrac{p_{XY}}{p_X}\right)$$

$$+ \Pr\left(U > p_X, \ p_Y < V \le \tfrac{p_Y - p_{XY}}{1 - p_X}\right),$$

and similarly,

$$\Pr(Y = 0, Y' = 1) = \Pr\left(U \le p_X, \ \tfrac{p_{XY}}{p_X} < V \le p_Y\right)$$

$$+ \Pr\left(U > p_X, \ \tfrac{p_Y - p_{XY}}{1 - p_X} < V \le p_Y\right).$$

Altogether we have

$$\Pr(Y \ne Y') = p_X \left| p_Y - \tfrac{p_{XY}}{p_X} \right| + (1 - p_X) \left| p_y - \tfrac{p_Y - p_{XY}}{1 - p_X} \right|$$

$$= 2 |p_{XY} - p_X p_Y|,$$

thus $eCov(X, Y) = 2 |p_{XY} - p_X p_Y|$.

Finally, $eVar(X) = 2 p_X (1 - p_X)$ is straightforward.

The absolute value of Pearson's correlation ρ for indicators is

$$|\rho(X, Y)| = \frac{|p_{XY} - p_X p_Y|}{\sqrt{p_X (1 - p_X) p_Y (1 - p_Y)}},$$

thus for indicators X and Y we have $|\rho(X, Y)| < eCor(X, Y)$ (and we have equality iff $p_X = p_Y$).

Based on this observation one can suspect that $|\rho(X, Y)| \le eCor(X, Y)$ for all real-valued random variables with finite variance. This conjecture is also supported by the fact that the independence of X, Y implies their un-correlatednes. In the other extreme case when $\rho(X, Y) = \pm 1$ we know that $Y = f(X)$ where f is a similarity (here a linear function) and by Theorem 23.2 below in this case we have $eCor(X, Y) = 1$.

This conjecture, however, can easily be disproved. The following theorem shows that if the joint distribution of X, Y is bivariate normal, the opposite inequality holds.

Theorem 23.4. *Let (X,Y) be bivariate normal with correlation $\rho(X,Y) = \rho$. Then*

$$eCor(X,Y) \le \left[1 - \sqrt{1-\rho^2}\,\right]^{1/2} \le |\rho|,$$

and the last inequality is strict unless $\rho = 0$ or $\rho = \pm 1$.

Actually we have the following:

Conjecture 23.2. *Let (X,Y) be bivariate normal with correlation $\rho(X,Y) = \rho$. Then $eCor(X,Y) = \left[1 - \sqrt{1-\rho^2}\,\right]^{1/2}$.*

Proof of Theorem 23.4. Let (X,Y) be bivariate normal with $Var(X) = \sigma_X^2$, $Var(Y) = \sigma_Y^2$ and $\rho(X,Y) = \rho$. We can suppose $EX = EY = 0$ and $\sigma_X^2 \ge \sigma_Y^2$. Let X' and Y' be independent zero mean normal with variances σ_X^2 and σ_Y^2, respectively. Finally, set $X = X'$ and $Y = (\sigma_Y/\sigma_X)\rho X' + \sqrt{1-\rho^2}\,Y'$. Then X,Y have the prescribed joint distribution, and $Y - Y'$ is normal with mean 0 and variance

$$\sigma_Y^2\left[\rho^2 + \left(1 - \sqrt{1-\rho^2}\,\right)^2\right] = 2\sigma_Y^2\left[1 - \sqrt{1-\rho^2}\,\right],$$

hence

$$eCov(X,Y) \le E|Y - Y'| = \frac{2\sigma_Y}{\sqrt{\pi}}\left[1 - \sqrt{1-\rho^2}\,\right]^{1/2}.$$

In the denominator of $eCor$ we have

$$eVar(Y) = \frac{2\sigma_Y}{\sqrt{\pi}} \le \frac{2\sigma_X}{\sqrt{\pi}} = eVar(X),$$

thus

$$eCor(X,Y) \le \left[1 - \sqrt{1-\rho^2}\,\right]^{1/2} \le |\rho|,$$

and the last inequality is strict unless $\rho = 0$ or $\rho = \pm 1$. □

Remark 23.2. Let us apply the Manhattan distance for pairs. Then by the triangle inequality for δ, we have

$$\delta(X,X') + \delta(Y,Y') \ge |\delta(X,Y) - \delta(X',Y')|,$$

thus

$$eCov(X,Y) \ge E\big|\delta(X,Y) - \delta(X',Y')\big| \ge \big|E\delta(X,Y) - E\delta(X',Y')\big|.$$

Concerning the lower bound of $eCor(X,Y)$, if $\sigma_X = \sigma_y$, then Remark 23.2 provides the following inequality: $|1 - \sqrt{1-\rho}\,| \le eCor(X,Y)$.

The arguments in the proofs support the next conjecture.

Conjecture 23.3. *In computing the infimum $eCov(X,Y) = \inf E\big[\delta(X,X') + \delta(Y,Y')\big]$, "under general conditions" we can suppose $X = X'$ or $Y = Y'$.*

This conjecture is not true without any conditions as the following counterexample shows.

Example 23.2. If X and Y are $1-1$ functions of each other, then the last conjecture would imply that $eCor(X, Y) = 1$ because Y is a function of $X = X'$ thus independent of Y'. Hence $eCov(X, Y) = \min\{eVar(X), eVar(Y)\}$. Thus in case of continuous marginals the empirical $eCor$ would always be 1 because then no vertical or horizontal lines can contain more than one sample point. This is, however, not true as is shown by the following sample of four elements: $(1, 4), (2, 2), (3, 3), (4, 1)$. Here $eVar = 5/4$ for both coordinates but $eCov = 1$.

Theorem 23.5. *The claim of Conjecture 23.2 implies Conjecture 23.3.*

Proof. The infimum in the theorem can be computed by applying conditional quantile transformations. Suppose $X = X'$. Let $F(x)$, $G(y)$ denote the cdf of X and Y, respectively, and $G(y|x) = \Pr(Y \le y \mid X = x)$, the conditional cdf of Y. Then the infimum of $E|Y - Y'|$ under the condition that $Y = Y'$ in distribution, but X', Y' are independent, equals

$$E\left(\int_{-\infty}^{\infty} |G(y|X) - G(y)| dy \right) = \int_{-\infty}^{\infty} E|G(y|X) - G(y)| \, dy.$$

Note that $G(y) = EG(y|X)$, thus the integrand on the right hand side is a kind of a mean absolute difference. An alternative formula for $eCov$ is

$$eCov(X, Y) = \int_{-\infty}^{\infty} \int_{-\infty}^{\infty} |G(y|x) - G(y)| \, dF(x) \, dy.$$

In the case of jointly normal X, Y the conditional quantile transformation leads to the same representation of Y as a linear combination of X' and Y' that we used in the proof of Theorem 23.4. Thus our Conjecture 23.3 would follow from Conjecture 23.2. ☐

23.5   Empirical Earth Mover's Correlation

The earth mover's metric suggests the following earth mover's distance definition between two sequences $x := (x_1, x_2, \ldots, x_n)$ and $y := (y_1, y_2, \ldots, y_n)$:

$$eCov_n(x, y) := \inf_{\pi} \sum_{i=1}^{n} \delta(x_i, y_{\pi(i)}),$$

where the infimum is taken for all permutation π on the integers $1, 2, \ldots, n$. One can easily see that for real-valued data, if the ordered sample is denoted by subscripts in brackets, then

$$eCov_n(x, y) := \sum_{i=1}^{n} |x_{(i)} - y_{(i)}|.$$

The empirical version of $eCov$ is the minimum transportation cost between the following two mass distributions or probability distributions:

(Q_1) $1/n$ mass at each point (x_i, y_i), $i = 1, 2, \ldots, n$;

(Q_2) $1/n^2$ mass at each point (x_i, y_j), $i, j = 1, 2, \ldots, n$.

The empirical version of $eCor$ is

$$eCor_n(x, y) = \frac{eCov_n(x, y)}{\min[eVar_n(x), eVar_n(y)]},$$

where $eVar_n(x)$ and $eVar_n(y)$ is simply the Gini's mean of x and y.

It is easy to see that the empirical $eVar$ is the arithmetic average of the distances $\delta(x_i, x_j)$ because the cost to transport $1/n^2$ mass from the point (x_i, x_j) to the main diagonal (x, x) is at least $\delta(x_i, x_j)/n^2$ and we can achieve this via "horizontal" transportation only. This is not the case if we want to transport to n general points, not necessarily on the main diagonal. The "naive" computational complexity of the empirical $eVar$, which is essentially Gini's mean difference is $O(n^2)$ but for real-valued random variables we can decrease it to $O(n \log n)$.

The complexity of the computation of the empirical $eCov$ is less obvious.

Our transportation problem can be reduced to an assignment problem between two sets of n^2 points, thus according to the "Hungarian algorithm" [Kuhn, 1955] this optimization can be solved in polynomial time. The algorithmic complexity of assignment problem for two sets of n points is $O(n^3)$ [Edmonds and Karp, 1972, Munkres, 1957, Tomizawa, 1971]. Thus in our case the complexity can be reduced to $O(n^6)$.

This is not very encouraging. A better complexity, namely $O(n^3 \log^2 n)$, is in Kleinschmidt and Schannath [1995]. Here the authors show that for the (linear) transportation problem with m supply nodes, n demand nodes and k feasible arcs, there is an algorithm which runs in time proportional to $m \log m(k + n \log n)$ assuming w.l.o.g. that $m \geq n$; still at least one order of magnitude worse than the algorithmic complexity, $O(n^2)$, of computing the distance covariance or the distance correlation. This is the price we need to pay for the generality of $eCov$ and $eCor$. The AMPL (A Mathematical Programming Language) code is easy to apply for computing empirical $eCov$ and then $eCor$. In Ajtai et al. [1984] it was shown that given n random blue and n random red points on the unit square, the transportation cost between them is typically $\sqrt{n \log n}$. Our problem is to find the optimal transportation costs when the distance is the Manhattan distance and the number of red points is different from the number of blue points (the total mass is the same).

A recent paper [Agarwal et al., 2017] suggests that our task of computing the earth mover's distance between two sets of size n^2 can be done with the first algorithm in the cited paper with $O(\log^2(1/\varepsilon))$ approximation error bound in $O(n^{2+\varepsilon})$ steps, for any $\varepsilon > 0$. See also Altschuler et al. [2017], Rigollet and Niles-Weed [2018], and Altschuler et al. [2019] on related topics.

23.6 Dependence, Similarity, and Angles

Heuristically it is clear that zero dependence = no similarity = independence and maximum dependence = perfect similarity. We can decrease dependence via increasing dissimilarity. This resembles the entropy–information duality: we can decrease entropy (uncertainty) via increasing information (see Maxwell's demon and Szilárd's explanation).

Let us make this idea of similarity and dependence more quantitative.

(i) *Angles.* Geometric similarity started with Thales of Miletus more than 2,500 years ago. According to Thales two triangles are similar if their angles ($A \geq B \geq C$ and $a \geq b \geq c$) are the same: $A = a$, $B = b$, $C = c$. But what are the least similar triangles? Let us measure the dissimilarity by $d = |A - a| + |B - b| + |C - c|$. Then triangles are least similar when d is maximum. These triangles are the regular and the "flat" triangles when $a = b = 0, c = \pi$. In this case $M = M(3) = 4\pi/3$ is the maximum of d. For n-gons the maximum is $m = M(n) = 4(n-2)\pi$, which tends to 4π increasingly. We can call $1 - d/M$ a measure of similarity. Unfortunately, for $n > 3$ the angles do not determine similarity so our first idea works for triangles only. The simple measure $1 - d/M$ of angle based similarity does not work for more general polygons than the triangle, at least not directly.

(ii) *Pearson's correlation*, Cor, can also be viewed as an angle, more precisely as the cosine of an angle. Based on n observations, $\text{Cor}((x_1, y_1), (x_2, y_2), \ldots, (x_n, y_n))$ is the cosine of the angle in the Euclidean space $\mathbb{R}^n$ between the centered vectors $x_1 - \bar{x}, x_2 - \bar{x}, \ldots, x_n - \bar{x}$ and $y_1 - \bar{y}, y_2 - \bar{y}, \ldots, y_n - \bar{y}$.

Cosine similarity is frequently applied in data mining. Cosine similarity is a simple measure of similarity between two non-zero vectors of an inner product space that measures the cosine of the angle between them. Pearson's correlation is essentially the same thing applied to centered vectors.

(iii) *Distance correlation* is also the cosine of an angle, but between two $n \times n$ double centered distance matrices in the Hilbert space of $n \times n$

matrices. There is no problem in defining angles in any Hilbert space via the inner product (x, y):

$$\cos(a) = (x/\|x\|, y/\|y\|),$$

where $\|x\| = (x, x)^{1/2}$.

As long as we want to define dissimilarity / dependence in metric spaces that are of strong negative type (embeddable into Hilbert spaces) we have no problem in defining dependence measures / similarities via cosine of angles.

(iv) *Earth mover's correlation.* For general metric spaces *eCor* can be a measure of similarity. In case of our triangle example in (i), $eCov = d/3$, thus d is essentially the Earth mover measure of dissimilarity. See also Deb et al. [2020], Deb and Sen [2021].

A

Historical Background

In 1927-28 two fundamental papers changed our views on matter and mind. Heisenberg's uncertainty principle [Heisenberg, 1927] led to a probabilistic description of matter (unlike Newton's deterministic laws), and von Neumann's minimax theorem of zero-sum games [von Neumann, 1928] led to a probabilistic description of our mind (see also *The Uncertainty Principle of Game Theory* in Székely and Rizzo [2007]). The energy perspective in this book reflects another type of duality, namely the duality between energy and mind. The name *energy of data* derives from Newton's gravitational potential energy, and there is an elegant relation to the notion of potential energy between statistical observations. Energy statistics are functions of distances between statistical observations in metrics paces. Thus even if the observations are complex objects, like functions, one can use their real-valued nonnegative distances for inference. In the "energy world" we do not work with the observations themselves, only with their distances or with their dissimilarities. Dissimilarities stored in our mind can be transformed to statistical energy for statistical inference. The history of statistical energy can be traced back to the non-statistical papers by Riesz [1938, 1949], who defined the α-energy of a probability measure μ as $E|X - X'|^\alpha$ where X, X' are independent and identically distributed with probability measure μ.

What makes Riesz α-energy applicable in statistics is the surprising property that if X, Y are independent and $0 < \alpha < 2$, then *double-centering* the Riesz energy, that is $2E|X - Y|^\alpha - E|X - X'|^\alpha - E|Y - Y'|^\alpha$, makes it always nonnegative and equal to 0 if and only if X and Y are identically distributed. This is what we exploited in this book many times. If we define the energy distance of independent random variables X and Y as $2E|X - Y|^\alpha - E|X - X'|^\alpha - E|Y - Y'|^\alpha$ whenever $0 < \alpha < 2$, then energy distance can be viewed as the potential energy of X with respect to Y or vice versa. This notion was first introduced in the Energy-Matter-Mind project (described in first chapter of this book) by G. J. Székely as a preparation for an invited lecture initiated by A. N. Kolmogorov in the early 1980s. The terms "statistical energy," "$\mathcal{E}$-statistic," and "energy of data" were coined in Székely's lectures at MIT, Yale, and Columbia University in 1985, where it was proved that for real-valued random variables the statistical energy is exactly twice Cramér's distance introduced in Cramér [1928]. In higher dimensions, however, the two distances are different because the energy distance is rotational invariant while Cramér's distance is not. The explanation of data energy with increasing details are found in e.g. Székely [2002], Székely and Rizzo [2004a, 2013b, 2017]. For statisticians it is comforting to see that the notion "energy of data" is not misleading for physicists; just the contrary, see e.g. Aslan and Zech [2005].

The history of Theorem 3.3 and especially the history of Lemma 2.1 goes back to Gel'fand and Shilov [1964] who showed that in the world of generalized functions, the Fourier transform of a power of a Euclidean distance is also a (constant multiple of a) power of the same Euclidean distance. See equations (12) and (19) [Gel'fand and Shilov, 1964, pp. 173–174] for the Fourier transform of $|x|^\alpha$. Thus, one can extend the validity of Lemma 2.1 using generalized functions, but Theorem 3.3 itself is not in Gel'fand and Shilov [1964]. The duality between powers of distances and their Fourier transforms is similar to the duality between probability density functions of random variables and their characteristic functions (especially of normal distributions whose probability density functions have the same form as their characteristic functions). This duality was called by Gauss a "beautiful theorem of probability theory" ("Schönes Theorem der Wahrscheinlichkeitrechnung" [Fischer, 2011, p. 46]). The proof of Theorem 3.2 in the univariate case appeared as early as 1989 [Székely, 1989]. An important special case of Theorem 3.2, namely $E|X + X'| \geq E|X - X'|$ for all real-valued X and X' with finite expectations, was a college level contest problem in Hungary in 1990 [Székely, 1996, p. 458]. On this inequality see more details in Buja et al. [1994]. This inequality and its statistical applications to testing diagonal symmetry was the topic of Székely's talk at the Lukacs Symposium organized by C. R. Rao with the title "Statistics for The 21st Century" (April, 1998). The conference volume is Rao and Székely [2000]. Székely and Móri [2001] has a more detailed version.

Russian mathematicians also published proofs of these theorems and their generalizations to metric spaces; see Klebanov [2005] and the references therein. See also Mattner [1997] and Morgenstern [2001] from the German school. On early applications see the test of bivariate independence of Feuerverger [1993], and the test of homogeneity of Székely [2002], Székely and Rizzo [2004a], and Baringhaus and Franz [2004]. For historical comments on "Hoeffding-type inequalities" and their generalizations see [Gneiting and Raftery, 2007, Section 5.2]. In 2000 Nail Bakirov joined the energy statistics research at Bowling Green State University for four months, and continued joint research on energy statistics until 2010, when he died suddenly after he was struck by an automobile while walking home from work in Ufa, Russia.

Distance correlation also has a fascinating history. Distance covariance of X and Y can be considered as a special case of energy distance when X is replaced by (X, Y) and Y is replaced by (X^*, Y^*) where X^* and Y^* are independent and have the same (marginal) distribution as X and Y, respectively.

But this is not how distance correlation was discovered. It was discovered as a combination of the Galton-Pearson covariance and the Brownian motion. Correlation and regression were discovered by Francis Galton in 1888 (see Stigler [1989]), the related mathematical formulae for covariance and correlation are due to Karl Pearson [Pearson, 1895]. These notions were invented to describe quantitatively the randomness in inheritance / mutations / evolution in biology described in the classical book "On the Origin of Species" in 1859 by Charles Darwin. Galton was Darwin's half-cousin. Correlation was very useful to describe linear or at least monotone type relationships like height and weight. It was clear from the beginning that non-linear, especially non-monotone type dependencies are typically not captured by the Galton-Pearson approach.

More than a hundred years later it turned out that we need to combine the Galton-Pearson approach with Brownian motions. In 1827 the botanist Robert Brown was studying pollen grains suspended in water under a microscope when he observed minute particles, ejected by the pollen grains, executing a jittery motion. By repeating the experiment with particles of inorganic matter he was able to rule out that the motion was life-related/biological in nature although its origin was yet to be explained. In 1905 Einstein wrote a paper with the title "Investigations on the theory of Brownian motion." This paper was one of his famous four papers published in 1905, in the "annus mirabilis." In layman's terms, in this paper Einstein explained Brownian motion with the help of atoms that hit the pollen from all directions; this idea supported the existence of hypothetical atoms. Einstein [1905] computed the variance of the Brownian motion at time t. The formula for this variance is Ct where the constant C helped to determine the size of invisible atoms and molecules indirectly from the visible Brownian motion diffusion. Remember that in 1905 outstanding scientists like Ernst Mach and Wilhelm Ostwald doubted the existence of atoms. (By the way, the other three 1905 papers of Einstein proved the existence of photons, described special relativity, and proved the famous $E = mc^2$ formula.) After Einstein, Brownian motion became a well-known concept in physics.

Almost twenty years later Brownian motion also became a fundamental notion in mathematics when Norbert Wiener constructed a mathematical model of Brownian motion in which the basic probabilities were the values of a measure defined on subsets of a space of continuous function realizations of Brownian motions. This measure has since been commonly called "Wiener measure" and the Brownian motion is called "Wiener process" denoted by $W(t)$. All this happened a decade before Kolmogorov's famous 1933 book was published on the axiomatization of probability theory based on measure theory.

In 2000 Székely raised the question of computing Pearson's covariance for $W(X)$ and $W'(Y)$ where W, W' are independent Wiener processes and the random variables X, Y are independent of them. Székely showed that if we combine Pearson's covariance (originally discovered to describe random biological processes of inheritance, mutation, and evolution) with Brownian/Wiener processes (originally applied to describe random diffusion in physics), then we get a revolutionary new type of dependence measure. Given W, W', the conditional covariance $cov(W(X), W'(Y))$, unlike the Galton-Pearson covariance, $cov(X, Y)$, can capture the independence of X and Y. Random variables X, Y with finite expectations are independent if and only if conditioned on W, W', $cov(W(X), W'(Y)) = 0$ with probability 1. This is equivalent to the equation $E[cov^2(W(X), W'(Y))] = 0$. The formula $E[cov^2(W(X), W'(Y))]$ was then called squared Brownian covariance.

It is almost unbelievable that Wiener processes, W, W' that are independent of X, Y can be relevant at all in measuring the dependence between X and Y. The heuristic explanation for this "miracle" is Paul Lévy's "forgery theorem": if you run a Brownian motion/Wiener process long enough, then it will write your signature (provided your signature is performed continuously) with a precision as high as you want it. Thus if the Brownian covariance is zero, then the covariances of all continuous functions of X, Y are also zero and this implies the independence of X and Y. In 2007 and 2009 two joint papers of Gábor Székely with Nail Bakirov and Maria Rizzo were published on distance correlation and Brownian correlation [Székely, Rizzo, and Bakirov, 2007, Székely and Rizzo, 2009a]. These papers explain technical and algorithmic details of these new notions including a very simple computing formula for empirical Brownian/distance correlations. The 2009 paper also explains why Brownian correlation and distance correlation are the same. Russell Lyons extended distance correlation to general metric space valued random variables, especially to Hilbert space valued variables, including Hilbert spaces of stochastic processes [Lyons, 2013]. The algorithmic complexity of the empirical distance correlation for bivariate data was decreased from the natural $O(n^2)$ to $O(n \log n)$ where n is the sample size [Huo and Székely, 2016]. Móri and Székely [2019] published a new "minimalist" set of four axioms for dependence measures; this system of axioms justifies the use of distance correlation and discourages the application of many other known measures of dependence.

In this century energy statistics and distance correlation have been extended and applied in nearly every direction of modern statistics and data science, thanks to the work of hundreds of contributors. It is truly inspiring to see all that has been accomplished and continues to develop. The story of energy statistics is still in its opening chapters.

B

Prehistory

CONTENTS

B.1 Introductory Remark

This Appendix is a copy of an 1985 unpublished manuscript of G.J. Székely, the starting point of applications of Thales' idea of *geometric similarities* in statistics, in data energy, especially in the axioms of dependence measures. Many classical statistical distances like Cramér's distance or Kolmogorov's distance are not rotation invariant. This observation led in the 1980s to the discovery of the rotation invariant energy distance, energy statistics, and data energy, a topic of Székely's lecture series in 1985 in Hungary, in Russia and in the U.S. The lectures started with the philosophy of similarity. We know similar triangles have the same angles, small circles are similar to large circles, circle and ellipse are less similar, etc., but what are the "least similar" objects? This led to the idea that "least similar" is the same as "independent." The classical correlation is the cosine of the angle between samples of size n in the n dimensional Euclidean space of samples. Distance correlation is the cosine of the angle between $n \times n$ matrices in the Hilbert space of $n \times n$ double centered distance matrices. Least similar is when the distance correlation is zero, which is the same as independence. Most similar or simply similar means that distance correlation is 1.

The history of statistics and in fact the history of exact sciences, started with the invention of number names about 6,000 years ago. Measurements of distances between cities and celestial bodies gave many excitements. Some of the measurements were easy, some are hard, e.g. the measurements of distance between ships in the ocean and between celestial bodies. Sophisticated measurements of distances via computing their ratios started with Thales of Miletus around 600 BC. He discovered the notion of similarity: when the shadow of a man has the same length as his height then the shadow of a pyramid also has the same length as its height thus, we can easily measure it. The corresponding triangles are similar. If we enlarge triangles, then the corresponding lengths of sides change but not their ratios. What remains intact or invariant is the angle. The notion of angle became fundamental in Thales' geometry. In statistics if instead of working with statistical data or statistical observations directly, e.g. adding, averaging, multiplying them as we do in classical statistics, if we want to understand the "soul" of data, then it is better to compute all distances between them and then work with these distances. From these distances we cannot tell the average height or weight, we cannot tell the average age but we can tell the dependencies of data, we can understand the "soul" of data because the forest of numbers does not mask the "soul." This is crucial in big data projects.

B.2   Thales and the Ten Commandments

Thales (624? BC –546? BC) is traditionally considered the first Western philosopher, the founder of abstract geometry and deductive mathematics. Thales is said to have attempted the calculation of the sizes of both sun and moon, and to have divided the year into 365 days. He is the first to be called Sage.

Thales lived in Miletus, Asia Minor, which belongs to Turkey today. Before the 6th century BC Miletus was the greatest Greek city in the east influenced by the neighboring Lydians who are the originators of gold and silver coins in the 7th century BC and the first people to establish permanent retail shops. The Greeks quickly adopted the metallic coinage, which played an important part in the commercial revolution that transformed the Greek civilization in the 6th century. According to Herodotus, Thales achieved his fame as a scientist for having predicted an eclipse of the sun on May 28, 585 BC. The eclipse stopped the six year war between the Lydians under Alyattes and the Medes under Cyaxares, one year after the occupation of Jerusalem and the destruction of Solomon's Temple by the Babylonian king Nebuchadnezzar in 586 BC. Diogenes Laertius states that ("according to Herodotus and Douris and Democritus") Thales was the son of Examyes and Cleobuline, who were, according to Herodotus and some authorities, of Phoenician origin. They lived after the destruction of Israel by the Assyrians in 722 BC, when all ten tribes of Israel were forced into exile. Even today nobody knows the fate of these tribes. Some of them might have settled from Assyria to Phoenicia where Aramaic was the spoken language. According to some researchers the word "Thales" comes from the Hebrew world "Thal" meaning "Dew" and the "-es" ending is typical in Greek, see e.g. Parmenides, Aristoteles, Euclides. According to some other researchers "Thales" comes from the Greek word "Thalla" meaning "to blossom." Diogenes Laertius connects Thales' family line back to the Phoenician prince Cadmos, brother of Phoenix. The root of the Hellenized word Cadmos is "qdm" which means "the East" in Phoenician, as well as in Hebrew. It is very likely that Thales had Semitic roots, and even if he himself was not Jewish, or Semitic (Phoenician), he must have heard a lot about the monotheism of the Israelis and of the Ten Commandments. He traveled a lot, visited Egypt where he learned many empirical geometric rules, and might have crossed Judea and Jerusalem, and finally, he himself, his parents and grandparents grew up in the period 722 BC–586 BC before Solomon's Temple was destroyed.

Why is this hypothetical Jewish background of Thales so interesting or important at all? In case of most scientists their nationality is almost completely irrelevant, what matters most are their scientific results. But in case of the first known Western scientist, it can be very important to see where his roots were. The roots or his educational sources might explain what he did and also why he did it. In this connection it is interesting to mention that, in his probably youthful work "On Piety," Theophrastus referred to the Jews as the "philosophical" race who at night prayed and observed the stars (the visible gods of Platonic theology). Clearchus also believed that in Syria, philosophers are called "Ioudaoi" because they inhabit the territory called "Ioudaia." In most cases, however, the Greeks naturally fused the various priests of the Aramaic-speaking East into a sacerdotal brotherhood of sages. The Jews were also placed among the sacerdotal sages of the Orient. The Greeks did not distinguish metaphysics from theology either. They compared the Egyptian priests, the Persian Magi to their own wise men. So it is natural to suppose that the peers of Thales were the Jewish prophets. Thales was a contemporary of Ezekiel, the Prophet. For Ezekiel, similarity in society was the basis for his prophecies, for Thales similarity in geometry was the basis for his predictions in nature.

This environment helps us to understand how mathematics might have been transformed from a collection of applicable empirical rules in Egypt and Babylonia to a deductive science. My claim is this: Thales viewed the whole empirical geometry of the Egyptians through the eyes of Jewish Prophets. *It was the example of the Jewish monotheism and the example of the Ten Commandments that directly influenced Thales' thinking in his approach to natural philosophy and geometry.* The continuation of Jewish theology is not the Greek theology (it has nothing to do with monotheism) but the Greek philosophy and the axiomatic science. Thales transplanted and transformed the supernatural and axiomatic approach of the Jewish religion to the natural philosophy of the Greeks. That is why Jewish scholars always felt Greek philosophers so close to their heart and at the same time so far from their religion. According to Aristotle, Thales was the founder, the "arkhegos" of Ionian natural philosophy. But there is nobody without predecessors.

Why was monotheism so important for the Jews that it was worth dying for? Many Jews sacrificed their lives in battles against Greeks in the 2nd century BC. All they wanted is to defend their only God. The importance of monotheism is this: if there is only one God and you do not follow the only God's will, then you cannot say that there is another God who actually approves or likes what you did, or committed (e.g. killings, say, sacrificing your own child in order to please your God, as it was common in Phoenicia, e.g. in Carthage). If there is only one God, then his will is an unquestionable "axiom." The first form of monotheism appeared in Egypt at the time of the Akhenaten (14th century BC). There are some researchers who say that Akhenaten is actually identical with Moses. In any case, some time between 1600 BC and 1200 BC monotheism became accepted by the Jewish people while in Egypt after Akhenaten the whole idea of monotheism disappeared. The idea of the only God was the first attempt for searching a unifying hypothesis for explaining our world. For the Jews this unifying hypothesis was supernatural. Many hundred years later Thales was the first to suggest that a single material substratum explains the whole universe; for Thales this substratum was the water or moisture. In geometry, Thales's unifying notion was the *angle* ("gonia"), a notion introduced by him. *The angle is Thales's most important contribution to geometry. With the help of angles he introduced a new geometry, which is based on similarity transformations that do not change angles.*

In this geometry two triangles are similar if they have the same angles but not necessarily the same sides. In similar triangles the *ratios* of their corresponding sides are the same. The Greek word for ratio is "logos," the equality of ratios is *"analogy"* thus in similar triangles the lengths of their corresponding sides form an "analogy." Following Thales, Pythagoras (570? BC-500? BC) studied other kinds of ratios, e.g. 12:8 = 9:6 is an "analogy." These "analogies" became the building blocks of Pythagoras' music theory and cosmogony (music of spheres), e.g. 12:8 = 9:6 = 3:2 is a "quint" ratio. The notion of angles and similarity applied to the shadows of gnomons helped Thales to compute astronomical distances and the distance of ships at sea from the shore. He also measured the Egyptian pyramids. All these results originated from Thales' abstract idea of geometric angles and similarity. He definitely deserved to be one of the Seven Wise Men of the Greeks.

Thales showed that

(i) angles at the base of a triangle having two sides equal length are equal,
(ii) opposite angles of intersecting straight lines are equal,
(iii) a triangle is determined if its base and their angles relative to the base are given,
(iv) the angle inscribed in a semicircle is a right angle = 'orthe gonia' (this claim is traditionally called "Thales' theorem").

The Sumerians, whose written code of laws dates back to 2500 BC were probably the first people to have a written code, but it lacked the passion of Justice of the Mosaic laws because they were not "democratic" they were not applicable to all without favoritism. Six hundred years later, the Sumerian code was augmented and incorporated by the Babylonians into the Code of Hammurabi, but again this Code did not have the "no exception" spirit of the Torah. The Ten Commandments is the first system of axioms of a written code, Torah, *applicable to all without exception.* This is exactly what mathematics is about: axioms and theorems with no exception. The prehistory of mathematics is in the Mosaic laws. The contents are different but the logical structures are the same. Accept a few axioms, do not question their validity, and prove or disprove everything with the help of these Commandments or axioms.

Indirect Proofs in mathematics were preceded by indirect proofs of judges. Just think of the following situation. According to the 6th Commandment "You shall not steal." Now if someone is accused of stealing something and he says "it was not me," then the judge must prove that the suspect is in fact a thief. He can prove it by showing that the claim that the suspect is innocent leads to a contradiction. Isn't this similar to the spirit of the proof of irrationality of the square root of 2? Doesn't the word "suspect" resemble the mathematical notion of "conjecture." If we can prove our claim, then the conjecture becomes a theorem, a suspect becomes a criminal.

In indirect proofs we need to deal with negations. The Ten Commandments lists 3 do-s and 7 don't-s. Don't-s were preferred by Jews because everything was allowed not specifically forbidden. Don't kill, Don't commit adultery, Don't steal, Don't bear false witness against your neighbor, etc. Otherwise you could do whatever you wanted. Compare Hillel who said 100 years before Jesus that "Do not do unto others what you don't want others to do unto you." And here is how Jesus said something similar: "Do unto others what you want others to do unto you."

Many scientists regard the Ten Commandments as an epitome of prophetic teachings, and they date the Ten Commandments some time after Amos, the first prophet having a book in the Bible, and king Hoshea (after 750 BC). Others say that the Ten Commandments are simply a summary of the legal and priestly traditions of Israel. As I see this is a complete misunderstanding of the situation. *The Ten Commandments is simply the first system of axioms*, not a summary. E.g. Euclid's axioms and postulates are not summaries of the book "Elements." The axioms are the starting points not the conclusions. The difference cannot be bigger. It is interesting to note that in the time of Euclid Alexandria was not only a center of Hellenism but, next to Jerusalem, was also home to the largest Jewish community in the world and it is likely that Jewish scientists contributed to the theory of geometry.

If both the Jewish monotheism and the Ten Commandments were known to Thales, it is easy to imagine that he wanted to build his natural philoso phy based on these ideas by "simply" transforming the supernatural beliefs of Jews to natural hypotheses, by transforming the commandments to axioms, suspects to conjectures, convincing juristic arguments to mathematical proofs. Thales himself could not finish the construction of a geometric system of axioms but he was the one who made the first steps in this direction. 300 years after Thales, Euclid's Elements became the Torah of geometers. Euclid has five postulates and five axioms ("common notions") as if they were commandments on two stone tablets. Euclid wanted to start the Elements with things that were self-evident and applicable without exception like the Ten Commandments.

Next to the Bible, the Elements may be the most translated, published and studied of all books produced in the Western world. Even today these books can show the right way to Justice and Truth in our society and in natural philosophy.

Addendum—many years later

According to Joong J. Fang, founding editor of Philosophia Mathematica, as his obituary in PM indicated, the driving force of his intellectual life was the question to which in this work I have given a completely new answer. Fang called it the Needham question because it was posed by the great scholar, Joseph Needham, of Chinese science: Why did science begin in the Greek civilization rather than in China? Fang was a Korean and deeply interested in and disturbed by this question. His answer was sociological and led him to develop the sociology of mathematics, a new field of his own invention. Mine is that the answer is divine.

Bibliography

J. Aaronson, R. Burton, H. Dehling, D. Gilat, T. Hill, and B. Weiss. Strong laws for L- and U-statistics. *Transactions of the American Mathematical Society*, 348:2845–2865, 1996.

G. Adelson-Velskii and E. M. Landis. An algorithm for the organization of information. *Proceedings of the USSR Academy of Sciences*, 146:263–266, 1962. (Russian) English translation by Myron J. Ricci in Soviet Math. Doklady, 3:1259–1263, 1962.

B. Afsari, A. Favorov, E. J. Fertig, and L. Cope. REVA: a rank-based multi-dimensional measure of correlation. *bioRxiv*, page 330498, 2018.

C. C. Agarwal, A. Hinnenburg, and D. A. Keim. On the surprising behavior of distance metrics in high dimensional space. In *International Conference on Database Theory*, pages 420–434, 2001. URL https://bib.dbvis.de/uploadedFiles/155.pdf.

P. K. Agarwal, K. Fox, D. Panigrahi, K.R. Varadarajan, and A. Xiao. Faster algorithms for the geometric transportation problem. In B. Aronov and M. J. Katz, editors, *33rd International Symposium on Computational Geometry (SoCG 2017)*, pages 7:1–7:16, 2017. URL https://doi.org/10.4230/LIPIcs.SoCG.2017.7.

A. Ajtai, J. Komlós, and G. Tusnády. On optimum matchings. *Combinatorica*, 4:259–264, 1984.

M. G. Akritas and S. F. Arnold. Fully nonparametric hypotheses for factorial designs I: Multivariate repeated measures designs. *Journal of the American Statistical Association*, 89(425):336–343, 1994.

A. V. Alekseyenko. Multivariate Welch t-test on distances. *Bioinformatics*, 32(23):3552–3558, 2016. doi: 10.1093/bioinformatics. PMID: 27515741.

J. Altschuler, J. Niles-Weed, and P. Rigollet. Near-linear time approximation algorithms for optimal transport via Sinkhorn iteration. *Advances in Neural Information Processing Systems*, 30, 2017. URL https://arxiv.org/pdf/1705.09634.pdf.

J. Altschuler, F. Bach, A. Rudi, and J. Niles-Weed. Massively scalable Sinkhorn distances via Nyström method. *Advances in Neural Information Processing Systems*, 32, 2019. URL https://arxiv.org/abs/1812.05189.

417

S. Amari and H. Nagaoka. *Methods of information geometry*, volume 191. American Mathematical Soc., 2007.

M. J. Anderson. A new method for non-parametric multivariate analysis of variance. *Austral Ecology*, 26(1):32–46, 2001.

M. J. Anderson. Permutational multivariate analysis of variance (PERMANOVA). *Wiley Statsref: Statistics Reference Online*, pages 1–15, 2014.

T. W. Anderson and D. A. Darling. A test of goodness of fit. *Journal of the American Statistical Association*, 49(268):765–769, 1954.

B. C. Arnold. *Pareto Distributions*. Chapman and Hall/CRC, 2015.

K. J. Arrow. *Social Choice and Individual Values*, volume 12. Yale University Press, 2012.

B. Aslan and G. Zech. Statistical energy as a tool for binning-free, multivariate goodness-of-fit test, two-sample comparison and unfolding. *Nuclear Instruments and Methods in Physics Research Section A: Accelerators, Spectrometers, Detectors and Associated Equipment*, 537(3):626–636, 2005.

K. Baba, R. Shibata, and M. Sibuya. Partial correlation and conditional correlation as measures of conditional independence. *Australian & New Zealand Journal of Statistics*, 46(4):657–664, 2004. doi: DOI10.1111/j.1467-842X.2004.00360.x.

N. K. Bakirov and G. J. Székely. Brownian covariance and central limit theorem for stationary sequences. *Theory of Probability & Its Applications*, 55 (3):371–394, 2011.

N. K. Bakirov, M. L. Rizzo, and G. J. Székely. A multivariate nonparametric test of independence. *Journal of Multivariate Analysis*, 97(8):1742–1756, 2006.

L. Baringhaus and C. Franz. On a new multivariate two-sample test. *Journal of Multivariate Analysis*, 88(1):190–206, 2004.

J. Beck. Sums of distances between points on a sphere — an application of the theory of irregularities of distribution to discrete geometry. *Mathematika*, 31(1):33–41, 1984.

M. G. Bellemare, I. Danihelka, W. Dabney, S. Mohamed, B. Lakshminarayanan, S. Hoyer, and R. Munos. The Cramér distance as a solution to biased Wasserstein gradients. *arXiv preprint arXiv:1705.10743*, 2017.

C. H. Bennett, P. Gacs, M. Li, P. M. B. Vitanyi, and W. Zurek. Information distance. *IEEE Transactions on Information Theory*, 44(4):1407–1423, 1998.

C. Berg. Stieltjes-Pick-Bernstein-Schoenberg and their connection to complete monotonicity. In J. Mateu and E. Porcu, editors, *Positive Definite Functions: From Schoenberg to Space-Time Challenges*. Department of Mathematics, University Jaume I, Castellon, Spain, 2008.

C. Berg, J. P. R. Christensen, and P. Ressel. *Harmonic Analysis on Semigroups. Theory of Positive Definite and Related Functions*, volume 100 of *Graduate Texts in Mathematics*. Springer-Verlag, Berlin-Heidelberg-New York, 1984.

G. Berschneider and B. Böttcher. On complex Gaussian random fields, Gaussian quadratic forms and sample multivariance. https://arxiv.org/abs/1808.07280, 2018.

A. Betken, H. Dehling, and M. Kroll. Block bootstrapping the empirical distance covariance, 2021.

P. J. Bickel and Y. Xu. Discussion of: Brownian distance covariance. *The Annals of Applied Statistics*, 3(4):1266–1269, 2009.

P. J. Bickel, F. Götze, and W. R. van Zwet. The Edgeworth expansion for U-statistics of degree two. *Annals of Statistics*, 14(4):1463–1484, 1986.

N. H. Bingham, A. Mijatović, and T. L. Symons. Brownian manifolds, negative type and geo-temporal covariances. *Communications on Stochastic Analysis*, 10(4):3, 2016. URL https://arxiv.org/abs/1612.06431.

P. E. Black. AVL tree. In Vreda Pieterse and Paul E. Black, editors, *Dictionary of Algorithms and Data Structures [online]*. National Institute of Standards and Technology (NIST), 2014. URL https://www.nist.gov/dads/HTML/avltree.html.

C. Blake and C. J. Merz. UCI repository of machine learning databases, 1998.

A. J. Blumberg, P. Bhaumik, and S. G. Walker. Testing to distinguish measures on metric spaces. *arXiv preprint arXiv:1802.01152*, 2018.

S. Bochner. Monotone funktionen, Stieltjes integrale und harmonische analyse. *Mathematische Annalen*, 108:378–410, 1933.

B. Böttcher, M. Keller-Ressel, and R. Schilling. Detecting independence of random vectors: generalized distance covariance and Gaussian covariance. *Modern Stochastics: Theory and Applications*, 5(3):353–383, 2018. ISSN 2351-6046. doi: 10.15559/18-vmsta116. URL http://dx.doi.org/10.15559/18-VMSTA116.

B. B. Böttcher, M. Keller-Ressel, and R. L. Schilling. Distance multivariance: New dependence measures for random vectors. *Annals of Statistics*, 47(5):2757–2789, 2019.

A. Bowman and A. Azzalini. *Applied Smoothing Techniques for Data Analysis: The Kernel Approach with S-Plus Illustrations.* Oxford University Press, Oxford, 1997.

A. W. Bowman and A. Azzalini. *R package sm: nonparametric smoothing methods (version 2.2-5.7).* University of Glasgow, UK and Università di Padova, Italia, 2021. URL http://www.stats.gla.ac.uk/~adrian/sm/.

R. C. Bradley. Central limit theorems under weak dependence. *Journal of Multivariate Analysis,* 11(1):1–16, 1981.

R. C. Bradley. A central limit theorem for stationary ρ-mixing sequences with infinite variance. *The Annals of Probability,* pages 313–332, 1988.

R. C. Bradley. *Introduction to Strong Mixing Conditions.* Kendrick Press, Heber City, Utah, 2007.

L. Breiman and J. H. Friedman. Estimating optimal transformations for multiple regression and correlation. *Journal of the American Statistical Association,* 80(391):580–598, 1985.

J. Bretagnolle, D. Dacunha Castelle, and J-L. Krivine. Lois stables et espaces Lp. *Annales de l'I.H.P. Probabilités et statistiques,* 2(3):231–259, 1966.

L. Brillouin. *Science and Information Theory.* Dover, second edition, 2004.

A. Buja, B. F. Logan, J. A. Reeds, and L. A. Shepp. Inequalities and positive definite functions arising from a problem in multidimensional scaling. *Annals of Statistics,* 22(1):406–438, 1994.

B. Böttcher. Dependence and dependence structures: estimation and visualization using the unifying concept of distance multivariance. *Open Statistics,* 1(1):1–48, Dec 2019. ISSN 2657-3601. doi: 10.1515/stat-2020-0001. URL http://dx.doi.org/10.1515/stat-2020-0001.

F. Cailliez. The analytical solution of the additive constant problem. *Psychometrika,* 48:343–349, 1983.

N. N. Cencov. *Statistical Decision Rules and Optimal Inference.* Number 53 in Translations of Mathematical Monographs. American Mathematical Society, 2000.

S. Chakraborty and X. Zhang. Distance metrics for measuring joint dependence with application to causal inference. *Journal of the American Statistical Association,* 114:1638–1650, 2019.

S. Chatterjee. A new coefficient of correlation. *Journal of the American Statistical Association,* 116:2009–2022, 2021.

X. Chen, X. Chen, and H. Wang. Robust feature screening for ultra-high dimensional right censored data via distance correlation. *Computational Statistics & Data Analysis*, 2017.

V. Chernozhukov, A. Galichon, M. Hallin, and M. A. Henry. Monge-Kantorovich depth, quantiles, ranks and signs. *Annals of Statistics*, 45(1): 223–256, 2017. arXiv:1412.8434v4 2015.

J. S. Cho and H. White. Generalized runs tests for the iid hypothesis. *Journal of Econometrics*, 162(2):326–344, 2011.

D. Christensen. Fast algorithms for the calculation of Kendall's τ. *Computational Statistics*, 20(1):51–62, 2005.

W. G. Cochran. The distribution of quadratic forms in a normal system, with applications to the analysis of covariance. *Mathematical Proceedings of the Cambridge Philosophical Society*, 30:178–191, 1934. doi: 10.1017/S0305004100016595.

L. Cope. Discussion of: Brownian distance covariance. *The Annals of Applied Statistics*, 3(4):1279–1281, 2009.

B. Cowley, J. Semedo, A. Zandvakili, M. Smith, A. Kohn, and B. Yu. Distance covariance analysis. In *Artificial Intelligence and Statistics*, pages 242–251, 2017.

T. F. Cox and M. A. A. Cox. *Multidimensional Scaling*. Chapman and Hall, second edition, 2001.

H. Cramér. On the composition of elementary errors: II, statistical applications. *Scandinavian Actuarial Journal*, 11:141–180, 1928.

M. W. Crofton. On the theory of local probability, applied to straight lines drawn at random in a plane; the methods used being also extended to the proof of certain new theorems in the integral calculus. *Philosophical Transactions of the Royal Society of London*, 158:181–199, 1868.

I. Csiszár. Eine informationstheoretische Ungleichung und ihre Anwendung auf den Beweis der Ergodizität von Markoffschen Ketten. *Magyar Tudományos Akadémia Matematikai Kutató Intézet Közleményei*, 8:85–108, 1963.

I. Csiszár. I-divergence geometry of probability distributions and minimization problems. *Annals of Probability*, 3(1):146–158, 1975.

I. Csiszár. Why least squares and maximum entropy? An axiomatic approach to inference for linear inverse problems. *Annals of Statistics*, 19(4):2032–2066, 1991.

S. Csörgő. Testing for normality in arbitrary dimension. *Annals of Statistics*, 14(2):708–723, 1986.

T. Daniyarov. *VaR: Value at Risk Estimation. R package version 0.2*, 2004. URL https://cran.r-project.org/src/contrib/Archive/VaR/.

R. A. Davis, M. Matsui, T. Mikosch, and P. Wan. Applications of distance correlation to time series. *Bernoulli*, 24(4A):3087–3116, 2018.

A. C. Davison and D. V. Hinkley. *Bootstrap Methods and their Application*. Cambridge University Press, Oxford, 1997.

B. de Finetti. La prévision: ses lois logiques, ses sources subjectives. *Annales de l'Institut Henri Poincaré*, 7:1–68, 1937.

N. Deb and B. Sen. Multivariate rank-based distribution-free nonparametric testing using measure transportation. *Journal of the American Statistical Association*, 2021.

N. Deb, P. Ghoshal, and B. Sen. Measuring association on topological spaces using kernels and geometric graphs. https://arxiv.org/abs/2010.01768, 2020.

P. Deheuvels. La fonction de dependance empirique et ses propriétés: Un test non paramétrique d'indépendance. *Bulletin de la Classe des Sciences, V. Serie, Académie Royale de Belgique*, 65:274–292, 1979.

H. Dehling, M. Matsui, T. Mikosch, G. Samorodnitsky, and L. Tafakori. Distance covariance for discretized stochastic processes. *Bernoulli*, 26(4):2758–2789, 2020. arXiv preprint arXiv:1806.09369.

M. M. Deza and E. Deza. *Encyclopedia of Distances, 4th revised edition*. Springer, New York, 2016.

M. M. Deza and M. Laurent. *Geometry of Cuts and Metrics*, volume 15 of *Algorithms and Combinatorics*. Springer-Verlag, Berlin, 1997.

P. Diaconis and D. Freedman. Finite exchangeable sequences. *The Annals of Probability*, 8(4):745–764, 1980.

D. Donoho. 50 years of data science. *Journal of Computational and Graphical Statistics*, 26(4):745–766, 2017.

L. E. Dor. Potentials and isometric embeddings in L1. *Israel Journal of Mathematics*, 24:260–268, 1976.

P. Dubey and H.-G. Müller. Fréchet analysis of variance for random objects. *Biometrika*, 106(4):803–821, 2019.

P. Duchesne and P. Lafaye de Micheaux. Computing the distribution of quadratic forms: Further comparisons between the Liu-Tang-Zhang approximation and exact methods. *Computational Statistics and Data Analysis*, 54:858–862, 2010.

J. Dueck, D. Edelmann, T. Gneiting, and D. Richards. The affinely invariant distance correlation. *Bernoulli*, 20(4):2305–2330, 2014.

J. Dueck, D. Edelmann, and D. Richards. Distance correlation coefficients for Lancaster distributions. *Journal of Multivariate Analysis*, 154:19–39, 2017.

B. Ebner and N. Henze. Tests for multivariate normality – a critical review with emphasis on weighted L^2-statistics. *TEST*, 29:845–892, 2020. doi: https://doi.org/10.1007/s11749-020-00740-0.

K. Eckerle. NIST circular interference transmittance study. *National Institute of Standards and Technology*, 1979.

D. Edelmann, K. Fokianos, and M. Pitsillou. An updated literature review of distance correlation and its applications to time series. *International Statistical Review*, 87(2):237–262, 2019.

D. Edelmann, D. Richards, and D. Vogel. The distance standard deviation. *Annals of Statistics*, 48(6):3395–3416, 2020.

D. Edelmann, T. F. Móri, and G. J. Székely. On relationships between the Pearson and the distance correlation coefficients. *Statistics & Probability Letters*, 169:108960, 2021.

F. Y. Edgeworth. Correlated averages. *Philosophical Magazine, 5th series*, 34:190–204, 1892.

J. Edmonds and R. M. Karp. Theoretical improvements in algorithmic efficiency for network flow problems. *Journal of the ACM*, 19:248–264, 1972. URL https://doi.org/10.1145/321694.321699.

B. Efron and R. J. Tibshirani. *An Introduction to the Bootstrap*. Chapman & Hall/CRC, Boca Raton, FL, 1993.

A. Einstein. Ist die trägheit eines körpers von seinem energieinhalt abhängig? *Annalen der Physik*, 18(13):639–643, 1905.

P. Erdős and A. Rényi. On a new law of large numbers. *Journal d'Analyse Mathématique*, 23:103–111, 1970.

P. A. Ernst, L. A. Shepp, and A. J. Wyner. Yule's "nonsense correlation" solved! *The Annals of Statistics*, 45(4):1789–1809, 2017. doi: 10.1214/16-AOS1509. URL https://projecteuclid.org/euclid.aos/1498636874.

Y. Escoufier. Le traitement des variables vectorielles. *Biometrics*, 29(4):751–760, 1973.

P. M. Esfahani and D. Kuhn. Data-driven distributionally robust optimization using the Wasserstein metric: Performance guarantees and tractable reformulations. *Mathematical Programming*, 171(1):115–166, 2018. URL https://arxiv.org/abs/1505.05116.

J. Fan, Y. Feng, and L. Xia. A projection-based conditional dependence measure with applications to high-dimensional undirected graphical models. *Journal of Econometrics*, 218(1):119–139, 2020.

Y. Fan, P. L. de Micheaux, S. Penev, and D. Salopek. Multivariate nonparametric test of independence. *Journal of Multivariate Analysis*, 153:189–210, 2017.

K.-T. Fang and Y. Wang. *Number Theoretic Methods in Statistics*. Chapman and Hall, London, 1994.

Y. Farjami, D. Rahbari, and E. Hosseini. Distance correlation between plaintext and hash data by genetic algorithm. *Pertanika Journal of Science & Technology*, 26(3), 2018.

G. Fejes-Tóth and L. Fejes-Tóth. Dictators on a planet. *Studia Scientiarum Mathematicarum Hungarica*, 15:313–316, 1980.

M. Fekete. Über die Verteilung der Wurzeln bei gewissen algebraischen Gleichungen mit ganzzahligen Koeffizienten. *Mathematische Zeitschrift*, 17(1):228–249, 1923.

A. Feragen and A. Fuster. Geometries and interpolations for symmetric positive definite matrices. In *Modeling, Analysis, and Visualization of Anisotropy*, pages 85–113. Springer, 2017.

A. Feragen, F. Lauze, and S. Hauberg. Geodesic exponential kernels: when curvature and linearity conflict. In *IEEE Conference on Computer Vision and Pattern Recognition, CVPR*, 2015.

T. Ferenci, A. Körner, and L. Kovács. The interrelationship of HbA1c and real-time continuous glucose monitoring in children with type 1 diabetes. *Diabetes Research and Clinical Practice*, 108(1):38–44, 2015.

A. Feuerverger. A consistent test for bivariate dependence. *International Statistical Review/Revue Internationale de Statistique*, 61:419–433, 1993.

A. Feuerverger. Discussion of: Brownian distance covariance. *The Annals of Applied Statistics*, 3(4):1282–1284, 2009.

H. Fischer. *A History of the Central Limit Theorem: From Classical to Modern Probability Theory*. Springer, 2011.

K. Fokianos and M. Pitsillou. Testing independence for multivariate time series via the auto-distance correlation matrix. *Biometrika*, 105:337–352, 2018.

J. Fox and S. Weisberg. *An R Companion to Applied Regression.* Sage, Thousand Oaks, CA, third edition, 2019. URL https://socialsciences. mcmaster.ca/jfox/Books/Companion/.

G. França, M. L. Rizzo, and J. T. Vogelstein. Kernel k-groups via Hartigan's method. *IEEE Transactions on Pattern Analysis and Machine Intelligence*, 43(12):4411–4425, 2020. URL https://ieeexplore.ieee.org/document/ 9103121.

M. Fréchet. Sur quelques points du calcul fonctionnel. *Rendiconti del Circolo Mathematico di Palermo*, 22:1–74, 1906.

M. Fréchet. Les éléments aléatoires de natures quelconque dans un espace distancié. *Annales de l'I.H.P.*, 10(4):215–310, 1948.

M. Fréchet. Sur la distance de deux lois de probabilité. *Comptes Rendus Hebdomadaires des Seances de l Academie des Sciences*, 244(6):689–692, 1957.

J. L. Freedman. *Crowding and Behavior.* WH Freedman, 1975.

P. Gács, J. Tromp, and P. M. B. Vitányi. Algorithmic statistics. *IEEE Transactions on Information Theory*, 47(6):2443–2463, 2001.

M. Galassi, J. Davies, J. Theiler, B. Gough, G. Jungman, P. Alken, M. Booth, and F. Rossi. *Reference Manual. Edition 1.4, for GSL Version 1.4*, 2003. URL https://inspirehep.net/files/ 6003fe0f9876d72f25d0db0aa6348248.

F. Galton. Co-relations and their measurement, chiefly from anthropometric data. *Proceedings of the Royal Society of London*, 45:135–145, 1888.

R. Gao and A. J. Kleywegt. Distributionally robust stochastic optimization with Wasserstein distance. *Mathematics of Operations Research*, 2022. URL https://arxiv.org/abs/1604.02199. Published Online:5 Aug 2022 https://doi.org/10.1287/moor.2022.1275.

I. M. Gel'fand and G. E. Shilov. *Generalized Functions, Volume I: Properties and Operations (translation by E. Salatan of the Russian edition of 1958).* Academic Press, New York, 1964.

C. Genest and B. Rémillard. Test of independence and randomness based on the empirical copula process. *TEST: An Official Journal of the Spanish Society of Statistics and Operations Research*, 13(2):335–369, 2004. URL https://EconPapers.repec.org/RePEc:spr:testjl:v:13: y:2004:i:2:p:335-369.

C. R. Genovese. Discussion of: Brownian distance covariance. *The Annals of Applied Statistics*, 3(4):1299–1302, 2009.

A. Genz and F. Bretz. *Computation of Multivariate Normal and t Probabilities*. Lecture Notes in Statistics. Springer-Verlag, Heidelberg, 2009. ISBN 978-3-642-01688-2.

C. Gini. Variabilità e mutabilità. *Reprinted in Memorie di metodologica statistica (Ed. Pizetti E, Salvemini, T)*, 1, 1912.

G. Girone and A. M. D'Uggento. About the mean difference of the inverse normal distribution. *Applied Mathematics*, 7(14):1504–1509, 2016.

T. Gneiting and A. E. Raftery. Strictly proper scoring rules, prediction, and estimation. *Journal of the American Statistical Association*, 102(477):359–378, 2007.

S. C. Goslee and D. L. Urban. The ecodist package for dissimilarity-based analysis of ecological data. *Journal of Statistical Software*, 22(7):1–19, 2007.

F. Götze. Asymptotic expansions for bivariate von Mises functionals. *Zeitschrift für Wahrscheinlichkeitstheorie und verwandte Gebiete*, 50(3): 333–355, 1979.

F. Q. Gouvêa. Was Cantor surprised? *American Mathematical Monthly*, 118: 198–209, 2011.

J. C. Gower. Some distance properties of latent root and vector methods used in multivariate analysis. *Biometrika*, 53:25–328, 1966.

U. Grenander. *Probabilities on Algebraic Structures*. Courier Corporation, 2008.

A. Gretton and L. Györfi. Consistent nonparametric tests of independence. *Journal of Machine Learning Research*, 11:1391–1423, 2010.

A. Gretton and L. Györfi. Strongly consistent nonparametric test of conditional independence. *Statistics and Probability Letters*, 82:1145–1150, 2012.

A. Gretton, K. Fukumizu, and B. K. Sriperumbudur. Discussion of: Brownian distance covariance. *The Annals of Applied Statistics*, 3(4):1285–1294, 2009.

P. Grunwald and P. Vitányi. Shannon information and Kolmogorov complexity. *arXiv preprint cs/0410002*, 2010. URL https://homepages.cwi.nl/~paulv/papers/info.pdf.

J. C. Guella. Generalization of the energy distance by Bernstein functions. *Journal of Theoretical Probability*, pages 1–28, 2022. URL https://arxiv.org/abs/2102.00633.

E. J. Gumbel. Multivariate exponential distributions. *Bulletin of the International Statistical Institute*, 39:469–475, 1961.

X. Guo, Y. Zhang, W. Hu, H. Tan, and X. Wang. Inferring nonlinear gene regulatory network from gene expression data based on distance correlation. *PLoS ONE*, 9(2), 2014. doi: 10.1371/journal.pone.0087446. e87446.

N. Gürtler and N. Henze. Goodness-of-fit tests for the Cauchy distribution based on the empirical characteristic function. *Annals of the Institute of Statistical Mathematics*, 52(2):267–286, 2000.

H. A. Güvenir, G. Demiröz, and N. Ilter. Learning differential diagnosis of erythemato-squamous diseases using voting feature intervals. *Artificial Intelligence in Medicine*, 13(3):147–165, 1998.

M. Hallin. On distribution and quantile functions, ranks, and signs in R^d: a measure transportation approach, 2017. URL https://ideas.repec.org/p/eca/wpaper/2013-258262.html.

M. Hallin, E. del Barrio, J. A. Cuesta-Albertos, and C. Matrán. Distribution and quantile functions, ranks, and signs in dimension d: a measure transportation approach. *Annals of Statistics*, 49(2), 2021.

J. Haman. *The energy goodness-of-fit test and E-M type estimator for asymmetric Laplace distributions*. PhD thesis, Bowling Green State University, Bowling Green, Ohio, 2018.

R. K. S. Hankin. Special functions in R: introducing the gsl package. *R News*, 6, October 2006.

D. P. Hardin and E. B. Saff. Discretizing manifolds via minimum energy points. *Notices of the American Mathematical Society*, 51(10):1186–1194, 2004.

J. A. Hartigan and M. A. Wong. Algorithm AS 136: A k-means clustering algorithm. *Applied Statistics*, 28(1):100–108, 1979.

T. Hastie, R. Tibshirani, and J. Friedman. *Elements of Statistical Learning*. Springer, New York, second edition, 2009. URL https://hastie.su.domains/ElemStatLearn/.

W. Heisenberg. Über den anschaulichen Inhalt der quantummmechanischen Kinematic und Mechanic. *Zeitschrift für Physik*, 43:172–198, 1927.

R. Heller, Y. Heller, and M. Gorfine. A consistent multivariate test of association based on ranks of distances. *Biometrika*, 100(2):503–510, 2013.

B. Hemmateenejad and K. Baumann. Screening for linearly and nonlinearly related variables in predictive cheminformatic models. *Journal of Chemometrics*, Volume 32 Issue 4 page numbers "e3009" Wiley Online Library, 2018.

N. Henze. On Mardia's kurtosis test for multivariate normality. *Communications in Statistics–Theory and Methods*, 23(4):1031–1045, 1994.

N. Henze and T. Wagner. A new approach to the BHEP tests for multivariate normality. *Journal of Multivariate Analysis*, 62:1–23, 1997.

N. Henze and B. Zirkler. A class of invariant and consistent tests for multivariate normality. *Communications in Statistics: Theory Methods*, 19: 3595–3617, 1990.

E. Herbin and E. Merzbach. The multiparameter fractional brownian motion. In G. Aletti, A. Micheletti, D. Morale, and M. Burger, editors, *Math Everywhere: Deterministic and Stochastic Modelling in Biomedicine, Economics and Industry. Dedicated to the 60th Birthday of Vincenzo Capasso*, pages 93–101. Springer, Berlin, Heidelberg, 2007. ISBN 978-3-540-44446-6.

H. Herwartz and S. Maxand. Nonparametric tests for independence: a review and comparative simulation study with an application to malnutrition data in India. *Statistical Papers*, 61(5):2175–2201, 2020. URL https://EconPapers.repec.org/RePEc:spr:stpapr:v:61: y:2020:i:5:d:10.1007_s00362-018-1026-9.

P. Hjorth, P. Lisonek, S. Markvorsen, and C. Thomassen. Finite metric spaces of strictly negative type. *Linear Algebra and Its Applications*, 270:255–273, 1998.

W. Hoeffding. Scale-invariant correlation theory (English translation). *Schriften des Mathematischen Seminars Institute und des Instituts für Angewandte Mathematik der Universität Berlin*, 5(3):181–233, 1940.

W. Hoeffding. A class of statistics with asymptotic normal distribution. *Annals of Mathematical Statistics*, 19(3):293–325, 1948.

Y. Hong. Consistent testing for serial correlation of unknown form. *Econometrica: Journal of the Econometric Society, Volume 64*, pages 837–864, 1996.

Y. Hong. Hypothesis testing in time series via the empirical characteristic function: a generalized spectral density approach. *Journal of the American Statistical Association*, 94(448):1201–1220, 1999.

H. R. Horn. On necessary and sufficient conditions for an infinitely divisible distribution to be normal or degenerate. *Z. Wahrscheinlichkeitstheorie v. Geb.*, 21:179–187, 1972.

A. M. Horst, A. P. Hill, and K. B. Gorman. *palmerpenguins: Palmer Archipelago (Antarctica) penguin data*, 2020. URL https:// allisonhorst.github.io/palmerpenguins/. R package version 0.1.0.

L. Horváth and P. Kokoszka. *Inference for Functional data with Applications.* Springer, New York, 2012.

L. Horváth, M. Hušková, and G. Rice. Test of independence for functional data. *Journal of Multivariate Analysis*, 117:100–119, 2013.

W. Hua, T. E. Nichols, and D. Ghosh. Multiple comparision procedures for neuroimaging genomewide association studies. *Biostatistics*, 16(1):17–30, 2015.

C. Huang and X. Huo. A statistically and numerically efficient independence test based on random projections and distance covariance. *ArXiv e-prints*, 2017. URL https://arxiv.org/abs/1701.06054v1.

C. Huang, V. Roshan Joseph, and Simon Mak. Population quasi-Monte Carlo. *Journal of Computational and Graphical Statistics*, pages 1–14, 2022. URL https://arxiv.org/abs/2012.13769.

J. Huber. Partial and semipartial correlation–a vector approach. *The Two-Year College Mathematics Journal*, 12(2):151–153, 1981.

X. Huo and G. J. Székely. Fast computing for distance covariance. *Technometrics*, 58(4):435–447, 2016.

J-P. Imhof. Computing the distribution of quadratic forms in normal variables. *Biometrika*, 48(3/4):419–426, 1961.

M. E. Jakobsen. Distance covariance in metric spaces: Non-parametric independence testing in metric spaces (master's thesis). *arXiv preprint arXiv:1706.03490*, 2017.

Sz. Jaroszewitz and Z. Lukasz. Székely regularization for uplift modeling. In *Studies in Computational Intelligence*, volume 605, pages 135–154. Springer, 2015.

Z. Jin and D. S. Matteson. Generalizing distance covariance to measure and test multivariate mutual dependence via complete and incomplete V-statistics. *Journal of Multivariate Analysis*, 168:304–322, 2018.

Z. Jin, S. Yao, D. S. Matteson, and X. Shao. *EDMeasure: Energy-Based Dependence Measures*, 2018. R package version 1.2.

M. E. Johnson. *Multivariate Statistical Simulation.* Wiley, New York, 1987.

N. L. Johnson, S. Kotz, and N. Balakrishnan. *Continuous Univariate Distributions*, volume 2. John Wiley & Sons, second edition, 1995. ISBN 0-471-58494-0.

V. R. Joseph and A. Vakayil. SPlit: An optimal method for data splitting. *Technometrics*, 64(2):166–176, 2022. URL https://www.tandfonline.com/doi/full/10.1080/00401706.2021.1921037.

J. Josse and S. Holmes. Measures of dependence between random vectors and tests of independence. Literature review, 2014.

J. Josse and S. Holmes. Measuring multivariate association and beyond. *Statistics Surveys*, 10:132, 2016. URL https://www.ncbi.nlm.nih.gov/pmc/articles/PMC5658146/.

L. Kantorovich. On the translocation of masses. *C.R. (Doklady) Proceedings of the USSR Academy of Sciences (N.S.)*, 31(37):199–201, 1942.

L. Kantorovich. On a problem of Monge (in Russian). *Uspekhi Matematicheskikh Nauk*, 3:225–226, 1948.

G. J. Kerns and G. J. Székely. De Finetti's theorem for abstract finite exchangeable sequences. *Journal of Theoretical Probability*, 19(3):589–608, 2006.

A. Y. Kim, C. Marzban, D. B. Percival, and W. Stuetzle. Using labeled data to evaluate change detectors in a multivariate streaming environment. *Signal Processing*, 89(12):2529–2536, 2009.

Seongho Kim. *ppcor: Partial and Semi-Partial (Part) Correlation*, 2015. URL https://CRAN.R-project.org/package=ppcor. R package version 1.1.

G. Kimeldorf and A. R. Sampson. Monotone dependence. *Annals of Statistics*, 6:895–903, 1978.

L. Klebanov. *$\mathcal{N}$-Distances and Their Applications*. Charles University, Prague, 2005.

P. Kleinschmidt and H. Schannath. A strongly polynomial algorithm for the transportation problem. *Mathematical Programming*, 68:1–13, 1995. URL https://doi.org/10.1007/BF01585755.

W. R. Knight. A computer method for calculating Kendall's tau with ungrouped data. *Journal of the American Statistical Association*, 61(314):436–439, 1966.

A. Koldobsky and Y. Lonke. A short proof of Schoenberg's conjecture on positive definite functions. *Bulletin of the London Mathematical Society*, 31(6):693–699, 1999.

A. N. Kolmogorov. Sulla determinazione empirica di una legge di distribuzione. *Giornale dell'Instituto Italiano degli Attuari*, 4:83–91, 1933.

A. N. Kolmogorov. Three approaches to the quantitative definition of information. *Problems Information Transmission*, 1(1):1–7, 1965.

V. Koltchinskii and E. Giné. Random matrix approximation of spectra of integral operators. *Bernoulli*, 6(1):113–167, 2000.

V. S. Koroljuk and Yu. V. Borovskich. *Theory of U-statistics.* Mathematics and its Applications. 273. Kluwer Academic Publishers Group, Dordrecht, 1994. (Translated by P. V. Malyshev and D. V. Malyshev from the 1989 Russian original ed.).

M. R. Kosorok. Discussion of: Brownian distance covariance. *The Annals of Applied Statistics*, 3(4):1270–1278, 2009.

S. Kotz, N. Balakrishnan, and N. L. Johnson. *Continuous Multivariate Distributions, Volume 1.* John Wiley & Sons, second edition, 2000.

S. Kotz, T. .J Kozubowski, and K. Podgórski. *The Laplace Distribution and Generalizations.* Springer, 2001.

J. Kowalski and X. M. Tu. *Modern Applied U-statistics*, volume 714. John Wiley & Sons, 2008.

H. W. Kuhn. The Hungarian method for the assignment problem. *Naval Research Logistics Quarterly*, 2:83–97, 1955. URL https://doi.org/10.1002/nav.3800020109.

S. Kullbach and R. A. Leibler. On information and sufficiency. *Annals of Mathematical Statistics*, 22:79–86, 1951.

H. H. Kuo. *Gaussian measures in Banach spaces*, volume 463 of *Lecture Notes in Mathematics*. Springer-Verlag, Berlin-New York, 1975.

H. J. Landau and L. A. Shepp. On the supremum of a Gaussian process. *Sankhyā: The Indian Journal of Statistics, Series A*, pages 369–378, 1970.

A. J. Lee. *U-statistics: Theory and Practice.* Routledge, 2019.

Y. Lee, C. Shen, C. E Priebe, and J. T. Vogelstein. Network dependence testing via diffusion maps and distance-based correlations. *Biometrika*, 106 (4):857–873, 2019.

P. Legendre. Comparison of permutation methods for the partial correlation and partial Mantel tests. *J. Statist. Comput. Simul.*, 67:37–73, 2000.

P. Legendre and L. Legendre. *Numerical Ecology.* Elsevier, third English edition, 2012.

E. L. Lehmann. *Theory of Point Estimation.* John Wiley, New York, 1983.

J. Lei. Convergence and concentration of empirical measures under wasserstein distance in unbounded functional spaces. *Bernoulli*, 26(1):767–798, 2020.

R. V. Lenth. Some properties of U-statistics. *The American Statistician*, 37 (4):311–313, 1983.

A. Leucht and M. H. Neumann. Degenerate U- and V-statistics under ergodicity: Asymptotics, bootstrap and applications in statistics. *Annals of the Institute of Statistical Mathematics*, 65:349–386, 2013.

P. Lévy. *Calcul des Probabilités*. Gauthier-Villars, Paris, 1925.

R. Li, W. Zhong, and L. Zhu. Feature screening via distance correlation learning. *Journal of the American Statistical Association*, 107(499):1129–1139, 2012. doi: 10.1080/01621459.2012.695654. PMID: 25249709.

S. Li. *K-groups: A Generalization of k-means by Energy Distance*. PhD thesis, Bowling Green State University, Bowling Green, Ohio, 2015a.

S. Li and M. L. Rizzo. K-groups: A Generalization of K-means Clustering. *ArXiv e-prints*, November 2017.

T. Li and M. Yuan. On the optimality of Gaussian kernel based non-parametric tests against smooth alternatives, 2019. https://arxiv.org/pdf/1909.03302.pdf.

W. Li, Q. Wang, and J. Yao. Eigenvalue distribution of a high-dimensional distance covariance matrix with application. https://arXiv.org/pdf/2105.07641.pdf, 2021.

Y. Li. *Goodness-of-Fit Tests for Dirichlet Distributions with Applications*. PhD thesis, Bowling Green State University, Bowling Green, Ohio, 2015b.

E. H. Linfoot. An informational measure of correlation. *Information and Control*, 1:85–89, 1957.

W. Liu and R. Li. Variable selection and feature screening. In *Macroeconomic Forecasting in the Era of Big Data*, pages 293–326. Springer, 2020.

Y. Liu, V. de la Pena, and T. Zheng. Kernel-based measures of association. *Wiley Interdisciplinary Reviews: Computational Statistics*, 10(2): e1422, 2018.

G. M. Ljung and G. E. P. Box. On a measure of lack of fit in time series models. *Biometrika*, 65(2):297–303, 1978.

S. Lloyd. Least squares quantization in pcm. *Information Theory, IEEE Transactions on*, 28(2):129–137, 1982.

J. Y. Lu, Y. X. Peng, M. Wang, S. J. Gu, and M. X. Zhao. Support vector machine combined with distance correlation learning for Dst forecasting during intense geomagnetic storms. *Planetary and Space Science*, 120:48 – 55, 2016. ISSN 0032-0633. doi: http://dx.doi.org/10.1016/j.pss.2015.11.004.

E. Lukacs. A characterization of the gamma distribution. *The Annals of Mathematical Statistics*, 26(2):319–324, 1955.

R. Lyons. Distance covariance in metric spaces. *Annals of Probability*, 41(5): 3284–3305, 09 2013. doi: 10.1214/12-AOP803. URL `http://dx.doi.org/10.1214/12-AOP803`.

R. Lyons. Hyperbolic space has strong negative type. *Illinois Journal of Mathematics*, 58(4):1009–1013, 2014. URL `http://projecteuclid.org/euclid.ijm/1446819297`.

R. Lyons. Errata to "distance covariance in metric spaces." *Annals of Probability*, 46(4):2400–2405, 2018.

R. Lyons. Strong negative type in spheres. *Pacific Journal of Mathematics*, 307:383–390, 2020.

S. Mak and V. R. Joseph. Projected support points: a new method for high-dimensional data reduction, 2018a.

S. Mak and V. R. Joseph. Support points. *The Annals of Statistics*, 46(6A): 2562–2592, 2018b. https://arxiv.org/abs/1609.01811.

N. Mantel. The detection of disease clustering and a generalized regression approach. *Cancer Research*, 27:209–220, 1967.

K. V. Mardia. Measures of multivariate skewness and kurtosis with applications. *Biometrika*, 57:519–530, 1970.

K. V. Mardia. Some properties of classical multidimensional scaling. *Communications in Statistics: Theory and Methods*, 7(13):1233–1241, 1978.

K. V. Mardia, J. T. Kent, and J. M. Bibby. *Multivariate Analysis*. Academic Press, London, 1979.

M. Matsui and A. Takemura. Empirical characteristic function approach to goodness-of-fit tests for the Cauchy distribution with parameters estimated by MLE or EISE. *Annals of the Institute of Statistical Mathematics*, 57(1): 183–199, 2005.

D. S. Matteson and N. A. James. A nonparametric approach for multiple change point analysis of multivariate data. *Journal of the American Statistical Association*, 109(505):334–345, 2014.

D. S. Matteson and R. S. Tsay. Independent component analysis via distance covariance. *Journal of the American Statistical Association*, 2016. doi: 10.1080/01621459.2016.1150851.

L. Mattner. Strict negative definiteness of integrals via complete monotonicity of derivatives. *Transactions of the American Mathematical Society*, 349(8): 3321–3342, 1997.

S. Mazur and S. Ulam. Sur les transformationes isométriques d'espaces vectoriels normés, 1932.

M. W. Meckes. Positive definite metric spaces. *Positivity*, 17(3):733–757, 2013.

K. Menninger. *Number Words and Number Symbols: A Cultural History of Numbers.* Dover, 2015.

O. Mersmann. *microbenchmark: Accurate Timing Functions*, 2021. URL https://CRAN.R-project.org/package=microbenchmark. R package version 1.4-9.

D. Meyer, E. Dimitriadou, K. Hornik, A. Weingessel, and F. Leisch. *e1071: Misc Functions of the Department of Statistics, Probability Theory Group (Formerly: E1071), TU Wien*, 2022. URL https://CRAN.R-project.org/package=e1071. R package version 1.7-11.

G. Monge. Mémoire sur la théorie des déblais et de remblais. *Histoire de l'Académie Royale des Sciences de Paris avec les Mémoires de Mathématique et de Physique pour la même année*, pages 666–704, 1781.

D. Morgenstern. Proof of a conjecture by Walter Deuber concerning the distance between points of two types in R^d. *Discrete Mathematics*, 226:347–349, 2001.

T. F. Móri. Essential correlatedness and almost independence. *Statistics and Probability Letters*, 15:169–172, 1992.

T. F. Móri and G. J. Székely. Representations by uncorrelated random variables. *Mathematical Methods of Statistics*, 26:149–153, 2017.

T. F. Móri and G. J. Székely. Four simple axioms of dependence measures. *Metrika*, 82(1):1–16, 2019.

T. F. Móri and G. J. Székely. The Earth Mover's correlation. *Annales Universitatis Scientiarum Budapestinensis de Rolando Eötvös Nominatae Sectio Computatorica*, 50:249–268, 2020.

T. F. Móri and G. J. Székely. Pseudorandom processes. *Stochastic Processes and their Applications*, 150:669–715, 2022.

T. F. Móri, G. J. Székely, and M. L. Rizzo. On energy tests of normality. *Journal of Statistical Planning and Inference*, 213:1–15, 2021. ISSN 0378-3758. URL https://doi.org/10.1016/j.jspi.2020.11.001.

J. Munkres. Algorithms for the assignment and transportation problems. *Journal of the Society for Industrial and Applied Mathematics*, 5(1):32–38, 1957.

A. Naor and G. Schechtman. Planar earthmover is not in L_1. *SIAM Journal on Computing*, 37:804–826, 2007. https://doi.org/10.1137/05064206X.

M. A. Newton. Introducing the discussion paper by Székely and Rizzo. *The Annals of Applied Statistics*, 3(4):1233–1235, 2009.

H. D. Nguyen. An introduction to approximate Bayesian computation. In *Research School on Statistics and Data Science*, pages 96–108. Springer, 2019.

H. D. Nguyen, J. Arbel, H. Lü, and F. Forbes. Approximate Bayesian computation via the energy statistic. *IEEE Access*, pages 131683–131698, 2020. URL https://arxiv.org/abs/1905.05884.

J. Niles-Weed and P. Rigollet. Estimation of Wasserstein distances in the spiked transport model. *arXiv preprint arXiv:1909.07513*, 2019.

J. P. Nolan. *Univariate Stable Distributions*. Springer, 2020.

P. Ofosuhene. *The Energy Goodness-of-Fit Test for the Inverse Gaussian Distribution*. PhD thesis, Bowling Green State University, Bowling Green, Ohio, 2020.

J. Oksanen et al. *vegan: Community Ecology Package*, 2022. URL https://CRAN.R-project.org/package=vegan. R package version 2.6-2.

M. Omelka and S. Hudecová. A comparison of the Mantel test with a generalised distance covariance test. *Environmetrics*, 24(7):449–460, 2013.

W. Pan, Y. Tian, X. Wang, and H. Zhang. Ball divergence: Nonparametric two sample test. *Annals of Statistics*, 46(3):1109–1137, 2018.

W. Pan, X. Wang, H. Zhang, H. Zhu, and J. Zhu. Ball covariance: a generic measure of dependence in Banach space. *Journal of the American Statistical Association*, 115(529):307–317, 2020.

S. Panagiotis. N-sphere chord length distribution, 2014. URL http://arxiv.org/abs/1411.5639.

S. Panda, C. Shen, R. Perry, J. Zorn, A. Lutz, C. E. Priebe, and J. T. Vogelstein. Nonparametric MANOVA via independence testing. arxiv.1910.08883, 2020.

M. Papadakis, M. Tsagris, M. Dimitriadis, S. Fafalios, I. Tsamardinos, M. Fasiolo, G. Borboudakis, J. Burkardt, C. Zou, K. Lakiotaki, and C. Chatzipantsiou. *Rfast: A Collection of Efficient and Extremely Fast R Functions*, 2022. URL https://CRAN.R-project.org/package=Rfast. R package version 2.0.6.

T. Park, X. Shao, and S. Yao. Partial martingale difference correlation. *Electronic Journal of Statistics*, 9(1):1492–1517, 2015.

K. R. Parthasarathy. *Probability Measures on Metric Spaces*. American Mathematical Society, 1967.

V. Patrangenaru and L. Ellingson. *Nonparametric Statistics on Manifolds and their Applications to Object Data Analysis*. CRC Press, 2015.

K. Pearson. Notes on regression and inheritance in the case of two parents. *Proceedings of the Royal Society of London*, 58:240–242, 1895.

M. Peligrad. An invariance principle for φ-mixing sequences. *The Annals of Probability*, 13:1304–1313, 1985.

N. Pfister, P. Bühlmann, B. Schölkopf, and J. Peters. Kernel-based tests for joint independence. *Journal of the Royal Statistical Society: Series B (Statistical Methodology)*, 80(1):5–31, 2018.

M. Pitsillou and K. Fokianos. dCovTS: Distance Covariance/correlation for time series. *R Journal*, 8:324–340, 2016.

M. T. Pratola, H. A. Chipman, E. I. George, and R. E. McCulloch. Heteroscedastic BART using multiplicative regression trees, 2017. http://apps.olin.wustl.edu/conf/SBIES/Files/pdf/2017/219.pdf.

A. P. Prudnikov, A. Brychkov, and O. I. Marichev. *Integrals and Series*. Gordon and Breach Science, New York, 1986.

M. L. Puri and P. K. Sen. *Nonparametric Methods in Multivariate Analysis*. Wiley, New York, 1971.

R Core Team. *R: A Language and Environment for Statistical Computing*. R Foundation for Statistical Computing, Vienna, Austria, 2022. URL https://www.R-project.org/.

C. R. Rao. *Linear Statistical Inference and its Applications*. Wiley, New York, second edition, 1973.

C. R. Rao and G. J. Székely, editors. *Statistics for the 21st century: Methodologies for Applications of the Future, Proceedings of the 1998 Eugene Lukacs Symposium*, New York, 2000. Marcel Dekker.

B. Rémillard. Discussion of: Brownian distance covariance. *The Annals of Applied Statistics*, 3(4):1295–1298, 2009.

A. Rényi. On measures of dependence. *Acta Mathematica Sci Hungarica*, 10 (3-4):441–451, 1959.

A. Rényi. On measures of information and entropy. *Proceedings of the fourth Berkeley Symposium on Mathematics, Statistics and Probability 1960*, pages 547–561, 1961.

D. Reshef, Y. Reshef, H. Finucane, S. Grossman, G. McVean, P. Turnbaugh, E. Lander, M. Mitzenmacher, and P. C. Sabeti. Detecting novel associations in large data sets. *Science*, 334:1518–1524, 2011.

M. Riesz. Intégrales de Riemann–Liouville et potentiels. *Acta Scientiarum Mathematicarum (Szeged)*, 9:1–42, 1938.

M. Riesz. L'intégrale de Riemann–Liouville et le problème de Cauchy. *Acta Mathematica*, 81:1–223, 1949.

P. Rigollet and J. Niles-Weed. Entropic optimal transport is maximum-likelihood deconvolution. https://arxiv.org/abs/1809.05572, 2018.

M. L. Rizzo. *A New Rotation Invariant Goodness-of-Fit Test*. PhD thesis, Bowling Green State University, Bowling Green, Ohio, 2002.

M. L. Rizzo. A test of homogeneity for two multivariate populations. In *2002 Proceedings of the American Statistical Association, Physical and Engineering Sciences Section*, Alexandria, VA, 2003. American Statistical Association.

M. L. Rizzo. New goodness-of-fit tests for Pareto distributions. *ASTIN Bulletin: Journal of the International Association of Actuaries*, 39(2):691–715, 2009.

M. L. Rizzo. *Statistical Computing with R*. The R Series. Chapman & Hall/CRC, second edition, 2019. ISBN 9781466553323.

M. L. Rizzo and J. T. Haman. Expected distances and goodness-of-fit for the asymmetric Laplace distribution. *Statistics & Probability Letters*, 117: 158–164, 2016. doi: http://dx.doi.org/10.1016/j.spl.2016.05.006.

M. L. Rizzo and G. J. Székely. DISCO analysis: A nonparametric extension of analysis of variance. *Annals of Applied Statistics*, 4(2):1034–1055, 2010.

M. L. Rizzo and G. J. Székely. *energy: E-Statistics: Multivariate Inference via the Energy of Data*, 2022. URL https://CRAN.R-project.org/package=energy. R package version 1.7-9.

P. Robert and Y. Escoufier. A unifying tool for linear multivariate statistical methods: The RV-coefficient. *Applied Statistics*, 25(3):257–265, 1976.

Y. Rubner, L. J. Guibas, and C. Tomasi. The earth mover's distance as a metric for image retrieval. *International Journal of Computer Vision*, 40 (2):99–121, 2000.

J. Rudas, J. Guaje, A. Demertzi, L. Heine, L. Tsibanda, A. Soddu, S. Lauareys, and F. Gomez. A method for functional network connectivity using distance correlation. In *Conference Proceedings, Annual International Conference of the IEEE Engineering in Medicine and Biology Society*, pages 2793–2796. IEEE Engineering in Medicine and Biology Society, 2014.

P. S. Rudman. *How Mathematics Happened: The First 50,000 Years*. Prometeus Books, 2007.

I. Z. Ruzsa and G. J. Székely. *Algebraic Probability Theory*. Wiley, New York, 1988.

E. B. Saff and A. B. J. Kuijlaars. Distributing many points on a sphere. *The Mathematical Intelligencer*, 19:5–11, 1997.

P. P. Saviotti. *Technological Evolution, Variety and Economy*. Edward Elgar: Cheltenham, 1996.

D. Schmandt-Bessarat. *How Writing Came About*. University of Texas Press, 1992.

W. M. Schmidt. Irregularities of distribution. IV. *Inventiones mathematicae*, 7:55–82, 1969.

I. J. Schoenberg. Remarks to Maurice Fréchet's article 'Sur la définition axiomatique d'une classe d'espace distanciés vectoriellement applicable sur l'espace de Hilbert.' *Annals of Mathematics*, 36:724–732, 1935.

I. J. Schoenberg. On certain metric spaces arising from Euclidean spaces by a change of metric and their imbedding in Hilbert space. *Annals of Mathematics, Second Series*, 38(2):787–793, 1937.

I. J. Schoenberg. Metric spaces and completely monotone functions. *Annals of Mathematics, Second Series*, 39(4):811–841, 1938a.

I. J. Schoenberg. Metric spaces and positive definite functions. *Transactions of the American Mathematical Society*, 44(3):522–536, 1938b.

B. Schölkopf and A. J. Smola. *Learning with Kernels*. MIT Press, Cambridge, MA, 2002.

E. Schrödinger. Zur Theorie der Fall-und Steigversuche an Teilchen mit Brownscher Bewegung. *Physikalische Zeitschrift*, 16:289–295, 1915.

E. Schrödinger. An undulatory theory of the mechanics of atoms and molecules. *Physical Review*, 28(6):1049–1070, 1926.

E. Schrödinger. *What is Life - the Physical Aspect of the Living Cell*. Cambridge University Press, Cambridge, England, 1944.

D. Sejdinovic, B. Sriperumbudur, A. Gretton, and K. Fukumizu. Equivalence of distance-based and RKHS-based statistics in hypothesis testing. *Annals of Statistics*, 41(5):2263–2291, 2013.

R. J. Serfling. *Approximation Theorems of Mathematical Statistics*. Wiley, New York, 1980.

G. Sferra, F. Fratini, M. Ponzi, and E. Pizzi. Phylo_dcor: distance correlation as a novel metric for phylogenetic profiling. *BMC Bioinformatics*, 18(1): 396, 2017.

M. Shamsuzzaman, M. Satish, and J. D. Pintér. Distance correlation based nearly orthogonal space-filling experimental designs. *International Journal of Experimental Design and Process Optimization*, 4(3/4):216–233, 2015.

X. Shao and J. Zhang. Martingale difference correlation and its use in high-dimensional variable screening. *Journal of the American Statistical Association*, 109(507):1302–1318, 2014.

C. Shen and J. T. Vogelstein. The exact equivalence of distance and kernel methods for hypothesis testing. *arXiv preprint arXiv:1806.05514*, 2018.

C. Shen, C. E. Priebe, and J. T. Vogelstein. From distance correlation to multiscale graph correlation. *Journal of the American Statistical Association*, 2019.

W. Sheng and X. Yin. Direction estimation in single-index models via distance covariance. *Journal of Multivariate Analysis*, 122:148–161, 2013.

W. Sheng and X. Yin. Sufficient dimension reduction via distance covariance. *Journal of Computational and Graphical Statistics*, 25(1):91–104, 2016. doi: 10.1080/10618600.2015.1026601.

N. Simon and R. Tibshirani. Comment on "Detecting novel associations in large data sets" by Reshef et. al., 2011. URL https://arxiv.org/abs/1401.7645.

G. L. Simpson. *permute: Functions for Generating Restricted Permutations of Data*, 2022. URL https://CRAN.R-project.org/package=permute. R package version 0.9-7.

Ian H Sloan. Quadrature methods for integral equations of the second kind over infinite intervals. *Mathematics of Computation*, 36(154):511–523, 1981.

S. Smale. Mathematical problems for the next century. *Mathematical Intelligencer*, 20:7–15, 1998.

P. E. Smouse, J. C. Long, and R. R. Sokal. Multiple regression and correlation extensions of the Mantel test of matrix correspondence. *Systematic Zoology*, 35(62):627–632, 1986.

P. Spector, J. Friedman, R. Tibshirani, T. Lumley, S. Garbett, and J. Baron. *acepack: ACE and AVAS for Selecting Multiple Regression Transformations*, 2016. URL https://CRAN.R-project.org/package=acepack. R package version 1.4.1.

T. Speed. A correlation for the 21st century. *Science*, 334(6062):1502–1503, 2011.

T. A. Stamey, J. N. Kabalin, J. E. McNeal, I. M. Johnstone, F. Freiha, E. A. Redwine, and N. Yang. Prostate specific antigen in the diagnosis and treatment of adenocarcinoma of the prostate: II. radical prostatectomy treated patients. *Journal of Urology*, 141(5):1076–1083, 1989.

C. Stein. A bound for the error in the normal approximation to the distribution of a sum of dependent random variables. In *Proceedings of the sixth Berkeley symposium on mathematical statistics and probability, volume 2: Probability Theory*, volume 6, pages 583–603. University of California Press, 1972.

W. Steutel, F and K. van Harn. *Infinite Divisibility of Probability Distributions on the Real Line*. Marcel Dekker, 2004.

S. M. Stigler. *The History of Statistics: The Measurement of Uncertainty before 1900*. Belknap Press, Cambridge, Massachusetts, 1986.

S. M. Stigler. Francis Galton's account of the invention of correlation. *Statistical Science*, 4:73–79, 1989.

K. B. Stolarsky. Sums of distances between points on a sphere. II. *Proc. Amer. Math. Soc.*, 41:575–582, 1973.

G. J. Székely. Why is 7 a mystical number? (in Hungarian). *MIOK Évkönyv, ed. S. Scheiber*, pages 482–487, 1981.

G. J. Székely. *Paradoxes in Probability Theory and Mathematical Statistics (Translated from Hungarian)*. D. Reidel, 1986.

G. J. Székely. Potential and kinetic energy in statistics (in Hungarian). Technical report, Budapest Institute of Technology (Budapest Technical University), 1989. Lecture Notes.

G. J. Székely. *Contests in Higher Mathematics*. Springer, New York, 1996.

G. J. Székely. E-statistics: Energy of statistical samples. Technical Report 02–16, Bowling Green State University, Department of Mathematics and Statistics, 2002. URL dx.doi.org/10.13140/RG.2.1.5063.9761. Also technical reports by same title, from 2000–2003 and NSA grant # MDA 904-02-1-0091 (2000–2002).

G. J. Székely. Student's t-test for scale mixture errors. In *2nd Lehmann Symposium – Optimality*, volume 49, pages 9–15. IMS Lecture Notes – Monograph Series, 2006.

G. J. Székely and N. K. Bakirov. Extremal probabilities for Gaussian quadratic forms. *Probab. Theory Related Fields*, 126:184–202, 2003.

G. J. Székely and N. K. Bakirov. Brownian covariance and CLT for stationary sequences. Technical Report 08-01, Bowling Green State University, Department of Mathematics and Statistics, 2008.

G. J. Székely and T. F. Móri. A characteristic measure of asymmetry and its application for testing diagonal symmetry. *Communications in Statistics – Theory and Methods*, 30(8 & 9):1633–1639, 2001.

G. J. Székely and M. L. Rizzo. Testing for equal distributions in high dimension. *InterStat*, 5, November 2004a. URL `https://citeseerx.ist.psu.edu/viewdoc/download?doi=10.1.1.598.3473&rep=rep1&type=pdf`.

G. J. Székely and M. L. Rizzo. Mean distance test of Poisson distribution. *Statistics & Probability Letters*, 67(3):241–247, 2004b.

G. J. Székely and M. L. Rizzo. Hierarchical clustering via joint between-within distances: Extending Ward's minimum variance method. *Journal of Classification*, 22(2):151–183, 2005a.

G. J. Székely and M. L. Rizzo. A new test for multivariate normality. *Journal of Multivariate Analysis*, 93(1):58–80, 2005b.

G. J. Székely and M. L. Rizzo. The uncertainty principle of game theory. *The American Mathematical Monthly*, 114(8):688–702, 2007.

G. J. Székely and M. L. Rizzo. Brownian distance covariance. *Annals of Applied Statistics*, 3(4):1236–1265, 2009a.

G. J. Székely and M. L. Rizzo. Rejoinder: Brownian distance covariance. *Annals of Applied Statistics*, 3(4):1303–1308, 2009b.

G. J. Székely and M. L. Rizzo. On the uniqueness of distance covariance. *Statistics & Probability Letters*, 82:2278–2282, 2012.

G. J. Székely and M. L. Rizzo. The distance correlation t-test of independence in high dimension. *Journal of Multivariate Analysis*, 117:193—213, 2013a. doi: 10.1016/j.jmva.2013.02.012.

G. J. Székely and M. L. Rizzo. Energy statistics: A class of statistics based on distances. *Journal of Statistical Planning and Inference*, 143(8):1249–1272, 2013b.

G. J. Székely and M. L. Rizzo. Partial distance correlation with methods for dissimilarities. *Annals of Statistics*, 42(6):2382–2412, 2014. doi: 10.1214/14-AOS1255. URL `http://projecteuclid.org/euclid.aos/1413810731`.

G. J. Székely and M. L. Rizzo. The energy of data. *The Annual Review of Statistics and Its Application*, 4:447–479, 2017. doi: 10.1146/annurev-statistics-060116-054026. URL `http://www.annualreviews.org/doi/abs/10.1146/annurev-statistics-060116-054026`.

G. J. Székely and M. L. Rizzo. Comments on: Tests for multivariate normality—a critical review with emphasis on weighted L^2–statistics. *TEST*, 29: 907–910, 2020. doi: 10.1007/s11749-020-00741-z.

G. J. Székely, M. L. Rizzo, and N. K. Bakirov. Measuring and testing independence by correlation of distances. *Annals of Statistics*, 35(6):2769–2794, 2007.

L. Szilárd. Über die Entropieverminderung in einem thermodynamischen System bei Eingriffen intelligenter Wesen. (On the decrease of entropy in a thermodynamic system by the intervention of intelligent beings). *Zetschrift für Physik*, 53:840–856, 1929. Based on Szilárd's 1922 Ph. D. dissertation in Berlin and on Szilárd's 1925 paper in *Zetschrift für Physik*, Volume 32, 753.

M. Talagrand. Small tails for the supremum of a Gaussian process. In *Ann. l'Institut Henri Poincaré (B) Probabilités et Statistiques*, volume 24, pages 307–315, 1988.

T. Tao. The spherical Cayley-Menger determinant and the radius of the Earth, 2019. URL https://terrytao.wordpress.com/2019/05/25/. Daily archive for blog post 25 May, 2019.

O. Thas and J-P. Ottoy. Some generalizations of the Anderson–Darling statistic. *Statistics & Probability Letters*, 64(3):255–261, 2003.

D. Tjøstheim and K. O. Hufthammer. Local Gaussian correlation: A new measure of dependence. *Journal of Econometrics*, 172(1):33–48, 2013.

D. Tjøstheim, H. Otneim, and B. Støve. Statistical dependence: Beyond Pearson's ρ. *Statistical Science*, 37(1):90–109, 2022.

N. Tomizawa. On some techniques useful for solution of transportation network problems. *Networks*, 1:173–194, 1971. URL https://doi.org/10.1002/net.3230010206.

W. S. Torgerson. *Theory and Methods of Scaling*. Wiley, New York, 1958.

M. Tsagris, M. Pitsillou, and K. Fokianos. *dCovTS: Distance Covariance and Correlation for Time Series Analysis*, 2021. URL https://CRAN.R-project.org/package=dCovTS. R package version 1.2.

J. W. Tukey. The future of data analysis. *The Annals of Mathematical Statistics*, 33(1):1–67, 1962.

M. C. K. Tweedie. Statistical properties of inverse Gaussian distributions. I. *The Annals of Mathematical Statistics*, 28(2):362–377, 1957a.

M. C. K. Tweedie. Statistical properties of inverse Gaussian distributions. II. *The Annals of Mathematical Statistics*, 28(2):696–705, 1957b.

S. M. Ulam. Adventures of a mathematician, 1976.

United States Bureau of the Census. Statistical Abstract of the United States, 1970.

E. Vaiciukynas, A. Verikas, A. Gelzinis, M. Bacauskiene, and I. Olenina. Exploiting statistical energy test for comparison of multiple groups in morphometric and chemometric data. *Chemometrics and Intelligent Laboratory Systems*, 146:10–23, 2015.

A W Van der Vaart. *Asymptotic Statistics*, volume 3. Cambridge University Press, Cambridge, England, 2000.

V. Vapnik. *The Nature of Statistical Learning Theory*. Springer, New York, 1995.

T. Varina, R. Bureaua, C. Muellerb, and P. Willett. Clustering files of chemical structures using the Székely–Rizzo generalization of Ward's method. *Journal of Molecular Graphics and Modelling*, 28(2):187–195, 2009.

W. N. Venables and B. D. Ripley. *Modern Applied Statistics with S*. Springer, New York, fourth edition, 2002. URL https://www.stats.ox.ac.uk/pub/MASS4/. ISBN 0-387-95457-0.

R. von Mises. On the asymptotic distributions of differentiable statistical functionals. *Annals of Mathematical Statistics*, 18:209–348, 1947.

J. von Neumann. Theorie der Gesellschaftspiele. *Mathematische Annalen*, 100:295–320, 1928.

G. Wahba. *Spline Models for Observational Data*, volume 59 of *CBMS-NSF Regional Conference Series in Applied Mathematics*. SIAM, 1990.

G. Wahba. Positive definite functions, reproducing kernel Hilbert spaces and all that, The Fisher Lecture at JSM 2014, August 6, 2014, Boston, Massachusetts, 2014.

A. Wald. *Sequential Analysis*. John Wiley, Hoboken, New Jersey 1947.

X. Wang, W. Pan, W. Hu, Y. Tian, and H. Zhang. Conditional distance correlation. *Journal of the American Statistical Association*, 110(512):1726–1734, 2015. doi: 10.1080/01621459.2014.993081. PMID: 26877569.

G. R. Warnes, B. Bolker, T. Lumley, and CRAN team. *gtools: Various R Programming Tools*, 2022. URL https://CRAN.R-project.org/package=gtools. R package version 3.9.2.2.

L. N. Wasserstein. Markov processes on countable space products describing large systems of automata. *Probl. Pered. Inform.*, 5(3):64–72, 1969. (L. N. Vasserstein) translated as 'Problems of Information Transmission' 5 (1969), no. 3, 47–52; MR0314115] and R. L. Dobrušin [Teor. Verojatnost. i Primenen. 15 (1970), 469–497; MR0298716].

N. Wermuth and D. R. Cox. Concepts and a case study for a flexible class of graphical Markov models. In *Robustness and Complex Data Structures*, pages 331–350. Springer, 2013.

H. Weyl. *Space–Time–Matter.* Dutton, 1922.

S. S. Wilks. On the independence of k sets of normally distributed statistical variables. *Econometrica, Journal of the Econometric Society*, pages 309–326, 1935.

G. Yang. *The Energy Goodness-of-Fit Test for Univariate Stable Distributions.* PhD thesis, Bowling Green State University, Bowling Green, Ohio, 2012.

S. Yao, X. Zhang, and X. Shao. Testing mutual independence in high dimension via distance covariance. *Journal of the Royal Society, Ser. B*, 3: 455–480, 2018.

S. Yitzhaki. Gini's mean difference: A superior measure of variability for non-normal distributions. *Metron*, 61(2):285–316, 2003.

G. Young and A. S. Householder. Discussion of a set of points in terms of their mutual distances. *Psychometrika*, 3:19–22, 1938.

K. Yousuf and Y. Feng. Partial distance correlation screening for high dimensional time series. https://arxiv.org/abs/1802.09116, 2018.

S. Zacks. *Parametric Statistical Inference: Basic Theory and Modern Approaches.* Pergamon, Oxford, 1981.

Z. Zhou. Measuring nonlinear dependence in time-series, a distance correlation approach. *Journal of Time Series Analysis*, 33(3):438–457, 2012.

C. Zhu, X. Zhang, S. Yao, and X. Shao. Distance-based and RKHS-based dependence metrics in high dimension. *Annals of Statistics*, 48(6): 3366–3394, 2020.

L. Zhu, K. Xu, R. Li, and W. Zhong. Projection correlation between two random vectors. *Biometrika*, 104(4):829–843, 2017.

S. Zucker. Detection of periodicity based on independence tests–III. Phase distance correlation periodogram. *Monthly Notices of the Royal Astronomical Society: Letters*, 474(1):L86–L90, 2018. URL https://academic.oup.com/mnrasl/article/474/1/L86/4693844.

Index